P9-AFZ-669

METHODS IN CELL BIOLOGY

VOLUME 21

Normal Human Tissue and Cell Culture

B. Endocrine, Urogenital, and Gastrointestinal Systems

Advisory Board

METHODS IN CELL BIOLOGY

Prepared under the Auspices of the American Society for Cell Biology

VOLUME 21

Normal Human Tissue and Cell Culture
B. Endocrine, Urogenital, and Gastrointestinal Systems

Edited by

CURTIS C. HARRIS
NATIONAL CANCER INSTITUTE
NATIONAL INSTITUTES OF HEALTH
BETHESDA, MARYLAND

BENJAMIN F. TRUMP
SCHOOL OF MEDICINE
UNIVERSITY OF MARYLAND
BALTIMORE, MARYLAND

GARY D. STONER
NATIONAL CANCER INSTITUTE
NATIONAL INSTITUTES OF HEALTH
BETHESDA, MARYLAND

1980
ACADEMIC PRESS
A Subsidiary of Harcourt Brace Jovanovich, Publishers
New York London Toronto Sydney San Francisco

ACADEMIC PRESS, INC.
111 Fifth Avenue, New York, New York 10003

United Kingdom Edition published by
ACADEMIC PRESS, INC. (LONDON) LTD.
24/28 Oval Road, London NW1 7DX

LIBRARY OF CONGRESS CATALOG CARD NUMBER: 64–14220

ISBN 0–12–564140–0

PRINTED IN THE UNITED STATES OF AMERICA

80 81 82 83 9 8 7 6 5 4 3 2 1

We dedicate these volumes to the memory
of our friend and colleague, Dr. Douglas Janss

Contents

LIST OF CONTRIBUTORS

Numbers in parentheses indicate the pages on which the authors' contributions begin.

NABIL ABAZA, Cancer Bioassay Laboratory, Department of Pathology, Medical College of Pennsylvania, Philadelphia, Pennsylvania 19129 (287)

THOMAS A. ADAMEC, Department of Pathology, University of North Carolina School of Medicine, Chapel Hill, North Carolina 27514 (1)

ARNE ANDERSSON, Department of Histology, University of Uppsala, Biomedicum, P.O. Box 571, S-751 23 Uppsala, Sweden (135)

HERMAN AUTRUP, Human Tissue Studies Section, Laboratory of Experimental Pathology, National Cancer Institute, Bethesda, Maryland 20205 (385)

MERRILL S. BABCOCK, Pasadena Foundation for Medical Research, Pasadena, California 91101 (195)

THERESA L. BEN,[1] Endocrine Carcinogenesis Section, Frederick Cancer Research Center, Frederick, Maryland 21701 (107)

R. J. BERRY, Department of Oncology, Middlesex Hospital Medical School, London W1 England (257)

CHARLES N. CARNEY, Department of Pathology, University of North Carolina School of Medicine, Chapel Hill, North Carolina 27514 (1)

J. W. COMBS, Department of Pathology, University of Pennsylvania at Hershey, Hershey, Pennsylvania, 17033 (79)

M. HALL-CRAGGS, Department of Pathology, Maryland Cancer Program and Maryland Institute for Emergency Medicine, University of Maryland School of Medicine, Baltimore, Maryland 21201 (309)

ROBERT M. DONALDSON, JR., Department of Internal Medicine, Yale University School of Medicine, New Haven, Connecticut 06510 (349)

B. HUGH DORMAN, Department of Pathology, University of North Carolina School of Medicine, Chapel Hill, North Carolina 27514 (1)

L. M. FRANKS, Department of Cellular Pathology, Imperial Cancer Research Fund, Lincoln's Inn Fields, London WC2A 3PX, England (153)

VALERIO M. GENTA, Department of Pathology, University of North Carolina School of Medicine, Chapel Hill, North Carolina 27514 (1)

JOE W. GRISHAM, Department of Pathology, University of North Carolina School of Medicine, Chapel Hill, North Carolina 27514 (1)

CURTIS C. HARRIS, Human Tissue Studies Section, Laboratory of Experimental Pathology, National Cancer Institute, National Institutes of Health, Bethesda, Maryland 20205 (79, 331)

BARRY M. HEATFIELD, Department of Pathology, University of Maryland School of Medicine, Baltimore, Maryland 21201 (171)

CLAES HELLERSTRÖM, Department of Histology, University of Uppsala, Uppsala, Sweden (135)

R. M. HICKS, School of Pathology, Middlesex Hospital Medical School, London W1, England (257)

ELIZABETH A. HILLMAN, Department of Pathology, University of Maryland School of Medicine, Baltimore, Maryland 21201 (79, 107, 331)

[1] *Present address:* Laboratory of Experimental Pathology, National Center Institute, National Institutes of Health, Bethesda, Maryland 20205.

J. C. IRWIN, Department of Hormone Biochemistry, Imperial Cancer Research Fund, Lincoln's Inn Fields, London WC2A 3PX, England (51)

DOUGLAS H. JANSS,[1] Department of Pathology, University of Maryland School of Medicine, Baltimore, Maryland 21201 and Endocrine Carcinogenesis Section, Frederick Cancer Research Center, Frederick, Maryland 21701 (79, 107)

RAYMOND T. JONES, Department of Pathology and Maryland Institute for Emergency Medical Services, University of Maryland School of Medicine, Baltimore, Maryland 21201 (429)

M. W. KAHNG, Department of Pathology, Maryland Cancer Program and Maryland Institute for Emergency Medicine, University of Maryland School of Medicine, Baltimore 21201 (309)

M. EDWARD KAIGHN, Pasadena Foundation for Medical Research, Pasadena, California 91101 (195, 253)

CYRUS R. KAPADIA, Department of Internal Medicine, Yale University School of Medicine, New Haven, Connecticut 06510 (349)

DAVID G. KAUFMAN, Department of Pathology, University of North Carolina School of Medicine, Chapel Hill, North Carolina 27514 (1)

DANA J. KESSLER, Department of Medical Microbiology, College of Medicine, University of Vermont, Burlington, Vermont 05405 (457)

D. KIRK,[2] Department of Hormone Biochemistry, Imperial Cancer Research Fund, Lincoln's Inn Fields, London WC2A 3PX, England (51)

M. A. KNOWLES, Department of Oncology, Middlesex Hospital Medical School, London W1 England (257)

JOHN F. LECHNER, Pasadena Foundation for Medical Research, Pasadena, California 91101 (195)

JOSEPH LEIGHTON, Cancer Bioassay Laboratory, Department of Pathology, Medical College of Pennsylvania, Philadelphia, Pennsylvania 19129 (287)

EDMUND LEIPUS, Department of Obstetrics and Gynecology, Rush Medical College, Rush Presbyterian St. Lukes Medical Center, Chicago, Illinois 60612 (29)

MAREEN MARNELL, Pasadena Foundation for Medical Research, Pasadena, California 91101 (195)

MARC J. MASS, Department of Pathology, University of North Carolina School of Medicine, Chapel Hill, North Carolina 27514 (1)

ROGER J. MAY, Department of Medicine, Harvard Medical School, and The Gastrointestinal Unit, Massachusetts General Hospital, Boston, Massachusetts 02114 (403)

SUSAN A. MELIN, Department of Pathology, University of North Carolina School of Medicine, Chapel Hill, North Carolina 27514 (1)

E. MILROY, Department of Urology, Middlesex Hospital Medical School, London W1, England (257)

K. SHANKAR NARAYAN, Pasadena Foundation for Medical Research, Pasadena, California 91101 (195)

GUY J. PHOTOPULOS, Department of Obstetrics and Gynecology, University of Tennessee

[1]Deceased.

[2]*Present address:* Pasadena Foundation for Medical Research, 99 North El Molino Avenue, Pasadena, California 91101.

School of Medicine, Memphis, Tennessee 38163 (1)

HENRY C. PITOT, Departments of Oncology and Pathology, McArdle Laboratory for Cancer Research, The Medical School, University of Wisconsin, Madison, Wisconsin 53706 (441)

JUDITH POWELL,[1] Department of Pathology, University of North Carolina School of Medicine, Chapel Hill, North Carolina 27514 (1)

ANDREA QUARONI, Department of Medicine, Harvard Medical School, and The Gastrointestinal Unit, Massachusetts General Hospital, Boston, Massachusetts 02114 (403)

T. ROBBINS, Department of Pathology, University of Maryland School of Medicine, Baltimore, Maryland 21201 (79)

NANCY T. RODGERS, Department of Pathology, University of North Carolina School of Medicine, Chapel Hill, North Carolina 27514 (1)

HAYATO SANEFUJI, Department of Pathology, University of Maryland School of Medicine, Baltimore, Maryland 21201 (79, 171)

T. SATO, Department of Pathology, Maryland Cancer Program and Maryland Institute for Emergency Medicine, University of Maryland School of Medicine, Baltimore, Maryland 21201 (309)

WARREN I. SCHAEFFER, Department of Medical Microbiology, College of Medicine, University of Vermont, Burlington, Vermont 05405 (457)

W. SCHÜRCH, Department of Pathology, Hotel-Dieu Hospital, Montreal, Quebec, Canada (331)

LOUISE B. MALAN-SHIBLEY, Endocrine Carcinogenesis Section, Frederick Cancer Research Center, Frederick, Maryland 21701 (107)

ALPHONSE E. SIRICA, Departments of Oncology and Pathology, McArdle Laboratory for Cancer Research, The Medical School, University of Wisconsin, Madison, Wisconsin 53706 (441)

M. W. SMITH, Department of Pathology, Maryland Institute for Emergency Medicine, University of Maryland School of Medicine, Baltimore, Maryland 21201 (309)

ROBERT STEIN, Cancer Bioassay Laboratory, Department of Pathology, Medical College of Pennsylvania, Philadelphia, Pennsylvania 19129 (287)

GARY D. STONER, Human Tissue Studies Section, Laboratory of Experimental Pathology, National Cancer Institute, National Institutes of Health, Bethesda, Maryland 20205 (429)

KURT STROMBERG, Viral Pathology Section, Laboratory of Viral Carcinogenesis, Division of Cancer Cause and Prevention, National Cancer Institute, Bethesda, Maryland 20205 (227)

RUY TCHAO, Cancer Bioassay Laboratory, Department of Pathology, Medical College of Pennsylvania, Philadelphia, Pennsylvania 19129 (287)

JERRY S. TRIER, Division of Gastroenterology, Department of Medicine, Peter Bent Brigham Hospital and Department of Medicine, Harvard Medical School, Boston, Massachusetts 02115 (365, 475)

A. TRIFILLIS, Department of Pathology, Maryland Cancer Program and Maryland Institute for Emergency Medicine, University of Mary-

[1] *Present address:* Department of Pathology, University of Kentucky School of Medicine, Lexington, Kentucky 40506.

land, School of Medicine, Baltimore, Maryland 21201 (309)

B. F. TRUMP, Department of Pathology, Maryland Cancer Program and Maryland Institute for Emergency Medicine, University of Maryland School of Medicine, Baltimore, Maryland 21201 (79, 171, 309, 327, 331, 429)

DAVID TSURUMOTO, Department of Obstetrics and Gynecology, Rush Medical College, Rush Presbyterian St. Lukes Medical Center, Chicago, Illinois 60612 (29)

M. J. VOCCI, Department of Pathology, University of Maryland School of Medicine, Baltimore, Maryland 21201 (79, 331)

LESLIE A. WALTON, Department of Obstetrics and Gynecology, University of North Carolina School of Medicine, Chapel Hill, North Carolina 27514 (1)

GEORGE D. WILBANKS, Department of Obstetrics and Gynecology, Rush Medical College, Rush Presbyterian St. Lukes Medical Center, Chicago, Illinois 60612 (29)

PREFACE

One of the central problems in biomedical research is the extrapolation of data from experimental animals to the human situation. This is especially a problem in studies of chronic diseases such as atherosclerosis, diabetes, and cancer. Extrapolation is made difficult by variation among species as well as variation among individuals within a species. Such variation may be wide, particularly in outbred populations such as humans. Finally, within a single individual one may find variations in response to exogenous agents among different tissues and within a tissue among different cell types.

The use of cultured human tissues and cells for investigations in biomedical research can be a logical bridge between experimental animals and the beneficiary of our research, humans. Volumes 21A and 21B of *Methods in Cell Biology* describe the current state of methodology in culturing human tissues and cells. The organization is on the basis of organ systems and both explant and cell culture approaches are included. A common strategy in this field of research is initially to develop the methodology with cells and tissues from experimental animals and then to adapt this methodology for use with human cells and tissues. Thus far, some investigations have been conducted only in cells and tissues from experimental animals. Studies with human material are still at an early stage of development. A few chapters are included that reflect the progression in methodology from experimental animals to humans. In general, this progression has been easier than most of us would have predicted.

The primary objective of these methodologic studies is to provide experimental tools for investigating the pathogenesis of human disease. In fact, as noted in many of the chapters, cultured human cells and tissues are already being used in conjunction with sophisticated methods to investigate a wide variety of important problems in biomedical research. The purpose of these volumes is to further encourage future investigations.

CURTIS C. HARRIS
BENJAMIN F. TRUMP
GARY D. STONER

Chapter 1

Studies of Human Endometrium in Organ Culture

DAVID G. KAUFMAN, THOMAS A. ADAMEC, LESLIE A. WALTON,[1] CHARLES N. CARNEY, SUSAN A. MELIN, VALERIO M. GENTA, MARC J. MASS, B. HUGH DORMAN, NANCY T. RODGERS, GUY J. PHOTOPULOS,[2] JUDITH POWELL,[3] AND JOE W. GRISHAM

Department of Pathology, University of North Carolina School of Medicine, Chapel Hill, North Carolina

---

[1]Department of Obstetrics and Gynecology.

[2]Current address: Department of Obstetrics and Gynecology, University of Tennessee School of Medicine, Memphis, Tennessee 38163.

[3]Current address: Department of Pathology, University of Kentucky School of Medicine, Lexington, Kentucky 40506.

ISBN 0-12-564140-0

I. Introduction

Approximately 37,000 new cases of endometrial carcinoma will be diagnosed in the United States this year. Thus, endometrial carcinoma is now the most prevalent cancer of the female genital tract in the United States, but only limited insight into the etiology of this disease has been elucidated by epidemiological studies. Consequently, we have attempted to learn more about the pathogenesis of endometrial cancer by studying this target tissue directly as organ cultures *in vitro*.

Efforts to study structurally intact endometrial tissue are not new. As early as the 1920s, attempts were made to maintain endometrial tissue *in vitro*. These investigations used menstrual effluent or tissue from endometrial curettings and evaluated the morphology of cellular outgrowths from explants maintained for up to 1 month on plasma clots. Randall, Stein, and Stuermer were the first group to systematically investigate the cytodynamic properties of human endometrial tissue derived from biopsies. In their first study, Randall *et al.* (1950) maintained tissue fragments immersed in fluid media for up to 35 days. Parameters investigated for their effect on culture maintenance included gas composition, pH, and temperature. Other studies (Stuermer and Stein, 1950) utilized the hanging drop and roller tube culture techniques. Later studies (Stuermer and Stein, 1952) evaluated variations in biochemical properties of the tissue including respiration, anaerobic and aerobic glycolysis, and representative enzyme pathways as a function of the menstrual cycle. Ehrmann *et al.* (1961) maintained endometrial fragments in organ culture on serum-agar plates. Although their cultures often evidenced extensive stromal necrosis, the epithelium was well maintained, in one case for as long as 12 weeks. Morphologic evidence of secretory activity was observed in the epithelium of the cultured tissue in the absence of supplemental hormones, as well as in the presence of estrogen, progesterone, or both hormones. Hughes *et al.* (1969) placed endometrial organ cultures on filter membranes supported by stainless-steel grids in organ culture dishes. Using fluid medium supplemented with serum, they observed biochemical and morphologic responses to estrogen and progesterone in endometrial cultures maintained for up to 18 days *in vitro*. Specifically, the morphology of cultures was better preserved when media were supplemented with both estrogen and progesterone at near physiologic levels as compared to media without both hormones. More recent studies by a number of investigators have dealt largely with the variations of macromolecular synthesis and hormone receptor levels as functions of the menstrual cycle state. In these studies tissue generally has been maintained *in vitro* for short periods of time to characterize biochemical properties or observe hormonal effects. These observations usually were made with short-term culture, typically within 1 or 2 weeks.

The ultimate goal of this project is to gain a greater understanding of the process of chemical carcinogenesis in human endometrium. One possible avenue for accomplishing this goal is to observe long-term toxic and carcinogenic effects of chemicals on this tissue. Such observations require that methods be developed for the long-term maintenance of this tissue *in vitro* as organ cultures. Furthermore, such methods should preserve, as far as possible, the morphologic and biochemical properties of the endometrium. In this chapter we describe such an *in vitro* model designed for long-term maintenance of human endometrial explants in organ culture.

II. Materials and Methods

A. Obtainment of Human Endometrial Tissue

The use of human tissue for experimental studies requires far more clinical judgment than would be required for use of animal tissues. Scientific investigations with human tissue must not compromise the clinical care of the patient or impair a complete and accurate pathological diagnosis. Our studies have utilized endometrial tissue obtained from hysterectomy specimens. Effort has been made to select carefully a small amount of tissue for experiments, in order not to interfere with pathological diagnosis. For this reason we have developed a detailed procedure for evaluating patients and their tissue prior to use in our studies. Obtainment of tissue from other human organs poses problems because operative specimens generally are removed for intrinsic disease. Obtainment of human endometrial tissue is in a different category because there are other gynecologic disorders in addition to endometrial disease that necessitate hysterectomy. Nonetheless, the procedures which have been utilized to obtain human endometrial tissue may benefit investigators who wish to use this or other human tissues.

Our tissue obtainment procedure requires a judgment to be made by the patient, the physician, and the laboratory investigator. Patients admitted for elective hysterectomy undergo preliminary screening to eliminate individuals who cannot freely offer informed consent (mentally incompetent persons, minors, prisoners, acutely ill patients, etc.) and those whose clinical history suggests the presence of intrinsic endometrial disease. The remaining patients are informed about the objectives of the project, the method of specimen collection, and the possible benefit to humanity. They are informed fully that no direct benefit will accrue to them as individuals. Possible risks related to their participation are outlined. Only those patients who freely offer their informed consent are considered as donors. (In fact, no patient so informed has refused to participate.) During the operation the surgeons re-evaluate the patient's condition to determine whether unsuspected findings mitigate against including the patient in the study. Providing that the operative findings do not interdict selection of the tissue, a pathologist on the staff of the project is summoned to the operating room, reviews the pelvic findings with the surgeon, receives the specimen, and, in a separate room, aseptically bisects the uterus from cervix to fundus. The specimen is next evaluated by a senior attending surgical pathologist for gross evidence of endometrial abnormality. This pathologist then judges whether the specimen can be used in the study and outlines the specific region of the endometrium which can be sampled for experimental purposes. Finally, the research staff may judge whether any features of the specimen preclude its use (endometrial atrophy, inflammatory exudate, gross hemorrhage, etc.). Only when all of these requirements have been met is the specimen accepted for experimental study. Up to one-half of the endometrial tissue is removed for experimental use; the remainder of the uterus is evaluated diagnostically by the surgical pathology service.

B. Sampling of Tissue and Transport to Laboratory

With the exception of sampling the tissue on an open bench top, every practical attempt is made to maintain aseptic conditions. For experimental study, a thin perpendicular slice of the selected endometrial tissue and superficial myometrium is removed with a sterile scalpel blade. This slice is used for morphologic evaluation and immediately is placed in ice cold 2.5% glutaraldehyde (Polysciences, Warrington, Pennsylvania) in 0.1 *M* sodium cacodylate buffer (Polysciences), pH 7.4, for fixation. The remainder of the endometrial tissue from the designated area is scraped from the uterus with a scalpel held perpendicular to the surface. This allows the separation of sheets of endometrial tissue from the underlying myometrium. The tissue immediately is placed in transport medium (see Section II, C) in a sterile container and is transferred to the tissue culture laboratory usually within 30 minutes after hysterectomy.

C. Organ Culture Procedures

Once the human endometrial tissue is received in the tissue culture laboratory, it is washed by swirling with three changes of transport medium. The sheets and fragments of tissue next are transferred to a 100-mm plastic tissue culture dish. The tissue is minced into approximately 2-mm cubes by manual slicing with two opposed scalpels. The tissue fragments are placed on Millipore filters (HA, 0.2 μm) and suspended by stainless-steel mesh in organ culture dishes (Falcon, Cockeysville, Maryland). Usually four explants are placed in each dish, and an average of 20 dishes is prepared from each specimen. The absorbant pad in the outer well of the organ culture plate is saturated with 2–3 ml of sterile water and sufficient medium is placed in the center well to barely lift the filter off the screen support. Medium is changed every 4 days by aspirating spent medium using sterile pasteur pipets. Organ culture assemblies are changed every 2 weeks. The cultures are incubated in a water-jacketed incubator at 37°C in 95% air and 5% CO_2.

The culture medium used routinely is CMRL 1066, although Eagle's basal medium (BME) with Earle's salts and Ham's F-12 were evaluated in preliminary experiments for their ability to support long-term maintenance of tissue. All three media are obtained as 10× concentrations from Grand Island Biological Company (GIBCO), Grand Island, New York and are diluted prior to use. The media are supplemented with 10% heat-inactivated fetal calf serum, 100 U/ml penicillin, and 100 μg/ml streptomycin (all obtained from GIBCO). The media are also buffered with 25 m*M* HEPES (Sigma Chemical Company, St. Louis, Missouri) and adjusted to pH 7.4. CMRL 1066 is further supplemented with 10 m*M* L-glutamine (GIBCO) for routine cultures.

The transportation and washing medium consists of supplemented CMRL 1066 with 10 times the normal concentrations of penicillin and streptomycin plus 0.5 mg/ml gentamicin (Schering Corp., Kenilworth, New Jersey).

Our studies include the use of various hormonal supplements. Insulin is used at a concentration of 4 mg/liter (crystalline bovine, 24 IU/mg, Sigma). Estrogen is in the form of 17β-estradiol (Calbiochem, San Diego, California), and when used alone is at a concentration of 10^{-7} *M*. When used with progesterone, the concentration of estradiol is 10^{-8} *M*. Progesterone (Sigma) is used at a concentration of 10^{-7} *M*. Both steroid hormones are prepared in 10^{-4} *M* stock solutions in 100% ethanol, then diluted to a 10^{-5} *M* working stock in sterile culture medium. Hormonally supplemented culture medium is prepared by adding these stock solutions to the basal medium in the amounts necessary to obtain the desired concentrations.

In experiments using serum-free medium, fatty acid-free crystalline human serum albumin (Sigma) is used at concentrations of 0.6, 1.5, and 3.0%. Albumin lots are assayed for the presence of estradiol, and only those lots with undetectable amounts (< 10 pg/ml) are used.

D. Methods for Morphologic Evaluation

Tissue specimens are fixed in 2.5% glutaraldehyde (Polysciences) or 4% paraformaldehyde (Fisher Scientific, Fairlawn, New Jersey) buffered with 0.1 *M* sodium cacodylate (Polysciences), pH 7.4. The tissue is then washed in 0.1 *M* sodium cacodylate buffer and dehydrated through graded concentrations of ethanol. For high-resolution light microscopy, the specimens are embedded in glycol methacrylate (Polysciences). Sections are cut at 1 μm on a Sorvall JB-4 microtome and stained with hematoxylin and eosin or periodic acid–Schiff (PAS) stains. When 1 μm sections are to be used for autoradiography, unstained sections are coated with undiluted NTB2 track emulsion (Eastman Kodak, Rochester, New York) in the dark. Slides are stored in light-tight slide boxes with dessicant for 2 weeks and then the emulsion is developed in D-19 developer (Eastman Kodak). Slides are finally stained with Giemsa stain through the emulsion.

Tissue to be evaluated by electron microscopy is fixed as described above, washed with 0.1 *M* sodium cacodylate buffer, pH 7.4, and postfixed with 2% osmium tetroxide (Fisher Scientific) in the same buffer. The tissue then is dehydrated through graded ethanols, cleared with propylene oxide (Polysciences), and embedded in Epon 812 (Polysciences). Sections are placed on grids and examined with a JEOLCO 100B electron microscope at 60 kV.

E. Methods for Biochemical Evaluation

Tritiated thymidine ([^{3}H]Thd, 16 Ci/mmole), uridine ([^{3}H]Urd, 40 Ci/mmole), or leucine ([^{3}H]Leu, 40 Ci/mmole), each obtained from New England Nuclear (Boston, Massachusetts), are added singly to medium at 50 μCi/ml concentrations and the corresponding unlabeled compound is omitted from the medium to minimize isotopic dilution with reduction of specific activity. Tissue specimens are cultured in these media for 4.5 hours and then removed and washed three times in Ca^{2+}–Mg^{2+}-free Hank's saline solution (GIBCO) to remove unincorporated tritiated precursors. When the objective is to study the incorporation of precursors using autoradiography as the analytic technique, the process previously specified is used. To determine the specific activity of precursor incorporation into macromolecules, the explants are first homogenized in a rough glass Dounce homogenizer with tight-fitting pestle in 2 ml of a solution containing 0.15 *M* NaCl, 0.1 *M* Tris–HCl, pH 7.4, and 5 m*M* Na_3EDTA. Because of the small size of the available tissue explants, large quantities of macromolecules are added to facilitate precipitation. Samples labeled with [^{3}H]Thd or [^{3}H]Urd receive 1.5 mg of bovine serum albumin (BSA, Sigma, St. Louis, Missouri); 400 μg of calf thymus DNA (Worthington Biochemical Co., Freehold, New Jersey) is added to samples labeled with [^{3}H]Leu. After mixing, two volumes of 1.2 *N* perchloric acid (PCA) is added to precipitate macromolecules. After 15 minutes at 4°C the precipitate is pelleted by centrifugation at 800 *g* for 15 minutes at 4°C. The pellet then is washed repeatedly with cold 0.2 *N* PCA and repelleted until no radioactivity is present in the supernatant. The pellet next is dissolved in 1 ml of 0.3 *N* NaOH and incubated at 37°C for 1 hour. This procedure results in approximately 95% of the RNA being acid soluble when macromolecules are reprecipitated with 1.2 *N* PCA. The RNA hydrolysate is the supernatant recovered following this reprecipitation of macromolecules. Radioactivity and UV absorbance at 260 nm are determined in this supernatant. From these data the specific activity (DPM per microgram RNA) of incorporation of uridine is calculated. After washing with 0.2 *N* PCA until only background radioactivity is found, the pellet is suspended in 1.2 *N* PCA and DNA is hydrolyzed during incubation at 60°C for 1 hour. The supernatant is the DNA hydrolysate used to evaluate [^{3}H]Thd incorporation. Radioactivity and UV absorbance (260 nm) in the hydrolysate are determined to yield a specific activity (DPM per microgram DNA) of incorporation of thymidine. The [^{3}H]Leu incorporation is determined in the pellet after dissolving in 0.3 *N* NaOH. The specific activity of incorporation of leucine is determined by quantitating the radioactivity and protein concentration in the solubilized pellet. Proteins are quantitated using an assay described by Schaffner and Weissman (1973), modified for use with smaller volumes.

A series of samples are included to determine the background incorporation of radioactivity and these samples are processed in parallel with each batch of specimens. Endometrial explants are dipped in labeled medium containing the radioactive precursors, then washed immediately, and processed in order to determine radioactivity adsorbed to endometrial specimens but not incorporated into macromolecules. Samples containing only BSA are processed to correct for UV absorbant material released after alkaline or acid hydrolysis. Samples containing only DNA carrier are processed to detect contaminants remaining after alkaline and acid hydrolysis which would interfere with the quantitation of proteins.

III. Results

A. Transportation Techniques

The hospital in which the operative specimens are obtained is near the tissue culture laboratory. Consequently, optimization of techniques for transporting tissue was simple. Nonetheless, we have investigated two main parameters concerned with obtaining and transporting tissue: (*a*) whether tissue preservation was best maintained when endometrial samples initially were scraped from the uterus or when they were removed *en bloc,* and (*b*) whether transportation of the tissue was best accomplished in medium at room temperature or in an ice bath. It appeared that all procedures provided acceptable morphologic preservation of the tissue during the brief period (10 minutes) required to reach the tissue culture laboratory. There was only marginal morphologic evidence to prefer the use of scraped tissue maintained in ice-cold media. Scraping the endometrium from the uterine cavity did not appear to affect the viability of the tissue; it did, however, minimize the amount of extraneous myometrium present in the specimen. Since a lower temperature decreases the metabolic rate and, hence, partially mitigates effects of oxygen deprivation, we chose to transport the tissue in ice-cold media.

Contamination of tissue specimens was an intermittent problem in the initial portion of our studies, perhaps because tissue sampling had to be done on an open bench top. To combat this problem, culture media used to transport tissue is supplemented with 10 times the normal concentrations of penicillin and streptomycin, plus 0.5 mg/ml gentamicin. Upon arrival in the laboratory, the tissue is washed with three changes of this high-antibiotic medium.

B. Organ Culture Support

The efficiency of a variety of culture techniques was evaluated for long-term maintenance of endometrial tissue as organ cultures. Organ cultures were plated directly onto tissue culture plates, placed on top of gelfoam sponges or on tissue culture plates within a ring of gelfoam sponge, and on a Millipore filter membrane lying on top of a stainless-steel grid supported on a Falcon organ culture dish. Organ cultures plated directly on plates with or without a ring of gelfoam tended to develop explant outgrowths. Also oxygen diffusion was insufficient for long-term survival of the explants. Cultures supported by gelfoam sponges initially survived well but after a period of time in culture the gelfoam sponge disintegrated leaving the tissue unsupported. The best system for long-term maintenance of organ cultures was the organ culture dish with tissue supported on a Millipore filter. This system appeared to provide the cultures with adequate oxygenation and a high humidity environment with good diffusion of nutrients.

C. Optimization of Culture Conditions

A number of parameters have been evaluated in an effort to determine culture conditions that best maintain human endometrial tissue as organ cultures for extended periods of time. The principal method used to determine the quality of tissue maintenance under various experimental conditions has been morphologic evaluation using both light and electron microscopy. The criteria for judging whether specific culture conditions were beneficial were the preservation of viable epithelial and stromal elements with organotypic interrelationship and with a minimum of cellular necrosis and tissue disorganization.

Three commercially available media were evaluated for their ability to support the long-term maintenance of human endometrial organ cultures. One medium used was Eagle's basal medium (BME) with Earle's salts, a medium containing salts, vitamins, and amino acids. Ham's F-12 is a medium further enriched with trace elements, additional amino acids, linoleic acid, and glucose. CMRL 1066 does not contain the same trace elements or linoleic acid but does contain additional amino acids as well as deoxynucleotides, enzyme cofactors, cholesterol, and a nonionic detergent. Evaluation of the various media was based on morphological properties of the organ cultures prepared from the same endometrial specimen following maintenance of organ cultures in the specific medium.

The histologic appearance of normal endometrial tissue is illustrated (Fig. 1) for purposes of comparison. Experiments were run for a maximum of 8 weeks and samples were taken for morphologic study at 2-week intervals. All three media appeared to be satisfactory for maintenance of organ cultures for periods of up to 4 weeks. However, after 8 weeks, organ cultures maintained in BME (Fig. 2) showed disintegration of the stroma and deterioration of the epithelium. The epithelial cells varied from flat to low cuboidal and discontinuities in the epithelium were numerous. Organ cultures maintained in Ham's F-12 for 8 weeks (Fig. 3) showed somewhat better preservation of the stroma, but the epithelium had extensive discontinuities and remaining cells evidenced necrosis. In comparison, organ cultures maintained in CMRL 1066 for 8 weeks (Fig. 4) had largely intact epithelium with cuboidal cells lining glands and surfaces and contained occasional mitotic cells. Although there was extensive deterioration of the stroma, a proportion of stromal cells survived. Thus, because of the somewhat better morphologic preservation of endometrial tissue in CMRL 1066, this medium was chosen for routine use in long-term maintenance of organ cultures.

Further efforts to establish a suitable medium for long-term survival of endometrial organ cultures next turned to a consideration of the effects of glutamine concentration. Previous studies had suggested that endocrine-responsive tissues required an increased concentration of glutamine (Steinberger *et al.*, 1970; George and Solomon, 1978). The results indicated that a high concentration of glutamine greatly increased morphologic preservation of both stroma and epithelium. When explants maintained in CMRL 1066 with a 10 m*M* concentration of glutamine (Fig. 5) were compared with explants maintained in CMRL 1066 with 1 m*M* concentration of glutamine, samples revealed better stromal viability and good epithelial preservation.

Previous studies with short-term cultures of endometrial tissue have either employed media which contained insulin (Shapiro and Forbes, 1978) or used insulin as a supplement to the medium (Hustin, 1975). Addition of 96 IU/liter of insulin to CMRL 1066 medium resulted in improvement of stromal viability comparable to that observed in the presence of high glutamine (Fig. 5). When CMRL 1066 was supplemented with both insulin (96 IU/liter) and a high concentration of glutamine (10 m*M*) (Fig. 6), both the stroma and epithelium of organ cultures were well maintained and were quite similar to the corresponding structures in the normal (uncultured) tissue (Fig. 1). Based upon these observations, all later organ culture studies were performed with both insulin and glutamine (at 10 m*M*) as supplements to the 1066 medium.

The necessity for the inclusion of serum albumin in the culture medium was evaluated in experiments in which three different concentrations of human serum albumin, 0.6, 1.5, and 3.0%, were substituted for fetal calf serum. When cultures were maintained for 4 weeks in media with 0.6 or 1.5% albumin extensive

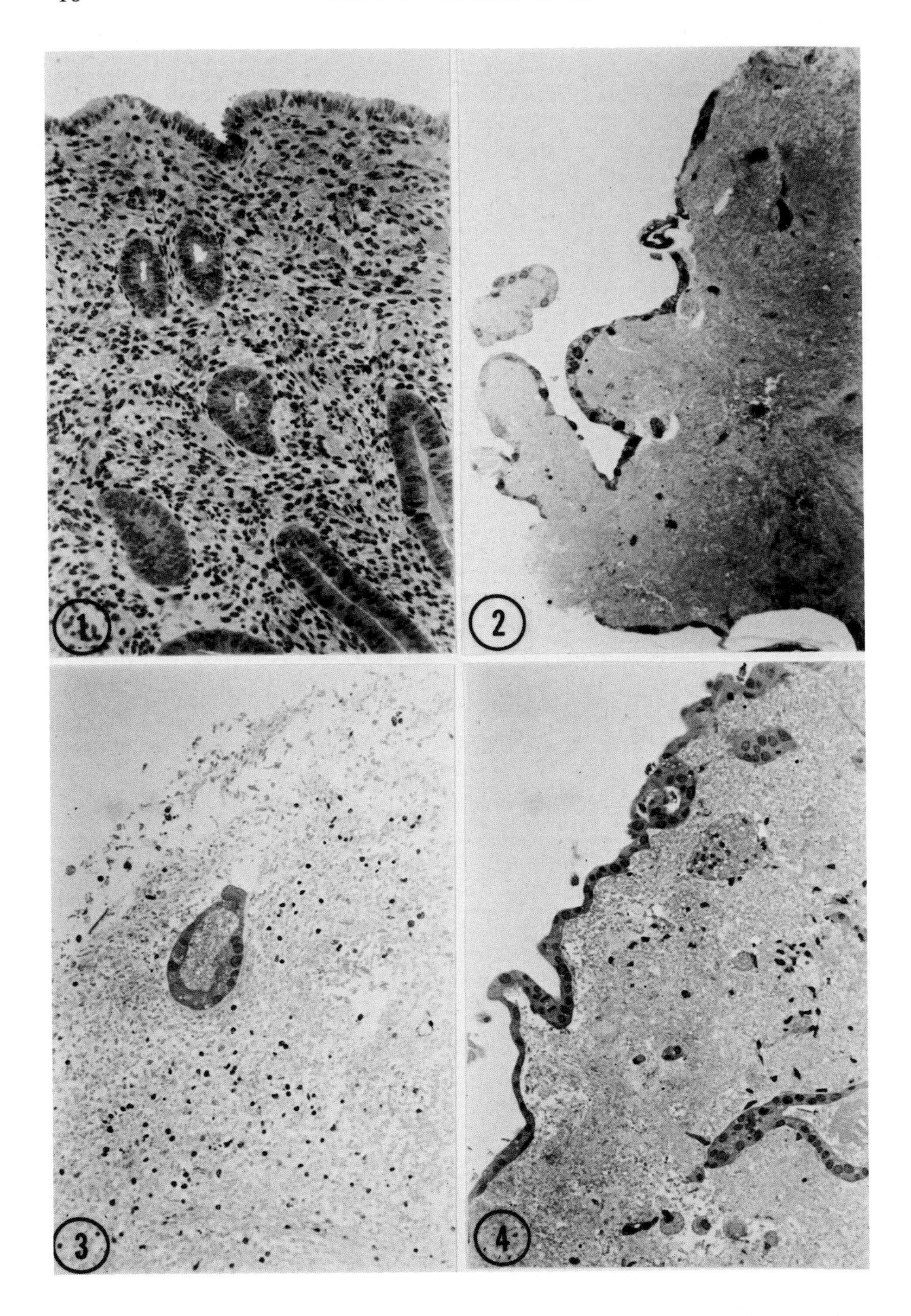
1
2
3
4

deterioration of both epithelial and stromal components was found. In contrast, even after 4 weeks, cultures maintained in 3.0% albumin (Fig. 7) appeared viable although there was stromal deterioration and epithelial simplification.

Female steroid sex hormone supplements have also been thought to benefit the maintenance of endometrial organ cultures. Because of the well-known proliferative influence of estrogens on the endometrium *in vivo,* studies were performed to evaluate the effects on organ culture maintenance of supplementing the medium with 17β-estradiol (10^{-7} *M*). Whereas estradiol improved the appearance of endometrial organ cultures during short intervals of maintenance, cultures continuously exposed to medium containing estradiol had decreased viability with necrosis of stroma and low attenuated epithelium (Fig. 8).

Since human endometrium is subjected to cyclic variations in steriod hormone concentrations during menstrual cycles *in vivo,* further studies were designed to evaluate the possibility that long-term maintenance of endometrial organ cultures might be improved by continuation of such cycling *in vitro*. Experiments were performed in which cultures were maintained in medium containing 17β-estradiol (10^{-7} *M*) alternating at 14-day intervals with medium containing 17β-estradiol (10^{-8} *M*) plus progesterone (10^{-7} *M*). These conditions were chosen to approximate roughly the hormonal milieu of the endometrium during the proliferative and secretory phases, respectively, of the menstrual cycle *in vivo*. The initial hormonal composition of the medium used for cultures and the date for changing to the alternate medium were determined from the date in the patient's menstrual cycle when the tissue is obtained. For example, endometrial tissue obtained at day 7 of the menstrual cycle, which is day 7 of the proliferative phase, would be cultured in medium with 10^{-7} *M* estradiol and no progesterone for 7 days prior to changing to the medium with both estradiol and progesterone (simulating the secretory phase). Histologic examination of organ cultures that had undergone repetitive hormonal cycling over prolonged intervals revealed that the cultures retained morphologic characteristics with well-preserved epithelium and viable stroma (Figs. 9–12).

FIG. 1. Uncultured human endometrium in the proliferative phase of menstrual cycle. There is loose cellular stroma surrounding glands composed of cuboidal to columnar epithelium. H + E, × 370.

FIG. 2. Human endometrium in organ culture for 8 weeks in BME. The epithelium is constituted by a discontinuous single layer of pyknotic cells and the stroma is largely acellular and degenerated. H + E, × 380.

FIG. 3. Human endometrium in organ culture for 8 weeks in Ham's F-12. The surface epithelium has degenerated and only an occasional remnant of glandular epithelium is seen. The stroma has undergone extensive degeneration but retains some cellularity. H + E, × 620.

FIG. 4. Human endometrium in organ culture for 8 weeks in CMRL 1066. The epithelium is intact and cuboidal, and displays mitotic activity. The stroma is largely necrotic but has foci of surviving cells. H + E, × 390.

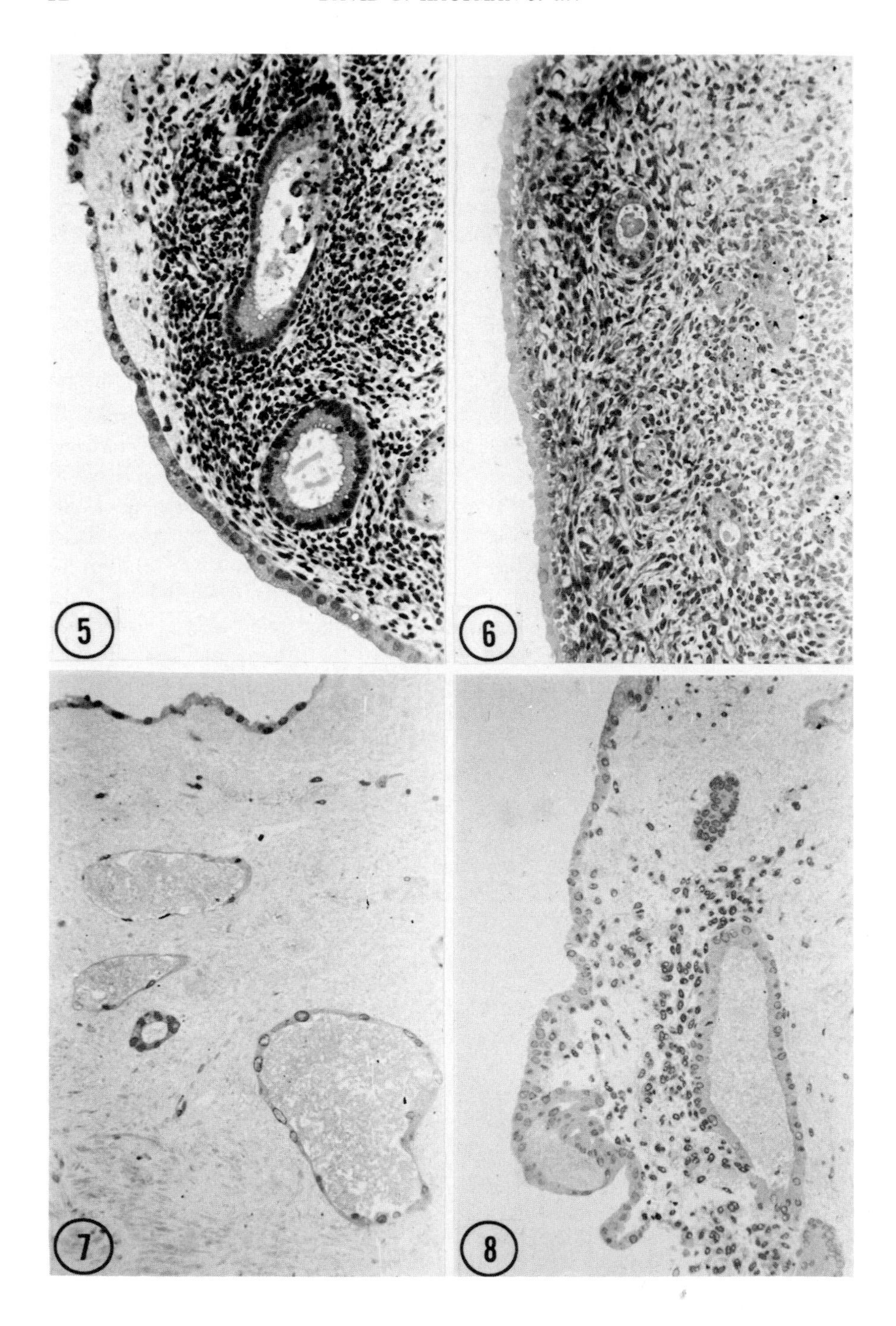
5
6
7
8

Organ cultures exposed to repetitive cycles of maintenance in media with alternating hormonal supplements were also evaluated to determine whether the tissue retained the ability to respond to hormonal stimuli after prolonged intervals in culture. Figure 9 illustrates an organ culture maintained with alternating hormonal supplements for 131 days and harvested during what corresponds to the "proliferative phase" when only estradiol (10^{-7} *M*) is added to the medium. In contrast, Fig. 10 illustrates an organ culture maintained for 126 days and harvested during the "secretory phase" when both estradiol (10^{-8} *M*) and progesterone (10^{-7} *M*) were present in the medium. In this case the cells have subnuclear vacuolization indicative of secretory activity and glands are tortuous and branched. Evidence of hormonal response is more apparent by transmission electron microscopy. Figure 11 is an electron micrograph of a culture maintained with alternating hormonal supplements for 226 days and harvested during the "proliferative phase." The cells are simple and show no evidence of secretory activity. The epithelium also has microvilli, tight cellular junctions, and desmosomes. In comparison, Fig. 12 is an electron micrograph of a culture from the same specimen maintained for 231 days and harvested during the "secretory phase." The cells have well-developed rough endoplasmic reticulum, secretory vacuoles, and more complex "active" nuclear configuration, indicative of a secretory response to progesterone.

D. Biochemical Observations

Biochemical studies of endometrium in organ culture involved the analysis of incorporation of radioactively labeled thymidine, uridine, and leucine as precursors of DNA, RNA, and protein. Two major experimental designs were employed to analyze this incorporation. To quantitate the extent of incorporation of the precursors that existed under the various experimental conditions, specific activities of macromolecules were determined following isolation of the DNA, RNA, and protein fractions. In addition, autoradiographic studies were performed under similar experimental conditions to obtain information on the rela-

---

FIG. 5. Organ culture maintained in CMRL 1066 with 10 m*M* glutamine. The epithelium is better differentiated and there is greatly improved stromal survival. H + E, × 380.

FIG. 6. Organ culture maintained in CMRL 1066 with 10 m*M* glutamine and 96 U/liter insulin. Note the marked improvement of stromal morphology; the stroma approximates the appearance of stroma in uncultured tissue. H + E, × 425.

FIG. 7. Organ culture maintained in CMRL 1066 with 10^{-7}*M* 17β-estradiol. After 35 days of continuous exposure to estrogen, the epithelium is low and atrophic, and the stroma is largely degenerated. H + E, × 400.

FIG. 8. Organ culture maintained in CMRL 1066 for 4 weeks with standard supplements but containing 3% human serum albumin in place of serum. The columnar epithelium is intact and although extensively degenerated the stroma retains foci of viable cells. H + E, × 390.

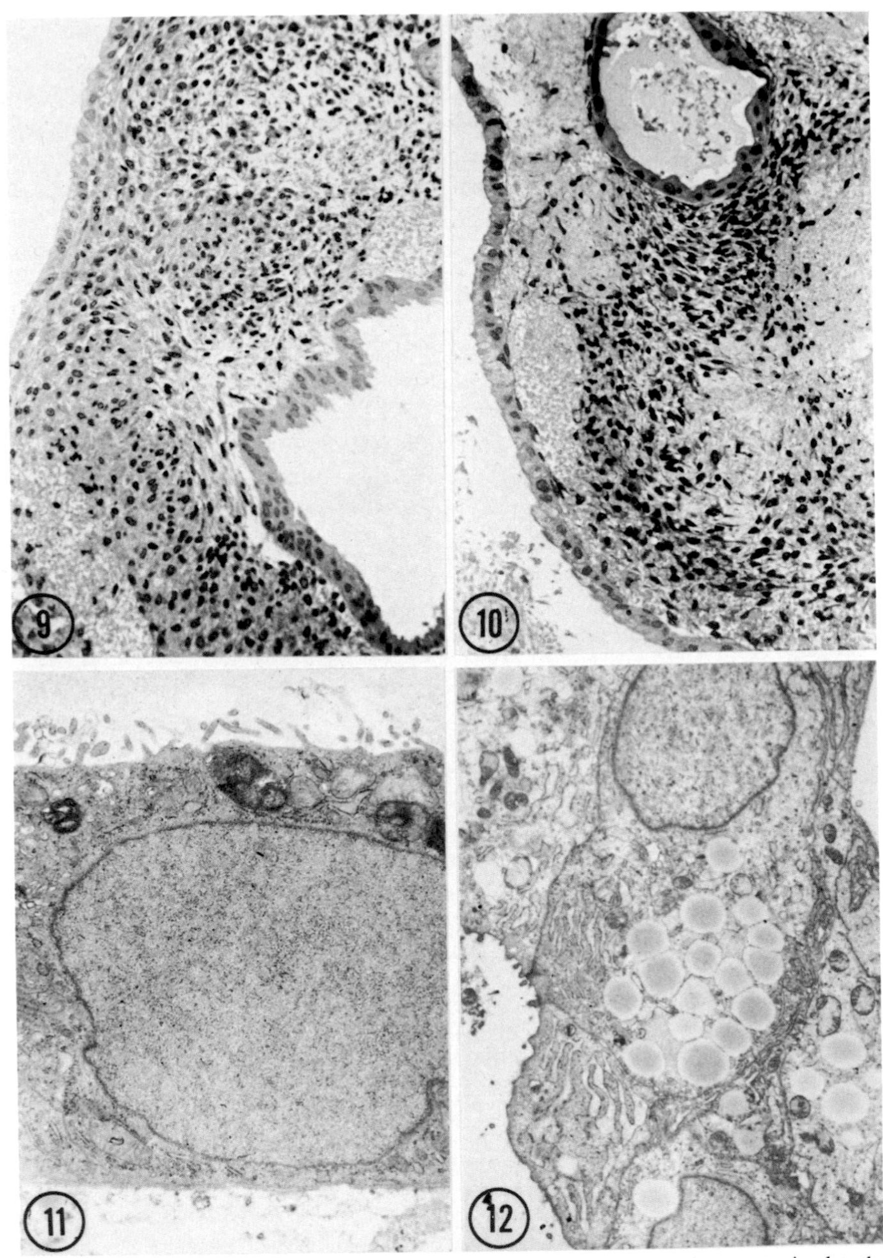

FIG. 9. Organ culture maintained in media with alternating hormonal supplements to simulate the menstrual cycle *in vitro;* harvested in the late "proliferative phase." The stromal elements are well maintained, and the epithelium is columnar and mitotically active. The epithelium has a "pseudostratified" appearance, similar to that seen at the corresponding menstrual cycle phase *in vivo*. H + E, × 380.

tive contribution of the various cell types and tissue structures within the culture to this incorporation.

Using the experimental approaches outlined, freshly obtained human endometrium which had been in organ culture for less than 5 hours was studied in relation to the menstrual cycle. Because the time between tissue obtainment and experimental analysis was short, it was presumed that endometrium placed in culture for this brief interval might be more representative of metabolic processes occurring *in vivo*. The specific activity of $[^3H]$Thd incorporated into endometrial DNA was minimal when determined in either the early proliferative phase or late secretory phase of the menstrual cycle. There was a well-defined peak of incorporation between days 12 and 16, corresponding to the late proliferative phase (Fig. 13). The overall extent of $[^3H]$Urd incorporation was greater than that for either $[^3H]$Thd or $[^3H]$Leu. The peak of $[^3H]$Urd incorporation (Fig. 14) occurred between days 14 and 20, slightly later in the menstrual cycle than the peak for $[^3H]$Thd. For early proliferative and late secretory phase tissue the specific activity of RNA was lower than at the peak, but remained above background levels. Incorporation of $[^3H]$Leu (Fig. 15) rose from near background levels in the middle of the proliferative phase to a peak in the later proliferative phase. Through the beginning and middle of secretory phase the level of $[^3H]$Leu incorporation remained elevated at approximately two-thirds of the peak value.

The incorporation of radioactive precursors into freshly obtained endometrial tissue in organ culture was also evaluated by autoradiography. Incorporation of $[^3H]$Thd largely was localized over the nuclei of cells (Figs. 16 and 17). At day 13 of the menstrual cycle, late in the proliferative phase, labeled nuclei were numerous and they were predominantly localized to epithelial cells (Fig. 16). In the middle of the secretory phase (day 21) labeled nuclei were far less numerous (Fig. 17). At this time there was virtually no labeling of the glandular epithelium; labeled cells were confined to the stroma. Under the conditions of labeling of organ cultures, autoradiographic grains from incorporated $[^3H]$Urd were most dense over nuclei but cytoplasmic labeling was also evident (Figs. 18 and 19). In

---

FIG. 10. Organ culture maintained under identical conditions as in Fig. 9 except that this explant was harvested in the late "secretory phase" of the simulated menstrual cycle. The epithelial component is columnar and contains vacuolated cells. The glands are dilated with intralumenal products. H + E, × 380.

FIG. 11. Electron micrograph of an organ culture maintained for 226 days with hormonal cycling and harvested in the "proliferative phase." There is little differentiation of cytoplasmic organelles and nuclei have a dispersed chromatin pattern. Note the microvilli and surface junctional apparatus typical of epithelial cells. × 4500.

FIG. 12. Electron micrograph of an organ culture maintained for 231 days with hormonal cycling and harvested in the "secretory phase" of the simulated menstrual cycle. There is marked cytoplasmic differentiation with secretory vacuoles and dilated endoplasmic reticulum indicating secretory activity. × 2500.

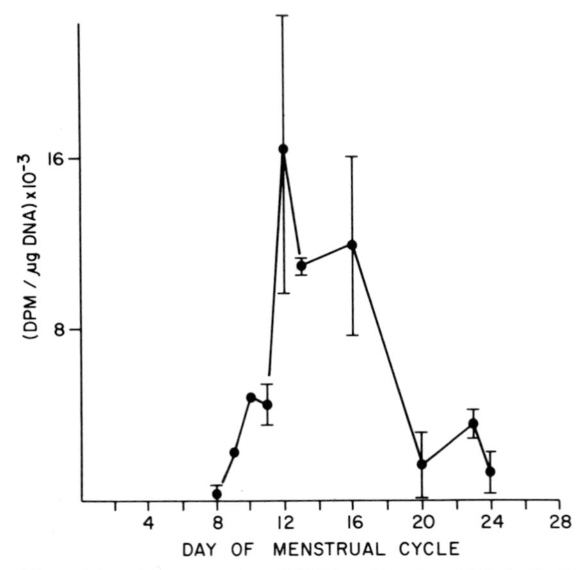

FIG. 13. Specific activity of incorporation of [^{3}H]thymidine into DNA in freshly obtained human endometrial tissue *in vitro* in relation to menstrual cycle.

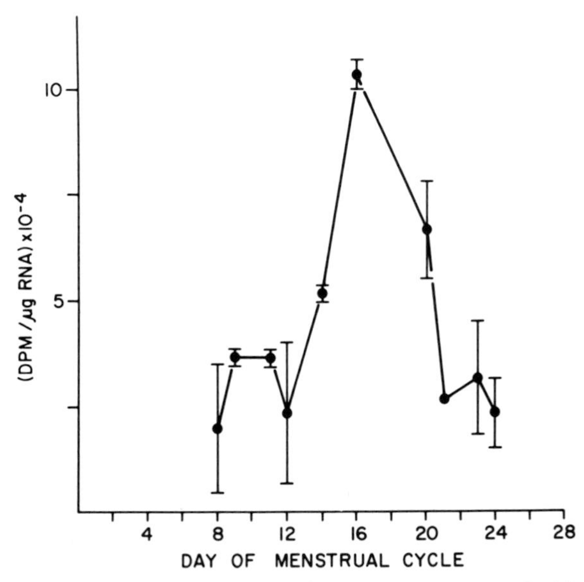

FIG. 14. Specific activity of incorporation of [^{3}H]uridine into RNA in freshly obtained human endometrial tissue in relation to menstrual cycle.

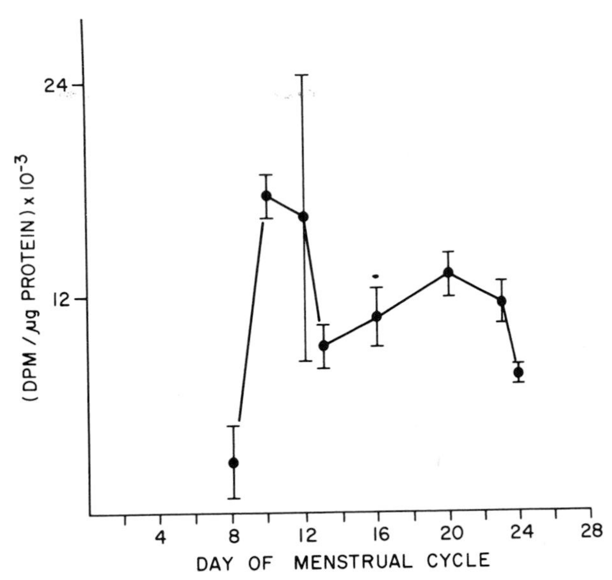

FIG. 15. Specific activity of incorporation of [^{3}H]leucine into protein in freshly obtained human endometrial tissue in relation to menstrual cycle.

the middle of the proliferative phase (day 8) there was a lower grain density over labeled cells (Fig. 18) than at day 13 in the late proliferative phase (Fig. 19). Uridine incorporation was largely localized to epithelial cells during the proliferative phase, whereas most labeled cells were in the stroma by the middle of the secretory phase (not illustrated). Autoradiographic grains from incorporated [^{3}H]Leu were evident over both nuclei and cytoplasm (Figs. 20 and 21). At day 8 in the middle of the proliferative phase the density of grains was low with only slightly greater labeling of epithelial cells than stromal cells (Fig. 20). In contrast, there was intense labeling of epithelial cells and some cells in the stroma at day 21 in the middle of the secretory phase (Fig. 21).

Endometrial organ cultures maintained *in vitro* for up to 200 days have also been evaluated for incorporation of radioactive precursors. Since the number of observations has been limited, these results should be regarded as preliminary. After prolonged intervals *in vitro,* organ cultures incubated with [^{3}H]Thd, [^{3}H]Urd, and [^{3}H]Leu for short intervals achieved specific activities in DNA, RNA, and protein that were quantitatively comparable to incorporation of the precursors in freshly obtained tissue. The extent of incorporation varied with the hormonal milieu in culture as it had with the menstrual cycle in the fresh tissue. For example, high levels of incorporation of [^{3}H]Thd were observed in organ cultures labeled during the middle and late portions of the simulated "proliferative phase" when estrogen but not progesterone was added to the medium.

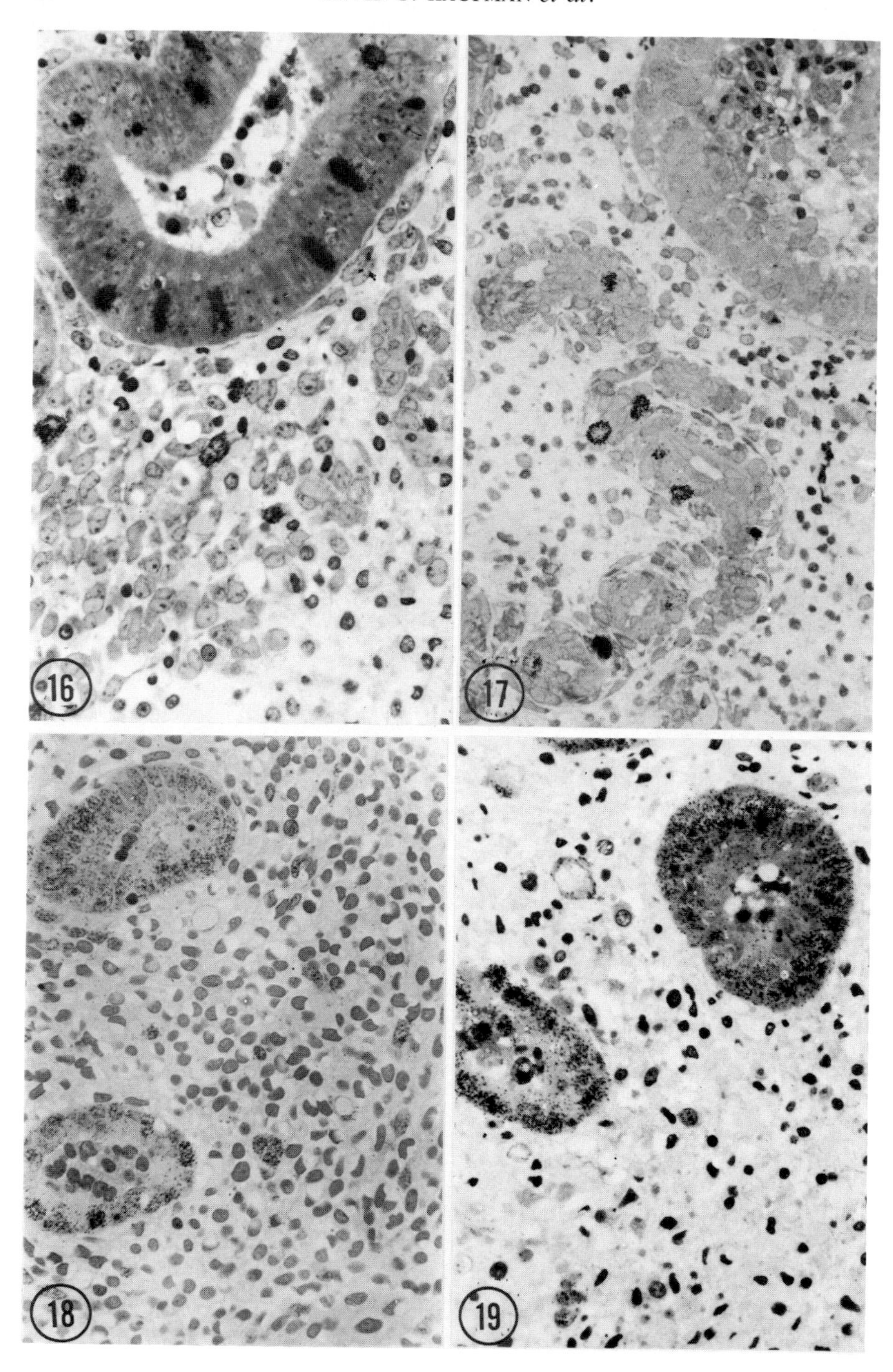
16
17
18
19

During the simulated "secretory phase" when the medium contained both progesterone and estrogen, incorporation of [^{3}H]Thd resulted in a lower DNA specific activity. Incorporation of radioactive precursors in organ cultures was also evaluated by autoradiography. Again, the extent of incorporation of radioactive precursors appeared comparable to that observed in the freshly obtained tissue. Autoradiographic grains were localized to cell nuclei when [^{3}H]Thd was incorporated, but overlay both nuclei and cytoplasm in cultures in which [^{3}H]Urd or [^{3}H]Leu had been incorporated. For each of the three precursors labeled cells were localized mainly to the epithelium during the simulated "proliferative phase" when the medium contained only estrogen as a hormone supplement. Again, the hormonal milieu in culture appeared to affect the extent of incorporation. Organ cultures incubated with [^{3}H]Thd during the simulated "proliferative phase" (Fig. 22) had far more labeled epithelial cell nuclei than did cultures incubated with [^{3}H]Thd during the simulated "secretory phase" (Fig. 23).

IV. Discussion

Development of techniques for long-term culture of endometrial tissue is crucial to the final realization of the goals of this project: the study of chemical carcinogenesis in human endometrium. The preceding results demonstrate that human endometrial tissue can be maintained as differentiated organ cultures *in vitro* for periods exceeding 6 months. Histologic appearance, macromolecular synthesis, and responsiveness to hormonal stimuli observed in organ cultures maintained for over 200 days *in vitro* were qualitatively similar to those of the freshly obtained tissue. Presumably these properties are like those of endometrial tissue *in vivo*.

In previous investigations endometrial tissue generally has been maintained in

FIG. 16. Autoradiogram of [^{3}H]thymidine labeling in freshly obtained human endometrium dated morphologically as day 13 of the menstrual cycle (late proliferative phase). Tissue was maintained as organ cultures for less than 5 hours after obtainment from the hysterectomy specimen. Autoradiographic grains are localized primarily over nuclei in the epithelium of a gland. Giemsa, × 1200.

FIG. 17. Autoradiogram of [^{3}H]thymidine labeling in freshly obtained human endometrium in the middle of the secretory phase (day 21) of the menstrual cycle. Labeled cells are confined to a blood vessel in the stroma. Epithelial cells in the gland (upper right) are not labeled. Giemsa, × 1300.

FIG. 18. Autoradiogram of [^{3}H]uridine labeling in human endometrial tissue in the middle of the proliferative phase (day 8). Grains are most dense over nuclei but the cytoplasm is also labeled. Label is localized to all epithelial cells in the glands but only a few of the stromal cells. Giemsa, × 1000.

FIG. 19. Autoradiogram of [^{3}H]uridine labeling in human endometrial tissue in late proliferative phase (day 13). Labeled cells are still primarily in the epithelium of glands but at this point in the cycle there is a considerably higher grain density. Giemsa, × 1000.

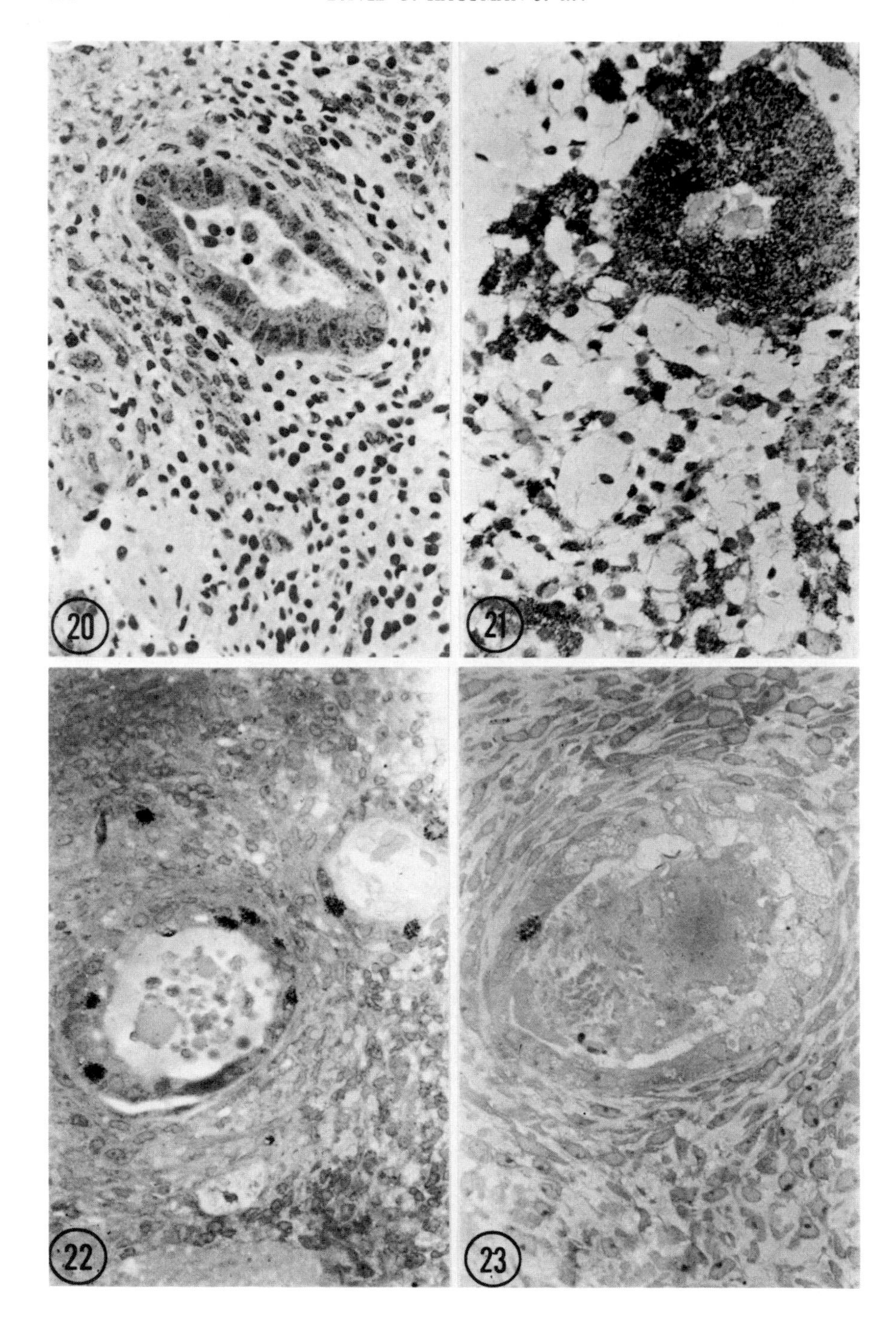
20
21
22
23

organ culture for short intervals, usually less than 2 weeks. This time interval probably sufficed to enable the research goals of those studies to be accomplished. Furthermore, the frequent occurrence of deterioration of cultures after a few days to 2 weeks *in vitro* (Ehrmann *et al.*, 1961; Hughes *et al.*, 1969) may have discouraged further attempts at prolonged maintenance. We have observed a similar initial deterioration of the endometrial tissue in culture. However, since the goals of our studies appeared to require far longer periods of organ culture, we maintained the cultures beyond this interval of early deterioration. We found renewed vitality of the tissue, presumably as a result of cellular regeneration. Maintenance of the culture for this additional period of time, however, does not suffice to explain the results reported, and other factors must be considered.

One feature of our methods which may have contributed to the sustained vitality of our organ cultures was the use of L-glutamine at concentrations much higher than those used in earlier studies with endometrial tissue. Most commercial tissue culture media, including CMRL 1066, contain glutamine at a concentration of 2 m*M* or less. Investigators culturing other steroid hormone target tissues *in vitro* have noted the need to add additional glutamine to raise concentrations to 4 m*M* or greater in order to improve differentiation of cultures and responsiveness to hormones (Steinberger *et al.*, 1970). Furthermore, recent experiments in rats have indicated that adult females metabolize glutamine at higher rates than males and newborns and that this difference is eliminated by ovariectomy (George and Solomon, 1978). These observations suggest that there is a critical interrelationship between glutamine and steroid hormones in hormone-responsive tissue which influences the differentiation of these tissues. Such a relationship may explain the apparent effect of high glutamine concentrations on endometrial organ cultures. The mechanism of such an effect is obscure, and the

---

FIG. 20. Autoradiogram of [^{3}H]leucine labeling in human endometrial tissue in the middle of the proliferative phase (day 8). Grains are distributed over both nucleus and cytoplasm and both epithelial and stromal cells are labeled. Giemsa, × 940.

FIG. 21. Autoradiogram of [^{3}H]leucine labeling in human endometrial tissue in the middle of the secretory phase (day 21) of the menstrual cycle. At this point in the menstrual cycle there is a considerably higher grain density. Giemsa, × 1200.

FIG. 22. Autoradiogram of [^{3}H]thymidine labeling in an organ culture of human endometrium maintained for 98 days in media with alternating hormonal supplements to simulate the menstrual cycle *in vitro* and harvested in the "proliferative phase." Labeled cells are located primarily in the epithelium of the glands and the grains are localized over cell nuclei. Giemsa, × 1000.

FIG. 23. Autoradiogram of [^{3}H]thymidine labeling in an organ culture of human endometrium maintained for 129 days in media with alternating hormonal supplements to simulate the menstrual cycle *in vitro* and harvested in the "secretory phase." There are few cells labeled with [^{3}H]thymidine under these conditions of culture which simulate the "secretory phase." Although not evident in this field, stromal cells are among those which are labeled under these culture conditions. Giemsa, × 1000.

validity of a critical role for high glutamine concentrations will require further study.

The enhancement of stromal viability observed when insulin was added to the culture medium occurred both in the presence and absence of high glutamine concentrations. This implies that at least part of the benefit of the presence of insulin derives from its isolated activity on stromal cells. Clearly insulin exerts an effect by enhancing the transport of glucose into cells, but in some systems it has also been shown to have an apparent direct stimulatory effect on cell proliferation in cells derived from tissue stroma (Prop and Hendrix, 1965). When supplementation with insulin is coupled with high glutamine concentrations there is enhancement of epithelial differentiation which suggests a possible combined effect of insulin and glutamine, rendering the epithelium more vital and hormonally responsive.

Considering the normal responsiveness of the endometrium *in vivo* to estrogen and progesterone, it was evident that the effects of these hormones should be evaluated on endometrium *in vitro*. Estrogen when unopposed by progesterone *in vivo* causes proliferation of endometrial cells. It was anticipated that treatment of cultures with 17β-estradiol would prolong survival and perhaps improve differentiation (Hughes *et al.*, 1969; Shapiro and Forbes, 1978). Whereas addition of estradiol (without progesterone) may have improved the cultures for brief intervals, it was evident that prolonged exposure to estradiol alone had an adverse effect on the organ cultures. The reasons for this effect are unclear, but similar results were observed in comparable tissue culture studies in the past. It has been conjectured that estrogen increased cellular metabolism faster than the culture systems could provide substrates (Rinard, 1972; Robertson *et al.*, 1961; Wilson and King, 1969).

The alternate approach taken for the study of effects of female steroid hormones on endometrial organ cultures was to reproduce an environment *in vitro* in which these hormones were cycled in a manner approximating the cyclic variations in these hormones during the menstrual cycle *in vivo*. It was hoped that the cyclic variations would increase the vitality of the cultures and previous results reported by other investigators suggested that such an effect could occur. Demers *et al.* (1970) showed that the condition of human endometrial explants improved when primed with estrogen followed by exposure to a combination of estrogen and progesterone. These studies, however, were of short duration (3–4 days) and involved only one hormonal cycle with approximately equimolar amounts of progesterone and estrogen. In our experiments, the molar ratio of progesterone to estrogen is 10, more closely reflecting the normally occurring decrease in concentration of estrogen relative to progesterone during the secretory phase of the menstrual cycle. With this more normal regimen, both morphologic and biochemical integrity were retained after repeated cycles during a 6-month interval.

Evaluation of the incorporation of radioactive precursors into macromolecules in endometrial organ cultures provides a separate means of judging the adequacy of the organ culture methods. Studies of incorporation of precursors into macromolecules in freshly obtained endometrial tissue were intended to provide an indication of the normal pattern of macromolecular synthesis in endometrium *in vivo*. These studies are limited by the fact that they were performed *in vitro* but in each case the incubation of tissue was completed within 5 hours of the hysterectomy. These results also are not strictly a measure of macromolecular synthesis because the incorporation of the precursors during the menstrual cycle may be influenced by possible variations in several cellular functions: uptake of the precursors into the cells, varying pools of precursors in the cells, activation to the proximate precursor for macromolecular synthesis, and intracellular transport to the site of synthesis. Nonetheless, results of incorporation of precursors in freshly obtained tissue do provide a standard for comparison for studies of incorporation in organ cultures maintained *in vitro* for prolonged times.

The results of specific activities of labeling of macromolecules in relation to menstrual cycle state in freshly obtained endometrial tissue show three distinct patterns. Incorporation of [^{3}H]Thd was low except during a peak in the late proliferative and early secretory phase which corresponds to the interval of maximal cell proliferation in the endometrium *in vivo*. For [^{3}H]Urd, incorporation was above background levels for all tested points in the menstrual cycle. The peak of [^{3}H]Urd incorporation in the early secretory phase corresponds to an interval during which an extensive change in differentiation occurs in endometrial cells. Perhaps this increased labeling of RNA relates to the reprogramming of the endometrial cells for this new state of differentiation. Although there is a peak of [^{3}H]Leu incorporation in the proliferative phase, the most distinctive feature of the pattern of [^{3}H]Leu incorporation is the sustained high level throughout the majority of the secretory phase. It is plausible that this latter feature corresponds to the period during which endometrial cells are actively synthesizing secretory glycoproteins.

These observations are in general agreement with results in previous studies with freshly obtained endometrial tissue. Nordqvist (1970) evaluated synthesis of DNA and RNA by quantitating the incorporation of [^{3}H]Thd and [^{14}C]Urd within 2 hours of tissue collection. Variations in incorporation of [^{3}H]Thd with the menstrual cycle in his study agree with those reported in our study. He observed smaller variations in Urd incorporation, but the time of peak incorporation corresponds to the peak observed in our studies. O'Grady *et al.* (1978) and Shapiro and Forbes (1978) examined the cytoplasmic protein contents of endometrial cells in relation to menstrual cycle. Both studies showed an increase in the content of protein in cells in the secretory phase as compared to the proliferative phase. These results are consistent with our observation of a sustained elevated level of incorporation of [^{3}H]Leu into protein throughout the early and middle portions of the secretory phase.

Autoradiography allows the incorporation of radioactive precursors into macromolecules to be evaluated with regard to location within cells and tissues. For example, [^{3}H]Thd was largely localized to nuclei, [^{3}H]Urd was most prominent in nuclei but also labeled the cytoplasm, and [^{3}H]Leu labeled both nuclei and cytoplasm. Incorporation of [^{3}H]Thd and [^{3}H]Urd was predominantly localized over epithelial cells in the proliferative phase and early in the secretory phase but was largely localized over stroma cells later in the secretory phase. In contrast, incorporation of [^{3}H]Leu occurred both over epithelium and stroma throughout the menstrual cycle. In all cases there was good qualitative agreement between grain densities in autoradiograms and the specific activities of incorporation for each precursor throughout the menstrual cycle.

The preceding results with freshly obtained endometrial tissue were used as a standard of comparison for studies of incorporation of precursors into macromolecules in endometrial organ cultures which had been maintained *in vitro* for prolonged intervals. Although the limited numbers of observations requires that conclusions be regarded as preliminary, the results with the organ cultures have generally paralleled the observations with fresh tissue. Specific activities of macromolecules following incorporation of radioactive precursors into organ cultures have been comparable to specific activities in similarly labeled freshly obtained tissue. Differences in extent of incorporation associated with hormonal supplementation of culture medium (i.e., simulated "proliferative phase" and "secretory phase") have generally agreed with variations in incorporation associated with the menstrual cycle in freshly obtained tissue. The localization of autoradiographic grains to nuclei and cytoplasm concurred for each radioactive precursor when fresh tissue and organ cultures were compared. Furthermore the distribution of labeled cells between epithelium and stroma appeared to relate to the hormonal milieu in culture medium at the time of labeling in a manner corresponding to menstrual cycle variations in freshly obtained tissue. It will require additional studies to confirm these conclusions further. Based upon the observations to date, however, the similarities in incorporation of precursors into macromolecules of newly obtained endometrial tissue and hormonally cycled endometrial organ cultures provide an additional validation of our organ culture techniques besides the similarities of morphologic appearance.

V. Perspectives

The results of this study demonstrate that human endometrial tissue can be maintained as organ cultures *in vitro* for many months. The tissue in culture retains differentiated morphology, responsiveness to hormones, and capacity for macromolecular synthesis. These properties provide opportunities for a wide variety of experimental studies.

Human endometrial organ cultures can be used to study several aspects of chemical carcinogenesis which cannot be contemplated *in vivo*. Organ cultures of human endometrium can be exposed to chemical carcinogens *in vitro* in an attempt to induce malignant transformation. Although it has never been accomplished previously, the relatively long period of time that this tissue remains viable in organ culture may provide favorable opportunity for the development of premalignant or malignant lesions in a human tissue in organ culture. After exposure to carcinogens, organ cultures can be harvested sequentially and examined for morphologic evidence of premalignant or malignant changes. If the interval during which organ cultures remain viable *in vitro* is not sufficient for malignant transformation, cultures can be xenotransplanted to extend the period of viability. Nude mice have a genetically determined immune deficiency which permits xenografting without rejection. Thus, human endometrial tissue exposed to carcinogens *in vitro* and then maintained in nude mice may have further opportunity to traverse the necessary steps in the process of malignant transformation. The determination of malignant transformation of the organ cultures would be made by evaluation of the morphology of xenografts recovered from the transplantation site or by evidence of tumorous growth and dissemination of transformed endometrial cells in nude mice. Since the organ cultures normally retain the relationship between epithelium and stroma, classic pathological criteria can be used to evaluate the tissue for evidence of malignancy. The opportunity to observe the sequential changes induced in human endometrial organ cultures by chemical carcinogens may offer a direct means of characterizing the morphologic stages in the natural history of endometrial cancer. Improvement of morphologic criteria for recognition of premalignant and endometrial lesions might facilitate diagnosis earlier in the natural history of endometrial cancer and thus improve survival of patients. Different carcinogens may be evaluated for their ability to transform human endometrial tissue in organ cultures. If they vary in effectiveness for malignant transformation the most effective carcinogens might be considered for epidemiologic studies to determine whether they are plausible etiologic agents for endometrial cancer.

The fact that endometrial organ cultures are responsive to the hormonal supplement in their culture medium offers the opportunity to evaluate the effects of these hormones on the process of carcinogenesis in this tissue. Estrogens, when present in excess and not balanced by appropriate amounts of progesterone, are thought to have a role in the etiology or pathogenesis of endometrial carcinoma. Experiments can be performed with endometrial organ cultures to evaluate the effect of estrogens on carcinogenesis *in vitro*. Such studies might be designed to determine whether various forms of estrogen act as carcinogens or promoters or both. Synergistic and antagonistic effects of both progesterone and estrogen could be evaluated in other studies in which modifications of the hormonal milieu are coupled with carcinogen administration *in vitro*.

The availability of human endometrial organ cultures that appear to maintain many properties of the human tissue *in vivo* offers opportunities for studying metabolism of carcinogenic and toxic compounds in this human tissue *in vitro*. Studies of this type can identify the typical route of metabolism of carcinogens and this information could facilitate comparison of metabolism between human endometrium and other human tissues. Comparisons of metabolic disposition could also be made between species. This would contribute to our estimation of carcinogenic risk in humans of compounds shown to be carcinogenic in animals. The ability to evaluate metabolism in endometrial tissue from a large number of patients may provide insight into the range of metabolic variability among a population of women. By comparing tissue from normal patients and patients with premalignant or malignant lesions of the endometrium, the importance of metabolic disposition as a risk factor in endometrial cancer might be determined.

In addition to direct studies of metabolism, the effects of carcinogens may also be evaluated by determining the extent of carcinogen binding to macromolecules within endometrial organ cultures. Since many investigators believe that carcinogens act by some critical effect on DNA within cells, the detection of carcinogen binding to the DNA within human endometrial cells may provide a better measure of the carcinogenic potential of the chemical in this human tissue. Presumably the ability to bind to DNA involves the net balance of carcinogen activation and elimination and, thus, comparisons of the extent of binding to DNA by various carcinogens may provide a measure of the relative effectiveness of the individual carcinogens in this tissue. Similarly, comparisons of the extent of carcinogen binding in endometrial organ cultures originating from different women may provide a measure of interindividual variability in the effects of carcinogens. If the extent of carcinogen binding in endometrial organ cultures is found to differ when patients with premalignant or malignant endometrial lesions are compared to normal women, then carcinogen binding studies in organ culture might eventually prove useful as a clinical means of assessing a woman's susceptibility to endometrial cancer.

Human endometrial organ cultures provide a promising system for attempting to transform human cells in organ culture. If transformation is achieved, this organ culture system could contribute to our insight into the sequence of stages in the progression to malignancy. Other studies could identify individual risk factors predisposing to the development of cancer and could identify xenobiotics or hormones that have a potential role in its etiology. Therefore, the availability of human endometrial organ cultures that can be maintained for prolonged intervals in culture provides valuable opportunities to learn more about the etiology and pathogenesis of endometrial cancer.

Acknowledgments

The authors acknowledge the excellent assistance of Sandra Murray, Carol Johnson, and Larry Gray in the preparation of this manuscript. We also thank the several gynecologists at North Carolina

Memorial Hospital who have participated in the obtainment of human endometrial tissue. This work was supported by a contract (NO1-CP-75956) and grants (T1-GM-0092 and R25-CA-17973) from the National Institutes of Health. Marc J. Mass is the recipient of a scholarship from the Stauffer Chemical Company. David G. Kaufman is the recipient of a Research Career Development Award (K04-CA-00431) from the National Cancer Institute.

References

Demers, L. M., Csermely, T., and Hughes, E. C. (1970). *Obstet. Gynecol.* **36,** 275-283.

Ehrmann, R. L., McKelvey, H. A., and Hertig, A. T. (1961). *Obstet. Gynecol.* **17,** 416-433.

George, J. P., and Solomon, S. (1978). *J. Endocrinol.* **79,** 271-276.

Hughes, E. C., Demers, L. M., Csermely, T., and Jones, D. B. (1969). *Am. J. Obstet. Gynecol.* **105,** 707-720.

Hustin, J. (1975). *Br. J. Obstet. Gynaecol.* **82,** 493-500.

Nordqvist, S. (1970). *J. Endocrinol.* **48,** 17-28.

O'Grady, J. E., Bell, S. C., Govan, A. D. T., and Black, W. (1978). *J. Endocrinol.* **77,** 21P-22P.

Prop, F. J. A., and Hendrix, S. E. A. M. (1965). *Exp. Cell Res.* **40,** 277-281.

Randall, J. H., Stein, R. J., and Stuermer, V. M. (1950). *Am. J. Obstet. Gynecol.* **60,** 711-720.

Rinard, G. A. (1972). *Biochim. Biophys. Acta* **286,** 416-425.

Robertson, G. L., Hagerman, D. D., Richardson, G. S., and Villee, C. A. (1961). *Science* **134,** 1986-1987.

Schaffner, W., and Weissman, C. (1973). *Anal. Biochem.* **56,** 502-514.

Shapiro, S. S., and Forbes, S. H. (1978). *Fertil. Steril.* **30,** 175-180.

Steinberger, E., Steinberger, A., and Ficher, M. (1970). *Recent Prog. Horm. Res.* **26,** 547-588.

Stuermer, V. M., and Stein, R. J. (1950). *Am. J. Obstet. Gynecol.* **60,** 1332-1338.

Stuermer, V. M., and Stein, R. J. (1952). *Am. J. Obstet. Gynecol.* **63,** 359-370.

Wilson, E. W., and King, R. J. B. (1969). *J. Endocrinol.* **43,** XL-XLI.

Chapter 2

Tissue Culture of the Human Uterine Cervix

GEORGE D. WILBANKS, EDMUND LEIPUS,
AND DAVID TSURUMOTO

*Department of Obstetrics and Gynecology,
Rush Medical College,
Rush Presbyterian St. Lukes Medical Center,
Chicago, Illinois*

I. Introduction

A. Scope

Most reports in the literature on the behavior of human uterine cervical tissues *in vitro* have focused mainly on abnormal material. For a review of cervical neoplasia *in vitro*, see Wilbanks and Fink (1976). These reports have generally lacked practical information on the tissue culture of normal cervical cells and have usually mentioned them briefly as controls for comparison with neoplastic cells. In addition, information on only a relatively small number of normal biopsies has been reported. This chapter will review the literature stressing normal human cervical culture and describe briefly the experience of this labora-

ISBN 0-12-564140-0

tory with normal human cervical cells in tissue culture in an attempt to fill in this apparent void. Results of routine cell culture, preliminary results on the effects of hormone addition, and preliminary results of organ cultures using porcine skin substrates will be presented.

B. Review of Methodology and Results

A summary of the literature on the tissue culture of normal human uterine cervix is presented in Table I. A variety of techniques have been used to initiate and maintain cultures from cervical biopsies of adult patients. Early cultural techniques (Gey *et al.*, 1952; Southam and Goettler, 1953; Grand and Ayre, 1954; Moore, 1955, 1956; Bromelow and Schaberg, 1957; Mellgren *et al.*, 1962; Mulligan, 1962; Simeckova *et al.*, 1962), which met with some success, utilized plasma clots and hanging drops on glass surfaces in several types of biological fluids. As they became available, chemically defined media with serum supplements (Mellgren *et al.*, 1962; Simeckova *et al.*, 1962; Cummins and Ross, 1963; Richart, 1964a,b; Nordbye, 1969; Wilbanks and Richart, 1966; Wilbanks, 1969; Wilbanks and Shingleton, 1970; Balduzzi *et al.*, 1972; Fink *et al.*, 1973; Schurch *et al.*, 1978) and various biologic substrates (Mellgren *et al.*, 1962; Auersperg and Worth, 1966; Balduzzi *et al.*, 1972; Fink *et al.*, 1973) were later used with resulting increased cell proliferation and enhanced maintenance of organ tissue architecture.

Accurate assessment or comparison of data from these studies is difficult because the number of possible experimental variables to explain divergent results from apparently similar studies is too great. The problems associated with the use of serum-supplemented media (Milo *et al.*, 1976; Keay, 1978), and the formation of toxic photoproducts from the breakdown of riboflavin and tryptophan in tissue culture media by ordinary fluorescent room light (Wang, 1976), further complicate interpretation. For example, some studies reported the growth of epithelial cells from normal areas of abnormal cervices having known neoplastic lesions, while others reported results with normal cervices from patients who were pregnant or who had other benign lesions of the uterus. In addition, some investigators separated the epithelial layer from the underlying stroma in an attempt to grow only epithelial cells and reported that cultures of pure epithelium or fibroblasts were obtained based on the standard histopathology or electron microscopy of only a small number of samples. Some investigators have subjectively graded colonial growth as poor, good, or excellent while others have measured the area of outgrowth around primary explants—even though in two areas of the same size the number of individual cells in one could be 3 to 10 times that in the other. Quantitation of the number of hormone-induced mitotic figures within an area of outgrowth has also been reported, though the stimulatory effect of background levels of hormones present originally in the serum supplement was not known and could not be compared between different studies. And finally, some studies have reported cell counts on primary cultures even though not all of the cells were of cervical origin (or perhaps not even of human origin) and were probably leukocytes and cells comprising blood vessels.

There has been little quantitative information on the routine cell culture of normal cervical tissue, the expected results, and the length of culture viability. One reason might be that no one has succeeded in obtaining a continuous line of either epithelial or fibroblast cells from the cervix. In most instances cultures were maintained for a few passages, and then the cells died. Thus long-term serial subculture has not been reported and practical methodology is lacking in this area.

Organ culture has recently been used to study endo- and ectocervical cells in normal tissues (Balduzzi *et al.*, 1972; Fink *et al.*, 1973; Schurch *et al.*, 1978). In these studies whole sections of tissue have been explanted with the stromal and epithelial layers intact. The tissue was placed either on a plastic or glass surface with the addition of growth medium, with or without mechanical rocking (Schurch *et al.*, 1978), or on a wire grid in specially designed organ culture dishes filled with growth media. In some studies a biological substrate, such as pigskin, collagen gel, or agar, was placed in between the tissue and the grid in an attempt to bring nutrients and moisture to, and remove wastes from, the tissues. These methods have met with limited success and relatively short-term studies (up to 6 months) have been performed with them to date (Schurch *et al.*, 1978).

II. Materials and Methods

A. Materials for Tissue Culture

Growth media used in this report consisted of the following: Eagle's basal medium (BME), RPMI-1640, or BioLab's modified Dulbecco's medium (BioLabs, Inc.) plus 15 to 20% fetal bovine serum (FBS) (BioLabs, Inc.), penicillin and streptomycin (P and S) [Grand Island Biological Co. (GIBCO)], each at 100 units and 100 μg/ml, respectively, and fungizone (Squibb) at 5.0 μg/ml.

Balanced salt solution (BSS) was composed of the following: Hank's balanced salt solution base, P and S or gentamicin (Schering) or kanamycin (Bristol) each at 100 units or 100 μg/ml, 10 μg/ml fungizone, glutamine at 0.292 mg/ml, and sodium bicarbonate to bring the pH to 7.2.

The trypsin–EDTA solution was composed of the following: 0.25% trypsin [1:250 stabilized for tissue culture (GIBCO)] and 0.1% ethylenediaminetetraacetic acid [EDTA (Sigma)] in Hank's balanced salt solution. The pH was adjusted to approximately 7.5 with sodium bicarbonate.

Hormones included insulin (Sigma 15500), β-estradiol (General Biochemicals

TABLE I

TISSUE CULTURE OF NORMAL HUMAN UTERINE CERVIX[a]

Investigator	Number of normal cervical biopsies	Number of cultures seeded	Culture methods or techniques employed	Growth characteristics	Additional comments
Gey *et al*. (1952)	No data given	No data given	Explants grown in chick plasma clot, bovine embryo extract, and human placental cord serum	Difficulty in initiating growth. Rapid differentiation	No quantitative data
Moore (1952)	No data given	11	Chick embryo plasma clot; hanging drop; under perforated cellophane; embryo extract, placental cord serum, and balanced salt solution	Limited growth in five cultures in 4–5 days; one could be subcultured; all died out at 40–45 days	Age range of patients was 23–42 years; two were pregnant
Southam and Goettler (1953)	No data given	3	Plasma clot; 50% human ascitic fluid, 45% BSS, and 5% chick embryo extract	1: epithelial, poor outgrowth; 1: fibroblastic; 1: no growth	
Grand and Ayre (1954)	No data given	No data given	Plasma clot; equal parts of 50% solution chick embryo extract, fowl plasma, and human placental serum	10% of cultures grew out in 10 days	

Moore (1955)	16	16	Chick embryo plasma clot	Four cultures showed small amount of epithelial cell growth; one culture had moderate epithelial growth	Eight biopsies from pregnant patients; eight from nonpregnant. Normal areas of Ca patients grew better than those from normal patients
Moore (1956)	51	51	Same as Moore (1955)	10 of 23 cultures from normal and pregnant patients grew; 23 of 28 normal areas from Ca or CIS patients grew well to excellent	
Bromelow and Schaberg (1957)	No data given	7	Organ culture; 25% human plasma, 16% horse serum, 12% human fetal brain press juice, in Hank's BSS	Poor growth or maintenance	
Mellgren *et al*. (1962)	No data given	No data given	On rat collagen; and in chick plasma clot; culture medium was Eagle's supplemented with human serum, adult or cord (26%)	Obtained 10^{2-3} cells after 10–20 days	Cultured normal areas of cervices with neoplasms; dissected epithelium from stroma
Mulligan (1962)	8	192 explants	Chicken plasma clot	Growth obtained in 68 explants for up to 15–29 days	Biopsies from parous women

(*continued*)

TABLE I (*continued*)

Investigator	Number of normal cervical biopsies	Number of cultures seeded	Culture methods or techniques employed	Growth characteristics	Additional comments
Simeckova *et al*. (1962)	40	No data given	Monolayer cultures in roller tubes or in plasma clot; medium 199 and 10% bovine serum	Poor to fair growth in all biopsies	
Cummins and Ross (1963)	At least 2	No data given	In Pyrex glass petri dishes; medium consisted of phosphite-buffered base containing fructose, glucose, Eagle vitamins, lactalbumin hydrolysate, or Medium 199, and 10% adult human serum and 10% fetal bovine serum	Maintained colonial growth for up to nine subcultures; fibroblasts mainly resulted	
Richart (1964a,b)	4	No data given	Dissection of epithelial layer from stroma; explantation under perforated cellophane in Eagle's MEM and 15% fetal bovine serum	Monolayer cultures of pure epithelial cells which could be subcultured one or more times with a trypsin versene solution	Unpredictable success rate

Auersperg and Worth (1966)	8	No data given	Explantation to glass, cellulose sponge, and collagen gel; culture medium was Waymouth's and 10% fetal bovine serum	Cell outgrowth began 4 to 14 days; ceased 9 to 35 days, with differentiation	"Secondary" epithelioid cells at 14 to 56 days with dying off at 21 to 100 days
Nordbye (1969)	11	No data given	Biopsies were minced and trypsinized to yield single-cell suspensions; seeded in Puck's Medium E2a	After 3 weeks, epithelial cells formed pavement-like sheet; fibroblasts were in lamellated pattern and swirls; at 5 weeks it was difficult to distinguish fibroblasts from epithelia	Number of subcultures were not reported
Wilbanks and Shingleton (1970); Shingleton and Wilbanks (1974)	8	40	Same as Richart (1964a)	Monolayers of mostly squamous epithelia; limited culture life; could not be subcultured	Showed that *in vitro* cells compared favorably with *in vivo* tissues by morphologic criteria for up to 21 days
Balduzzi *et al*. (1972)	7	7	Organ culture of endo- and ectocervix with underlying stroma; on 8% Noble agar containing Ham's F-12 basal medium and 10% newborn bovine serum	Five cultures had good morphology for up to 23 days; stromal degeneration at 11 to 12 days	
Vesterinen *et al*. (1980)[b]					

(continued)

TABLE I (*continued*)

Investigator	Number of normal cervical biopsies	Number of cultures seeded	Culture methods or techniques employed	Growth characteristics	Additional comments
Fink *et al*. (1973)	No data given	No data given	Organ culture of endocervix; epithelium and underlying stroma placed on thin slab of agar-gelled medium (1.4% agar in Eagle's BME, no serum supplement) supported by a wire grid	Held up well for 8 days. Columnar cell pyknosis at 10 days; observed progression of changes of squamous metaplasia seen *in vivo*	
Chaudhuri *et al*. (1974)	Not known; mixed data on normal with abnormal lesions; 25 abnormal patients reported	92	Epithelium was separated from underlying stroma in 28 cultures; in 64 cultures, they were left intact; grown in MEM or McCoy's 5A plus 20% fetal bovine serum and 2 m*M* glutamine; subcultured using trypsin	In 26/28 cultures of epithelium alone, epithelial cells grew but with fibroblast contamination at 5–7 weeks; with intact biopsies, 33/64 had epithelial growth but were overgrown with fibroblasts after 5 weeks. Epithelial cells survived 6–16 weeks and could not be subcultured whereas fibroblasts could be subcultured several times	Optimum pH for growth of epithelial cells was 7.2 and that for fibroblasts 7.6

Wilbanks (1975)	166	No data given	Method of Richart (1964a)	Good growth obtained in 140/166 biopsies. Three spontaneous cell transformations occurred	
Brdar *et al.* (1977)	No data given	No data given	Growth media consisted of Dulbecco modified Eagle's medium supplemented with 10% fetal bovine serum	Monolayers; could be subcultured at least five times	
Ishiwata *et al.* (1978)	15	484	Epithelial layer planted in culture; Ham's F-12, 10% fetal bovine serum, with and without 17β-estradiol	69% or 333/484 explants produced squamous cells; 21/484: mixture of squamous cells contaminated with fibroblasts; 14/484 were pure fibroblasts; 97 did not grow out (20%)	Measured colonial outgrowth and mitotic index. No cell line established
Schurch *et al.* (1978)	20	No data given	Organ culture of upper one-third of endocervix along with 2–3 mm underlying stroma; culture medium: CMRL-1066, hydrocortisone hemisuccinate, insulin, glutamine, and 5% heat-inactivated fetal bovine serum, on rocker platform	Cultures held up well for 24 weeks, by morphologic criteria	Metaplastic epithelia covered all exposed surfaces within 12 weeks

[a] Ca, carcinoma; MEM, minimal essential medium; CIS, carcinoma *in situ*.
[b] Reference published after submission of manuscript.

80200), progesterone (Schering), testosterone (Matheson, Coleman & Bell TX 95), hydrocortisone (Sigma H-4001), and cortisone (Sigma C-2755). Porcine skin substrates were tissue culture quality porcine skin (Burn Treatment Skin Bank of Phoenix, Arizona). Activated charcoal powder was "Norit A" (Matheson, Coleman, & Bell).

Instruments and equipment included #11 surgical knife blade with holder, plastic petri dishes, 100 × 15 mm and 60 × 15 mm (Lux), plastic screw cap tubes, 16 × 125 mm (Falcon), culture flask, 25 cm^2 (Falcon), organ tissue culture dishes, 60 × 15 mm (Falcon), organ culture grids (Falcon), dissecting microscope (American Optical, Model #56M-1), dissecting forceps (Matheson), eye scissors (E. Weck & Co.), B-D Cornwall Pipettor (Becton, Dickinson, and Co.), and a water-jacketed CO_2 incubator (National Appliance Co., model # 3331).

B. Methods for Tissue Culture

1. Human Uterine Cervical Tissue

All human cervical tissues used in the authors' study were derived from patients who had undergone hysterectomies for various benign conditions not related to cervical neoplasia, usually symptomatic pelvic relaxation, uterine fibroids, etc. Papanicolaou smears were negative. After a portion of the tissue was removed for histologic examination, adjacent areas of the cervix were selected for culture, and the epithelial layer as well as the underlying stroma were cut out with sharp surgical scissors. The pieces, approximately 0.5 × 0.5 × 0.3 cm (thickness), were placed in a plastic screw cap tube and then washed seven times with BSS. After the last wash, the media was removed and the pieces placed in a 100 × 15-mm plastic petri dish. While viewing each piece of tissue through a dissecting microscope (magnification at 70 ×) and holding the specimen in place with dissecting forceps, the epithelial layer, a light yellowish, cheese-like material, was gently separated from the stroma, using light scraping movements, with a surgical knife. Soaking the tissue in media containing at least 10% FBS for several hours or overnight in the refrigerator sometimes aided in separating the epithelial layer, but often reduced its growth potential.

After dissection, the pieces of epithelium were transferred to a 60 × 15-mm plastic petri dish where they were minced with sharp eye scissors, in 1.0 ml of growth medium. When the pieces were 1 mm or less in size, the suspension was further diluted with growth media and 1.0 ml was transferred to each 25-cm^2 culture flask. The optimal number of explants per flask ranged from 80 to 100 and up to six flasks were seeded from a single cervix. The cultures were incubated in a humidified atmosphere of 5% CO_2–95% air at 37°C.

The use of small volumes of media when culturing primary cervical tissues was very critical. The explants were not submerged in medium, but bathed on all sides by it, otherwise attachment and outgrowth did not occur. After attachment and commencement of growth, the cultures were fed two times per week, with increasing amounts of growth media added as the cells filled the flasks. Feeding of large numbers of cultures was accomplished using a Cornwall pipettor instead of time-consuming hand pipetting.

Subculturing of cells was accomplished by removing the growth media, adding 2.0 ml of trypsin–EDTA solution per 25-cm^2 flask for 30 seconds, and then incubating the flask at 37°C in a 5% CO_2–95% air, humidified atmosphere until the cell sheet flowed from the flask. The cells were then resuspended in growth medium, diluted 1:2, and planted in fresh flasks.

2. Organ Culture

Organ cultures of human cervical epithelium on pigskin substrates were maintained using RPMI-1640 and 15% fetal bovine serum as growth medium. The method used was that reported by Freeman *et al.* (1976). Briefly, pigskin, dermis side up, was placed on a wire grid in an organ culture dish (Falcon). Pieces of stripped epithelium with a thin underlying stromal layer were usually oriented stromaside down on the pigskin. Medium was added until the level touched the underside of the pigskin and was changed twice weekly. Incubation was in a humidified atmosphere of 5% CO_2–95% air at 37°C.

3. Marmoset Uterine Cervical Tissues

Cervical tissue from the uterus of a marmoset monkey was explanted into tissue culture and subcultured using basically the same method as for the human tissue except that, because of its small size (1 mm), no stripping off of the epithelial layer was done and the entire cervix was cultured. Because this type of tissue was heavily contaminated at the outset, the whole cervix was soaked in growth medium containing two to three times the regular amount of antibiotics for several hours at room temperature. Extra washings before mincing were also effective in controlling microbial outgrowth.

4. Preparation of Hormone Stock Solutions

Stock hormone solutions were made by dissolving the hormone in either absolute alcohol or acetone depending upon hormone solubility. Aliquots of the stock solutions were added to growth media to yield the working concentrations, over and above that already present in the FBS, listed in Table III. These concentrations also depended upon the solubility of dissolved hormone in the aqueous growth media.

5. Charcoal Treatment of Fetal Bovine Serum

Fetal bovine serum, 500 ml, control No. R 60804 was absorbed for 16 hours at room temperature with powdered charcoal (0.25% w/v). After removal of charcoal the serum was filter sterilized. The volume after this treatment was 400 to 450 ml and the pH was 7.03.

6. Radioimmunoassay of Steroid Hormones

The amount of steroid hormones in FBS before and after charcoal treatment was determined by radioimmunoassay (Abraham, 1974).

III. Results and Discussion

In a review of the literature on the tissue culture of normal cervical tissues, the problem of what constitutes growth in the cultures becomes readily apparent. Terms such as "poor," "fair," "good," and "excellent," and later, "confluent monolayer," have been used to describe the way the tissues grow *in vitro* (Wilbanks, 1975). In this report, growth refers to both small and large patches of cells, with or without explants, or colonial growth, to formation of confluent monolayers, all detected by visual scanning of culture vessels with low power, light microscopy and medium power, phase microscopy.

From Table II, slightly less than 50% of attempted cultures showed growth. Of these (282 cultures), 85% grew or could be maintained for up to 1 year, while 15% could be grown for greater than 1 year. In contrast, Ishiwata *et al.* (1978) obtained growth in 80% of 484 individual explants. This apparently higher success rate could be accounted for by chance alone since their sampling was considerably smaller than ours. The results in the present report were obtained from approximately 48,000 to 60,000 individual explants. On the other hand, their cultural conditions may have been more suitable, though such detailed information was not presented in their report.

Extended viability in culture could not be correlated with a particular age group since almost identical age groups were represented in each culture interval. The age range was 23 to 71 years.

Likewise, once overtly toxic lots of fetal bovine serum were eliminated there was no difference in growth-promoting ability of various lots of FBS. For instance, those lots of FBS that were used in cultures that grew for greater than 1

TABLE II

OVERALL GROWTH PERFORMANCE OF NORMAL, HUMAN CERVICAL CELLS IN TISSUE CULTURE

Culture interval	Number of patient biopsies	Total number of cultures (% of total)	Number of subcultivations	Age range of patients biopsied, in years
<1 Month	2	4 (1.0%)	1	27–42
1–6 Months	47	167 (28.0%)	1–10	23–57
6–12 Months	25	69 (12.0%)	1–8	26–71
>12 Months	23	42 (7.0%)	1–19	25–71
No growth	63	315 (52.0%)	—	25–71
Total	100	597 (100%)		

year were also used in cultures with no growth at all. Thus the growth potential of human cervical cells *in vitro* may vary greatly from cervix to cervix and from explant to explant from a single cervix.

Primary explants usually began growing out at 1–3 days to 1 month (Figs. 1 and 2). The reader is referred to Wilbanks (1975) for a detailed account of the fine structure of normal cervical cells in culture. Cultures of pure epithelium were rare and most were contaminated with fibroblasts from the stroma, the degree depending on the dissection technique. In most cultures these contaminants would eventually overgrow the epithelium. Epithelial cells had a tightly packed cobblestone appearance, while fibroblasts grew in parallel arrays and whorls. In most cultures that grew from 6 months to greater than 12 months, the fibroblasts would die out after three or four passages and a new type of large, bipolar cell would emerge. In very dense, confluent cultures, these cells would become epithelioid, while they would appear as "fat" fibroblasts in newly planted cultures (Fig. 3). Total viable counts performed on these cells ranged from 10^4 to 10^6 per confluent 25-cm^2 flask. The longest interval a culture survived was for 3.5 years and 19 subcultivations.

The effects of insulin and steroid hormones were tested in a small number of cultures selected at random. The preliminary results are presented in Table III. The treated cultures grew or were maintained in culture from less than 1 month to 12 months. Since no treated cultures grew for more than a year, this group of cultures did less well than untreated controls which grew for more than 1 year. There were no obvious toxic effects of the hormones. There were eight cultures in which there was no outgrowth. The small number of cultures tested have not allowed us to arrive at any firm conclusions about the effects of hormones on cervical cells at this time.

Background levels of steroid hormones in one lot of fetal bovine serum were determined by standard radioimmunoassay (Abraham, 1974). Insulin determina-

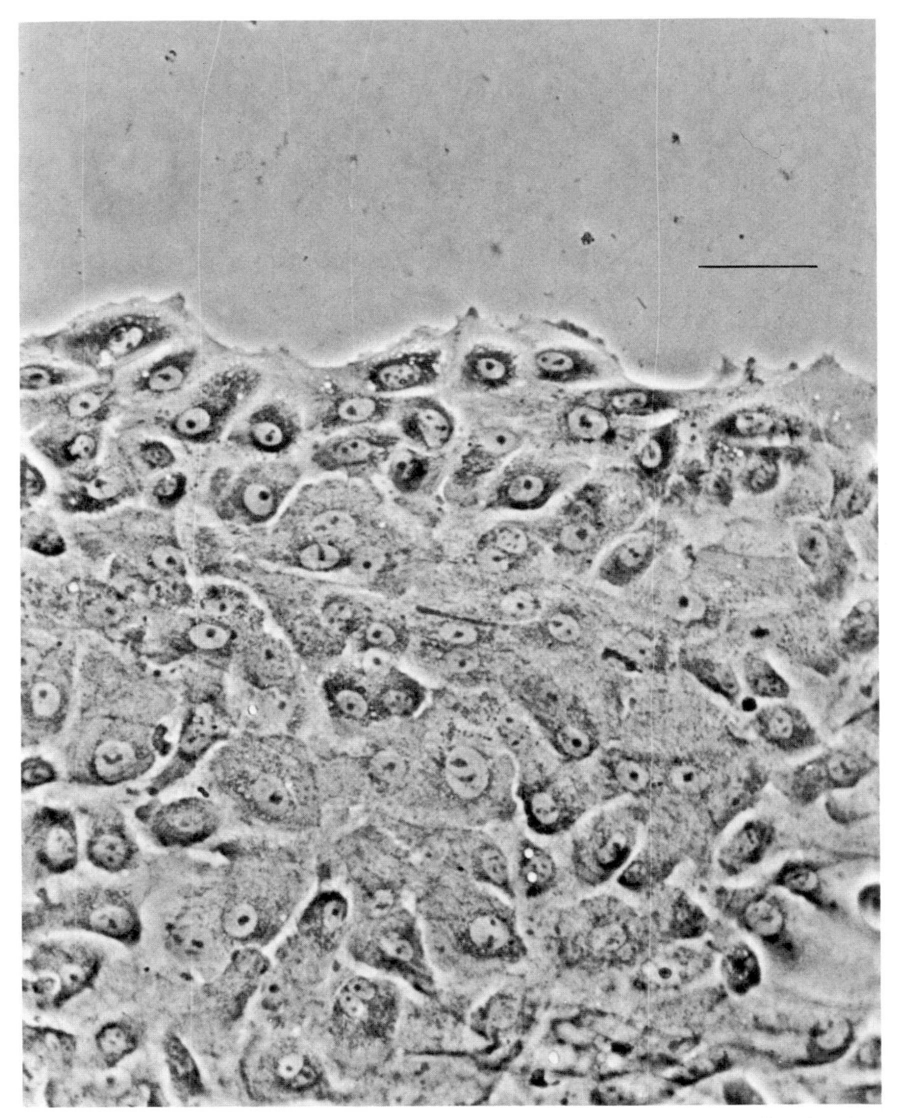

FIG. 1. Phase contrast photomicrographs of primary normal human cervical epithelium in culture for 5 days. Bar represents 40 μm.

tions were not performed. Progesterone, hydroxyprogesterone, estradiol-17β, androstenedione, dehydrotestosterone, testosterone, hydroxycortisone, and dehydroepiandrostane sulfate were assayed and their concentrations in FBS are listed in Table IV.

Attempts were made to remove these steroid hormones from FBS with char-

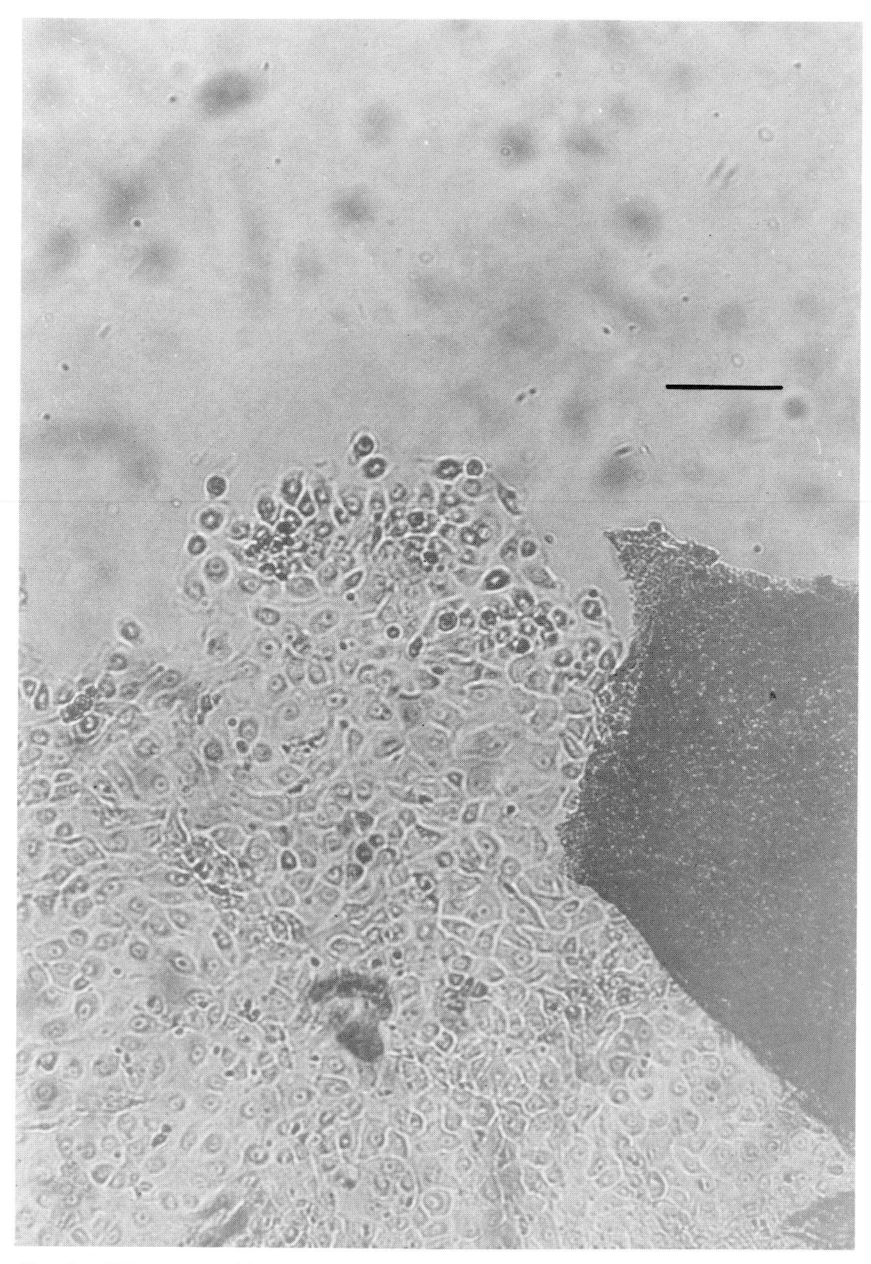

FIG. 2. Primary normal human cervical epithelium in culture for 7 days. Bar represents 80 μm.

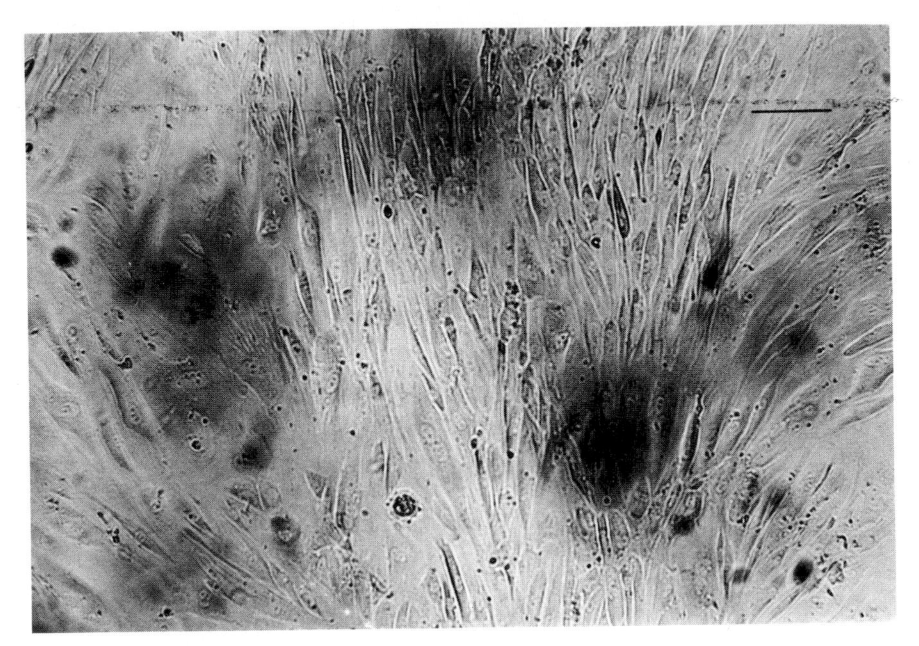

FIG. 3. Norman human cervical cells two weeks after second subcultivation. Note possible epithelioid replacement cells (see text). Bar represents 80 μm.

coal treatment (Heyns *et al.*, 1967). The results of these attempts are presented in Table IV. The charcoal was effective in removing most of the progesterone, all of the dehydrotestosterone, all of the testosterone, and a considerable amount of the hydroxyprogesterone. Heat inactivation of FBS caused all of the dehydrotestosterone to be converted to testosterone. The effects of this "stripped" serum, with or without hormones added back at known concentrations, on cervical cell cultures are now being investigated.

It was not known whether the background amounts of hormones in our serum were responsible for the effects of the added hormones. Milo *et al.* (1976) found inhibitory levels of progesterone in untreated FBS and these levels could explain the generally short longevity of an entire series of cultures from three patients biopsied. The lack of a striking difference between hormone-treated and untreated cultures suggests that inhibitory levels were already present in the FBS and that the effect could not be further increased because of already saturated hormone receptor sites. On the other hand, the one untreated culture that survived for greater than 12 months (Table III) and three subcultivations may be significantly different, suggesting that the levels of added hormones were detrimental. Nevertheless, the small number of biopsies and cell cultures tested does not allow any definitive conclusions.

Unlike human cells, the marmoset cells formed an apparently continuous line which is presently at passage 50 (Figs. 4 and 5). After a 1:2 dilution at the time of

TABLE III

OVERALL EFFECT OF HORMONE TREATMENT ON THE GROWTH PERFORMANCE OF NORMAL, HUMAN CERVICAL CELLS IN TISSUE CULTURE[a]

Hormone and concentration	Number of cultures treated with hormone (number of serial subcultivations after hormone treatment) Culture interval				
	<1 Month	1-6 Months	6-12 Months	>12 Months	No growth
Insulin					
5 μg/ml	0	0	3 (2p)	0	0
5 mg/ml	1 (2p)	2 (2p)	3 (2-3p)	0	1
β-Estradiol					
1-2 μg/ml	0	2 (1p)	2 (1p)	0	4
Progesterone					
1-2 μg/ml	0	6 (1p)	5 (1-2p)	0	1
Testosterone					
1-2 μg/ml	4 (1p)	2 (1p)	0	0	2
Hydrocortisone					
$1 \times 10^{-4}\ M$	0	3 (1p)	0	0	0
Cortisone					
$1 \times 10^{-6}\ M$	0	3 (1p)	0	0	0
No Hormone					
Biopsy 113	0	0	2 (2p)	1 (3p)	0
Biopsy 129	1 (1p)	0	0	0	0
Biopsy 274	0	5 (1p)	1 (7p)	0	0

[a] Cells were at subculture one or two at the beginning of treatment. Hormones were added at the time of explantation or subculture or up to 10 days later. Those receiving hydrocortisone or cortisone were switched to growth medium without added hormone after 2 months. Untreated control cultures were derived from the same patient biopsies as those cultures receiving added hormone.

subcultivation, confluent monolayers were formed within 3 to 5 days. These monolayers could be maintained for up to 1 month, if refed twice weekly with fresh growth media containing 10% FBS. This cell line appears to be purely epithelial without any contaminating fibroblasts. Unlike the human cells, it can be frozen down in liquid N_2 using 10% glycerol or dimethyl sulfoxide (DMSO). However, neither the human nor the marmoset cells could be cloned in soft agar.

In organ cultures of separated cervical epithelium, the epithelial cells migrated over, and often digested, the porcine skin, leaving circular cleared, translucent areas in their wake. These cleared areas served to monitor cell outgrowth. Preliminary histologic observations revealed that the layered growth of squamous cells was similar to that found on H and E sections of normal tissues *in vivo*. Ectocervical epithelium was maintained from 5 days to 1 month before degeneration and sloughing of the tissue layers were observed (see Figs. 6, 7, and 8).

TABLE IV

STEROID HORMONE CONTENT OF FETAL BOVINE SERUM USED IN THE TISSUE CULTURE OF CERVICAL CELLS

Hormone	Hormone concentrations (ng/ml) in fetal bovine serum		
	Untreated	Charcoal treated	Untreated; heat-inactivated at 56°C for 1 hour
P: Progesterone	0.148	0.06–0.088	0.202
17P: Hydroxyprogesterone	0.104	0.080	0.110
E_2: Estradiol-17β	0.145	0	0.166
A: Androstenedione	0.069	0.063	0.088
DHT: Dehydrotestosterone	0.049	0	0
T: Testosterone	0.096	0	0.169
CpF: Cortisol	0	0	0
D-S: DHEA sulfate[a]	0	0	0

[a] Dehydroepiandrosterone sulfate.

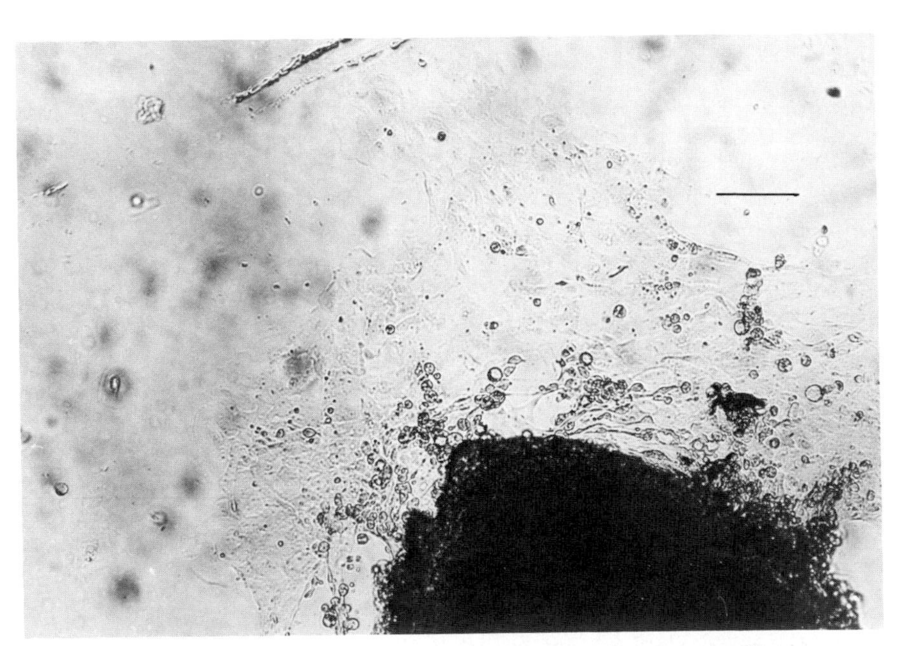

FIG. 4. Seven-day-old primary marmoset cervical cell culture. Bar represents 80 μm.

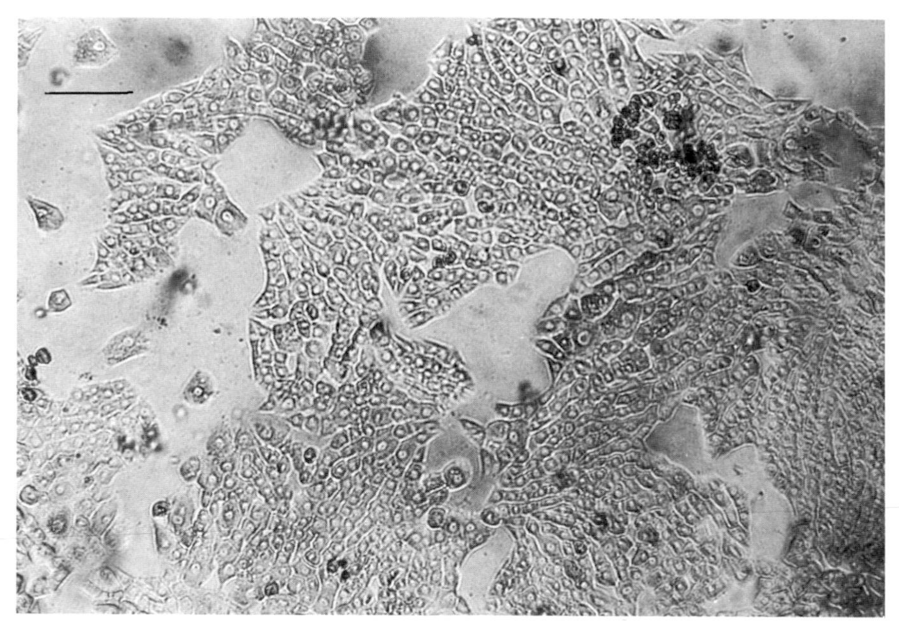

FIG. 5. Five-day-old culture of marmoset cervical cells at fortieth subculture. Bar represents 80 μm.

Certain lots of porcine skin did not support outgrowth and the resulting clearing effect in the substrate; these were actually toxic and rapid degeneration of the cervical tissues occurred. It is suggested that lots be pretested before large-scale experimentation is launched using porcine substrates.

IV. Perspectives

A. Current Applications and Findings

Human uterine cervical tissues in cell or organ culture have been used to study a diverse group of biologic problems. Many of these studies center around carcinogenesis in this organ, as well as its normal physiology.

Balduzzi *et al.* (1972) and Birch *et al.* (1976) used organ cultures of normal endocervix and ectocervix with and without the underlying stroma to show that herpes simplex virus type 2 (HSV-2) could infect and replicate in these tissues. Fink *et al.* (1975) used organ culture to study cellular events of metaplasia in the uterine cervix. Wilbanks and Fink (1976) and Marczynska *et al.* (1980) reported unsuccessful attempts to transform normal adult cervical epithelium of both human and marmoset origin in cell culture using UV-irradiated or unirradiated HSV-2, and/or incubation at 42°C, although primary hamster embryo and rat embryo cell

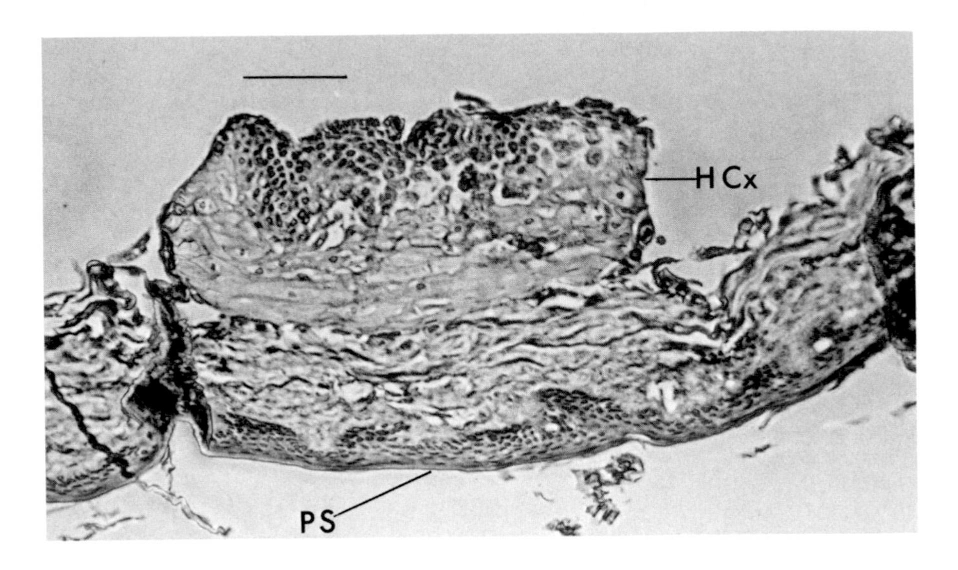

FIG. 6. Hematoxylin and eosin-stained section of normal human cervical epithelium and stroma (HCx) on porcine skin (PS) maintained in organ culture for 2 weeks. Bar represents 40 μm.

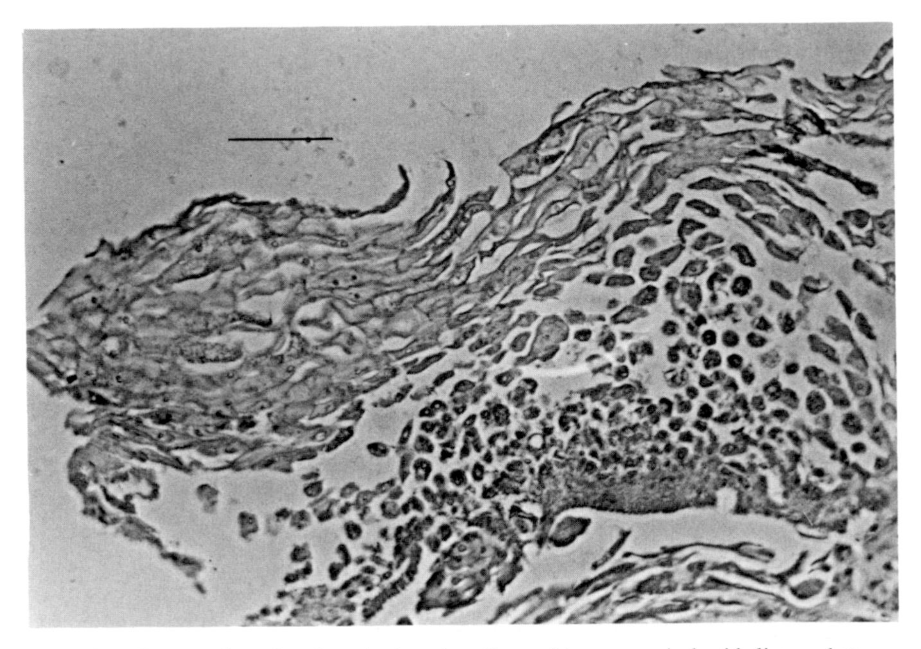

FIG. 7. Hematoxylin and eosin-stained section of normal human cervical epithelium and stroma (same biopsy as in Fig. 6) on organ culture grids for 2 weeks. Note lack of cohesiveness of epithelial cells. Bar represents 40 μm.

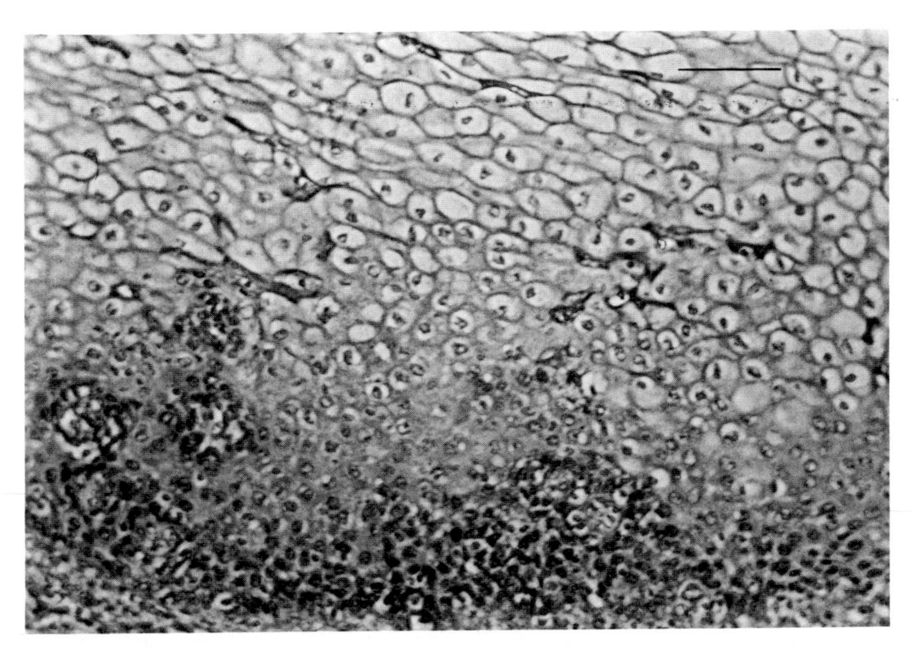

FIG. 8. Hematoxylin and eosin-stained histological section of a biopsy of normal human cervical tissue (same biopsy as in Figs. 6 and 7). Bar represents 40 μm.

cultures became transformed under these conditions. Vesterinen *et al.* (1977)induced a persistent latent herpesvirus infection in human ecto- and endocervical epithelial cells using adenine arabinoside (ara-A). The inhibiting effects of human leukocyte interferon on normal cervical cell cultures have recently been reported by Brdar *et al.* (1977). Ishiwata *et al.* (1978) have shown the stimulatory effect of low concentrations of estradiol-17β on growth and differentiation of normal cervical squamous cells in culture. Rowinski *et al.* (1978) cultured normal fibroblasts of the human uterine cervix with various lesions and examined their nuclear features using image analysis. They found a wide variation (as much as 10-fold) in the size of the nucleus which could not be correlated with the type of lesion involved.

B. Future Studies

Studies involving long-term effects on cervical tissues *in vitro,* e.g., carcinogen studies, should be performed in cell culture since only by this method can cells be maintained long enough for possible effects to be seen. Even then, a large number of biopsies must be sampled in order to obtain cultures that can be grown for a year or more. Organ culture should be used for relatively short-term studies. Short-term effects of drugs, physical, or biological agents, etc. seem to be suitable topics for study in organ cultures.

References

Abraham, G. E. (1974). *Acta Endocrinol.* (Suppl.) **183,** 1–42.
Auersperg, N., and Worth, A. (1966). *Int. J. Cancer* **1,** 219–238.
Balduzzi, P. C., Nasello, M. A., and Amstey, M. S. (1972). *Cancer Res.* **32,** 243–246.
Birch, J., Fink, C. G., Skinner, R. B., Thomas, G. H., and Jordan, J. A. (1976). *Br. J. Exp. Pathol.* **57,** 460–471.
Brdar, B., Krusix, J., Jusic, D., Soos, E., Ban, J., and Nagy, B. (1977). *Period. Biol.* **79,** 121–127.
Bromelow, J. H., and Schaberg, A. L. (1957). *Exc. Med.* (Spec.) **16,** 49.
Chaudhuri, S., Koprowska, I., Putong, P. B., and Townsend, D. E. R. (1974). *Cancer Res.* **34,** 1335–1343.
Cummins, G. T. M., and Ross, J. D. (1963). *Cancer Res.* **23,** 1581–1592.
Fink, C. G., Thomas, G. H., Allen, J. M., and Jordan, J. A. (1973). *Obstet. Gynaecol. Br. Commonw.* **80,** 169–175.
Freeman, A. E., Igel, H. J., Herrman, B. J., and Kleinfeld, K. L. (1976). *In Vitro* **12,** 352–362.
Gey, G. O., Coffman, W. D., and Kubicek, M. T. (1952). *Cancer Res.* **12,** 264–265.
Grand, C. G., and Ayre, J. E. (1954). *Obstet. Gynecol.* **4,** 411–417.
Heyns, W., Van Baelen, H., and DeMoor, P. (1967). *Clin. Chem. Acta* **18,** 361–370.
Ishiwata, I., Okumura, H., Nozawa, S., Kurihara, S., and Yamada, K. (1978). *Acta Cytol.* **22,** 556–561.
Keay, L. (1978). *Methods Cell Biol.* **20,** 169–209.
Marczynska, B., McPheron, L., Wilbanks, G. D., Tsurumoto, D. M., and Deinhardt, F. (1980). *Exp. Cell Biol.* **48,** 114–125.
Mellgren, J., Boeryd, B., and Hagman, M. (1962). *Cancer Res.* **22,** 139–146.
Milo, G. E., Malarkey, W. B., Powell, J. E., Blakeslee, J. R., and Yohn, D. S. (1976). *In Vitro* **12,** 23–30.
Moore, J. G. (1952). *Am. J. Obstet. Gynecol.* **64,** 13–24.
Moore, J. G. (1955). *W. J. Surg. Obstet. Gynecol.* **63,** 1–9.
Moore, J. G. (1956). *Surg. Forum* **7,** 507–510.
Mulligan, R. M. (1962). Abstr. *Annu. Meet., 13th, Tissue Culture Assoc.*
Nordbye, K. (1969). *Acta Obstet. Gynecol. Scand.* **48,** (Suppl. 3), 102–109.
Richart, R. M. (1964a). *Am. J. Obstet. Gynecol.* **88,** 710–714.
Richart, R. M. (1964b). *Cancer Res.* **24,** 662–669.
Rowinski, J., Koprowski, I., Chaudhuri, S., and Swenson, R. (1978). *Mater. Med. Pol.* **2,** 94–97.
Schurch, W., McDowell, E. M., and Trump, B. F. (1978). *Cancer Res.* **38,** 3723–3733.
Shingleton, H. M., and Wilbanks, G. D. (1974). *Cancer* **33,** 981–989.
Simeckova, M., Nichols, E. E., and Lonser, E. (1962). *Obstet. Gynecol.* **20,** 251–255.
Southam, C. M., and Goettler, P. J. (1953). *Cancer* **6,** 809–827.
Vesterinen, E., Leinikki, P., and Saksela, E. (1977). *Acta Pathol. Microbiol. Scand. B* **85,** 289–295.
Vesterinen, E. M., Nedrud, J. G., Collier, A. M., Walton, L. A., and Pagano, J. S. (1980). *Cancer Res.* **40,** 512–518.
Wang, R. J. (1976). *In Vitro* **2,** 19–22.
Wilbanks, G. D. (1969). *Obstet. Gynecol. Surv.* **24,** 804–837.
Wilbanks, G. D. (1975). *Am. J. Obstet. Gynecol.* **121,** 771–788.
Wilbanks, G. D., and Fink, C. G. (1976). *In* "The Cervix" (J. A. Jordan and A. Singer, eds.), pp. 429–441. Saunders, London.
Wilbanks, G. D., and Richart, R. M. (1966). *Cancer Res.* **26,** 1641–1647.
Wilbanks, G. D., and Shingleton, A. M. (1970). *Acta Cytol.* **14,** 182–186.

METHODS IN CELL BIOLOGY, VOLUME 21B

Chapter 3

Normal Human Endometrium in Cell Culture

D. KIRK[1] AND J. C. IRWIN

Department of Hormone Biochemistry,
Imperial Cancer Research Fund,
Lincoln's Inn Fields,
London, England

[1]Present address: Pasadena Foundation for Medical Research, 99 North El Molino Avenue, Pasadena, California 91101.

ISBN 0-12-564140-0

I. Introduction

The human endometrium, the mucous membrane lining the uterine cavity, is an epithelial/mesenchymal structure of mesodermal origin. It is of central importance in studies of reproductive physiology, representing a classic target tissue for ovarian sex steroid hormonal effects, a feature that has made endometrial structure a reliable indicator of ovarian function.

Initiation of *in vitro* studies of the human endometrium in 1926 (Mayer and Heim, 1926) has been followed by several investigators (Papanicolaou and Maddi, 1958, 1959; Figge, 1960, 1963; Notake, 1963; Hiratsu, 1968), all of whom are in general agreement about the following empirical observations of human endometrial cell cultures:

1. Two morphological cell types are observed in culture, epitheloid and fibroblastic;
2. Epitheloid cells are short-lived, being replaced by the fibroblastic type;
3. Proliferative endometrial cells grow better than secretory endometrial cells; and,
4. Premenopausal endometrial cells grow better than postmenopausal endometrial cells.

A major drawback of these early studies was the failure to separate the epithelial and mesenchymal components; thus the use of such mixed cultures (Hiratsu, 1968; Chen *et al.*, 1973) as an *in vitro* model for studying hormone action has made it impossible to interpret the results with respect to cell type.

Recent studies (Liszczak *et al.*, 1977; Kirk *et al.*, 1978a,b) have demonstrated, using different enzymic separation procedures, the relative ease with which normal human endometrium can be separated into its epithelial and stromal components which can then be identified morphologically at both the cellular and ultrastructural level in culture.

Even with the availability of such separation procedures, the normal human endometrial cell cultures have still proved disappointing as an *in vitro* model for studying hormone action because of (*a*) the short-term nature of the epithelial cultures, and (*b*) the total lack of response of such short-term cultures to estrogens (Figge, 1963; Liszczak *et al.*, 1977; Kirk *et al.*, 1978a, b). It has been our aim to study the growth processes in culture of this highly specialized cell type with a view to improving the culture conditions (including chemical and physical requirements) in an attempt to realize the potential that human endometrial epithelial cells may have for *in vitro* studies.

II. Materials

A. Source of Human Endometrium

Normal human endometrium was obtained from patients (ranging from 23 to 44 years of age) under sterile conditions, as curettings either at hysterectomy or as an out-patient procedure, the tissue used being surplus to that needed for standard histopathological examination. Samples were collected in Dulbecco's modified Eagle's medium (DMEM) supplemented with 5% fetal bovine serum (FBS, Flow Laboratories). Histology was checked for normalcy and the menstrual data assessed according to the standard schedule of Noyes *et al.* (1950).

B. Growth Medium

The growth medium (GM) for plating and maintaining epithelial and stromal cells was DMEM supplemented with 10% FBS, penicillin (100 units/ml), and streptomycin (100 μg/ml). The cultures, grown at 37°C, were buffered at pH 7.2 using a bicarbonate–CO_2 system supplemented with added HEPES buffer (10 m*M*; *N*-2-hydroxyethylpiperazine-*N*′-2-ethanesulfonic acid, Sigma).

C. Hormones

Dexamethasone (Sigma, London) progesterone, and estradiol-17β (Searle, High Wycombe) were all added as concentrated ethanolic solutions never exceeding a final ethanol concentration in the medium of 0.1% (v/v). Insulin (crystalline bovine, BDH, Poole, England) was dissolved in dilute hydrochloric acid, the final concentration in the medium being 0.05 m*M*.

Effects of hormones on cell growth were studied in DMEM supplemented with

FBS previously extracted with dextran-charcoal (Armelin *et al.*, 1974) to remove endogenous steroids.

III. Methods

A. Separation of Epithelium from Its Stroma

Curettings were maintained at ambient temperature during transport and then rinsed several times with collection medium to remove excess red blood cells. Approximation of sample size was determined from its displacement volume (DV) in the collection medium using a 10-ml graduated cylinder before it was minced into cubes (approximately 1 mm^3) with a tissue sectioner (Mickle Laboratory Engineering Co., Gomshall, Surrey). The minced sample was then subjected to a mild collagenase digestion (180 units/ml DMEM; Type I, Sigma Chemical Co. Ltd., London) overnight (15–20 hours) at room temperature (22–23°C) using a sample DV to digestion medium ratio of 1:50 (v/v). The digestion was carried out without shaking in 75-cm^2 Falcon flasks gassed with an 80% air–20% CO_2 atmosphere and the cap screwed tight. After digestion the stroma was reduced to single cells whereas the epithelial structures remained intact. These two components were then separated on a size basis using gravity sedimentation in a 20-ml universal bottle. After settling for 5 minutes the supernatant (stromal-rich fraction) was removed using a pasteur pipet. The pellet (epithelial-rich fraction) was further rinsed with DMEM, the supernatants being pipetted off and pooled as the stromal fraction.

Residual stromal cells and undigested clumps were substantially removed from the epithelial structures as follows: when the partially purified epithelium was placed in 75-cm^2 Falcon flasks in DMEM containing 0.5% FBS and then incubated at 37°C in a 95% air–5% CO_2 atmosphere, the residual stromal cells were selectively attached to the plastic within 30–45 minutes, leaving the epithelial structures floating in the medium. After repeating this procedure on the opposite side of the flask the purified epithelial structures were collected and their number estimated as follows: a known volume of a dilute suspension of epithelial clumps was added to a sterile Leighton tube. Using an eye-piece graticule (E_{35}, Graticule Limited, Tonbridge, Kent) and an inverted microscope, the mean number of epithelial clumps per unit area was determined from 10 random medial samples along the Leighton tube. The total number of epithelial clumps was thus easily calculated.

The pooled stromal fraction was further purified from contaminating epithelial clumps by filtration through two layers of a 27-μm pore-size polyester filter (PES 27, Schweizerische Seidengazefabrikag).

B. Cell Culture

Cells were plated, in growth medium, into 24-hole Linbro trays (Linbro Scientific Co., New Haven, Conn.) containing 13-mm glass coverslips.

C. Determination of Stromal Cell Growth

Cells were plated in 6-hole Linbro trays (35 mm diameter) at a seeding density of 4–6 $\times$ $10^3/cm^2$ in DMEM supplemented with 1% FBS extracted with dextran charcoal (DC, see Section II, C). Hormones were added 24 hours after plating in DMEM supplemented with 1 or 5% FBS (DC). Medium was changed every 3 days and after 8–10 days the final cell number of triplicate cultures was determined using a Model B Coulter electronic cell counter.

D. Scanning (SEM) and Transmission (TEM) Electron Microscopy

Fresh tissue or coverslip cultures were fixed and processed using the standard methods described previously (Kirk *et al.*, 1978b).

E. Scoring for Mitoses

Coverslip cultures, rinsed twice in cold (4°C) phosphate-buffered saline (PBS, Dulbecco's modification without Mg^{2+} and Ca^{2+} salts), were fixed in 95% methanol, hydrolyzed in 5 *N* hydrochloric acid at room temperature for 25 minutes, and stained by the Feulgen reaction. Mitotic indices (percentage cells in mitosis) were determined by examination of 10^3 randomly selected nuclei per coverslip culture.

F. Feulgen Microspectrophotometry

DNA values of Feulgen-stained interphase nuclei (see Section III, E) were determined using conventional single wavelength microspectrophotometry at 560 n*M* with a Barr and Stroud integrating microdensitometer. Net absorbance determinations (arbitrary units) were made on 50 randomly selected nuclei per coverslip culture. The arbitrary values for nuclear DNA content were converted to logarithms to the base 2 and delineated into nuclei with the three major DNA classes 2C, 4C, and intermediate (C = DNA content of a haploid chromosome set). Nuclei with the 2C amount of DNA were classed as G_1, those with twice this amount as G_2, while nuclei with intermediate values were regarded to be in DNA synthesis (S). Nuclei with DNA values greater than 4C were grouped into 8C, 16C, 32C, and 64C with all intermediate values again taken as nuclei in the S phase. No distinction was made between tetraploid G_1 and diploid G_2 nuclei and likewise for all adjacent ploidy levels. Calibration using bull sperm as a standard has confirmed that the 2C DNA value represents that of a diploid human cell. Although polyploidy strictly refers to cells with three or more sets of chromosomes, lack of karyotypic data has necessitated polyploidy to be defined here as nuclei with DNA content greater than the 4C level and the polyploid index (PI) of a culture as the percentage of nuclei with DNA content greater than the 4C level.

G. Determination of Ploidy from Measurements of Nuclear Diameters

The mean diameter of 50 randomly selected, Feulgen-stained interphase nuclei was obtained from microscopic measurement of two diameters at right angles for each nucleus using an eyepiece graticule (E_{17}, Graticules Ltd., Tonbridge, Kent). Assuming nuclei to be spherical and ploidy to be directly proportional to nuclear volume (Meek and Harbison, 1967), nuclear ploidy levels were determined from the calculated relative diameters of 2C (1.000), 4C (1.26), 8C (1.58), 16C (2.00), 32C (2.52), and 64C (3.17) nuclei. Estimates of polyploid indices (percentage nuclei $>$ 4C) using this method were in close agreement with those obtained by Feulgen microspectrophotometry.

H. Determination of DNA Synthesis by Incorporation of Tritiated Thymidine ([^{3}H]TdR)

1. Liquid Scintillation Counting

Coverslip cultures were continuously labeled with [^{3}H]TdR in the growth medium (0.5 μCi/ml; sp ac 20 Ci/mmole) and were counted for tritium in Aquasol using standard liquid scintillation methods.

2. Autoradiography

Cultures counted for tritium in Section III, H, 1 were rinsed in absolute methanol, stained with aceto-orcein, and processed for autoradiography using Kodak AR10 stripping film.

I. Check for PPLO Contamination

Although the cells used for these studies had not been tested for mycoplasma contamination, routine testing for PPLO, using the aceto-orcein method (Fogh

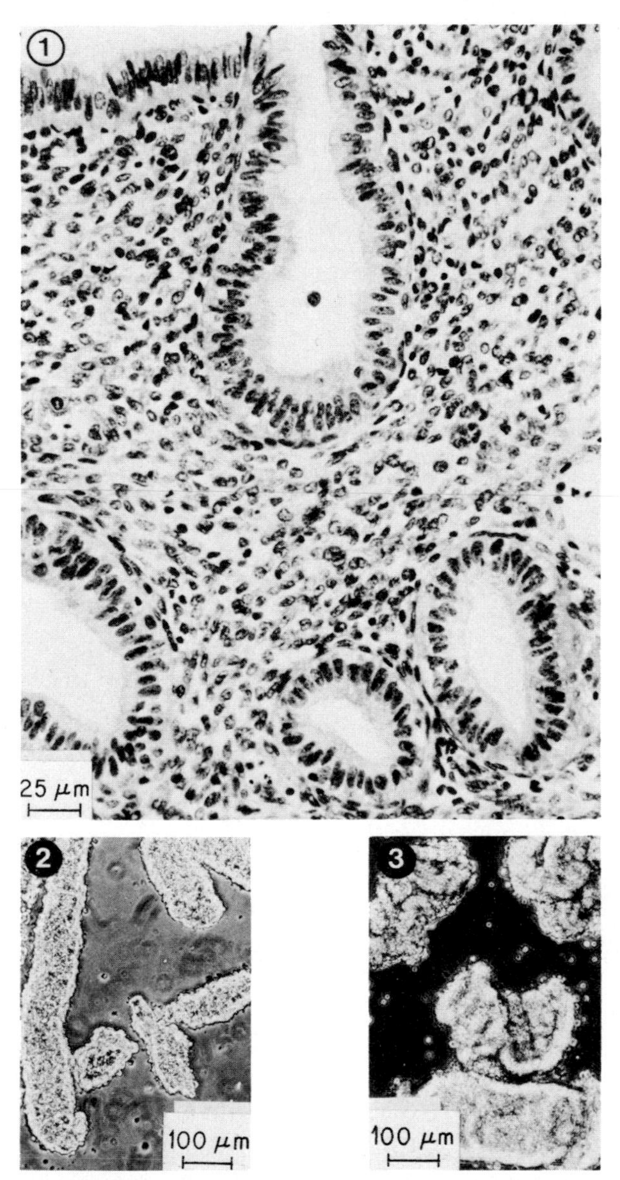

FIG. 1. Section of proliferative endometrium. Note the tall columnar and pseudostratified epithelial glands. H and E staining, × 400.

FIG. 2. Epithelial glandular structures after separation from a proliferative endometrium. Note the long narrow glands characteristic of the proliferative phase. Phase-contrast, × 100.

FIG. 3. Epithelial glandular structures after separation from a secretory endometrium. Note the tortuous, wide glands characteristic of the secretory phase. Phase-contrast, × 100.

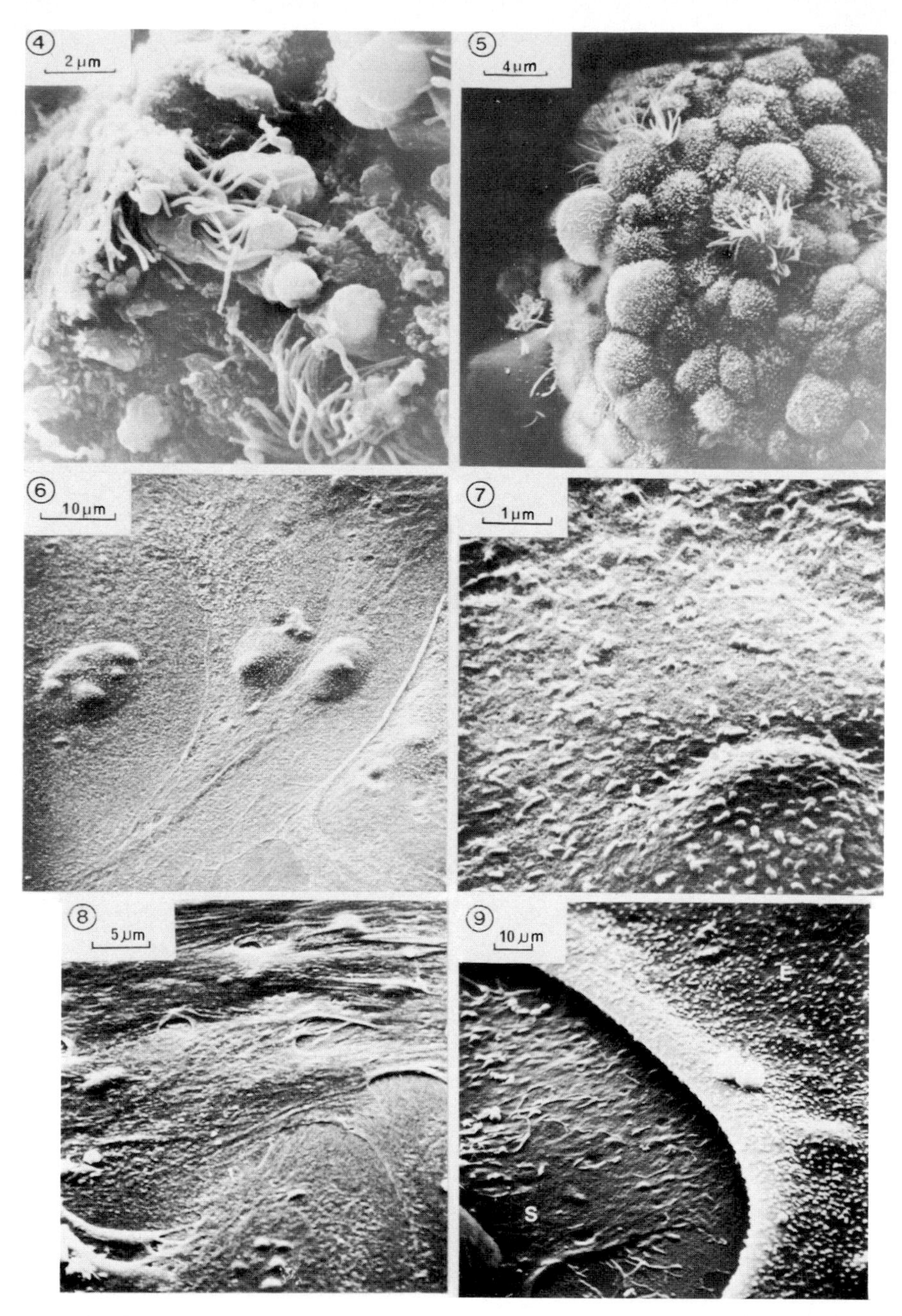

FIG. 4. Scanning electron micrograph (SEM) of intact luminal epithelium before collagenase digestion. Mucus and red blood cells are present. × 10,000.

FIG. 5. SEM of luminal epithelium after collagenase digestion. Cilia intact. × 4,000.

FIG. 6. SEM of migrating epithelial cells in culture 3 days after explanting. Note regular cell-to-cell borders. × 2000.

and Fogh, 1964), has subsequently shown mycoplasma-free cultures to give similar results.

IV. Results

A. Epithelial/Stromal Separation Procedure

The procedure for separating epithelium from stroma, outlined in Section III, A, was found to be the optimum method for collagenase digestion and produced cleanly separated, recognizable, glandular structures. Digestion of a proliferative endometrium (Fig. 1) yields straight and narrow glandular structures typical of the proliferative phase (Fig. 2) whereas separated secretory glands are typically wide and tortuous (Fig. 3). Transport on ice caused delays in cell outgrowth whereas digestion at 37°C and mincing to smaller fragments resulted in atypical epithelial colonies. Comparison of epithelial surface structures such as cilia and microvilli before (Fig. 4) and after (Fig. 5) collagenase treatment showed no deleterious effects on structural integrity.

Assuming that all cells in the epithelial glandular fragments had a diploid DNA content of 5 pg (Sober and Harte, 1970), determination of DNA content per clump has demonstrated that the present separation procedure yielded clumps containing 10^3–6×10^3 cells irrespective of the stage of the endometrium in the menstrual cycle. Efficiency of epithelial/stromal separation varied with the menstrual cycle; proliferative material yielded epithelia upward of 90% pure, whereas secretory epithelia was upward of 70% pure (as judged by visual observation).

B. Characterization of Cell Cultures

1. SEM

Outgrowths from explanted epithelial clumps in GM (Section II, B) consisted of flat cells with reduced microvilli and regular cell-to-cell borders typical of epithelium (Figs. 6 and 7).

FIG. 7. High power of Fig. 6 illustrating surface microvilli. × 20,000.

FIG. 8. SEM of primary stromal cell cultures 3 days after collagenase separation and plating. Note irregular cell to cell borders. × 3000.

FIG. 9. SEM comparing surface structures of primary stromal (S) and epithelial (E) cells 3 days after plating. Note regular microvillus structures on the epithelial cell (E) and the irregular surface structures on the stromal cell (S). × 10,000.

In contrast, stromal cultures showed irregular cell-to-cell borders (Fig. 8), no microvilli, and the presence of irregular surface blebs (Fig. 9).

2. TEM

Epithelial cultures possessed microvilli (with a central core and glycocalyx typical of epithelia) and an epithelial-type junctional complex (Fig. 10). TEM further confirmed the flat morphology of epithelial cultures [cf. *in vivo* 15 to 20 μm high (Fig. 1), *in vitro* 2 to 5 μm high (Fig. 10)]. A well-defined basal lamina was not observed.

Stromal cultures were also flat but did not have microvilli and showed mesenchymal-type cell to cell junctions (Fig. 11).

C. Behavior of Endometrial Cells in Monolayer Culture

Growth Form

a. Stromal Cultures. Initially the stromal cultures presented as a heterogeneous assemblage of stellate, fibroblastic, and macrophagic cell morphologies. Essentially they behaved like human diploid fibroblasts in that they could be passaged as monolayer cultures which showed a limited *in vitro* life span.

b. Epithelial Cultures. About 12 hours after the epithelial clumps had attached, cells migrated out and formed a typical epithelial colony (Figs. 12 and 13). The colony continued to spread and grow for up to 10–12 days when degeneration occurred. In contrast to the stromal cultures, endometrial epithelium survived only as short-term primary cultures which could not be successfully passaged using either the standard trypsin/versene method or the nonenzymic K^+ passage of Lechner *et al.* (1978). However, if the cultures were seeded at higher plating densities, where confluence occurred 2 to 3 days after seeding, then the cultures remained viable for up to 6 weeks (Figs. 14 and 15).

D. Cytodynamics of Primary Epithelial Cultures

In an attempt to determine why epithelial cultures were short-term, a systematic study of cell growth in terms of mitotic activity and the nuclear DNA cycle was undertaken.

1. Mitotic Activity

Epithelial cultures seeded at a range of densities from subconfluent (I) to sparse (IV) were monitored for mitotic activity (see Section III, E) during the

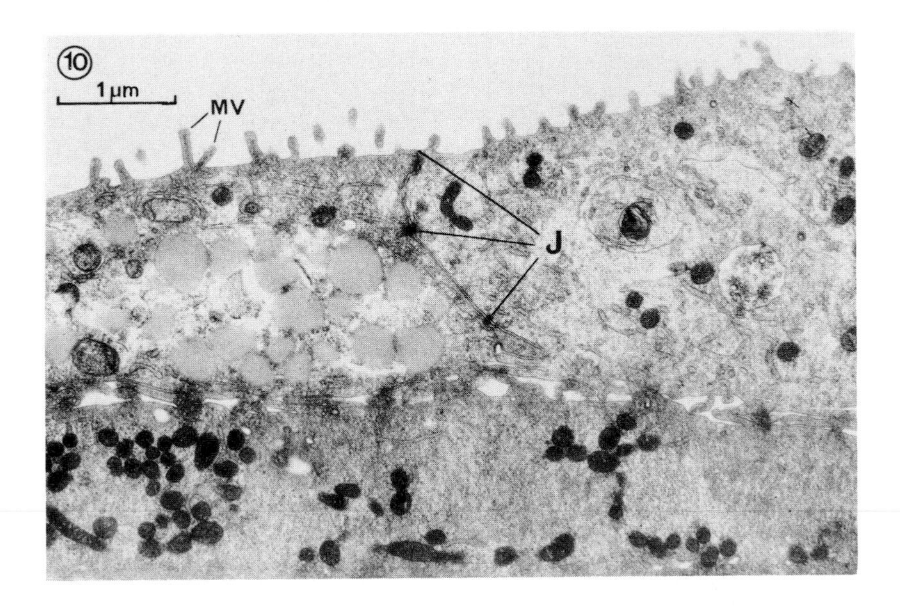

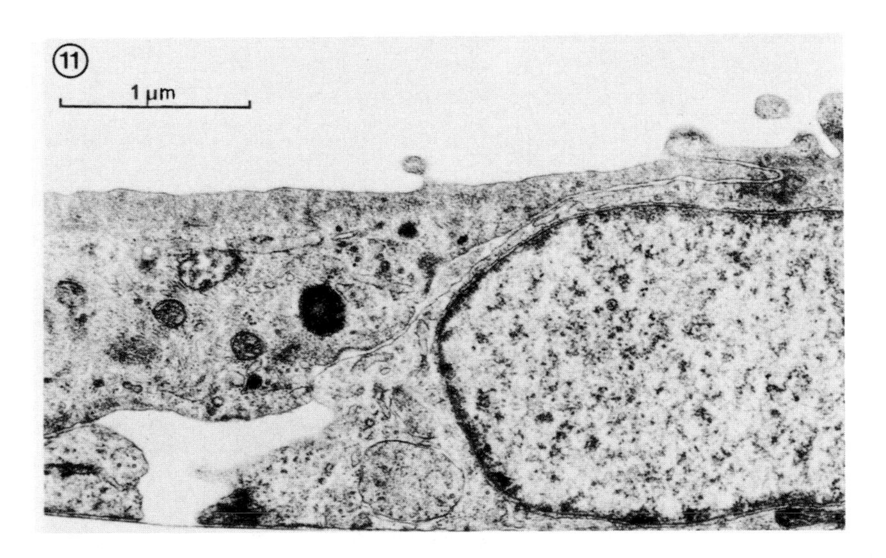

FIG. 10. Transmission electron micrograph (TEM) of 28-day-old epithelial culture derived from a proliferative endometrium and plated at confluent density. Note the microvilli (MV) and the junctional complex (J) typical of epithelia. × 25,000.

FIG. 11. TEM of stroma after 20 culture passages. Microvilli and the epithelial type of junctional complex are absent. × 40,000.

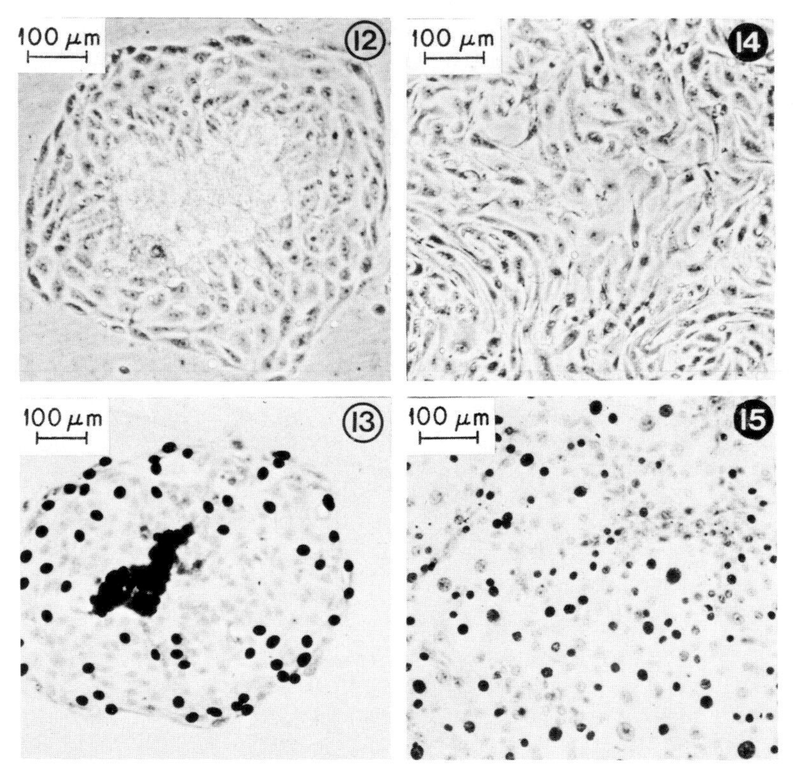

FIG. 12. Migrating cells from epithelial explant, derived from a proliferative endometrium, 24 hours after plating. Phase-contrast, × 100.

FIG. 13. Autoradiograph of migrating cells from epithelial explant, derived from a late proliferative endometrium, 40 hours after plating. Culture was pulsed with [^{3}H]TdR (Section III, H, 2) for 24 hours. H and E staining, × 100.

FIG. 14. Epithelial culture 20 days after plating at high (confluent) density. Note the characteristic whorling appearance. Phase contrast, × 100.

FIG. 15. Autoradiograph of 6-day-old epithelial culture (derived and pulsed with [^{3}H]TdR as described for Fig. 13) plated at subconfluent density (see density I, legend to Fig. 16). Note wide range in the size of nuclei synthesizing DNA. H and E staining, × 100.

16-day culture period. The results (Fig. 16) showed that essentially the cultures go through a wave of mitosis which can be shunted laterally by alterations in the plating density.

2. MICROSPECTROPHOTOMETRIC STUDY OF THE NUCLEAR DNA CYCLE

Mitosis is only a small window on the cell cycle and supplementary information on cell population dynamics can be obtained from a knowledge of the distribution of the nuclear DNA contents (see Section III, F). Nuclear DNA amounts were determined using Feulgen microspectrophotometry (Section III, F) on interphase nuclei in the cultures already scored for mitosis (Section IV, D, 1). The distribution of nuclear DNA contents of 3-day cultures (Fig. 17) showed a tight diploid–tetraploid distribution with very few nuclei with DNA content greater than 4C. However, the DNA profiles of 16-day cultures (Fig. 18) showed a marked increase in the number of nuclei with DNA contents greater than 4C, i.e., polyploid nuclei.

By plotting the percentage of nuclei with DNA content greater than 4C against time in culture a measure of the rate of polyploidization is obtained (Fig. 19a). Plating density affects the rate of polyploidization in epithelial cultures, higher plating densities being associated with higher rates of polyploidization. Under identical conditions of cell density and culture, homologous stromal cells do not show this polyploidization process (Fig. 19b). Although this analysis of polyploidy utilized a normal secretory endometrium, polyploidy was found to occur in

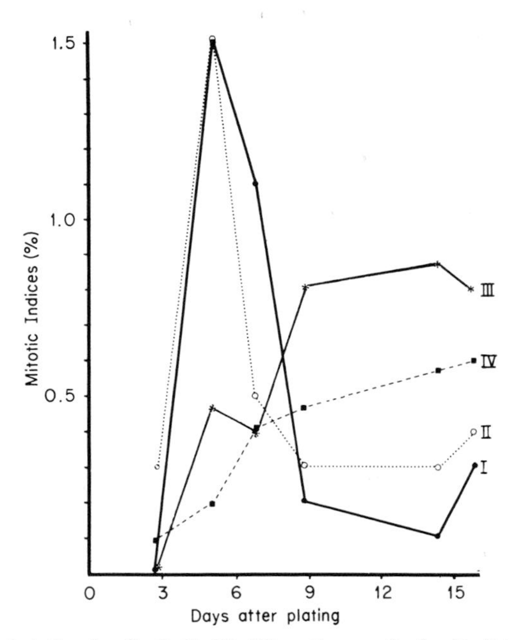

FIG. 16. Effect of plating density (I, II, III, IV) on the growth of epithelial cells migrating from explants. Each point represents the mean of triplicate cultures. Error bars, which have been omitted for clarity never exceeded $\pm$ 0.15 for densities I and II and $\pm$ 0.20 for densities III and IV. Epithelial clumps (= 1000 cells/clump) were plated at 2×10^3 (I), 10^3 (II), 5×10^2 (III), and 10^2 (IV) clumps per well in 2 ml of DMEM containing 15% FBS and incubated at 37°C in an atmosphere of 95% air–5% CO_2. Cultures were refed with fresh medium every 3 days. The highest density cultures (I) reached confluency after 5–6 days whereas the lowest density cultures (IV) never reached confluency.

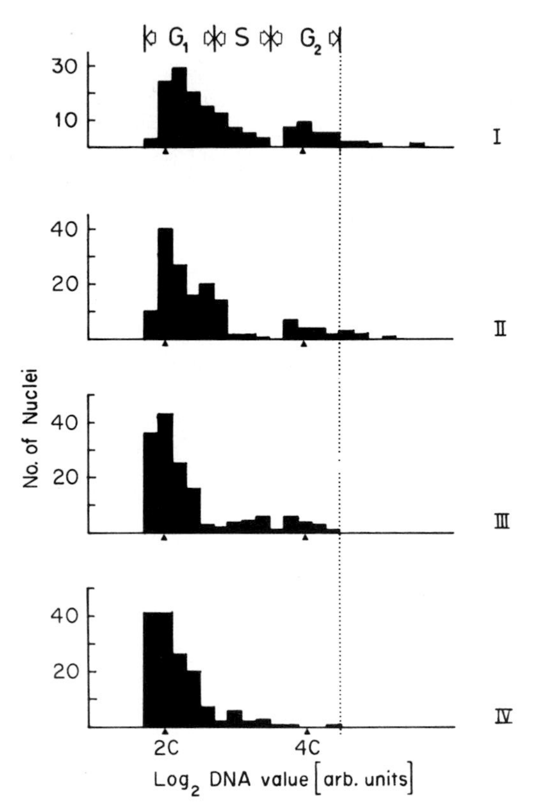

FIG. 17. Nuclear DNA profiles of 3-day epithelial cultures plated at densities I–IV (see Fig. 16). Data were pooled from triplicate cultures (n = 150) and ploidy levels assigned as described in Section III, F.

epithelial cultures from functionally different normal human endometria (Table I). The use of richer medium, such as Ham's F12 (or the MCDB series), Medium 199, Waymouth's, or supplementing with human serum, cortisol, and/or insulin, did not alter the nature of the polyploidization process. Cortisol and insulin did, however, improve the appearance of epithelial cultures by producing tighter colonies.

A clearer understanding of the processes involved in this polyploidization process is obtained from looking at the changes in the proportion of the different ploidy levels with increasing time in culture (Fig. 20). In density I the diploid (2C) population showed a significant drop from 63 to 9% over the 16-day culture period. The tetraploid (4C) population showed a less pronounced but also significant decline. Concomitant with this decline of the 2C and 4C nuclei was the increase in the number of 8C, 16C, and, to a lesser extent, 32C and 64C nuclei *in this specific order of appearance*. These changing patterns of ploidy distributions illustrate a flow of nuclei up the ploidy series resulting in a fall of the 2C population. Lowering the plating density altered the rate but not the overall nature of polyploid formation.

A prediction of the "flow" concept of polyploidization would be that DNA synthesis is necessary to achieve higher ploidy levels. It is therefore interesting that DNA values intermediate between all the ploidy classes occur (Fig. 20, ΣS), these intermediate values being taken to represent DNA synthesis. However, microspectrophotometric determinations of DNA synthesis are static measurements and, in order to try and firmly assess the role of DNA synthesis in the polyploidization process, the active incorporation of [^{3}H]TdR into DNA was studied.

3. Determination of Nuclear DNA Synthesis Using [^{3}H]TdR Autoradiography

Primary epithelial cultures were continuously labeled with [^{3}H]TdR and processed for autoradiography (Section III, H, 2). Cultures were then monitored for

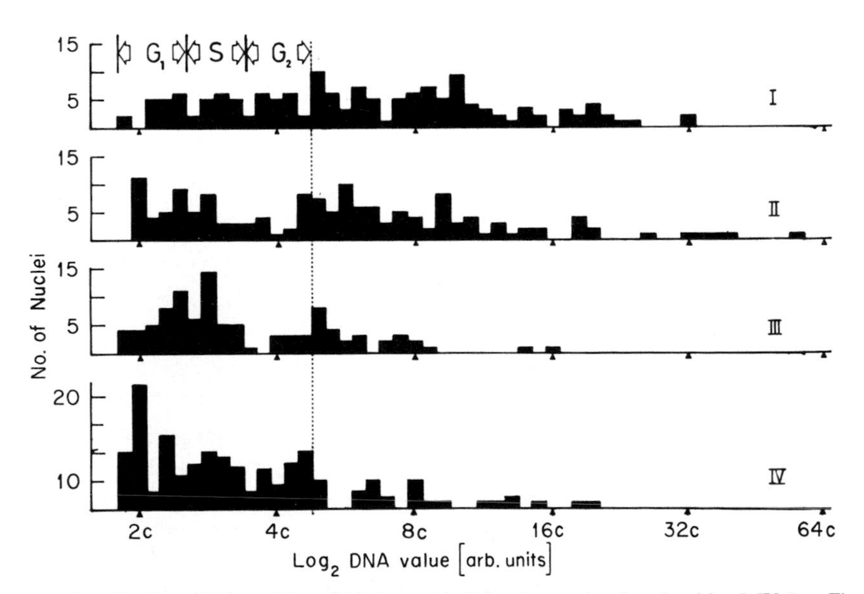

FIG. 18. Nuclear DNA profiles of 16-day epithelial cultures plated at densities I–IV (see Fig. 16). Data were pooled from triplicate cultures ($n = 150$).

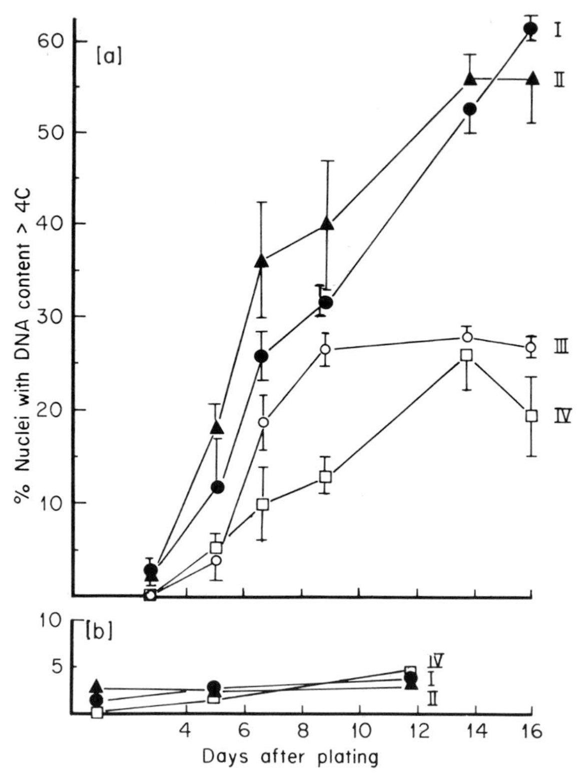

FIG. 19. Rate of polyploidization in endometrial cell cultures. (a) Primary epithelial cultures plated at densities I–IV (see Fig. 16). Each point represents the mean of triplicate cultures ± SE (vertical line). (b) Secondary stromal cultures plated at 7×10^4 (I), 3.5×10^4 (II), and 1.4×10^4 (III) cells per cm^2. Each point represents the mean of duplicate cultures. Epithelial and stromal cultures were derived from the same endometrium, the stroma having undergone four passages in culture.

polyploidy (using the method of nuclear diameters, Section III, G) and the measured nuclei simultaneously scored as labeled or unlabeled with [^{3}H]TdR (Fig. 21). As the cultures become progressively more polyploid (Fig. 21a,b,c) it is seen that all the polyploid cells are labeled with [^{3}H]TdR. This process is illustrated in autoradiographs of 24 hour cultures (Fig. 13) where nuclei synthesizing DNA are of uniform diploid–tetraploid size, and also in 120 hour cultures (Fig. 15) where polyploid nuclei are synthesizing DNA (note, culture in Fig. 15 was pulse-labeled with [^{3}H]TdR for 24 hours and not continuously labeled).

In conclusion the combined microspectrophotometric and autoradiographic data are in full agreement and show polyploid formation in endometrial epithelial cultures to be dependent on DNA synthesis.

TABLE I

SUMMARY OF POLYPLOID INDICES OF EPITHELIAL CULTURES FROM NORMAL HUMAN ENDOMETRIA

Endometrial type	Donor age (years)	Growth medium	Days in culture	Plating density[a]	Polyploid index[b]
Secretory				I	63
				II	56
(a)	38	DMEM + 15% FBS	16	III	27
				IV	19
(b)	36	DMEM + 10% FBS	8	II	30
(c)	42	DMEM + 20% FBS	5	III	18
(d)	44	DMEM + 20% FBS	5	IV	10
Proliferative					
		Ham's F_{12} + 10% FBS	16	III	63[c]
(a)	37	Ham's F_{12} + 10% FBS + hydrocortisone (1 μg/ml) + insulin (10 μg/ml)	16	III	65[c]
				III	50
(b)	23	DMEM + 15% FBS	8	IV	50
(c)	33	DMEM + 10% FBS	8	III	32

[a] Defined in legend of Fig. 1.
[b] Percentage of nuclei with DNA contents >4C.
[c] Data derived from measurements of nuclear diameters (see Sections II and III).

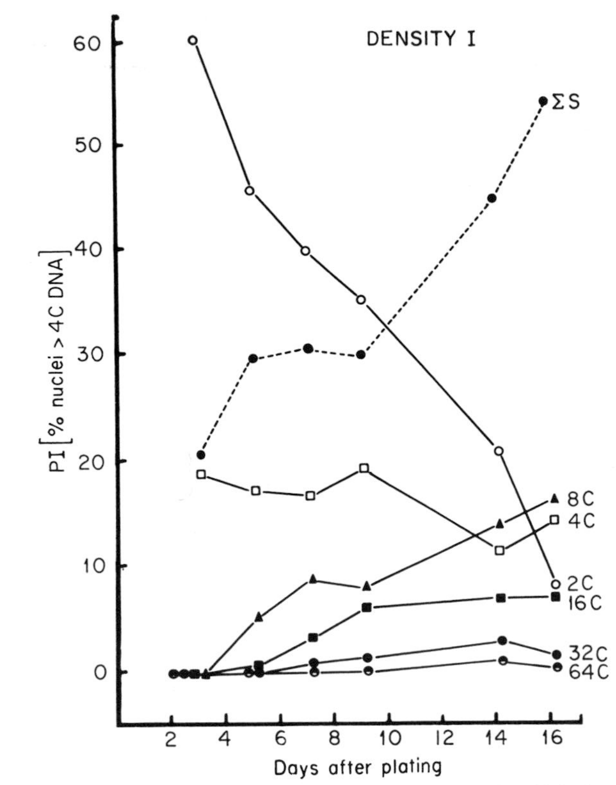

FIG. 20. Distributions of 2C, 4C, 8C, 16C, 32C, 64C nuclei and also all the intermediate values (ΣS) with increasing time in culture for epithelial cultures plated at density I (see legend to Fig. 16). Each point represents the mean of triplicate cultures.

E. Hormone Sensitivities

1. Epithelial Cultures

The effect of hormones on epithelial cell proliferation, as measured by mitotic activity and [^{3}H]TdR incorporation was studied. It was found that 10^{-8} M estradiol-17β, 10^{-7} M progesterone, or a combination of these hormones produced no consistent effects on spreading epithelial cultures.

2. Stromal Cultures

When the effects of hormones on the final cell numbers of uncloned stromal cultures were investigated, strains derived from two normal proliferative endometria (Table II) showed variable responses ranging from 3 to 62% for proges-

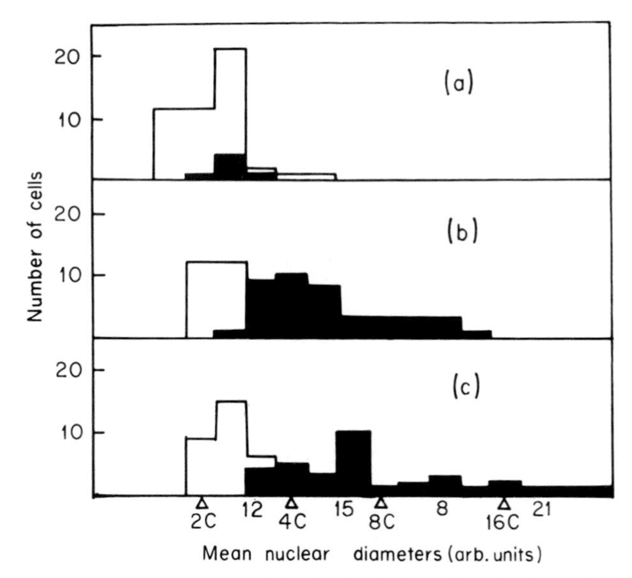

FIG. 21. Distribution of nuclei synthesizing DNA during polyploidy formation in epithelial cultures derived from a late proliferative endometrium. Cultures were plated at density II (see legend to Fig. 16), continuously labeled with [^{3}H]TdR, and processed for [^{3}H]TdR autoradiography (see Section III, H, 2) at 17 hours (a), 92 hours (b), and 140 hours (c) after plating; □, unlabeled; ■, labeled.

TABLE II

SUMMARY OF EFFECTS OF HORMONES[a] ON GROWTH OF HUMAN ENDOMETRIAL STROMAL CELLS *in Vitro*

		Mean increase (%) in final cell number[b] compared to controls										
					+ I					+ D		
Source of stroma	Experiment	E_2	P	E_2 + P	E_2	P	E_2 + P	I	D	E_2	I	E_2 + I
Proliferative	1[c]	60	62									
	2[c]	−13	3	−4								
Proliferative	3 a[c]	−3	8	−3	17	22	29	6				
	3 b[d]	21	26	7	65	43	42	21				
	4[d]	7			22			15	30	65	62	83

[a] Hormones: E_2, 10^{-8} *M* estradiol-17β; P, 10^{-7} *M* progesterone; D, 10^{-8} *M* dexamethasone; I, 2.5 μg/ml insulin.

[b] Standard errors of the means ranged between 5 and 15%.

[c] Growth medium containing 1% (DC) FBS.

[d] Growth medium containing 5% (DC) FBS.

terone and from −13 to 60% for estradiol as compared to control cultures. Combination of estradiol and progesterone produced no significant response in these stromal strains.

However, it is clear from the data (Table II, Experiments 3, 4) that the proliferative response to estradiol is substantially increased through synergistic effects with insulin and/or dexamethasone. Furthermore, maximal response is affected by the concentration of the DC serum used. The optimum concentration was found to be 5% (Table II; cf. Experiments 3a, 3b) whereas 10% resulted in no significant proliferative hormonal effects (unpublished observation).

F. Serum Dependence of the Plating Efficiency of Primary Epithelial and Stromal Cells

Incorporation of [^{3}H]TdR into DNA (Section III, H, 1) was taken as a measure of plating efficiency. Primary epithelial and first passage stromal cultures from normal secretory endometria (nonhomologous cultures) were plated in growth

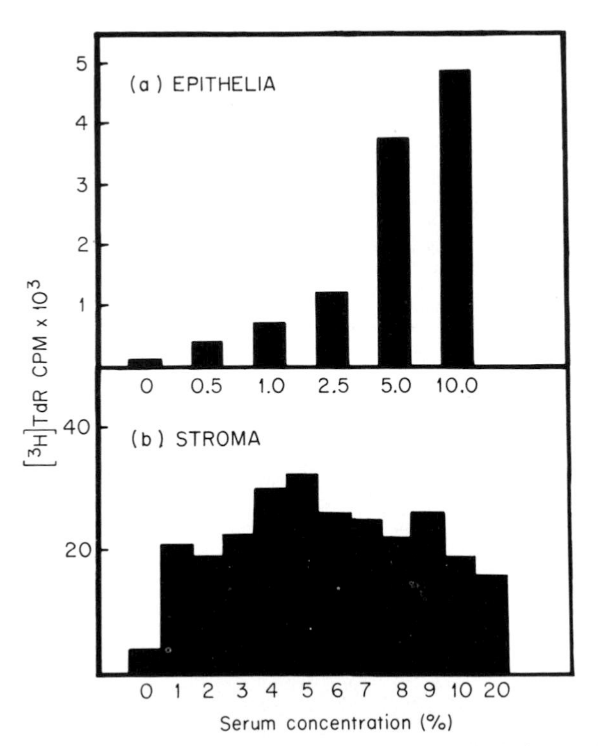

FIG. 22. Serum dependence of epithelial (a) and stromal (b) cells during attachment and early growth in culture. Epithelial cells were plated at density III (see legend to Fig. 16) and stromal cells at $5 \times 10^4/cm^2$.

medium (Section II, B) containing [^{3}H]TdR (0.5 μCi/ml) and varying amounts of FBS. Incorporation of [^{3}H]TdR was determined on coverslip cultures after 1 day for stromal cultures and after 3 days for epithelial cultures. During this period of attachment, spreading, and early growth, the stromal cells (Fig. 22b) showed a lower serum requirement than epithelial cells (Fig. 22a).

G. Effect of High Cell Density on Epithelial Cultures

Plating epithelia at high cell density (confluent density) increased culture viability for up to 6 weeks. Three-dimensional structures resembling domes in mouse mammary cultures (Das *et al.*, 1974) were a consistent feature of these high-density endometrial epithelial cultures.

V. Discussion

Human endometrium lends itself well to separation of its epithelial and stromal components with the use of gentle collagenase digestion. Although neither component is a homogeneous cell population in itself, the cultures obtained can be clearly categorized ultrastructurally as either of epithelial or mesenchymal origin. Recognition of epithelia in this study has relied heavily on their colony growth form with their characteristic intercellular junctional complexes and surface microvilli structures with glycocalyx (Franks and Wilson, 1977). However, Liszczak *et al.* (1977), using a hyaluronidase/trypsin-EDTA digestion of intact normal endometria, were also able to identify glandular epithelial cells in culture by the presence of secretory organelles.

Mesenchymal stromal cells grew in culture as single cells typical of human diploid fibroblasts and lacked both the epithelial-type microvilli and junctional complexes. Time-lapse cinematography further contrasted these two cell types in culture, the stroma appearing sedentary compared to the active whorling movements of the epithelia. There are also important functional differences between epithelial and stromal cultures.

A. Serum Dependence of Epithelial and Stromal Cultures

In contrast to the previous findings that epithelial cells are less serum dependent than fibroblasts (Dulbecco and Elkington, 1973), endometrial stromal cells showed a lower serum requirement than endometrial epithelial cells for attachment to the substratum and early growth. However, Dulbecco and Elkington (1973) used murine 3T3 cells, whereas other studies on normal human cells have shown foreskin fibroblasts to have about half the serum requirement of prostatic epithelial cells (J. F. Lechner, personal communication). This differential serum requirement has been exploited by us as a means of increasing the efficiency of the initial epithelial/stromal separation procedure by carrying out the differential attachment of the stroma in a low serum concentration (see Section III, A).

B. Stromal and Epithelial Hormone Responses *in Vitro*

The results of the hormone responses of stromal cultures are in general agreement with previous findings (Kirk *et al.*, 1978b), where progesterone and estrogen produced variable proliferative responses. In addition, the present study demonstrates that there are positive synergistic effects between estrogen, insulin, and/or dexamethasone and to a lesser extent between progesterone and insulin.

Since the epithelium is a target cell for estradiol *in vivo,* it was surprising that epithelial cultures showed essentially no response to this hormone. This is in agreement with previous studies on normal human endometrium where estradiol elicited no proliferative, histochemical (Figge, 1963) or ultrastructural changes (Liszczak *et al.*, 1977). A clue as to why sex steroids are impotent *in vitro* may be seen from the striking similarities between epithelial cells that migrate from explants *in vitro* and those that resurface the endometrium after menstrual desquamation (Table III). Both cell populations are flat with a fusiform morphology and have reduced surface microvilli with ciliated cells being absent. Furthermore, initially both ''repairing'' populations are composed exclusively of cells containing the diploid (2C) DNA amount, and cell migration continues for up to 48 hours before any mitoses are observed (Ferenczy, 1976). Since the resurfacing *in vivo* occurs under a very low estradiol concentration, it is possible that the *in vitro* migration is similarly an estrogen-independent repair program which renders the cells refractory to further anabolic stimulation by steroids.

An additional reason for the inefficacy of sex steroid function in the epithelial cultures could be the absence of its supporting stroma because Kratochwil and

TABLE III

SIMILARITIES BETWEEN THE BEHAVIOR OF EPITHELIAL CELLS IN THE RESURFACING ENDOMETRIUM AFTER MENSTRUAL DESQUAMATION *in Vivo*[a] AND IN THE MIGRATION FROM EXPLANTS *in Vitro*

1. No mitoses observed during the first 48 hours
2. Epithelial populations contain cells exclusively with the diploid (2C) DNA amount representing cells in the G_1 phase of the cell cycle. No nuclei synthesising DNA or higher ploidy levels are observed at this stage
3. Cells are very flat
4. Surface microvilli are markedly reduced in size
5. Cells present a fusiform morphology
6. Complete absence of ciliated cells

[a] See Ferenczy (1976).

Schwartz (1976) have demonstrated the androgenic response of mouse mammary epithelium to be mediated through the supporting mesenchyme.

C. Growth Behavior of Endometrial Cells *in Vitro*

Although stromal cultures showed an *in vitro* life span typical of human diploid fibroblasts (Hayflick and Moorhead, 1961) the epithelial cultures survived only as short-term primary cultures (10 to 15 days). It is generally found that normal, adult human epithelial cells cannot be serially cultivated like human diploid fibroblasts (Owens *et al.*, 1976). In agreement with previous workers (Figge, 1960, 1963; Papanicolaou and Maddi, 1958, 1959; Hiratsu, 1968) it was noted that epithelia from proliferative phase endometrium showed a better growth and a more typical epithelial colony structure than secretory-phase epithelia. The short life span of endometrial epithelial cultures very likely is related to the rapid accumulation of polyploid cells, a process not observed in stromal cultures.

Polyploidy, which is widespread in the animal kingdom (Brodsky and Uryvaeva, 1977) and commonly associated with animal differentiation (Geitler, 1939, 1953; Walker, 1958) can be produced in several ways including endoreduplication, failure of cyto- or karyokinesis, nuclear fusion, and selective gene amplification. The stepwise increase in ploidy levels argues against either nuclear fusion or selective gene amplification as being the primary mechanism. Although we do not have the cytological observations to specify the mechanism further, it must be concluded that both the densitometric and autoradiographic data have shown DNA synthesis to play an integral role in the polyploidization process.

Whereas it is true that polyploidy is usually associated with termination and degeneration of cells in culture (Miller *et al.*, 1977) there are two pertinent observations that attribute a possible physiological significance to endometrial epithelial polyploidization *in vitro*. First, the pattern of ploidy formation in endometrial epithelia *in vitro* is strikingly similar to that of the liver (Carriere, 1969) and the bladder (Walker, 1958). There is a steady depletion of 2C nuclei with subsequent increases in the higher ploidy levels, although time scales varied from days in endometrial cultures and mouse bladder, to years in the liver. Second, polyploid nuclei (up to 16C) are observed in the glandular epithelium of the endometrium during pregnancy, especially late pregnancy (Wagner and Richart, 1968). This occurrence of *in vivo* endometrial polyploidy, referred to as the Arias-Stella Reaction (ASR), has been experimentally induced in rat and human endometria with gonadotrophin and/or sex steroid combinations (Brown and Bradbury, 1947; Arias-Stella, 1955).

This association between growth stimulation and polyploidy in intact endometrium is found also in the mammalian liver, where repeated hepatectomies progressively enhance the polyploidy (Brodsky and Uryvaeva, 1977). More generally, polyploidy has also been associated with hyperplasia in the human thyroid (Beierwaltes and Al-Saadi, 1966), heart muscle (Kompmann *et al.*, 1966), and mouse liver and pancreas (Leuchtenberger *et al.*, 1954). Such observations suggest that the proportionality of mitotic rate and polyploidy observed in our endometrial epithelial cultures represents more than a degenerative culture phenomenon.

Although we believe the *in vitro* endometrial polyploidy described here to have some physiological meaning, it is conceivable that using optimal methods of separation and cell culture the terminal polyploidization process could be substituted for a different cellular programme. It is surely of great significance, though, that this nuclear behavior of epithelial cultures is markedly different from the stable diploid cycle of homologous stromal cells maintained under similar culture conditions.

Das *et al.* (1974) have previously reported a similar progressive polyploidization in mouse mammary epithelial cultures. In addition, they noted that very high-density cell cultures helped retain a diploid nuclear phenotype. Preliminary experience with very high-density endometrial epithelial cultures showed similar results although it was not clear whether the decreased polyploidy was due to recycling of tetraploid cells and/or a slowing down of polyploidization due to density inhibition of growth. The relative stability of these high-density epithelial cultures in being able to survive for longer periods *in vitro* would implicate high cell density as a useful physiological, stabilizing factor. Use of such high-density cultures might prove more useful for studying hormonal effects on growth and differentiation than the low-density "repairing" cultures have done to date.

VI. Perspectives

A. Endometrial Cell Cultures as Models for Studying the Mechanism of Hormone Action

1. Stromal Cells

The use of uncloned stromal cells for purely mechanistic studies of sex hormone action is frustrated by the inherent variation of the hormone responses observed. A uniformity of response might be obtained from studies on cloned cells. However, stromal cultures do promise to be a useful tool for studying growth factor interactions *in vitro*.

2. Epithelial Cells

It was thought that the use of the relatively more stable high-density endometrial cultures would prove more suitable for functional studies, since high cell density is important in maintaining the *in vivo* multicellular structure and functional properties of cultured thyroid epithelial cells (Fayet *et al.*, 1971). Disappointingly, such high-density cultures also failed to respond to hormones and perhaps this was due to the absence of a physiological substrata such as collagen which is known to maintain not only normal cytodifferentiation of mouse mammary epithelia *in vitro,* but also hormonal stimulation of casein secretion in these cells (Emerman and Pitelka, 1977; Emerman *et al.*, 1977; Katiyar *et al.*, 1978).

B. Use of Endometrial Cultures for Studying Epithelial-Mesenchymal Interactions *in Vitro*

Although epithelial/mesenchymal interactions are firmly established as important for normal embryonic development (Kratochwil, 1972), their role in the control of adult tissue homeostasis has received little attention (Tarin, 1972). The human endometrium provides a useful model system for studying such interactions by remixing separated epithelial and stromal cells. Epithelial cells can be either plated on or within its homologous stroma, as mixed cultures, or physically separated from its stroma and maintained as cocultures sharing the same culture medium.

Preliminary studies of coculturing coverslip cultures of epithelia and stroma showed the epithelia to inhibit stromal growth (unpublished observation).

C. Special Advantages of Endometrial Cell Cultures for Use in Cancer Research

Endometrial cancer is fairly common and shows a similar epidemiology to breast cancer. The use of exfoliative cytology as a routine investigative procedure in gynecologic disorders provides the unique opportunity of obtaining endometria of differing neoplastic potential ranging from normal, through hyperplastic, to invasive carcinoma. Potentially, this affords the *in vitro* comparison of the functional properties of epithelial and stromal components at the different stages of carcinogenesis.

D. Future Studies

It is our aim to develop functionally responsive epithelial cultures on collagen-coated nucleopore filters. By placing such filter cultures on stromal cultures, the role of cell contact on epithelial/mesenchymal interactions could be studied using different pore sizes.

Using such recombinations of heterologous epithelium and stroma from normal, hyperplastic, or neoplastic endometria, it is hoped to use epithelial/mesenchymal interactions to probe the normal → neoplastic transition in the human endometrium *in vitro*.

REFERENCES

Arias-Stella, J. (1955). *Arch. Pathol.* **60,** 49-58.

Armelin, H. A., Nishikawa, K., and Sato, G. H. (1974). *In* "Control of Proliferation in Animal Cells" (B. Clarkson and R. Baserga, eds.), pp. 97-104. Cold Spring Harbor Laboratory, Cold Spring Harbor, New York.

Beierwaltes, W. H., and Al-Saadi, A. A. (1966). *J. Clin. Endocrinol.* **26,** 729-734.

Brodsky, W. Ya., and Uryvaeva, I. V. (1977). *Int. Rev. Cytol.* **50,** 275-332.

Brown, W. E., and Bradbury, F. T. (1947). *Am. J. Obstet. Gynecol.* **53,** 749-757.

Carriere, R. (1969). *Int. Rev. Cytol.* **25,** 201-277.

Chen, L., Lindner, H. R., and Lancet, M. (1973). *J. Endocrinol.* **59,** 87-97.

Das, N. K., Hosick, H. L., and Nandi, S. (1974). *J. Natl. Cancer Inst.* **52,** 849-861.

Dulbecco, R., and Elkington, J. (1973). *Nature (London)* **246,** 197-199.

Emerman, J. T., and Pitelka, D. R. (1977). *In Vitro* **13,** 316-328.

Emerman, J. T., Enami, J., Pitelka, D. R., and Nandi, S. (1977). *Proc. Natl. Acad. Sci. U.S.A.* **74,** 4466-4470.

Fayet, G., Michel-Béchet, M., and Lissitzky, S. (1971). *Eur. J. Biochem.* **24,** 100-111.

Ferenczy, A. (1976). *Am. J. Obstet. Gynecol.* **124,** 64-74.

Figge, D. C. (1960). *Obstet. Gynecol.* **16,** 269-277.

Figge, D. C. (1963). *Acta Cytol.* **7,** 245-251.

Fogh, J., and Fogh, H. (1964). *Proc. Soc. Exp. Biol. Med.* **117,** 899-901.

Franks, L. M., and Wilson, P. D. (1977). *Int. Rev. Cytol.* **48,** 55-139.

Geitler, L. (1939). *Chromosoma (Berlin)* **1,** 1-22.

Geitler, L. (1953). *Protoplasmatologia* **6C,** 1-89.

Hayflick, L., and Moorhead, P. S. (1961). *Exp. Cell Res.* **25,** 585-621.

Hiratsu, T. (1968). *Kobe J. Med. Sci.* **14,** 29-48.

Katiyar, V. N., Enami, J., and Nandi, S. (1978). *In Vitro* **14,** 771-774.

Kirk, D., and King, R. J. B. (1979). *In Vitro* **5,** 374-382.

Kirk, D., King, R. J. B., Heyes, J., and Taylor, R. W. (1978a). *In* "Endometrial Cancer" (M. G. Brush, R. J. B. King and R. W. Taylor, eds.), pp. 385-394. Baillière Tindall, London.

Kirk, D., King, R. J. B., Heyes, J., Peachey, L., Hirsch, P. J., and Taylor, R. W. (1978b). *In Vitro* **14,** 651-662.

Kompmann, M., Paddags, I., and Sandritter, W. (1966). *Arch. Pathol.* **82,** 303-308.

Kratochwil, K. (1972). *In* "Tissue Interactions in Carcinogenesis" (D. Tarin, ed.), pp. 1-47. Academic Press, New York.

Kratochwil, K., and Schwartz, P. (1976). *Proc. Natl. Acad. Sci. U.S.A.* **73,** 4041-4044.

Lechner, J. F., Shankar Narayan, K., Ohnuki, Y., Babcock, M. S., Jones, L. W., and Kaighn, M. E. (1978). *J. Natl. Cancer Inst.* **60,** 797-801.

Leuchtenberger, C., Helweg-Larsen, H. F., and Murmanis, L. (1954). *Lab. Invest.* **3,** 245-260.

Liszczak, T. M., Richardson, G. S., MacLaughlin, D. T., and Kornblith, P. L. (1977). *In Vitro* **13,** 344-356.

Mayer, A., and Heim, K. (1926). *Zentralbl. Gynaekol.* **50,** 2688-2696.
Meek, E. S., and Harbison, J. F. A. (1967). *J. Anat.* **101,** 487-489.
Miller, R. C., Nichols, W. W., Pottash, J., and Aronson, M. M. (1977). *Exp. Cell Res.* **110,** 63-73.
Notake, Y. (1963). *J. Jpn. Obstet. Gynecol. Soc.* **15,** 695-705.
Noyes, R. H., Hertig, A. T., and Rock, J. (1950). *Fertil. Steril.* **1,** 3-25.
Owens, R. B., Smith, H. S., Nelson-Rees, W. A., and Springer, E. L. (1976). *J. Natl. Cancer Inst.* **56,** 843-849.
Papanicolaou, G. N., and Maddi, F. V. (1958). *Am. J. Obstet. Gynecol.* **76,** 601-618.
Papanicolaou, G. N., and Maddi, F. V. (1959). *Am. J. Obstet. Gynecol.* **78,** 156-173.
Sober, H. A., and Harte, R. A. (1970). *In* "Handbook of Biochemistry" (H. A. Sober and R. A. Harte, eds.) 2nd Ed., p. H-113, Chemical Rubber Co., Cleveland.
Tarin, D. (1972). *In* "Tissue Interactions in Carcinogenesis" (D. Tarin, ed.), pp. 81-93. Academic Press, New York.
Wagner, D., and Richart, R. M. (1968). *Arch. Pathol.* **85,** 475-480.
Walker, B. E. (1958). *Chromosoma* **9,** 105-118.

METHODS IN CELL BIOLOGY, VOLUME 21B

Chapter 4

Human Breast Organ Culture Studies[1]

E. A. HILLMAN AND M. J. VOCCI

Department of Pathology,
University of Maryland School of Medicine,
Baltimore, Maryland

J. W. COMBS

Department of Pathology,
University of Pennsylvania at Hershey,
Hershey, Pennsylvania

H. SANEFUJI, T. ROBBINS, AND D. H. JANSS

Department of Pathology,
University of Maryland School of Medicine,
Baltimore, Maryland

C. C. HARRIS

Human Tissue Studies Section,
National Cancer Institute,
Bethesda, Maryland

B. F. TRUMP

Department of Pathology and Maryland Institute for Emergency Medicine,
University of Maryland School of Medicine,
Baltimore, Maryland

[1]This is contribution #808 from the Cellular Pathobiology Laboratory University of Maryland School of Medicine, Baltimore, Maryland. This work was supported in part by NCI NO1-CP-43237.

ISBN 0-12-564140-0

I. Introduction

A review of the literature reveals that until now principally malignant human breast tissues have been grown using the organ culture technique. There are a few reports in which normal human breast tissue from various sources was grown, but the cultures survived for relatively short periods of time.

Most normal human organ cultures were established from biopsy specimens removed for pathological diagnostic purposes and found to be non-neoplastic (Barker *et al.*, 1964; Archer, 1968; Ceriani *et al.*, 1972; Wellings and Jentoft, 1972; Dilley and Kister, 1975; Kleinberg, 1975). Other sources of tissues have been from reduction mammoplasties (Van Bogaert, 1976) or autopsy specimens (Flaxman and Van Scott, 1972). There is one report on explant culture of human mammary tissue obtained from the second trimester of pregnancy (Flaxman *et al.*, 1976).

These previous studies were based primarily on short-term organ culture—3–6 days. There are only three reports in which organ cultures were analyzed for over 1 week. Barker *et al.* (1964) described the cultural conditions and role of insulin played in maintaining human organ cultures for up to 12 days. Flaxman (1974) and his colleagues have further extended the period of time in culture in tissue obtained from autopsy tissues. At 16 days ductal and myoepithelial cells were viable by morphological and autoradiographic criteria. Finally, the longest time that we are aware of that normal human mammary epithelium has been grown was reported by Flaxman *et al.* (1976). The tissue that they obtained from a female at the end of the second trimester of pregnancy survived for 21 days.

II. Materials and Methods

A. Tissue Obtainment

Human breast tissues were obtained from reduction mammoplasties or immediate autopsy cases. Breast tissue was received as quickly after removal as possible by a representative of our group and immediately was placed in cold L-15 medium for transport to the laboratory. Zerotime baseline tissue was placed simultaneously in 4F-1G fixative (McDowell and Trump, 1976) for light and electron microscopy. Upon arrival in the laboratory, epithelial elements were dissected free from the surrounding fat and stroma with the aid of a dissecting microscope and trimmed into fragments 0.5 to 1.0 cm^2 and four to six fragments were placed in each 60-mm plastic petri.

B. Culture Conditions

1. Baseline Culture Conditions

Control human breast organ cultures were grown in CMRL-1066 medium with the addition of 0.1 μg/ml hydrocortisone, 1 μg/ml bovine recrystalized insulin, 5% heat-inactivated fetal calf serum, 2 m*M* L-glutamine, and antibiotics (300 U/ml penicillin, 300 μg/ml streptomycin, 100 μg/ml gentamicin, and 0.5 μg/ml amphotericin B). A volume of 3 ml of culture medium was used for each culture and the medium was changed twice a week. The culture dishes were placed in environmentally controlled chambers maintained at 36.5°C, in an atmosphere of 45% O_2, 50% N_2, 5% CO_2 and rocked 10 cycles/minute.

2. Hormone Studies

The role of hormones in this organ culture system was investigated based on the well-established hormone sensitivities of human breast. Our studies utilized CMRL-1066 with varying insulin concentration, 5% charcoal absorbed heat-inactivated fetal calf serum, and various concentrations of hormones as shown in Table I.

The prolactin, progesterone, and estradiol levels used in the 28-day cycle

TABLE I

	Type of hormone culture conditions (ng/ml)				
		28 Day cycling regimen			
Hormone	Constant level	Week 1	Week 2	Week 3	Week 4
Progesterone	0.3	0.6	0.6	0.6	15
Aldosterone	50	50	50	50	50
Estradiol	0.30	0.034	0.35	0.034	0.034
Prolactin	10	8	8	15	8
Insulin			50 or 500		

regimen were based on studies by Mishell *et al.* (1971) and Reyes *et al.* (1977). Constant hormone levels were very similar to the hormone requirements developed for the rat mammary epithelial monolayer system (Janss *et al.*, 1980). Stock solutions of progesterone, aldosterone, and estradiol (Calbiochem) were dissolved in absolute ethanol and refrigerated at 4°C. Ovine prolactin (Hormone Distribution Program, NIAMDG, NIH) was dissolved in 0.05 *N* NaOH and stored at −70°C. Prior to use, the hormones were diluted in culture medium. Before hormone treatment, the human breast organ cultures were grown in control breast culture medium. After 2 weeks in culture, a few dishes were randomly selected and transferred to either the cycling or constant hormone regimen. Two levels of insulin (Table I) were evaluated. Cultures undergoing a 28-day hormone cycle had their hormone levels adjusted weekly. Cultures at constant levels had their hormone level maintained for the duration of the experiment. The medium was changed twice weekly. Cultures destined for autoradiography were incubated in 1.0 μCi/ml [^{3}H]TdR (specific activity 6.7 Ci/m*M*, New England-Nuclear, Boston, Ma.) added to bulk medium and given in place of a routine feeding 24 hours prior to harvest.

C. Light and Electron Microscopy

Prior to culture zerotime, the sample mentioned above, and at selected time intervals thereafter, explants were fixed for high-resolution light and electron microscopy in 4F-1G fixative (McDowell and Trump, 1976). Samples were processed following standard protocols. Light microscopic examinations were performed on paraffin sections stained with hematoxylin and eosin (H & E) and periodic acid–Schiff (PAS), and semithin plastic sections were stained with toluidine blue. Areas for examination by electron microscopy were selected from examination of semithin sections.

D. Autoradiography

After fixation in 4F-1G, tissues for autoradiography were paraffin embedded and sectioned at 5 μm. The slides from each experiment were dipped by machine (Kopriwa, 1967) in one session in Ilford K5 emulsion diluted 1:1 by weight with sterile water under conditions that produced a continuous monolayer of silver halide crystals over the tissue (Neely and Combs, 1976). Emulsion quality and proper exposure were determined on appropriate control and test slides. After 11 days exposure at 5°C, the slides were batch developed with Rodinal (Agfa-Gaevert) and stained with hematoxylin–eosin (Neely and Combs, 1976). The labeled and unlabeled cells and labeled and unlabeled mitotic figures were tallied in a sample of 1000 mammary epithelial cells under oil immersion and the labeling index (LI) and mitotic index (MI) were calculated as a percentage. Because of the relatively small amount of epithelium in each explant, the number of mitotic figures counted was small and the percentage labeled mitosis was calculated from the aggregate cell counts in each group.

III. Results

A. Baseline Studies

We felt that it was imperative to study the morphological characteristics of the breast tissue obtained from immediate autopsy (Trump *et al.*, 1973) and reduction mammoplasties prior to culture. Although the ultrastructural characteristics of normal mammary gland have been previously described (Tannenbaum *et al.*, 1969: Stirling and Chandler, 1976, 1977), several controversial points still remain; for instance, the significance of light and dark mammary epithelial cell (Toker, 1967). Rijsbosch and Meyer (1976) have raised the question of whether mammary epithelium obtained from reduction mammoplasties is truly normal. To date, all 22 cases we have studied were diagnosed by surgical pathologists as microscopically normal.

Our investigations have been focused primarily on the terminal ductules within the lobule. Several terminal ductules empty into a single central lumen. The ductule was composed of two types of epithelial cells: (*a*) secretory cells (mammary epithelium) lining the lumen, and (*b*) myoepithelial cells which are adjacent to the basal lamina. At the light microscope level toluidine blue stained semithin plastic sections (Fig. 1); the secretory cells and myoepithelial cells were easily distinguished. The secretory cells contained apical vesicles of varying size and bordered the lumen. Myoepithelial cells had a scalloped basal profile resting against the basement membrane. Typical fibroblasts of the dense intralobular stroma abutted the outer side of the basement membrane.

These features and additional details were visible by scanning (Fig. 2) (SEM) and transmission (Figs. 3 and 4) (TEM) electron microscopy. The secretory cells had delicate apical microvilli and were joined on the luminal surface by prominent junctional complexes. Secretory vesicles in the apical cytoplasm, predominantly free ribosomes, and an active Golgi apparatus were characteristic of these cells. Within the lumen of the ductule smooth droplets of secreted material were present (SEM, Fig. 2). By TEM (Fig. 3) this material was primarily granular, but fibrillar material was often present as well.

The scalloped basal border of myoepithelial cells, adjacent basement membrane, and limiting stromal cells were easily recognized in the SEM. In the TEM (Fig. 4), the myoepithelial cells were most often seen as slender cell processes with hemidesmosomes that accentuated the basal lamina and numerous myofilaments oriented mainly parallel to the basal surface. The tissue seen in Fig. 3 was

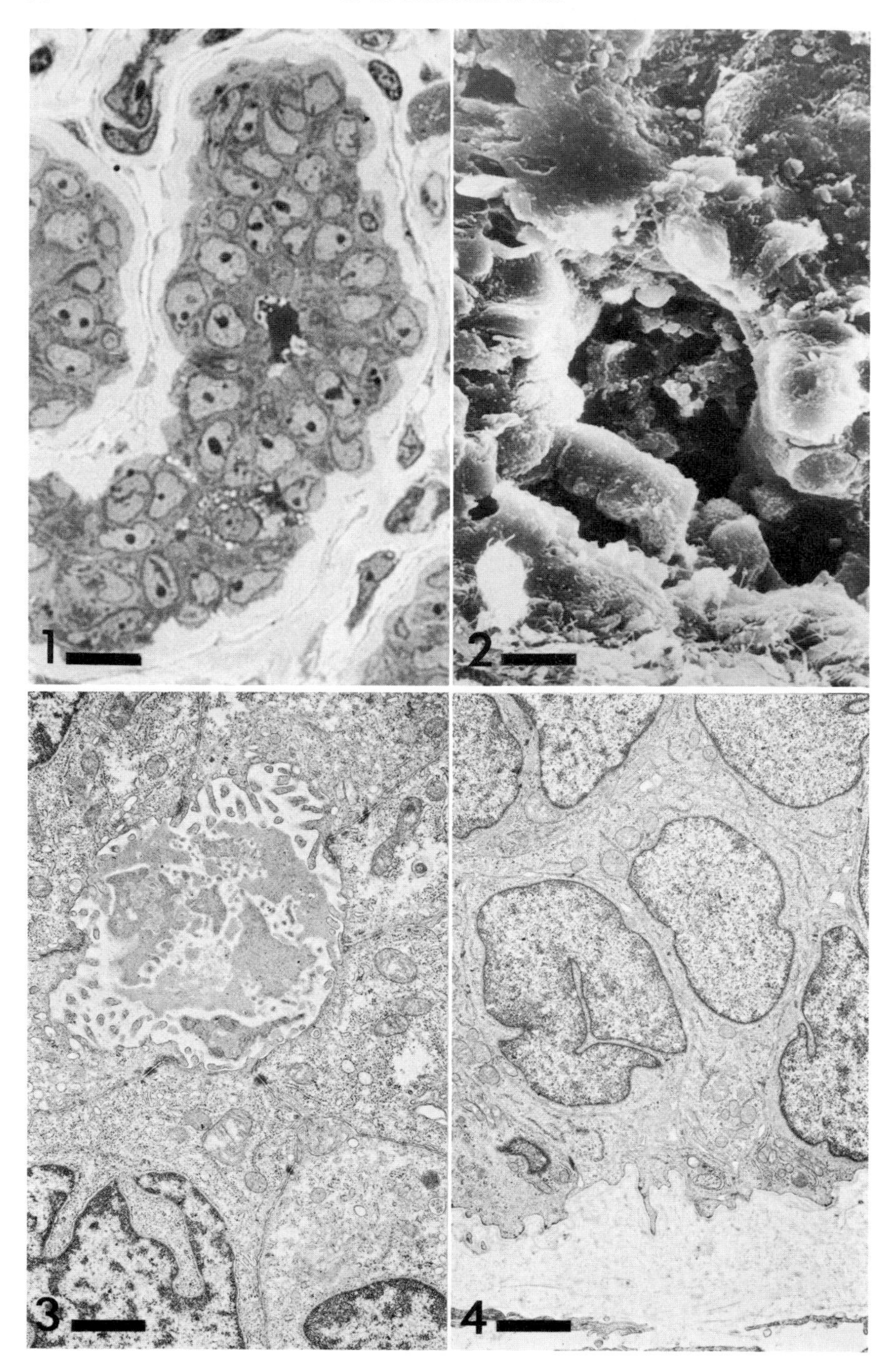
1
2
3
4

stored in cold (4°C) L-15 medium for 2 days and then processed for TEM. No ultrastructural differences were observed between these and zerotime specimens. In fact tissue was stored as isolated small fragments in cold L-15 medium for up to 2 weeks and still appeared viable, and indeed has been successfully cultured.

B. Organ Culture Studies

To date, we have studied a total of 22 cases, 17 cases obtained from reduction mammoplasties and 5 from immediate autopsies. The longest time interval that human breast epithelium has been maintained in our culture system was 26 weeks. Routinely, normal breast epithelium was maintained for 3–4 months. The length of time in culture that the epithelium remained viable appeared to have some relationship to the age of the patient. Premenopausal epithelium grew for a longer period. Also, the amount of the epithelium/unit area varied greatly; many breast specimens contained mostly fat and stroma. In these cases it was difficult to establish the numerous dishes required for the comparative long-term experiments. Human breast organ cultures were sampled at weekly time intervals. Emphasis was placed on correlative light and TEM.

1. Control Organ Culture

For the first 2 weeks in culture, the normal lobular structure was readily maintained. It was difficult to distinguish the zerotime sample from epithelium cultured for 1 week (Fig. 5). Within the central lumen, secretory material was often present. The microvilli on the apical surface of mammary epithelium were long and slender. The epithelial cells contained apical secretory vesicles, numerous free ribosomes, short segments of rough endoplasmic reticulum, and multiple Golgi apparatuses. The nucleus contained primarily euchromatin and a single nucleolus. These epithelial cells were joined by typical junctional complexes, and myoepithelial cells and basal lamina remained continuous around the duc-

---

FIG. 1. Epon semithin section of normal mammary ductule stained with toluidine blue. Note the appearance of secretory material in the lumen, small secretory droplets in the apices, the irregularity of the myoepithelium, and the thick basement membrane. Bar = 14.8 μm.

FIG. 2. Low-magnification scanning electron micrograph of ductule surrounded by fibroblasts and connective tissue. Bar = 30.0 μm.

FIG. 3. Transmission electron micrograph of ductular lumen stored for 2 days showing secretory material within the lumen. Note the apical microvilli and junctional complexes joining the cells. Bar = 1.30 μm.

FIG. 4. Basal aspect of ductule at zerotime. Note the appearance of the myoepithelial cells with irregular basalar protrusions extending in the basal lamina. Hemidesmosomes can be distinguished and the limiting fibroblastic cellular processes of the epithelial stromal junction are demonstrable. Bar = 3.0 μm.

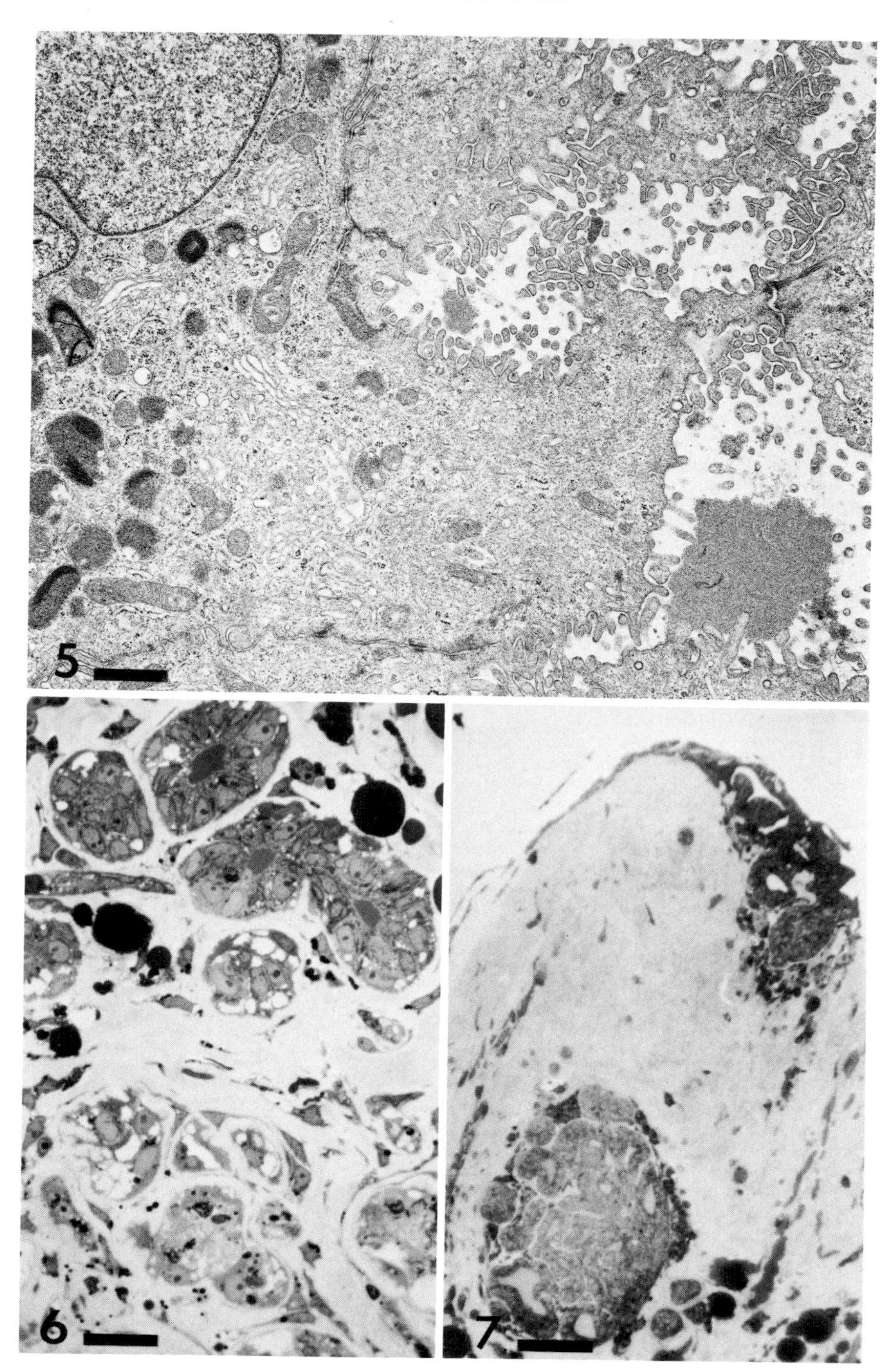
5
6
7

tule. After 2 weeks in culture (within the range of 1.5 to 3 weeks) certain ductules appeared necrotic as determined by the criteria of karyolysis and increased eosinophilia of the cytoplasm (Fig. 6). These autolysing ductules were often adjacent to ductules that appeared viable. The stroma surrounding both types of ductules was morphologically normal.

After 4 weeks in culture (3–6 week variation from case to case) the acinar epithelium within the central portion of the explant (Fig. 7) remained morphologically normal, but the epithelial cells were beginning to migrate from superficial acini out onto the surface of the explant. Occasionally the migrating epithelium originated in acini in the deeper portions of the explant. The stroma was still morphologically viable and contained numerous lipid droplets.

At the ultrastructural level (Fig. 8), the advancing edge of the epithelial outgrowth was primarily myoepithelium. The secretory epithelial cells overlying the migratory myoepithelial cells were often necrotic. The basal lamina, although present, was patchy and less dense. In some areas, the outgrowth was three or more cells thick. The migrating cells had numerous bundles of randomly oriented filaments that were associated with dense plaques typical of myofilaments (Fig. 9). In some instances hemidesmosomes were seen in the intercellular space (Fig. 10). In ductules in the central portion of the explant, typical secretory epithelium was usually present (Fig. 11). Typically, the luman contained secretory material that was finely granular and fibrillar. Numerous slender microvilli and secretory vesicles were observed in the apical portion of the cells and typical junctional complexes with prominent desmosomes joined the epithelial cells.

With continued time in culture, most of the epithelial cells maintained their secretory characteristics. Within one area, however, both glandular and epidermoid characteristics, including intercellular alveoli with secretory material and bundles of tonofilaments, were sometimes simultaneously present (Fig. 12). After 5–6 weeks in culture (Figs. 13 and 14) the epithelial cells became elongated and flattened (Fig. 13). The microvilli were short, stubby and randomly oriented, and apical. Smooth secretory droplets were present. The epithelial components usually retained their polarity in areas in which both the myoepithelial and epithelial cells were present (Fig. 14). The basal lamina was present, and both cell types contained numerous lysosomes. The epithelial cells were more plate-

---

FIG. 5. Ductular lumen of tissue cultured for 1 week. Note secretory material still present within the lumen and the slender, relatively long microvilli. Junctional complexes are along the sides of the cell and a few fine filaments can be seen. Bar = 1.20 μm.

FIG. 6. Semithin section of explant after 2 weeks in culture demonstrating viable and adjacent necrotic acinar epithelium. Bar = 80 μm.

FIG. 7. Survey micrograph of semithin section demonstrating both central acinar lobule and epithelial outgrowth onto the surface of the explant. Bar = 133.3 μm.

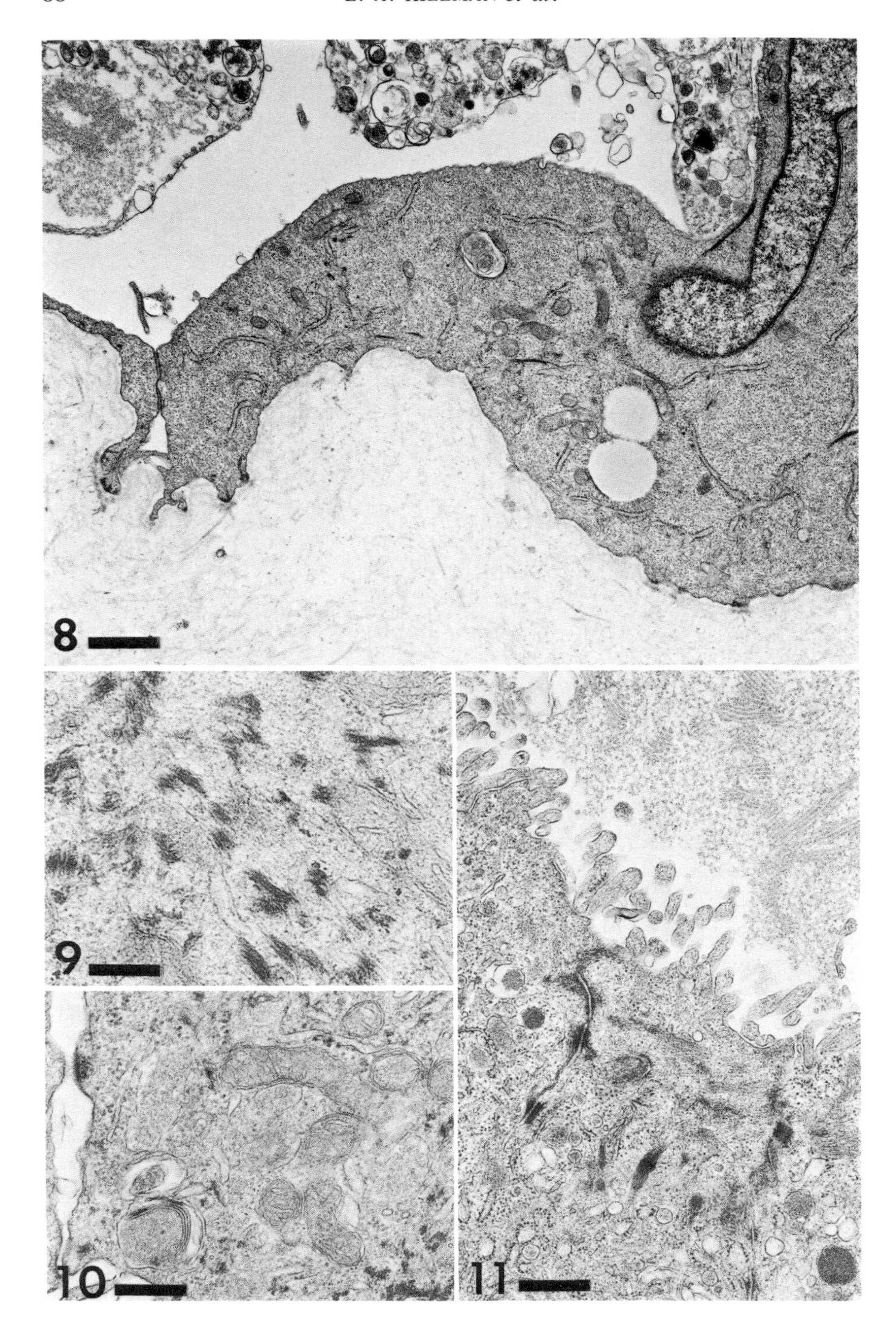
8
9
10
11

like and had numerous interdigitations of their lateral plasmalemmas. The nucleoli were extremely large and morphologically active.

After several months in control culture media, the epithelial components were still viable and appeared metabolically active. However, the number of intracellular alveoli (Fig. 15) increased and they were often multiple within the same cell. In areas where the basal lamina was present, it was adjacent to cells containing numerous filaments associated with dense plaques (Fig. 16). These myoepithelial cells also contained multiple Golgi apparatuses and myelin-like whorls. The epithelial cells were now primarily on the explant surface and one to three cell layers thick. The basal lamina was sparse and no longer continuous.

In summary, we have demonstrated that (*a*) normal human mammary epithelium can be maintained in organ culture for several months, (*b*) the mammary epithelium retained its secretory characteristics and epidermoid potential, and (*c*) the cellular outgrowth observed was primarily myoepithelial in origin.

2. Hormone Studies

Recently we have studied the role of hormones in maintaining human breast epithelium in our long-term organ culture system. In previous experiments, the hormones present in culture media were insulin, hydrocortisone, and those present in the fetal calf serum. We pursued two approaches: (*a*) a constant regimen of breast trophic hormones present at uniform levels throughout the culture period, and (*b*) a cycling hormone regimen, i.e., varying levels of hormones present for weekly periods in an attempt to generally mimic the *in vivo* menstrual cycle (see Table 1). It was our hope that hormone treatment would (*a*) maintain the central glandular structure of the epithelium, i.e., that secretory cells, myoepithelial cells, and basal lamina would be retained throughout the culture period, (*b*) delay if not prevent myoepithelial outgrowth, (*c*) enhance secretory activity of the epithelial cells, and (*d*) induce waves of DNA synthetic activity, similar to the *in vivo* synchronization of mammary epithelial replication in cycling female humans and rodents (Meyer, 1977; Purnell *et al.*, 1974; Purnell and Kopen, 1976).

FIG. 8. Micrograph of advancing edge of outgrowth epithelium showing cells with junctions and hemidesmosomes typical of myoepithelial cells. Note basal lamina forming beneath the myoepithelial cell and the necrotic cell debris at the top of the picture. Bar = 1.66 μm.

FIG. 9. High magnification of outgrowth cells showing clusters and plaques of dense filaments. These probably represent contractile myofilaments though some may be keratin. Bar = 1.0 μm.

FIG. 10. Micrograph showing hemidesmosomes in central portion of an area of outgrowth. Note preservation of organelles. Bar = 1.0 μm.

FIG. 11. Explant cultured for 4 weeks showing lumen containing secretory material which have paracrystalline arrangements. Note apical microvilli, junctional complexes, fine filaments, and small secretory droplets within the apical cytoplasm. Bar = 0.71 μm.

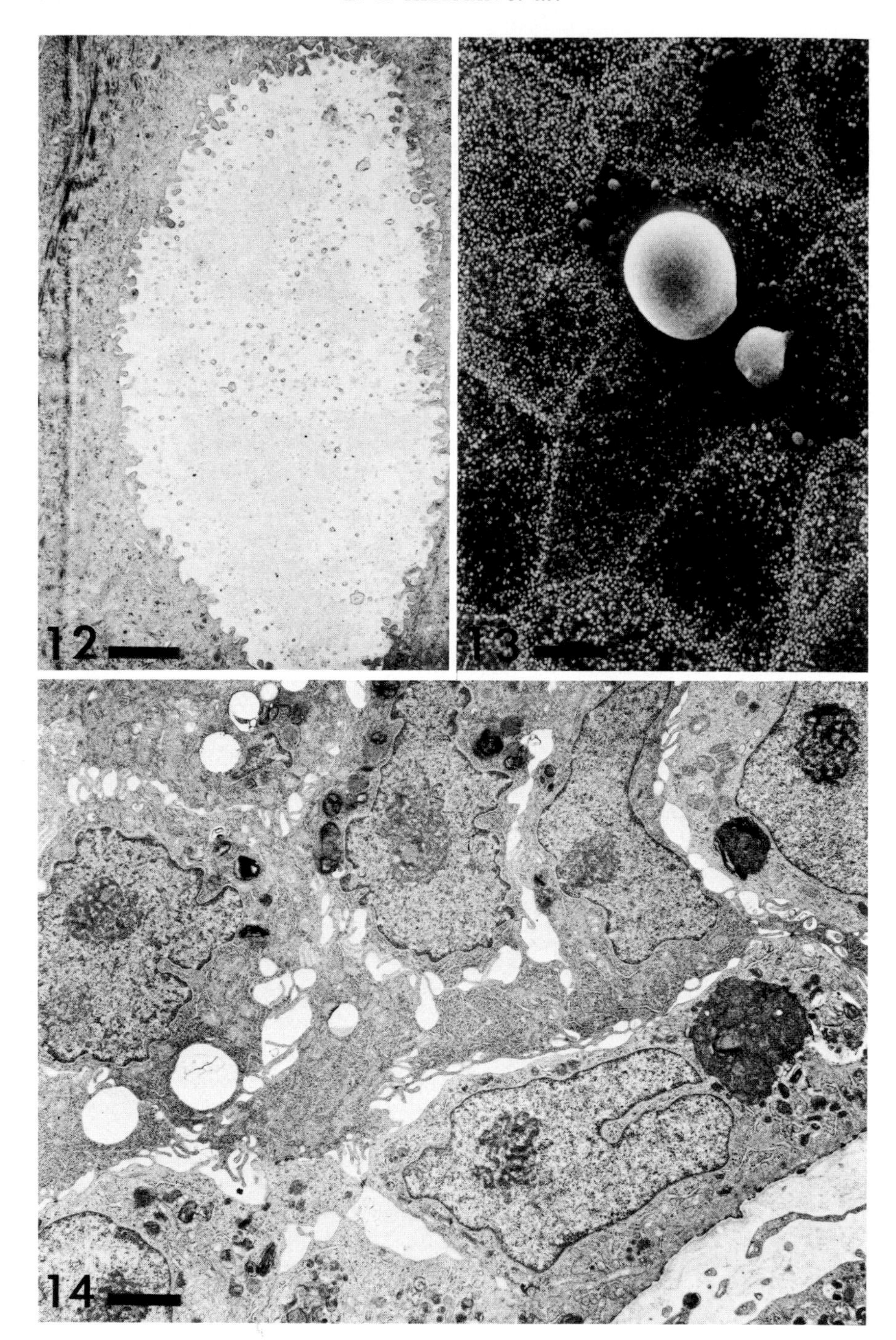
12
13
14

In each experiment, in addition to the cycling and constant hormone treatment, control cultures using our previously defined culture system were evaluated. In the earlier experiment only the low level of insulin, 0.05 μg/ml, was present in the cycling and constant hormone treatments. In the later experiments two levels of insulin, 0.05 and 0.5 μg/ml, were evaluated to determine if the insulin level might have been modifying the secretory activity.

There was a marked improvement in the maintenance of lobular architecture following either hormone treatment. After 6 weeks in the cycling hormone regimen (Fig. 17), the normal structure of the ductules, intralobular ducts, and supporting stroma was well preserved. Strongly PAS-positive secretory product was present within the lumena. In this particular experiment after 5–6 weeks in culture the explants grown in the presence of control breast medium displayed epithelial migration onto the surface of the explant with both low and high level insulin levels (Fig. 18). Epithelial cells were beginning to migrate in the cycling hormone regimen also but to a lesser degree.

Excellent viability of the ductules following 5 weeks of cycling hormone treatment was demonstrated (Fig. 19). Both the epithelial and myoepithelial cells were retained in culture. The secretory cells were metabolically active, judging from the plug of material present in the intercellular lumen. The hormone-treated epithelial cells were extremely electron lucent and contained numerous free ribosomes, mitochondria, and Golgi apparatuses. On the apical surfaces, the cells were sealed by typical prominent junctional complexes. Both the epithelial and myoepithelial cells contained large active nucleoli and numerous lysosomes. The basal lamina was intact and was part of a typical epithelial stromal junction. At higher magnification (Fig. 20), the characteristic myofibrils and hemidesmosomes characteristic of the myoepithelial cells were also present. These cells were actively synthesizing DNA (Fig. 21). Numerous developed [^{3}H]thymidine autoradiographic grains were localized primarily over the epithelial nuclei. However, only a small fraction of mitotic figures was labeled (Fig. 22) and few stromal cells were labeled. The very low stromal cell replication rate confirmed the morphologic finding that the culture conditions and/or the presence of epithelium prevented stromal overgrowth.

The viable morphology and [^{3}H]thymidine incorporation of some of the lobules was also retained after 6 weeks on the constant hormone regimen at the

---

FIG. 12. Cells in explant for 33 days showing both intercellular lumen and tonofilaments characteristics of epidermoid differentiation. Bar = 1.33 μm.

FIG. 13. Scanning micrograph of surface outgrowth showing the regular arrangement of the cells. Two blebs probably represent secretory activity. Bar = 4.0 μm.

FIG. 14. Transmission electron micrograph of a 6 week explant culture demonstrating the retention of both the myoepithelial and epithelial cells. Both cell types contain nuclei with morphologically active nucleoli and numerous cytoplasmic lysosomes. Bar = 2.22 μm.

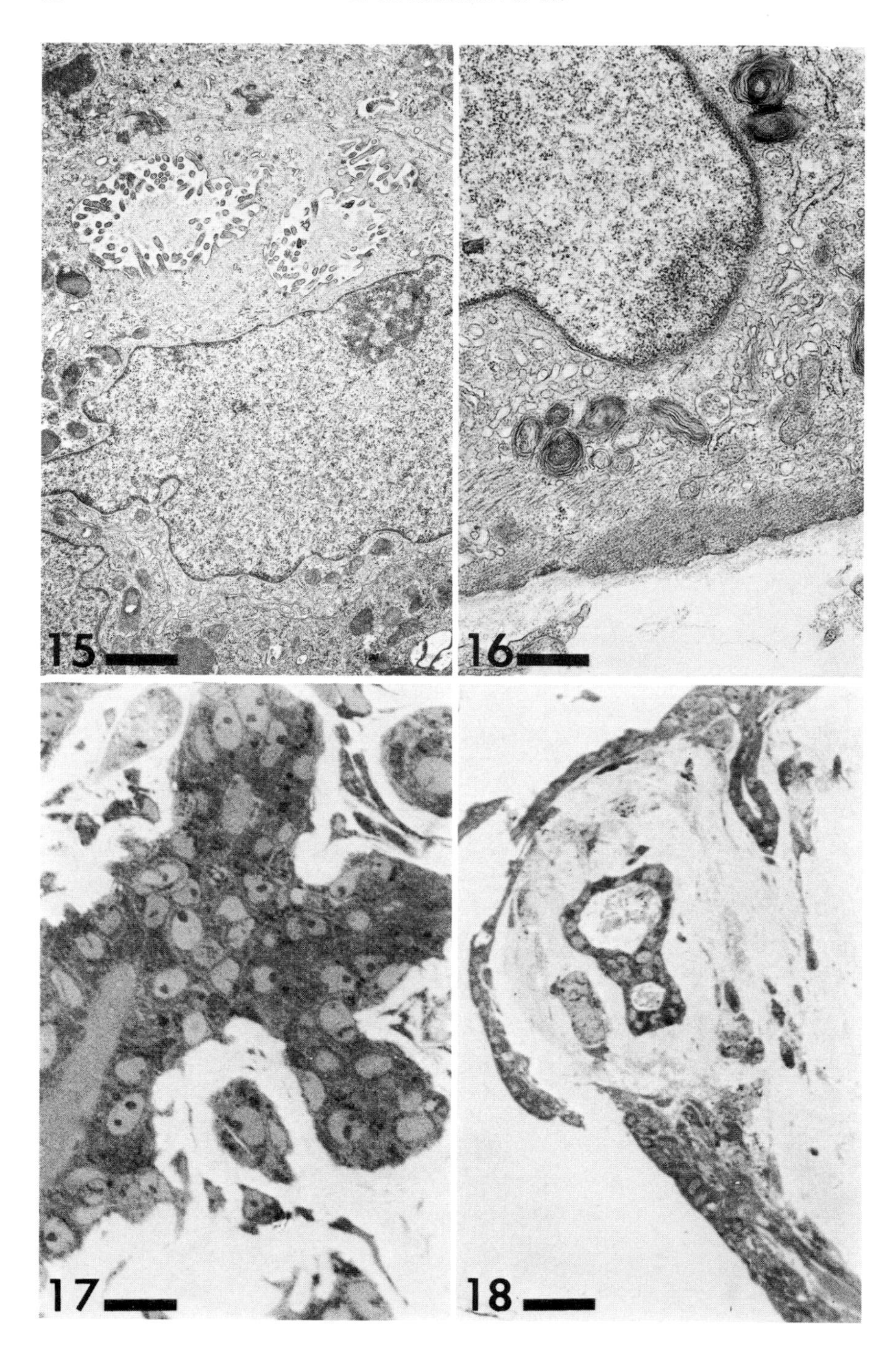
15
16
17
18

higher insulin level (Fig. 23). Both the epithelial and myoepithelial cells were extremely electron lucent. The normal acinar architecture was retained and the basal lamina was also present. The lumen of the ductules contained PAS-positive secretory material on paraffin sections. After 7 weeks on the constant hormone treatment (Fig. 24) the myoepithelial cells were still vigorously viable, but they contained numerous lysosomes, mitochondria, and Golgi apparatuses. The mammary epithelial cells in this area were necrotic.

From our morphological studies we concluded the following: (*a*) the lobular architecture of the normal human mammary gland was retained better in organ cultures following hormone treatment. The cycling hormone regimen appeared to provide the best growth, however, viable acinar epithelium was observed to a lesser degree in both constant and control culture media; (*b*) higher insulin level promoted a greater viability in all treatments; (*c*) in the cycling and constant hormone treated cultures, there was stimulation of typical secretory activity; and (*d*) hormone treatment delayed but did not prevent epithelial outgrowth onto the explant surface. The cycling hormone was superior in this regard.

The [^{3}H]thymidine labeling index data supported our morphological findings. The labeling index (LI) data are presented in Figs. 25–27. The details presented in the graphs indicated that insulin, other hormones, and the constituents of control organ culture medium strongly influenced the fraction of cells in the DNA synthesis phase and mitotic phase of the cell replication cycle. The values for LI indicated that the higher insulin level at least doubled the proportion of cells in these phases of replication. In the case of control human breast medium, the fraction of cells synthesizing DNA was more than five times greater, measured as averages, with the higher insulin level. Because of the strong time dependence of these data, further statistical analysis of the LI was not appropriate.

The plots in Figs. 25–27 revealed additional differences in the cell kinetic response between the groups. Because of the small size necessary for the organ explants, and the resulting sparsity of grandular epithelium, too few mitotic figures were available for the counts from individual slides to yield meaningful mitotic indices for individual explants, consequently, only the labeling indices

FIG. 15. Cells in explant culture for 119 days demonstrating intracellular lumens containing secretory product. Bar = 1.66 μm.

FIG. 16. On the basal surface, cells with numerous myofilaments, hemidesmosomes, lysosomes, and multiple Golgi apparatuses are observed. The basal lamina is sparse or absent in some areas. Bar = 0.75 μm.

FIG. 17. Explant cultured for 6 weeks using the cycling hormone regimen. Note increased preservation of ductular structures with an enhancement of secretory material in the lumen which is PAS-positive. Bar = 19.3 μm.

FIG. 18. Explant cultured in control breast media after 6 weeks. Note difference in the migration and secretory material. Bar = 66.6 μm.

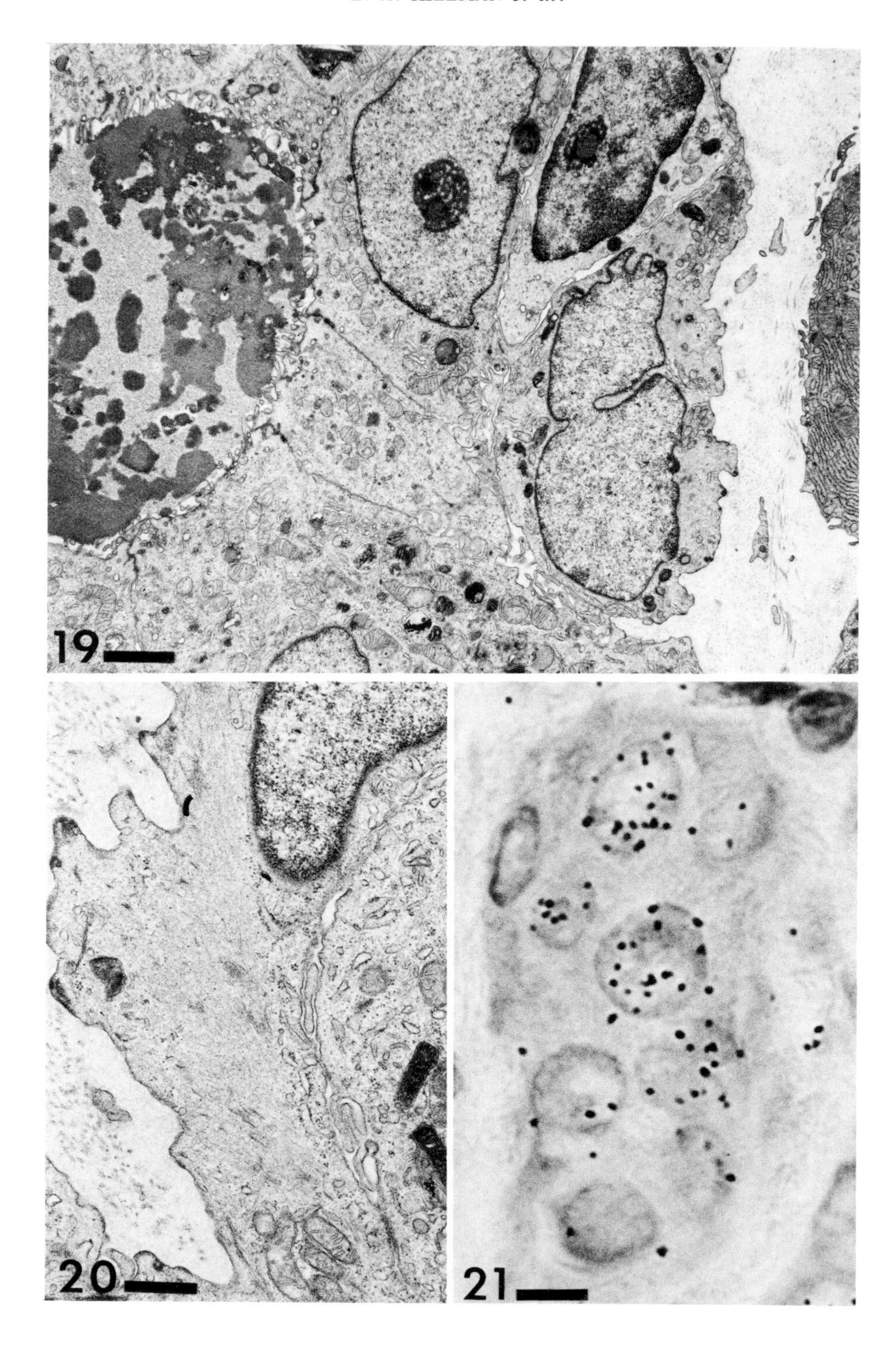
19
20
21

were plotted. In Figs. 25–27, the LI for cultures with low and high insulin media and differing ovarian hormone regimens are plotted for the 8 week interval of the experiment. Low insulin levels produced low levels of mammary epithelial cell replication irrespective of the ovarian hormone regimen, while the higher insulin level dramatically changed the pattern of the LI in all three hormone regimens.

The cycling regimen (Fig. 25) appeared to suppress replication during weeks 1, 2, and 3 followed by a low but distinct peak at 4 and 5 weeks and a rise from 6–8 weeks. The trace suggested that the LI was cyclic, but the curve did not extend far enough to confirm the 28-day period of the hormone cycle in the medium. Morphologically the last rise was accompanied by noticeable proliferation in the surviving glands, and did not appear to be related to cells streaming to the surface.

Control organ culture medium with high insulin (Fig. 26) also caused the LI to fluctuate, but much more dramatically than low insulin. The peak at 5 weeks towered over the other regimens. Judging from the morphologic appearance of the explants, this corresponded to a burst of replication related to the streaming migration of epithelial cells from glands to cover the surface of the explant with a two to four cell thick epithelium. At the same time many glands within the central portion of the explant also appeared to become necrotic.

The constant hormone regimen (Fig. 27) produced a qualitatively similar trace at the low level insulin level, but the suppression was delayed to week 3, and the LI rose steadily but slowly to week 8. (At that point, the cycling regimen produced a labeling index about twice as great as the other two media.) The constant hormone regimen with the high insulin level appeared to have a definite peak at 3 weeks, and fell thereafter. No tissue samples were available at 7 and 8 weeks due to contamination of these dishes, but by week 6, the LI in the constant hormone high and low insulin were similarly low. The microscopic appearance of the explants grown in constant hormone media was similar also, with most of the glandular elements consisting of necrotic "ghosts."

The fraction of labeled mitosis data was incongruous (Table II).

Owing to the small number of mitotic figures in each explant, the percentage labeled mitoses was reported as the aggregate for all the counts in each culture regimen. The results indicated that only a minor fraction of the mitotic figures observed were in fact labeled.

FIG. 19. TEM of explant cultured during 5 weeks of the cyclic hormone treatment. Note good preservation of the ductule with abundant PAS-positive secretory material in the lumen. Myoepithelial cells rest on a continuous basal lamina. Bar = 2.22 μm.

FIG. 20. Same conditions as Fig. 19 showing preservation of myoepithelial cell. Prominent hemidesmosomes and numerous intracytoplasmic myofilaments. Bar = 0.75 μm.

FIG. 21. Similar treatment as Figs. 19 and 20. High-resolution light micrograph demonstrating numerous autoradiographic grains over epithelial nuclei after labeling with [^{3}H]thymidine. Bar = 8.0 μm.

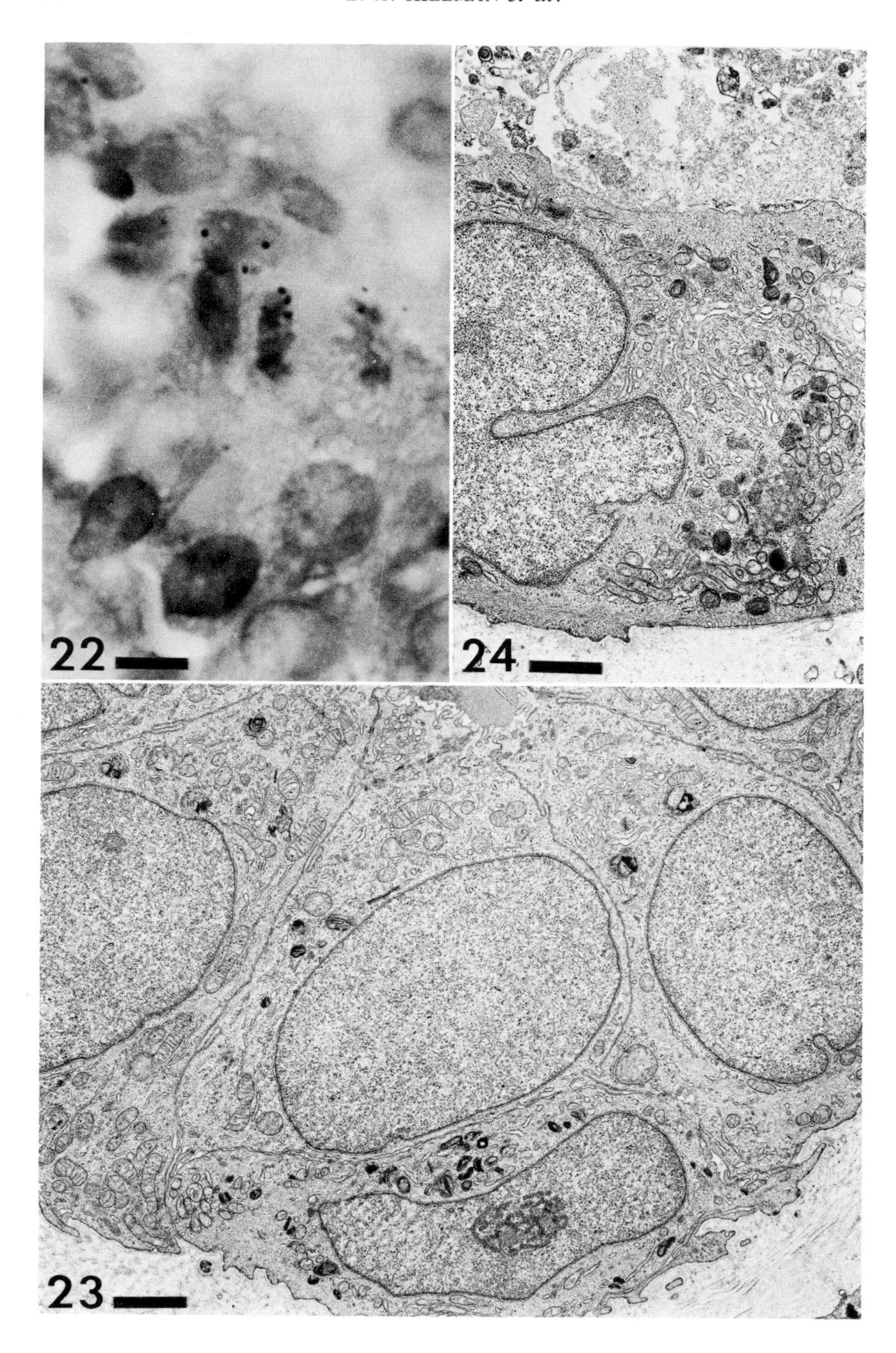
22
24
23

IV. Discussion

Our success in developing a long-term organ culture system for growing normal human breast epithelium was probably due to a combination of several factors including the tissue source, transport, and cultural conditions. For both control and hormone treatment studies, the explanted breast epithelium was obtained from relatively young (17–39 year, average age 32) premenopausal, healthy individuals. This tissue may have had different potential from the normal appearing areas from mastectomy or biopsy specimens from older women explanted by most previous investigators. The tissue used in our studies was still under cycling hormone control *in vivo,* and there was a greater proportion of epithelial nests within the supporting stroma and fat than in most surgical specimens. The specimens thus yielded more viable explants which in turn allowed greater freedom in designing long-term experimental protocols.

The conditions initially used for maintaining normal human breast epithelium in long-term organ culture (Hillman *et al.*, 1980) were similar to those which have successfully maintained several types of epithelium in long-term organ culture at the University of Maryland: bronchial epithelium (Barrett *et al.*, 1976), pancreatic ductal epithelium (Jones *et al.*, 1977), and endocervix epithelium (Schürch *et al.*, 1978). After evaluating several specimens using these cultural conditions it became apparent that we possibly were selecting for a certain cellular type as it appeared that the hormone responsive epithelial cell was dropping out of the cultural system.

After several weeks in control CMRL-1066 organ culture medium, squamous metaplasia was also often observed (Hillman *et al.*, 1980). Cells contained numerous bundles of filaments resembling typical tonofilaments. This phenomena in human mammary organ cultures has been described by several investigators (Barker *et al.*, 1964; Archer, 1968; van Bogaert, 1977). van Bogaert (1977) reported that the addition of 5 μg/ml insulin to M199 with Earle's salts enhanced squamous metaplasia. Barker *et al.* (1964) also observed that in the older insulin-treated cultures, hyperplasia of ductal cells suggestive of squamous metaplasia occurred. However, these investigators used a 5-fold greater level of insulin in their culture systems then used in our experiments. Elias and Armstrong (1973) working with much higher concentration of insulin

---

FIG. 22. Similar preparation as Fig. 20 showing labeled mitotic figure. Bar = 8.0 μm.

FIG. 23. Explant culture maintained for 6 weeks in the presence of constant hormone regimen. Note preservation of myoepithelial and epithelial cells and a small portion of a duct lumen containing secretory material. Bar = 2.0 μm.

FIG. 24. Preparation of explant culture after 7 weeks on constant hormone regimen. Myoepithelial cells after thie treatment are still viable, however, the epithelial cells appear necrotic. Bar = 1.6 μm.

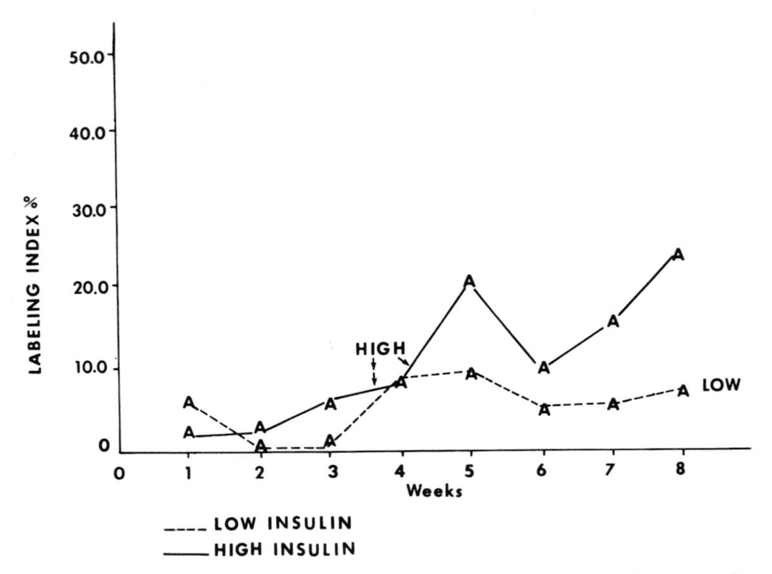

FIG. 25. [^{3}H]TdR labeling index in human mammary epithelium grown in organ culture in low and high insulin level cycling ovarian hormone regimens.

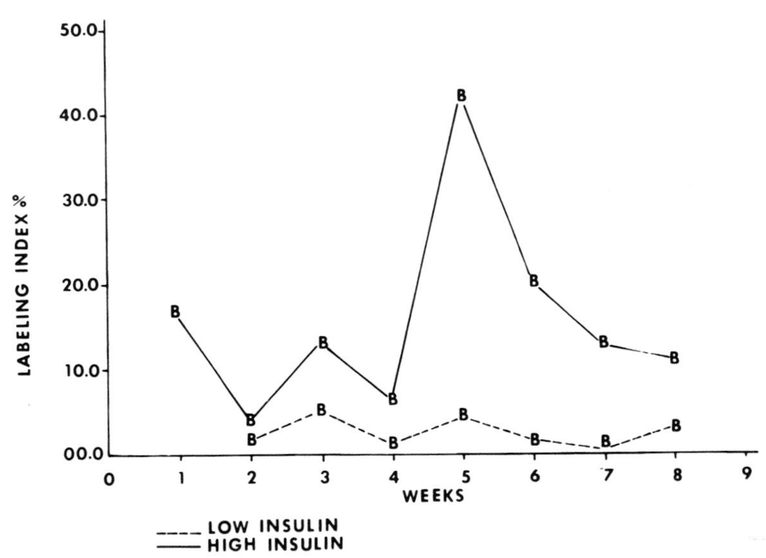

FIG. 26. The LI of mammary epithelium growing in low and high insulin level control organ culture medium.

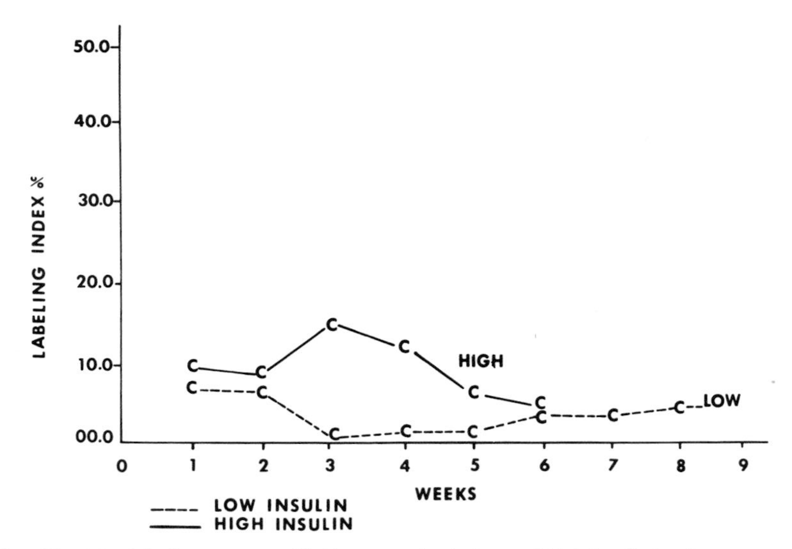

FIG. 27. The LI of mammary epithelium growing in low and high insulin level constant ovarian hormone regimens.

(5–50 μg/ml) induced hyperplastic and squamous metaplastic changes in the epithelium in cultured explants of fibroadenomas and human breast dysplasias. The lower level of insulin in concert with hydrocortisone could have had the same effect in our cultural system. The significance of this metaplasia is by no means clear. It may represent enhanced division and epithelial repair as it does in the tracheobronchial epithelium.

With increasing time in control organ culture medium, in addition to the squamous metaplasia, there was an increase in the numbers of intracellular aveoli. These structures have been described by Battifora (1975) as pathonomoic for malignant breast epithelium. In normal human breast organ culture system,

TABLE II

AGGREGATE PERCENTAGE LABELED MITOSIS AFTER 24 HOURS INCUBATION IN [^{3}H]THYMIDINE

	Low insulin (%)[a]	High insulin (%)
Cycling hormone media	20	30
Control organ culture media	15	23
Constant hormone media	20	5

[a] Calculated as the percentage of the labeled mitotic figures from all the specimens (1 to 8 weeks) in each of the different culture conditions. In the low insulin cultures, about 50 mitotic figures contributed to the percentages. In the high insulin cultures, about 100 mitotic figures were tallied.

however, they may have merely been a continuation of the normal secretory function of the epithelial cells. The homogeneous, granular, and/or fibrillular material usually present within the central lumina favored this postulate. They are seen in a variety of regenerating and neoplastic epithelia. In this regard Dermer and Sherwin (1975) have demonstrated using autoradiography at the light and electron microscopic level that explants derived from human infiltrating ductal carcinomas were capable of synthesizing glycoproteins that were secreted into intracellular and extracellular ductular structures. We also noted that with continued time in culture, the basal lamina was no longer continuous and in some areas was totally lacking. It appeared that it was primarily the myoepithelial cell that was secreting the basal lamina. As the myoepithelial cell migrated out onto the surface of the explant, the basal lamina was patchy and absent in most areas. At this time, the epithelial cells lost thier polarity and the myoepithelial cells sometimes became multilayered. Within the central portion of areas of myoepithelial proliferation, cells with typical myofilaments and hemidesmosomes were sometimes observed.

The stromal elements present in the explants undoubtedly contributed to their extended viability. The importance of the mammary epithelial–stromal junction has been well documented (Ozello, 1974; Gould and Battifora, 1976). In particular, Heuson *et al.* (1975) found that collagenase treatment of explants of infiltrating ductal carcinoma of the scirrhous type resulted in excellent survival and removed the requirement for insulin, prolactin, and hydrocortisone in the medium. They observed the stroma in treated explants to be loosened, and postulated that better nutrient penetration might have been responsible for the nonhormone-dependent growth. Collagenase might have had many other effects, however, particularly on cell surface interactions. In addition to total hormone supplementation, more recently, Wicha *et al.* (1979) demonstrated that human mammary epithelial cell growth was stimulated by unsaturated free fatty acids.

In any event, in our experiments the stroma remained viable and morphologically normal, the epithelial–stromal junction preserved its usual relationship, and lipid droplets remained numerous throughout the culture period. As the stromal elements degenerated with continued culture, it appeared that the epithelial elements migrated out onto the surface of the explant. The constant or the cycling hormone regimen preserved the typical lobular architecture for an extended period of time. Although epithelial migration did occur following hormone treatment, it was definately delayed. Further, both the epithelial and myoepithelial cells were retained and judging from their ultrastructure they were metabolically active. The cells had large prominent nucleoli, and an electron lucent cytoplasm which contained numerous free ribosomes, multiple Golgi apparatuses, and lysosomes. We believe this was due to the fact that the basal lamina was present and continuous throughout the culture period analyzed. After

PAS staining, the basement membrane was very prominent in hormone-treated cultures, but was patchy or absent in control cultures. Banerjee *et al.* (1977) have demonstrated that basal lamina was required for the maintenance of salivary gland lobular morphology and that it could be synthesized *in vitro*. Our observations parallel and reinforce these findings. Following either type of hormone treatment there was also a marked enhancement of secretory activity. This product was strongly PAS-positive primarily in normal intercellular aveoli.

The role of hormones in maintaining normal lobuloalveolar architecture in normal human breast organ cultures is quite controversial and has been recently reviewed by Osborne and Lippman (1978). Morphologic studies by Dilley and Kister (1975) indicated that insulin, and human prolactin, but not ovine prolactin, stimulated growth. In some instances, hormones were required while, in other instances, they seemed to have no effect; however, previous investigations used much higher levels of hormones which may have been toxic. In our studies, human hormones at physiological concentrations were utilized.

In addition to the role that hormones play in the normal maturation and metabolism of human mammary epithelium in organ culture, they are undoubtedly mitogenic. Insulin alone (van Bogaert, 1976) insulin or prolactin (Flaxman, 1974), or insulin plus prolactin (Flaxman and Lasfargues, 1973) have been demonstrated to increase DNA synthesis and enhance mitotic activity in normal human breast organ cultures. Similarly, their mitogenic role in organ cultures derived from benign and malignant human breast epithelium has also shown the importance of insulin, prolactin, and human placental lactogen in this system. Welsch *et al.* (1976) demonstrated that insulin (5 μg/ml) consistantly increased the incorporation of [^{3}H]thymidine into DNA of organ cultures of human mammary carcinoma explants. Although the addition of ovine prolactin to the insulin-containing medium did not routinely significantly enhance the insulin stimulation, however, a very small fraction of the breast tumor explants did respond to prolactin with increased DNA synthesis. More recently Welsch and McManus (1977) and Welsch (1978) have demonstrated that insulin and human placental lactogen can directly stimulate DNA synthesis of the epithelium in benign human breast tumors grown in organ culture and in athymic nude mice.

The complexity of the effects of insulin levels on major aspects of cell replication in our system was apparent. High insulin levels increased epithelial cell replicative activity in a manner strongly conditioned by the culture regimen. This correlated well with the microscopic morphology of the explants in the autoradiographic sections. The cycling hormone regimen appeared to maintain the mammary epithelium in reasonably normal condition for at least 8 weeks, although some lobules were necrotic. In the control organ culture medium, however, most of the glandular elements were at least partially necrotic. Similarly, the constant hormone medium did not appear to maintain the mammary epithelium to any greater extent than the control organ culture medium. Thus, cycling hormone treating with high insulin level was unique in the preservation of normal breast tissue from both the kinetic and morphological standpoint.

In our experiments, the [^{3}H]thymidine labeling index was proportional to the fraction of epithelial cells in DNA synthesis during the labeling pulse. Both the LI and the MI in native mammary epithelium of rodents were synchronized with the estrus cycle (Purnell *et al.,* 1974; Purnell and Kopen, 1976). The sparse evidence available indicates this may be true in human breast as well (Meyer, 1977). Based on these results, the cycling hormone regimen should have synchronized replication in the explant cultures. The LI results in conjunction with the morphology established that the cycling regimens, in the presence of high insulin levels, produced significant fluctuations in cell replication (see Fig. 25). The peak in LI at 5 weeks may have reflected the adaptation of the cultures to the fluctuating hormones in the medium. However, the drop in the LI at 6 weeks and the rise in weeks 7 and 8 could reflect synchronous replication of breast epithelial cells on the 28 day hormone cycle. However, data from subsequent weeks would obviously be required to confirm replication synchrony in the cycling hormone explant cultures. These studies are presently ongoing.

The low fraction of labeled mitosis counts raised several questions. The failure of all mitoses to be labeled by the 24 hour incubations [^{3}H]thymidine must have been due to either mitotic arrest or delay of mammary epithelial cells in metaphase. On the basis of the customary four phase model of cell replication, cells enter the replication cycle at the onset of DNA synthesis (S phase), pass, after DNA synthesis is complete, to the premitotic phase (G_2), and thence into mitosis (M). The customary duration of the G_2 phase is in the range of a few hours, so that the 24-hour [^{3}H]thymidine incubation followed by harvest employed in these experiments should have labeled all the mitotic figures.

The failure of mitotic cells to label in our data may have resulted from either an extraordinarily long premitotic gap (G_2) or outright arrest of mitotic figures. The latter explanation was favored by the common occurrence of metaphase mitotic figures in degenerating ductules. Many of these mitotic figures were in eosinophilic, vacuolated, and clearly nonviable epithelial cells. However, both mechanisms may be operating, since temporary mitotic delay for more than 24 hours has been observed in normal hamster mammary epithelium (Purnell *et al.,* 1974; Purnell and Saggers, 1974), and mitotic arrest has been observed in experimental rat mammary tumors (Combs *et al.,* 1970).

The kinetic significance of mitotic delay and/or arrest was enhanced by the demonstration by Kleinfeld and Sisken (1966) that cells arrested in mitosis by colcemid or colchicine for more than 6 hours eventually die. The more direct significance of mitotic delay/arrest in terms of organ culture vis-à-vis native breast was related to the problem of epithelial necrosis. Cell death in the mitotic step along with ductular necrosis has not been observed in normal breast.

A further incongruity in our data is not presented in the tables or figures. In most *in vitro* systems, a labeling pulse of 1 μCi/ml [^{3}H]thymidine for 24 hours results in adequately exposed autoradiographs within 24 to 48 hours. In these experiments, an exposure of 11 days provided only moderate grain density over labeled nuclei. This result was verified in independent labeling experiments done at a single time point. The low incorporation rate of [^{3}H]thymidine in these tissue specimens could be most easily explained on the basis of a greatly prolonged DNA synthesis phase in mammary tissue growing under the influence of cycling ovarian hormones. This interpretation fits with the low fraction of labeled mitoses if the entire cell cycle in cycling human mammary epithelium was dramatically longer than in other epithelia.

V. Perspectives

A. Current Applications and Findings

The overall long-term goal of the project is to study mammary carcinogenesis *in vitro*. Using our conditions for maintaining normal human breast epithelium in organ culture, the explants have been exposed to the direct acting carcinogen MNNG or the indirect acting carcinogen, DMBA. At designated time intervals, the explants were analyzed using light and electron microscopy. Parallel samples were xenografted into the athymic nude mice to determine whether the morphological alterations that were observed were indeed oncogenic. Following carcinogen exposure, there was a marked alteration in the morphology of the breast cultures. The nuclei were enlarged, contained active nucleoli, and had a bizarre shape. The normal intercellular architecture was altered. Areas of hyperplasia and epidermoid metaplasia were observed. In some areas, foci of cells were observed which were reminiscent of lobular carcinoma *in situ*. In this regard, the epithelial stromal junction was altered. There was a marked reduplication of the basal lamina. The myoepithelial cells on their basal surface were deeply serrated. After 1 week exposure to carcinogens, numerous labeled cells were observed primarily on the basal portion of the hyperplastic ductules. Comparable tissue samples were evaluated 1 month postxenotransplantation in nude mice. The epithelium was capable of incorporating [^{3}H]thymidine given 2 hours prior to retrieval of the graft; however, no differences between control and treated cultures were demonstrable. The remaining mice are still being held 1 year post-transplantation. One DMBA-treated transplant analyzed after 275 days in the nude mouse still had viable ductal epithelium. No demonstrable tumors have developed at this time.

Normal human mammary organ cultures, using our control culture conditions, derived from reduction mammoplasty cases have also been xenografted into athymic nude mice. After 1 week in culture, explants were transplanted into female or male athymic nude mice. The tissue was transplanted in various sites subcutaneously, into cleared fat pads, or into the shoulder fat pad. At designated intervals, the xenografts were retrieved and examined histologically. It appeared that the site of transplantation or the sex of the host did not interfere with the ability of the nude mouse to accept successfully control human breast organ cultures. Ductular epithelium capable of secretory function has been observed after 2 months in the nude mouse. At this time, viable ductular epithelium has been observed up to 190 days (Valerio, 1979). These studies thus demonstrate the feasibility of using the athymic nude mouse in our xenotransplantation studies.

B. Future Studies

In the future, we will continue to refine the hormonal requirements for normal human breast epithelium in organ culture. In addition, quantitative markers for metabolic activity of the epithelial cells will be employed. Radioimmunoassays for human 2-lactalbumin synthesis (Kleinberg *et al.,* 1977) and microspectrofluorimetric assay for secretory activity and basement membrane synthesis will be performed (Changaris *et al.,* 1978). Also immunospecific markers for prekeratin and myosin filaments will be employed in these studies.

In the future experiments, a stathmokinetic agent (colcemid) will be employed, with [^{3}H]thymidine and controls to determine the fraction of viable mitotic figures and their geographic relationship to ductular necrosis. Higher doses of [^{3}H]thymidine will be employed and the extended viability of the cultures after labeling will be studied. The accumulation of statistically significant counts of mitoses will require reading multiple autoradiographs. By using serial sections and histotopologic counting methods (Purnell *et al.,* 1974; Purnell and Kopen, 1976), the geometric association of successful and unsuccessful mitosis with ductular viability can be tested.

One of the difficulties in studying morphological data at several different levels of examination (light and EM) is that only correlative observations can be described. This is particularly difficult in analyzing breast tissue as asynchronly of mitotic activity and hormone responsiveness are known; therefore, methods will be developed whereby one will be able to study the same area at all levels. Namely, the whole mount technique will be modified so that one can select the area, photograph it, and then process that area for histological analysis. Then the same area can be selected and processed for ultrastructural observation. Thus one describes the morphological characteristics of one area throughout the total range of magnification.

Finally, the major emphasis of our future studies will be devoted to extending further our *in vitro* carcinogenesis studies. Using the cycling hormone regimen, at the time of maximal DNA synthesis, the cultures will be exposed to multiple and/or constant levels of carcinogens known to be organotrophic for mammary epithelium. At designated time intervals, the explants will be examined morphologically, and the labeling and mitotic indices determined. Randomly chosen explants will be transplanted into the athymic nude mice. Periodically the transplants will be harvested and evaluated using light, EM, and autoradiographic techniques. A certain portion of the explants will be held for extended periods of time to see if tumors indeed do develop. It may be required that the nude mouse be primed by hormonal innoculation (Welsch *et al.*, 1979) in order for these tumors to develop.

REFERENCES

Archer, F. L. (1968). *Arch. Pathol.* **85,** 62–71.

Banerjee, S. D., Cohn, R. H., and Bernfield, M. R. (1977). *J. Cell Biol.* **73,** 445–463.

Barker, B. E., Fanger, H., and Farnes, P. (1964). *Exp. Cell Res.* **35,** 437–448.

Barrett, L. A., McDowell, E. M., Frank, A. L., Harris, C. C., and Trump, B. F. (1976). *Cancer Res.* **36,** 1003–1010.

Battifora, H. (1975). *Arch. Pathol.* **99,** 614–617.

Ceriani, R., Contesso, G., and Nataf, B. M. (1972). *Cancer Res.* **32,** 2190–2196.

Changaris, D., Shengrund, C-L., and Combs, J. W. (1978). *J. Histochem. Cytochem.* **26,** 267–276.

Combs, J. W., Bennington, J. L., and Mackay, B. (1970). *Am. J. Pathol.* **59,** 90a.

Dermer, G. B., and Sherwin, R. P. (1975). *Cancer Res.* **35,** 63–67.

Dilley, W. G., and Kister, S. J. (1975). *J. Natl. Cancer Inst.* **55,** 35–36.

Elias, J. J., and Armstrong, R. C. (1973). *J. Natl. Cancer Inst.* **51,** 1341–1343.

Flaxman, B. A. (1974). *J. Invest. Dermatol.* **63,** 48–57.

Flaxman, B. A., and Lasfargues, E. Y. (1973). *Proc. Soc. Exp. Biol. Med.* **143,** 371–374.

Flaxman, B. A., and Van Scott, E. J. (1972). *Cancer Res.* **32,** 2407–2412.

Flaxman, B. A., Dyckman, J., and Feldman, A. (1976). *In Vitro,* **12,** 467–471.

Gould, V. E., and Battifora, H. (1976). *Pathol. Ann.* **11,** 353–386.

Heuson, J.-C., Pasteels, J.-L., Legros, N., Heuson-Stiennon, J., and Leclercq, G. (1975). *Cancer Res.* **35,** 2039–2048.

Hillman, E. A., Halter, S. A., Barrett, L. A., Trump, B. F., and Harris, C. C. (1980). (In preparation).

Janss, D. H., Malan-Shibley, L. B., Ben, T. L., and Hillman, E. A. (1980). Endocrinology (in preparation).

Jones, R. T., Barrett, L. A., van Haaften, C., Harris, C. C., and Trump, B. F. (1977). *J. Natl. Cancer Inst.* **58,** 557–565.

Kleinberg, D. L. (1975). *Science* **190,** 276–278.

Kleinberg, D. L., and Todd, J. (1978). *Cancer Res.* **38,** 4318–4322.

Kleinberg, D. L., Todd, J., and Groves, M. L. (1977). *J. Clin. Endocrinol. Metab.* **45,** 1238–1250.

Kleinfeld, R. G., and Sisken, J. E. (1966). *J. Cell Biol.* **31,** 369–379.

Kopriwa, B. F. (1967). *J. Histochem. Cytochem.* **14,** 923–928.

McDowell, E. M., and Trump, B. F. (1976). *Arch. Pathol.* **100,** 405–414.

Meyer, J. S. (1977). *Hum. Pathol.* **8,** 67–81.

Mishell, D. R., Nakamura, R. M., Crosigani, P. G., Stone, S., Kharma, K., Nagata, Y., and Thorneycroft, I. H. (1971). *Am. J. Obstet. and Gynecol.* **111,** 60–65.

Neely, J. E., and Combs, J. W. (1976). *J. Histochem. Soc.* **24,** 1057–1064.
Orr, M. F., and McSwain, B. (1954). *Texas Rep. Biol. Med.* **12,** 916–920.
Osborne, C. K., and Lippman, M. E. (1978). *In* "Breast Cancer. 2. Advances in Research and Treatment" (W. L. McGuire, ed.), pp. 103–154. Plenum, New York.
Ozzello, L. (1974). *J. Invest. Dermatol.* **63,** 19–26.
Purnell, D. M., and Kopen, P. (1976). *Anat. Rec.* **186,** 39–48.
Purnell, D. M., and Saggers, G. C. (1974). *J. Natl. Cancer Inst.* **53,** 825–828.
Purnell, D. M., Combs, J. W., and Saggers, G. C. (1974). *J. Natl. Cancer Inst.* **53,** 1691–1697.
Reyes, F. I., Winter, J. S. D., and Faiman, C. (1977). *Am. J. Obstet. and Gynecol.* **129,** 557–564.
Rijsbosch, J. K. C., and Meyer, A. H. (1976). *Arch. Chir. Neerl.* **28,** 271–280.
Schürch, W., McDowell, E. M., and Trump, B. F. (1978). *Cancer Res.* **38,** 3723–3733.
Stirling, J. W., and Chandler, J. A. (1976). *Virchows Arch. Pathol. Anat. Histol.* **372,** 205–226.
Stirling, J. W., and Chandler, J. A. (1977). *Virchows Arch. Pathol. Anat. Histol.* **373,** 119–132.
Tannenbaum, M., Weiss, M., and Marx, A. J. (1969). *Cancer* **23,** 958–978.
Toker, C. (1967). *J. Ultrastruct. Res.* **21,** 9–25.
Trump, B. F., Valigorsky, J. M., Dees, J. H., Mergner, W. J., Kim, K. M., Jones, R. T., Pendergrass, R. E., Garbus, J., and Cowley, R. A. (1973). *Hum. Pathol.* **4,** 89–109.
Valerio, M. (1979). Personal communication.
van Bogaert, L. J. (1976). *Cell Tissue Res.* **171,** 535–541.
van Bogaert, L. J. (1977). *Experientia* **33,** 1450.
Wellings, S. R., and Jentoft, V. L. (1972). *J. Natl. Cancer Inst.* **49,** 329–338.
Welsch, C. W. (1978). *Cancer Res.* **38,** 4054–4058.
Welsch, C. W., and McManus, J. M. (1977). *Cancer Res.* **37,** 2257–2261.
Welsch, C. W., de Iturri, G. C., and Brennan, M. J. (1976). *Cancer* **38,** 1272–1281.
Welsch, C. W., McManus, J. M., DeHoog, J. V., Goodman, G. T., and Tucker, H. A. (1979). *Cancer Res.* **39,** 2046–2050.
Wicha, M. S., Liotta, L. A., and Kidwell, W. R. (1979). *Cancer Res.* **39,** 426–435.

Chapter 5

Methods for the Isolation and Culture of Normal Human Breast Epithelial Cells[1]

DOUGLAS H. JANSS

Department of Pathology,
University of Maryland School of Medicine,
Baltimore, Maryland
and
Endocrine Carcinogenesis Section,
Frederick Cancer Reserach Center,
Frederick, Maryland

ELIZABETH A. HILLMAN

Department of Pathology,
University of Maryland School of Medicine,
Baltimore, Maryland

LOUISE B. MALAN-SHIBLEY AND THERESA L. BEN[2]

Endocrine Carcinogenesis Section,
Frederick Cancer Research Center,
Frederick, Maryland

---

[1]Supported by NCI Contracts NO1-CO-75380 and NO1-CP-43237.
[2]Present address: Laboratory of Experimental Pathology, National Cancer Institute, National Institutes of Health, Bethesda, Maryland.

ISBN 0-12-564140-0

I. Introduction

Studies of mammary gland growth, differentiation, and carcinogenesis have been limited mainly to nonhuman species especially the mouse. *In vitro* studies using organ cultures or cell cultures derived from the mouse, where a viral etiology of breast cancer seems well established (Lyons and Moore, 1965), have provided evidence that hormonal and other factors provide important modulation of cancer expression. While breast cancer is the most common neoplasm in women of the Western world, its pathogenesis remains unknown (MacMahon *et al.*, 1973). It is, therefore, important to consider whether *in vitro* studies on the mouse are relevant to an understanding of carcinogenesis in the human. Furthermore, comparative studies of human and mouse mammary gland in organ culture have demonstrated important differences in cell behavior *in vitro* (Elias, 1959; Rivera and Bern, 1961; Lockwood *et al.*, 1967; Dilley, 1971; Barker *et al.*, 1964; Wellings and Jentoft, 1972; Flaxman and Lasfargues, 1973). Thus, recent research has been focused on attempts to study human breast tissue in culture.

Numerous attempts have been made to study benign and malignant human breast tissue *in vitro* (Orr and McSwain, 1954; Lasfargues and Ozzello, 1958; Foley and Aftonomos, 1965; Whitescarver *et al.*, 1968; Ceriani *et al.*, 1972; Stoker *et al.*, 1976; Hallowes *et al.*, 1977a), but few studies have examined the *in vitro* growth of explants or isolated cells of the normal human breast (Orr and McSwain, 1955; Buehring, 1972; Flaxman, 1974; Gaffney *et al.*, 1976a; Taylor-Papadimitriou *et al.*, 1977). Such studies have been restricted for: (*a*) most available tissue specimens are obtained from patients who have undergone treatment for cancer, (*b*) surgical specimens are usually small and from nonlactating tissue which contain only small numbers of epithelial cells, and (*c*) procedures for the dissociation of large masses of tissue with subsequent isolation and maintenance of epithelial cells have not been adequately perfected.

A brief summary of the various techniques used to isolate epithelial cells from normal human breast tissues is presented schematically in Fig. 1. Under sterile conditions, the tissue specimen is trimmed to remove extraneous fat and washed in medium. After several washings, the tissue is placed in a culture dish with medium, minced into small pieces, and the mince is incubated at 37°C with gentle shaking. The tissue fragments are removed and cells "spilled" during mechanical separation are collected by centrifugation, washed, and cultured (Lasfargues and Ozzello, 1958). Remaining tissue fragments can be dissociated by treatment with various enzymes, cells collected and cultured (Lasfargues and Moore, 1971; Buehring and Williams, 1976). Numerous enzymatic dissociation methods have been used to obtain cells from human breast tissues (Pitelka *et al.*, 1969; Lasfargues and Moore, 1971; Prop and Wiepjes, 1973; Pretlow *et al.*, 1974; Gaffney *et al.*, 1976a). Minced tissue is incubated for various times at 37°C with trypsin, collagenase, hyaluronidase, pronase, or combinations of these enzymes with rather vigorous shaking. Once the tissue mince has become dissociated, cells are collected by centrifugation, washed, and cultured. Flaxman (1974) has combined dissection of mammary ducts with short-term incubation with trypsin–EDTA which then permitted the epithelium to be scraped from the connective tissue. The stripped epithelium is gently dissociated into cell clumps and single cells which are then cultured. Milk from lactating women or breast fluid from nonlactating women has been a source from which cell aggregates or single cells have been obtained for culture (Buehring, 1972; Gaffney *et al.*, 1976b; Taylor-Papadimitriou *et al.*, 1977). The cell aggregates attach slowly to

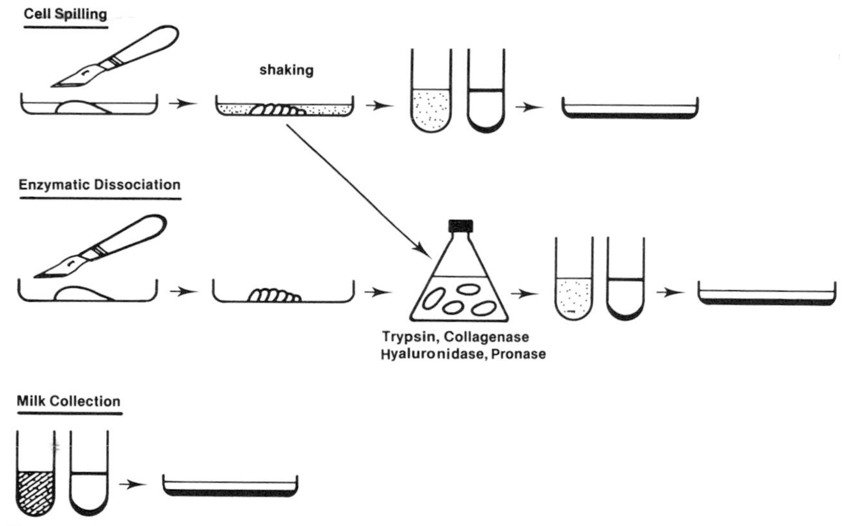

FIG. 1. Schematic representation of the various techniques which have been used in an attempt to isolate human breast epithelial cells for tissue culture studies.

either glass or plastic substrates but eventually give rise to colonies of slowly proliferating epithelioid cells. In contrast, single cells referred to as colostrum bodies, Donne cells, or "foam cells" attach readily to the substrate but soon degenerate.

While these techniques yield breast cells which when cultured give rise to colonies of epithelium, they are not particularly useful. Human breast epithelial cell cultures derived from the spilling and enzymatic treatment methods are usually slow growing and are rapidly overgrown by fibroblasts. Human milk does not yield a homogeneous cell population and cultures from milk contain colony-forming epithelial cells and a large number of "foam cells" which are either fully differentiated or macrophages (Holquist and Papanicolaou, 1956; Gaffney *et al.*, 1976b). While the viability of breast cells obtained by these methods is high, the low yield of epithelial cells coupled with the slow growth rate and rapid overgrowth of the cultures by fibroblasts represent distinct disadvantages.

Based upon our work with rat mammary tissue (Janss *et al.*, 1980; Malan-Shibley *et al.*, 1980), we have attempted to develop methods for tissue dissociation, for epithelial cell isolation, and for cellular proliferation which can be used routinely to establish cultures of normal human breast epithelium for studies of growth, differentiation, and carcinogenesis.

II. Materials and Methods

A. Tissue

Normal human breast tissue from reductive mammoplasty or immediate autopsy of young premenopausal women (14 to 30 years of age) was obtained from the Pathology Department of the University of Maryland School of Medicine. All procedures from the collection to the examination of the material, prior to its release, were performed under sterile conditions. Tissue was transported to the cell culture laboratory in L-15 medium containing antibiotics and maintained at 4°C. Once received, the tissue was rinsed with Medium 199 Hanks' salts (M199-H) containing 1% charcoal-extracted fetal bovine serum (FBS) and 50 μg/ml gentamicin (Medium 199, Grand Island Biological Co., Grand Island, NY; FBS, KC Biological, Lenexa, KS; gentamicin, Schering Corp., Port Reading, NJ). Rinsed tissue was cut into pieces approximately 1–2 cm^3, placed in fresh culture medium, and stored at 4°C until used. Stored in this manner, tissue could be used for cell isolation for at least 4 days and still yield viable epithelial cells.

B. Medium and Reagents

1. Collagenase

The dissociation mixture consisted of 0.1% collagenase (w/v) in M199-H containing 10% charcoal-extracted FBS and 50 μg/ml gentamicin (CLS collagenase, Type II, Worthington Biochemical, Freehold, NJ).

2. Ficoll

A stock solution of 30% ficoll (w/v) (Ficoll 400, Pharmacia Fine Chemicals, Piscataway, NJ) was prepared in calcium and magnesium-free Earle's balanced salt solution (CMF-EBSS; Flow Laboratories, Rockville, MD) and sterilized by autoclaving at 15 lb pressure for 20 minutes at 120°C. Working solutions of 1, 2, 8, and 20% ficoll were made by diluting the stock with M199-H containing 5% charcoal-extracted FBS and 50 μg/ml gentamicin.

3. Culture Medium

The basal culture medium consisted of Medium 199-Earle's salts (M199-E; Grand Island Biological Co., Grand Island, NY) with 100 m*M* glutamine. The medium was purchased in powder form, rehydrated in deionized, glass-distilled water, and sterilized by filtration through 0.22-μm filters (Millipore Corp., Bedford, MA). Supplements, including 5% charcoal-extracted FBS, hormones, and 50 μg/ml gentamicin, were added before use. Hormone supplements were added to the culture medium in various concentrations and combinations after Millipore filtration (0.22 μm) of 1 mg/ml stock solutions. Insulin (bovine pancreas: 25.9 U/mg, Schwarz/Mann, Orangeburg, NY) was prepared in (5×10^{-3} *N*) HCl. Prolactin (NIH-P-S-13-Ovine: 30.0 IU/mg, gift of the Pituitary Hormone Distribution Program, NIAMDD, Bethesda, MD) was dissolved in (5×10^{-4} *M*) NaOH. Aldosterone, 17β-estradiol, and progesterone (Sigma Chemical Co., St. Louis, MO) were dissolved in absolute ethanol. The final concentration of ethanol in the culture medium did not exceed 0.1%. All stock solutions were stored at 4°C until use (within 1 week) except for prolactin which was stored at −20°C.

4. Fetal Bovine Serum

Endogenous hormones present in fetal bovine serum have been reported to vary significantly among different lots and to be of sufficient concentration to influence cell growth (Esber *et al.*, 1973). To reduce variability and to establish the lowest possible baseline to which specific hormone supplements could be made, the serum was stripped of hormones by charcoal extraction. Charcoal extraction of endogenous fetal bovine serum hormones was carried out with 1% (w/v) activated charcoal (Mallinckrodt, Inc., St. Louis, MO) at room temperature. The mixture was stirred for 15 minutes at moderate speed. Charcoal was removed by centrifugation at 10,000 *g* (max) for 15 minutes at 4°C, and the adsorption procedure was then repeated. After the final centrifugation, adsorbed serum was sterilized by Millipore filtration (0.22 μm). Samples were taken to test for sterility and the adsorbed serum was stored at −20°C until use. For each new serum lot, aliquots of untreated and charcoal-extracted serum were analyzed by radioimmunoassay for insulin, cortisol, aldosterone, testosterone, estrone, 17β-estradiol, estriol, and progesterone by Herner Laboratories, Inc., Rockville, MD and Hazelton Laboratories, Vienna, VA. After extraction, hormone content was undetectable, or reduced to insignificant levels as shown in Table I.

C. Cell Isolation and Separation

Tissue pieces were removed from the storage medium, weighed, and then minced with a razor blade on a sterile glass plate or in a petri dish. Approximately 2.5 g of minced tissue was transferred to a sterile 50-ml screw-cap Fernback flask containing 12 ml of the collagenase dissociation mixture. Incubation was carried out at 37°C without agitation for at least 16 hours. At the end of the incubation period, the tissue mince still appeared intact but it could be dissociated by gentle swirling and pipetting the mixture until it had an opalescent

TABLE I

RANGE OF HORMONE LEVELS IN FETAL BOVINE SERUM BEFORE AND AFTER CHARCOAL ADSORPTION

Hormone[a]	Untreated	Adsorbed[b]
Insulin (μU/ml)	7–24	<5
Aldosterone (ng/ml)	0.1–2.0	<0.02
Cortisol (ng/ml)	1–5	<0.5
Testosterone (pg/ml)	114–500	<50
Progesterone (pg/ml)	100–595	<10
Estrone (pg/ml)	17–68	<10
Estradiol (pg/ml)	160–340	<10
Estriol (pg/ml)	<10–20	<10

[a] Radioimmunoassay for each hormone performed by Herner Laboratories, Inc., Rockville, MD and Hazelton Laboratories, Vienna, VA. Values represent the range of concentrations determined for seven different lots of fetal bovine serum.

[b] After charcoal adsorption, serum was concentrated 10-fold and values reported represent determinations performed on the 10× serum.

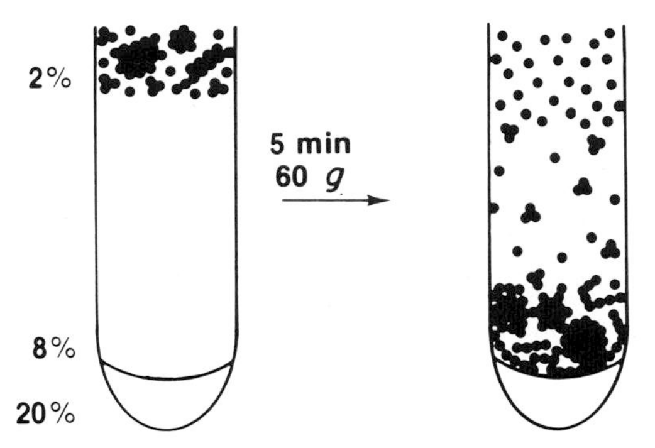

FIG. 2. Schematic representation of the separation of mammary epithelial elements from fibroblasts using ficoll gradient centrifugation.

appearance. This mixture was then filtered through one to two layers of 157-mesh nylon (Martin Supply Co., Baltimore, MD) to trap any undissociated fragments, and the mesh was thoroughly rinsed with M199-H. To ensure proper filtration, the dissociated tissue from no more than four flasks was strained through a single filter. The filtrate was transferred to Falcon 50-ml conical tubes and centrifuged at 500 *g* for 15 minutes at 4°C. The fat layer was removed from the top of the tube and discarded along with the remainder of the supernatant above the cell pellet. Each cell pellet was then dissociated in 10 ml M199-H, the volume brought to 50 ml, and the suspension centrifuged at 400 *g* for 10 minutes. Depending upon the amount of tissue originally used, the cells from 10 to 15 g of breast tissue were pooled and resuspended in a total volume of 30 ml M199-H. Sterile deoxyribonuclease I (bovine pancreas, Sigma Chemical Co., St. Louis, MO), 40,000 units in 0.4 ml, was added to the cell suspension which was then incubated at 37°C for 5 minutes with occasional gentle shaking. After incubation the cells were again centrifuged at 400 *g* for 10 minutes and the supernatant was decanted. Following the addition of 10 ml of a 1% ficoll solution containing 5% charcoal-extracted FBS and 50 μg/ml gentamicin, the cell pellet was again completely dissociated. The cell suspension was cooled at 4°C and layered over a continuous 2–8% ficoll gradient (80 ml) resting on a 5 ml cushion of 20% ficoll. The gradient had been allowed to equilibrate overnight or for a minimum of 4 hours at 4°C before layering. The gradients were gradually accelerated to 600 rpm (60 *g* max) and centrifuged 5 minutes at 4°C in a Sorvall RC-3 centrifuge equipped with an HL-8 rotor (DuPont Company, Sorvall Operations, Newtown, CT). Erythrocytes, connective tissue fibroblasts, and debris remained in the top half of the gradient while the epithelial elements sedimented to the lower half to third of the gradient (Fig. 2). Gradients were fractionated by removing successive 10 ml volumes, starting from the top. Each fraction was examined by phase contrast microscopy to check the distribution of cell types throughout the gradient. The parenchyma was not completely dissociated into single cells but appeared as epithelial aggregates or hollow balls and tubes of cells in the bottom half of the gradient. A second gradient centrifugation of the intermediate fractions (containing a mixture of all cell types) was used to increase the final cell yield.

The epithelial, or fibroblast, fractions were individually combined, diluted with M199-H plus 5% charcoal-extracted FBS, and centrifuged at 400 *g* for 15 minutes. The supernatant was decanted, and a uniform cell suspension was prepared for counting. Hollow balls and tubes of cells or single cells were checked for viability by trypan blue exclusion and were enumerated using a hemocytometer. An average yield of 1×10^6 epithelial elements (hollow balls and tubes) and 5×10^6 fibroblasts per 20 g of starting breast tissue generally was obtained. The cells were again pelleted, and the cell pellet was prepared for immediate use or for viable frozen storage. Tests for sterility were made routinely by inoculating an aliquot of the final supernatant into the following media: tryptic soy agar containing 5% sheep blood, tryptose phosphate broth, and sabouraud dextrose broth (England Laboratories, Beltsville, MD). Samples were incubated for 1 month at 25 and 3°C and periodically examined for the growth of bacteria, fungi, or yeast.

Cells to be frozen were resuspended in M199-H to give a concentration of 4×10^6 epithelial elements or 2×10^7 fibroblasts per ml. An equal volume of 2× freezing solution (M199-H, 40%; charcoal-extracted FBS, 40%; dimethyl sulfoxide, 20%; penicillin, 200 U/ml; and streptomycin, 200 μg/ml) was added with moderate agitation. Approximately 3×10^6 epithelial elements or 1×10^7 fibroblasts in 1.5 ml were distributed per freezing vial. After a slow-rate freeze to −60°C, the vials were then stored at −100 to −150°C in the vapor phase of a liquid nitrogen freezer.

D. Culture Conditions

For high-resolution light and electron microscopy or hormonal influence on labeling index studies, the required number of cells was thawed rapidly at 37°C, added to warmed M199-H containing 5% charcoal-extracted FBS and centrifuged at 110 *g* for 15 minutes. This cell pellet obtained from cells frozen prior to use or the cell pellet of freshly isolated elements was gently resuspended in M199-E, and the cells were counted and added to the basal medium containing the test hormones. Human breast epithelial elements were planted at a seeding density of 4.5×10^3 elements/cm^2 in Lab-Tek chamber slides (Miles Laboratories, Inc., Naperville, IL) for autoradiography or 35- and 60-mm Falcon culture dishes for microscopic evaluation. Control human breast fibroblast or mouse dermal fibroblast (obtained from Dr. Stuart H. Yuspa of the Laboratory of Experimental Pathology, NCI, Bethesda, MD) cultures were seeded at 1×10^5 cells/cm^2. Cultures were maintained at 37°C in a humidified incubator (Forma

Scientific, Marietta, OH) with 5% CO_2, 95% air. Medium was replaced (slides, 3.5 ml; 35-mm culture dishes, 2 ml; 60-mm culture dishes, 4 ml) on day 2 after planting and at 2- to 3-day intervals thereafter, but always 24 hours before a [^{3}H]thymidine incorporation period.

E. Autoradiography and Labeling Index

Cells were grown for 5 days in medium with or without hormones, and then exposed to 1.0 μCi/ml of [*methyl*-^{3}H]thymidine (New England Nuclear, Boston, MA, specific activity 6.7 Ci/mmole) for 1 hour. Cells were washed twice with cold medium followed by cold phosphate-buffered saline and then fixed with glacial acetic acid–methanol (1:3) for 10 minutes. After fixation, the cells were washed with water and the slides were allowed to air dry. Slides were processed for autoradiography by the dip-coating technique with Kodak NTB-2 emulsion (Eastman Kodak, Rochester, NY) as described by Rogers (1973). After exposure for 2 weeks at 4°C in total darkness, the slides were developed for 2 minutes in Kodak Dektol developer, fixed for 8 minutes in 30% sodium thiosulfate, rinsed in tap water for 20 minutes, and stained with Giemsa. The labeling index was based on counts of at least 2000 cells per slide performed by two individuals. Every fourth field was counted during a systematic scanning of the slide provided that the field showed no evidence of sloughing emulsion and was a minimum of 50% confluent. A nucleus was considered labeled when covered by five or more developed silver grains. A minimum of four slides for each hormone variable was prepared from cells isolated from each tissue sample. The number of labeled nuclei per 1000 cells counted was recorded and an average of the six or more readings was used for each value reported.

F. Preparation of Samples for Light and Electron Microscopy

Cell pellets of human breast epithelial elements obtained either prior to or after ficoll gradient centrifugation and cultures of epithelial elements were fixed for high-resolution light and electron microscopy in 4% formaldehyde, 1% glutaraldehyde (McDowell and Trump, 1976). After primary fixation, specimens were stored or rinsed several times in 0.1 *M* sodium cacodylate buffer pH 7.4, containing sucrose. Tissue was postfixed in 1% OsO_4 buffered with 0.1 *M* collidine buffer, stained *en bloc* with uranyl acetate, dehydrated through a graded series of alcohol, and embedded in Epon (Luft, 1961). Thick 0.5- to 1-μm plastic sections were stained with toluidine blue (Trump *et al.*, 1961) and examined with the light microscope. Representative areas were selected for ultrastructure examination. Thin sections were doubly stained with 2% uranyl acetate in water and lead citrate (Venable and Coggeshall, 1965) and examined in a Jeol 100B electron microscope.

III. Results

A. Morphology

Human breast tissue obtained for epithelial cell culture studies contains a large amount of adipose tissue interlaced with a loose network of fibrous connective tissue. In each sample, areas of dense fibrous connective tissue are found. Within the connective tissue small islands of resting ductules (terminal ducts and alveoli) one epithelial cell layer thick and surrounded by myoepithelial cells can be seen. These parenchymal islands appear to be more abundant in the dense connective tissue (Fig. 3). Phase contrast microscopic examination of the cell mixture obtained after collagenase dissociation reveals the presence of numerous cell aggregates (hollow balls and tubes of cells), single cells, erythrocytes, and much cell debris (Fig. 4). Following centrifugation of the cell mixture through a continuous ficoll gradient, single cells, erythrocytes, and cell debris remain in the upper third of the gradient. Cell cultures arising from these single cells have a typical fibroblast-like morphology (Fig. 5). The lower third of the ficoll gradient contains an enriched mixture of cellular aggregates (Fig. 6) in which single cells are rarely found. If single cells are present, they are removed by a second ficoll gradient centrifugation. Cultures arising from the cellular aggregates have a pavement-mosaic pattern characteristic of epithelial-like cells (Fig. 7).

Dissociation of human breast tissue with a weak collagenase solution effectively frees the epithelial tree from the fat and connective tissue stroma and reduces it to cellular aggregates which are isolated as hollow balls and tubes of cells. The organization of epithelial cells into ductal and ductular structures, however, is not altered. In semithin Epon section (Fig. 8), entire ductules composed of epithelial cells surrounded by a thin layer of myoepithelial cells are observed. Figure 9 is a survey electron micrograph of normal human breast used as starting material illustrating the morphology of a ductular unit. Epithelial cells containing numerous free ribosomes and mitochondria, an inconspicous Golgi apparatus, and a small amount of smooth endoplasmic reticulum within a relatively small cytoplasmic volume are seen. Large nuclei occupy the major fraction of each cell volume. These cells are joined on their lateral surfaces by characteristic desmosomes. Myoepithelial cells are joined to the basal surface of the

FIG. 3. Lobular unit surrounded by dense connective tissue in normal human breast specimen prior to dissociation. Stained with H & E. Bar = 150 μm.

FIG. 4. Cell mixture following collagenase dissociation of normal human breast tissue. Phase-contrast. Bar = 72 μm.

FIG. 5. Typical fibroblast culture arising from single cells isolated by ficoll gradient centrifugation from cell dissociation mixture. Phase-contrast. Bar = 40 μm.

FIG. 6. Ficoll gradient fraction containing cellular aggregates (balls and tubes of cells). Note absence of single cells. Phase-contrast. Bar = 40 μm.

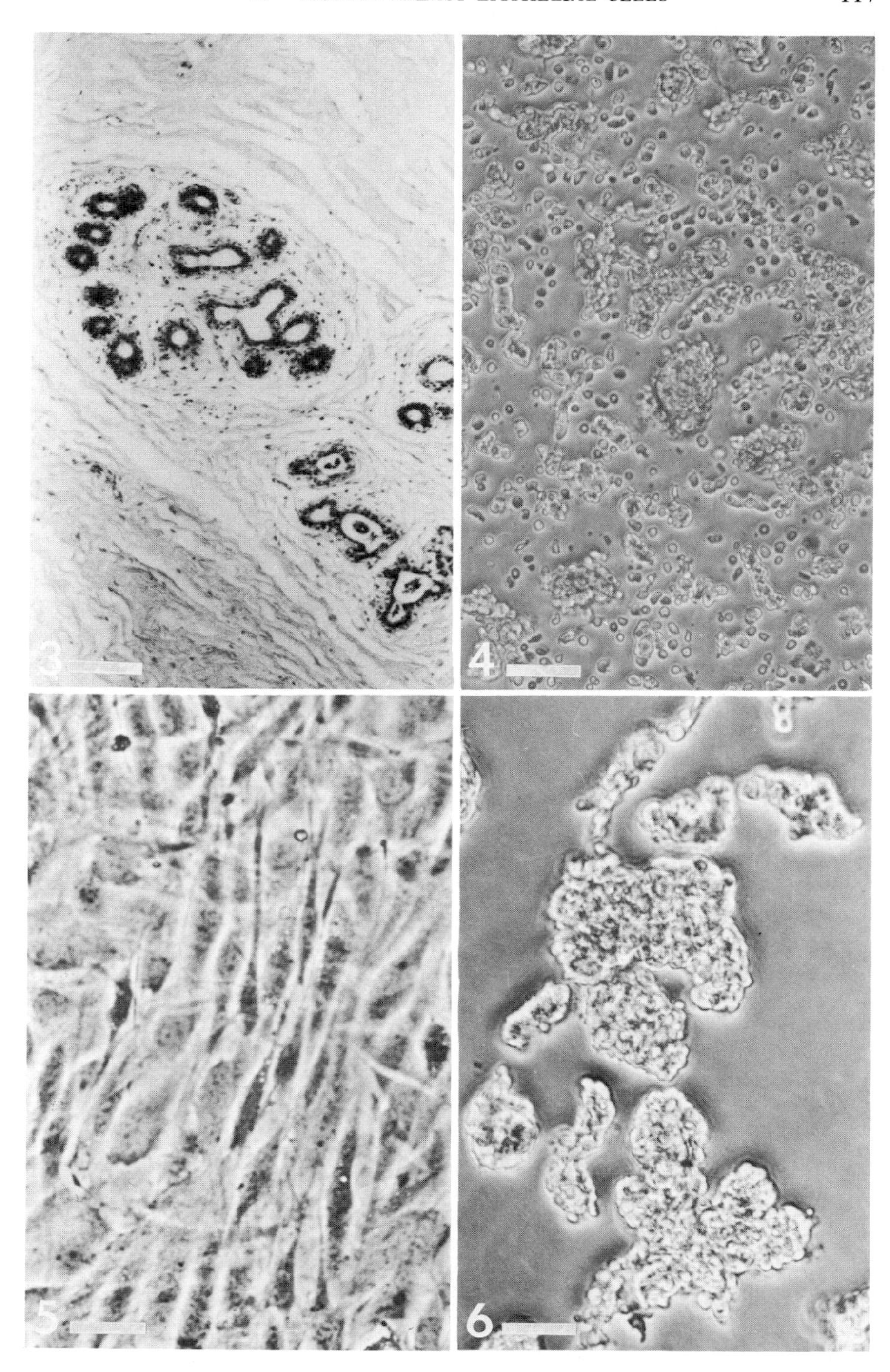
3
4
5
6

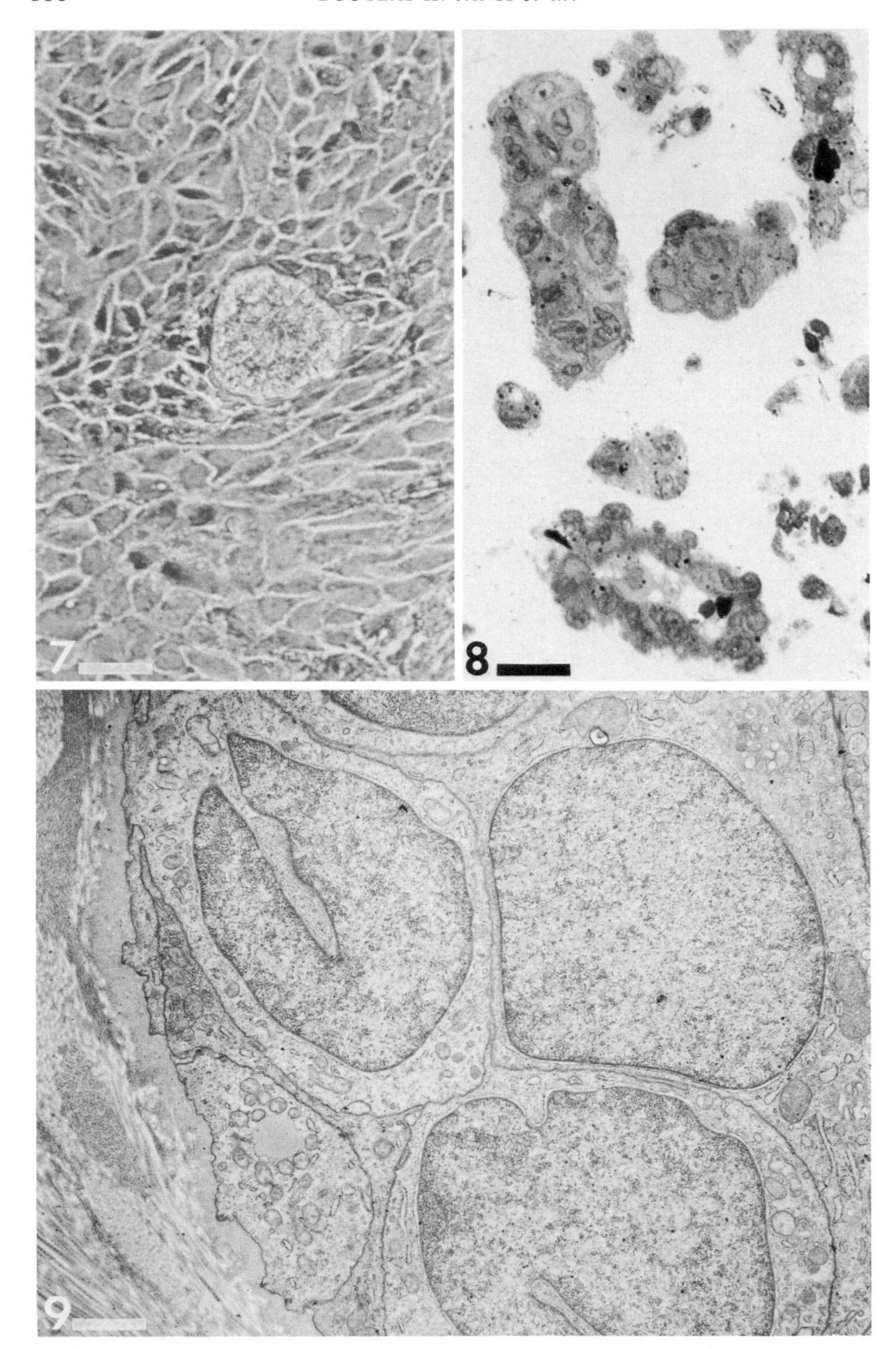
7
8
9

breast ductule cells by typical junctional complexes. These myoepithelial cells have characteristic hemidesmosomes on their basal surface. A well-formed epithelial–stromal junction with all the typical components is present.

When plated in medium containing hormones, the isolated ductal and ductular (hollow tubes and balls of cells) elements rapidly attach and, within 2 days, islands of cells can be observed. Within 4–5 days, confluent cultures of polygonal cells whose morphology is similar to that observed in freshly isolated elements and breast tissue *in vivo* are obtained. While not conclusive proof of the absence of fibroblast-like cells, the only morphology observable is that typical of epithelial cells. Semithin Epon sections of human breast epithelial cultures oriented perpendicular to the plane of growth are present in Figs. 10 and 11. In Fig. 10, a slender outgrowth of epithelial cells can be seen. The attachment site of an epithelial element from which cellular outgrowth advances is observed in Fig. 11. After 4–5 days, epithelial outgrowth ranges from a monolayer to a multilayer culture. Within the multilayered areas, intracellular alveoli surrounded by numerous epithelial cells can be seen (Figs. 10 and 11). Electron microscopic examination (Fig. 12) of an area of slender outgrowth such as that seen in Fig. 10 shows that the epithelial cells are attached to each other and to the myoepithelial cells by typical junctional complexes. The outgrowth seen in this area is composed primarily of cells which have short slender microvilli on their upper surface which is in contact with the culture medium. Components characteristic of breast epithelial cells which are readily observed include numerous free ribosomes and small mitochondria, the active Golgi apparati, the large nucleus with a single prominent nucleolus, and a small amount of smooth endoplasmic reticulum. Prominent myofilaments are present in the myoepithelial cells.

Since freshly isolated breast parenchymal elements contain both myoepithelial and epithelial cells with distinct organizational polarity, it was important to demonstrate similar structures and organization in cultured cells. In this tangentional electron micrograph of a 5 day culture, characteristic myofilaments of myoepithelial cells running parallel to the petri dish are seen (Fig. 13). While a fine homogeneous material is observed on the basal surface of these cells, periodic acid–Schiff (PAS) staining of semithin Epon sections failed to reveal whether this is indeed basal lamina. On the basal surface of the myoepithelial cells prominent hemidesmosomes are easily identified (Fig. 13, inset). As the epithelial cells grow out from the attached elements, they are joined by desmo-

---

FIG. 7. Characteristic pavement–mosaic pattern of epithelial-like cell outgrowth arising from a centrally located cellular aggregate. Phase-contrast. Bar = 40 μm.

FIG. 8. Ficoll gradient isolated ductular elements illustrating viable epithelial and myoepithelial cells. Semithin plastic section stained with T & P. Bar = 22 μm.

FIG. 9. Survey electron micrograph of the basal surface of a normal resting breast ductular unit found in an immediate autopsy specimen. Bar = 1.25 μm.

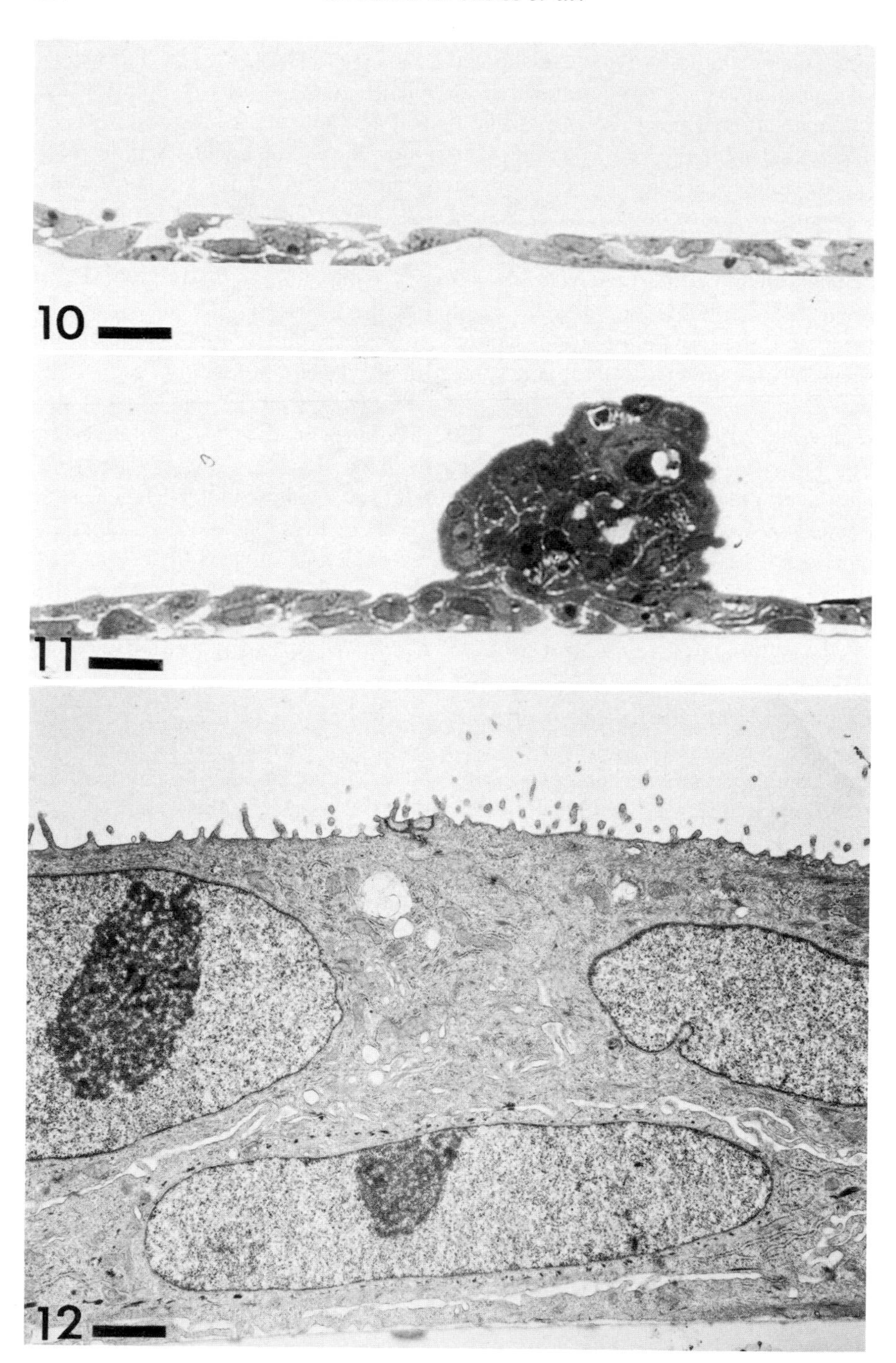
10
11
12

somes and contain numerous fine filaments oriented parallel to the plane of the petri dish (Fig. 14). In some instances, these fine filaments tend to form bundles.

A survey electron micrograph of these cells illustrates the typical epithelial cell morphology of a preponderance of free ribosomes and small mitochondria, numerous fine filaments, short segments of organized endoplasmic reticulum, and abundant microvilli present on the apical surfaces (Fig. 15). The large nuclei have a single nucleolus and the chromatin is dispersed. Intracellular alveoli containing secretory material are often observed.

Within the central portion of attached clumps of epithelial cells from which outgrowth originates (as seen previosly in Fig. 11) secretory material is frequently seen within intercellular ascini (Fig. 16). Cells surrounding the ascinus contain active Golgi apparatuses. Epithelial cells in contact with the growth medium (luminal surface of the outgrowth) have numerous microvilli (Fig. 17). Many slender cytoplasmic filaments are also present in these cells. Epithelial cells on the luminal surface as well as within the central portion of the outgrowth have enlarged, active nucleoli (Figs. 16 and 17).

B. Hormone Requirements for Growth

When human breast epithelial elements were plated in medium without hormones or with insulin and aldosterone alone, minimal cell attachment and epithelial cell outgrowth was observed. These cultures exhibited small islands of epithelial cell outgrowth which never coalesced to form confluent cultures. Epithelial cell outgrowth required a minimum hormone combination of insulin, aldosterone, and prolactin. With medium containing insulin, aldosterone, prolactin, 17β-estradiol, and progesterone, confluent cultures were obtained within 5 to 7 days after cell plating. Once epithelial cell cultures had reached confluency in the hormone-supplemented medium, they could be maintained for as long as 60 days with medium changes every 2 to 3 days in which the hormone supplementation was reduced to insulin and aldosterone. If cells were then released from the culture dish with 0.05% trypsin plus 0.02% EDTA and reseeded as a 1:2 dilution in medium supplemented with all five hormones, cell replication resumed and confluent cultures were obtained. Human mammary fibroblasts and mouse dermal fibroblasts grew well in culture medium without hormone additions if the

---

FIG. 10. Semithin plastic section oriented perpendicular to the plane of growth and demonstrating typical polarity of the outgrowth of epithelial cells. Stained with T & P. Bar = 22 μm.

FIG. 11. Semithin plastic section with the same orientation as Fig. 10. Note attachment site of cellular aggregate with advancing epithelial cell outgrowth. Stained with T & P. Bar = 22 μm.

FIG. 12. Survey electron micrograph of human breast cell culture demonstrating epithelial and myoepithelial cells of the outgrowth. Section oriented perpendicular to the plane of growth. Bar = 1.66 μm.

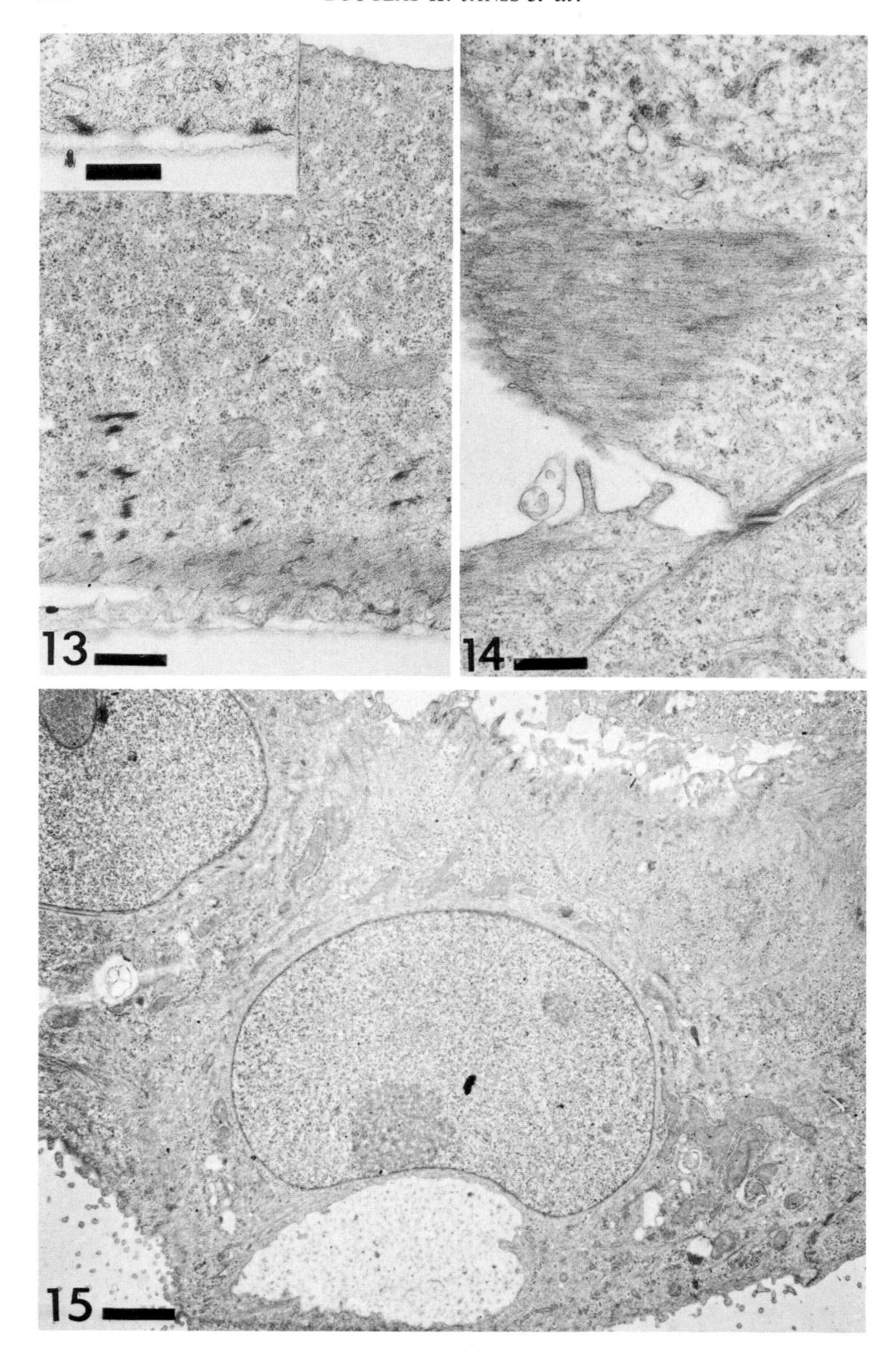
13
14
15

serum concentration was increased to 15 to 20%. Confluent cultures were obtained within 4 to 5 days after seeding.

Based upon our studies with rat mammary epithelial cell cultures (Malan-Shibley *et al.*, 1980), the influence of hormone addition to the culture medium on the DNA synthesis of human breast epithelial cells, human breast fibroblasts, and mouse dermal fibroblasts was examined by determining the labeling index of cells after 5 days in culture (Table II). The addition of insulin and aldosterone to the culture medium had little influence on the labeling index of human breast epithelial cells. The greatest increase in the labeling index was found when 0.5 μg/ml insulin, 0.5 μg/ml aldosterone, and 0.5 μg/ml prolactin were present in the culture medium. Although 0.0003 μg/ml 17β-estradiol and 0.003 μg/ml progesterone added in succession to the insulin, aldosterone, and prolactin supplemented medium resulted in a somewhat lower labeling index, epithelial cell growth was still maintained. Human breast epithelial cells in the later stages of mitosis could be readily identified by phase-contrast microscopy when grown and maintained in medium supplemented with all five hormones (Figs. 18 and 19). Both human breast and mouse dermal fibroblasts had a labeling index much greater than that determined for the human epithelial cells. However, hormone additions had no influence on the labeling index of fibroblasts of human breast or mouse dermis origin.

C. Cell Viability and Cryopreservation

The viability of freshly isolated breast epithelial elements ranged from 85 to 95% as determined by trypan blue exclusion. Breast fibroblasts did not survive the collagenase dissociation as well and viability varied between 65 and 75%. Only 20–30 g of tissue has been obtained from some reductive mammoplasty surgery and all immediate autopsy cases. The expected number of epithelial elements which could be isolated from a specimen of this size would preclude *in vitro* time kinetic, steroid receptor assay, casein synthesis, and transformation studies requiring multiple culture dishes and replicate samples. However, these types of studies might be possible if breast cells obtained from women of similar age and endocrine history could be pooled. Thus, we have examined the cryopreservation

---

FIG. 13. Electron micrograph demonstrating morphological characteristics of myoepithelial cells which include prominent myofilaments, dense plaques, and hemidesmosomes (see inset). Bar = 0.8 μm. Inset bar = 0.5 μm.

FIG. 14. Illustration of typical fine filaments present in epithelial cells joined by desmosome. Bar = 0.66 μm.

FIG. 15. Survey electron micrograph demonstrating secretory activity of human breast epithelial cells in culture. Secretory material is evident within the intracellular alveolus. Bar = 2.4 μm.

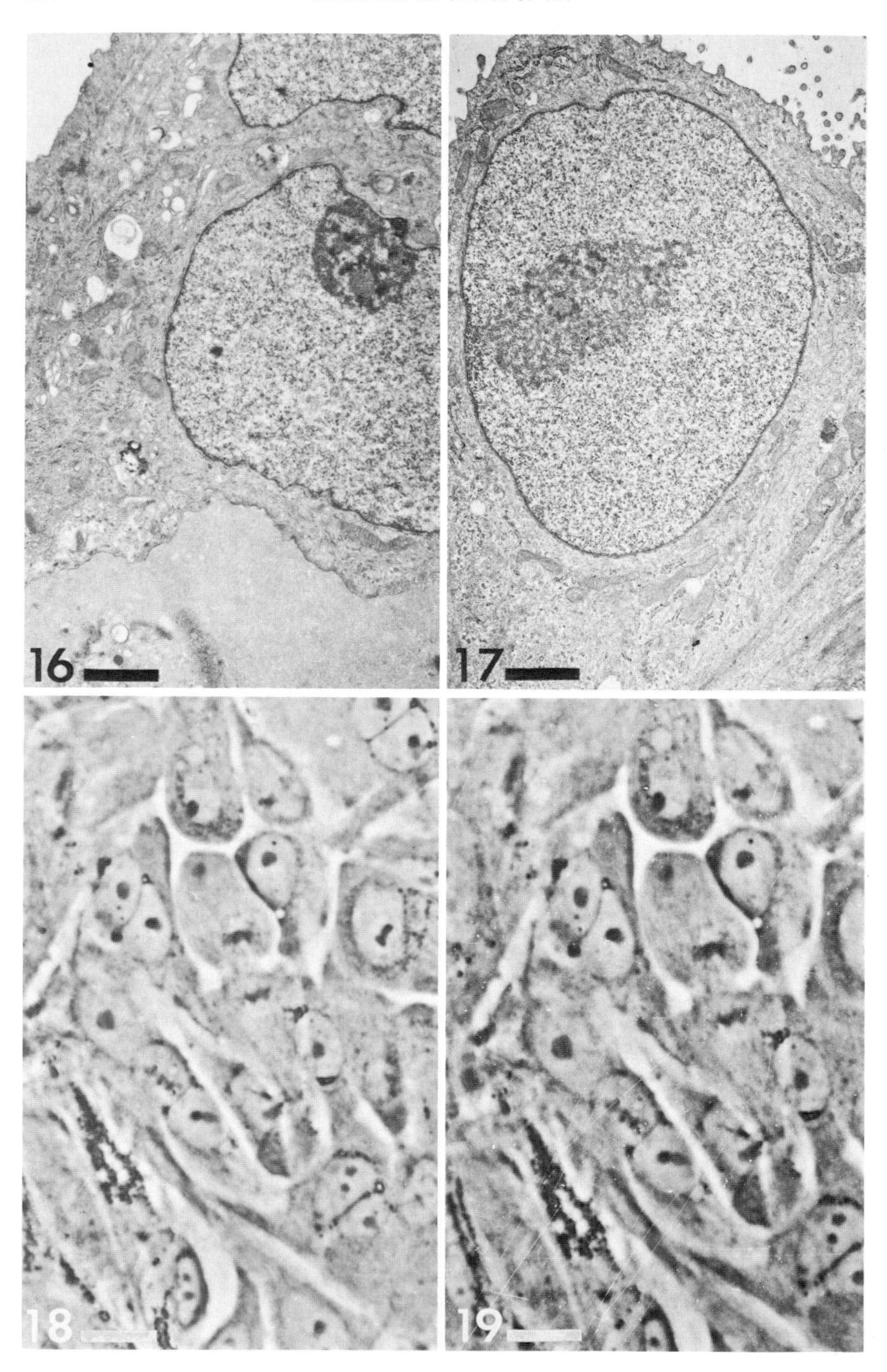
16
17
18
19

TABLE II

INFLUENCE OF HORMONES ON THE LABELING INDEX OF PRIMARY CELL CULTURES OF NORMAL HUMAN BREAST EPITHELIAL CELLS, HUMAN BREAST FIBROBLASTS, AND MOUSE DERMAL FIBROBLASTS

	Medium additions[b]				
Cell type[a]	NH	IA	IAP1	IAP1E	IAP1EPg
Breast epithelium	20 ± 2[c]	28 ± 2	98 ± 5	89 ± 3	72 ± 3
Breast fibroblasts	270 ± 24	292 ± 29	284 ± 21	301 ± 33	265 ± 35
Dermal fibroblasts	281 ± 32	300 ± 25	295 ± 30	289 ± 35	268 ± 35

[a] Cells were cultured for 5 days in various types of medium and then exposed to 1.0 μCi/ml [*methyl*-^{3}H]thymidine (6.7 Ci/mmole) for 1 hour.

[b] Additions to Medium 199 containing 5% charcoal-extracted fetal bovine serum and 50 μg/ml gentamicin: NH, no hormones; I, insulin 0.5 μg/ml; A, aldosterone 0.5 μg/ml; P1, prolactin 0.5 μg/ml; E, 17β-estradiol 0.0003 μg/ml; Pg, progesterone 0.003 μg/ml.

[c] Number of labeled nuclei per 1000 cells expressed as the mean ± SE. A minimum of four slides for each hormone variable was prepared for each cell type and at least 2000 cells per slide were counted independently by two individuals.

of human breast cells. When the total cell pellet isolated by collagenase dissociation is subjected to cryopreservation prior to ficoll gradient separation of cell types, the viability of cells is greatly reduced. Furthermore, the epithelial elements tend to adhere to one another and to form strands of cells throughout the ficoll gradient, rather than sedimenting as a distinct fraction. If the breast epithelial elements were subjected to cryopreservation after isolation by ficoll gradient centrifugation, viability ranged from 80 to 85% in elements stored in liquid N_2 for as long as 6–9 months.

IV. Discussion

The basis for the paucity of studies using cell cultures of normal human breast epithelium, can be traced to the technical problems associated with the dissocia-

FIG. 16. Electron micrograph of multilayer area of cellular outgrowth illustrative of secretary activity. Intercellular secretary material is evident at the lower portion of the cell. Bar = 1.5 μm.

FIG. 17. Electron microscopic demonstration of surface epithelial cell with numerous microilli, prominent active nucleolus, and abundant microfilaments. Bar = 1.66 μm.

FIG. 18. Example of mitotic cell frequently observed in human breast epithelial cell cultures grown in hormone-supplemented medium. Note cell in early to mid-telophase in center of field with densely packed chromosomes at the two poles. Cell in late telophase is evident in upper right portion of the field. Phase-contrast. Bar = 16.6 μm.

FIG. 19. Same field as in Fig. 18 seen at a later time. Cell eventually completed mitosis. Phase-contrast. Bar = 16.6 μm.

tion of breast material to yield sufficient numbers of viable epithelial cells for culture. A variety of methods involving enzymes, singly or in combination, as well as mechanical disruption which are used to dissociate tissues into individual cells (Waymouth, 1974), have been adapted for normal human breast. Although the prolonged exposure to strong proteolytic enzymes and the harsh mechanical disruption treatments dissociated human breast tissue into individual cells, epithelial cell yield, cell viability, and growth rate were severely compromised. These individual cells when cultured gave rise to slowly growing epithelial cell islands which were rapidly overgrown by fibroblasts. Ceriani *et al.* (1978) have shown that these methods destroyed cell-to-cell junctional complexes and caused extensive damage to the mammary cell surface which altered membrane structure and function. This type of damage would account for the reduction in cell viability and growth rate as reported. While the addition of hormones to the culture medium briefly maintained or restored the synthesis of some specific proteins and carbohydrates, these processes were not sustained (Larson, 1969).

In an attempt to reduce the degree of trauma to which human breast tissue is subjected prior to culturing, we have developed a simple and rapid tissue dissociation technique. This method reduces enzyme concentration, buffers enzymatic action, and minimizes the mechanical agitation required to dissociate human breast tissue. The dense connective tissue and adipose cells are effectively stripped from the islands of epithelium. While the connective tissue is dissociated into individual cells, the parenchyma remains as cellular aggregates which retain their organization as ductular and ductal elements surrounded by a layer of myoepithelial cells. Based upon the report of Banerjee *et al.* (1977), the use of a collagenase dissociation technique without the addition of hyaluronidase should result in the retention of the integrity of the basal lamina of the cellular aggregates. Cell surface damage and disruption of cell junctional complexes are reduced leading to a greater cell viability. Furthermore, the dissociation to parenchymal elements facilitates the separation of epithelial and fibroblast fractions. Ficoll gradient centrifugation parameters can be formulated to isolate the individual fibroblasts from the cellular aggregates of epithelial cells based upon differences in cell size and density (Pretlow and Boone, 1969; Pretlow, 1971). High-resolution light and electron microscopic evaluation of the isolated epithelial elements revealed that the parenchyma has survived the dissociation and separation procedures with no alteration in morphology or cellular organization.

Cellular aggregates attach and grow rapidly when cultured with medium supplemented with 5% charcoal-adsorbed FBS and physiological levels of hormones. While insulin and aldosterone singly or in combination maintains the DNA synthetic activity of human breast epithelial cells, prolactin increases the labeling index of these cells. Progesterone and 17β-estradiol when added in combination with the other hormones maintains this high level of DNA synthetic activity. Previous reports have indicated that the addition of insulin, aldosterone, prolactin, 17β-estradiol, or progesterone to the medium bathing human breast cell cultures is not required for maintenance of mitotic activity (Barker *et al.*, 1964; Wellings and Jentoft, 1972; Flaxman and Lasfargues, 1973; Flaxman, 1973). In some instances, DNA synthetic activity was suppressed by hormone addition. However, the hormone concentrations used in these previous studies were pharmacological and, when tested in our system, resulted in severe cellular toxicity and subsequent cell death. Human breast epithelium grew rapidly when supplemented with quantitatively and qualitatively physiological hormone levels (compare values in Table III) and could be maintained for long periods once confluency was reached. When grown and maintained with fixed levels of hormones, ultrastructural characteristics normally associated with secretory epithelial cells were induced (Tobon and Salazar, 1975).

Using this dissociation and isolation procedure, normal human breast tissue obtained from reductive mammoplasty and immediate autopsy can be rapidly and routinely processed to yield from the same specimen sufficient numbers of either fibroblasts or epithelial elements to establish multiple cultures of both cell types. Although not conclusive evidence, light and electron microscopic examination of the epithelial elements after ficoll isolation and of the cultures of these isolated epithelial elements failed to reveal contaminating fibroblasts. It is important to stress that using this procedure both myoepithelial and epithelial cells are isolated intact and are still joined by characteristic epithelial junctional complexes. When cultured, the epithelial cells maintain their normal polarity in that myoepithelial cells are attached to the culture dish with the epithelial cells growing on the luminal or uppermost surface in contact with the culture medium. This characteristic epithelial cell polarity may be the result of the retention of the basal

TABLE III

REPORTED HUMAN PLASMA CONCENTRATIONS OF SELECTED HORMONES COMPARED WITH *in Vitro* SUPPLEMENTS

	Concentration	
Hormone	Physiological (per ml plasma)	*In vitro* (per ml medium)
Insulin	0.5–5.0 ng[a]	5–50 ng
Glucocorticoid	20–60 ng	0.5–50 ng
Prolactin	8–27 ng	5–50 ng
Estradiol	20–350 ng	30 pg
Progesterone	0.3–22 ng	3–30 ng

[a] Values reported for normal women (Ross *et al.*, 1970; Mishell *et al.*, 1971; Abraham *et al.*, 1974; Baird and Fraser, 1974; England *et al.*, 1974; McNeilly and Chard, 1974; Williams, 1974; Sherman and Korenman, 1975; Reyes *et al.*, 1977).

lamina following collagenase dissociation and/or the synthesis of new structural components of the basal lamina during cellular outgrowth. Both the myoepithelial and epithelial cells are joined to adjacent cells by prominent desmosomes and the luminal surface of the epithelial cells are covered by slender microvilli projecting into the culture medium. Furthermore, the intracellular organization and morphology of these cells in culture as determined by both high-resolution light and electron microscopy is characteristic of human breast epithelium as seen *in vivo* (see Chapter 4 of this volume; Murad and van Haan, 1968; Tannenbaum *et al.*, 1969; Ozzello, 1971, 1974; Stirling and Chandler, 1976, 1977). It is important to note that cell cultures derived from normal human breast resemble those derived from virgin rat mammary gland both in morphology (Hollmann, 1974; Russo *et al.*, 1976; Janss *et al.*, 1980) and in hormonal responsiveness (Dilley, 1971; Hallowes *et al.*, 1977b; Malan-Shibley *et al.*, 1980). The rat mammary epithelial cell culture system may, thus, be used as a model for human breast to develop and define further conditions for optimal cell growth and maintenance that would otherwise require large numbers of normal human breast specimens. Culture parameters such as serum-free medium, pH maintenance with organic buffers, temperature requirements, and substrate (collagen versus plastic) specificity can be evaluated initially using rat mammary cells. In addition, techniques and procedures to examine normal cell growth and differentiation as well as carcinogenesis can be developed using the rat mammary epithelial cell system and then applied to normal human breast epithelium *in vitro*.

V. Perspectives

While it is important to devise methods for the routine isolation of large numbers of epithelial elements from normal human breast for the subsequent growth and maintenance of such cells *in vitro,* a critical characterization of the system must be undertaken before it can be used for studies of cell growth, differentiation, and carcinogenesis. It is imperative that not only morphological characteristics of cells in culture be documented as epithelial and similar to their *in vivo* counterparts but that they retain normal hormonal dependency and growth regulatory pathways. If these cell culture systems are to be used for studies of transformation and mechanisms of carcinogenesis, it is mandatory that the procedures used for tissue dissociation and cellular maintenance in culture have not altered the ability of the cells to recognize and respond to hormonal stimuli for growth and differentiation. Thus, cultured cells when transplanted into an appropriate host animal should give rise to outgrowths which are morphologically normal and which respond to the fluctuations in the hormonal milieu during pregnancy, lactation, postlactional regression, and old age involution as does the host's mammary parenchyma.

It is significant that human breast cells *in vitro* are dependent upon and proliferate in response to hormones which have been shown to be active *in vivo*. We have previously defined hormone combinations and levels which give optimal rat mammary epithelial cell growth in culture and which mimic serum fluctuations during the estrus cycle (Malan-Shibley *et al.*, 1980; DeAngelo *et al.*, 1979). The findings of these studies indicate that it will be necessary to investigate the influence of hormonal fluctuations on the proliferative and secretory phases of the menstrual cycle, and on the morphology and synthetic activity of normal human breast cells in culture.

It is well documented that mechanisms whereby hormones control cellular activity are initiated by the association of a hormone with its specific receptor protein. The synthesis of the receptor protein may be dependent upon another hormone(s), e.g., prolactin stimulates the synthesis of an estrogen receptor, and one of the consequences of estrogen interaction with its specific receptor is the induction of the synthesis of a specific cytoplasmic progesterone receptor. To ensure that hormone regulatory pathways are retained in rat mammary epithelial cell cultures, we have measured estrogen and glucocorticoid receptor levels *in vitro* and related receptor content to estradiol and aldosterone induction of DNA synthesis (Malan and Janss, 1978; Malan-Shibley *et al.*, 1979). Methods developed for these types of cell culture studies can now be used to demonstrate and quantify steroid receptors in human breast cells *in vitro* and to evaluate the retention of steroid hormone regulatory pathways. Another indication of the retention of hormonal pathways and the establishment of the epithelial nature of rat mammary cells in culture has been the recent demonstration of differentiation as measured by casein and α-lactalbumin synthesis (A. B. DeAngelo, unpublished observation). *In vitro* differentiation was found to be dependent upon the growth of cells on a collagen matrix and the cycling of hormones in the medium to mimic the *in vivo* changes associated with pregnancy and lactation. Such findings lend support to attempts to induce cellular differentiation of human breast epithelium *in vitro*.

The morphological characterization of our rat mammary epithelial cell culture system has demonstrated the presence of alveolar, ductal, and myoepithelial cells which respond to and require a defined set of hormones for growth as does the mammary parenchyma *in vivo*. However, it was imperative to characterize the system further to ensure that long-term growth in culture had not altered the capability of these cells to grow "normally" and differentiate when transplanted into an isologous host. Given the proper mesenchymal background, both freshly isolated and cultured cells would be expected to give rise to outgrowths with morphologically identifiable components concomitant with the host mammary parenchyma when transplanted into the "cleared" or epithelium-free mammary fat pad of recipient animals. Furthermore, these outgrowths should respond to hormonally mediated growth and differentiation cycles associated with pregnancy, lactation, postlactional regression, and old age involution.

To evaluate this requirement, the fourth and fifth abdominal mammary fat pads on both sides of 21-day-old female Lewis rats were cleared of mammary rudiments by a modification of the technique of DeOme *et al.* (1959). Animals were allowed to recover from surgery and received transplants of freshly isolated or cultured mammary epithelial cells when 50–65 days old. Epithelial elements for transplantation and for growth in culture prior to transplantation were obtained from 50-day-old virgin Lewis females as previously described (Janss *et al.*, 1980). Primary cell cultures were maintained for 5, 7, 14, or 20 days before removal from the culture vessel by treatment with trypsin–EDTA (0.05–0.02%). Freshly isolated mammary epithelial elements or cultured epithelial cells were washed several times with serum-free medium and, following the final centrifugation, the medium was carefully removed. From this cell pellet, a uniform number of cells could be routinely removed with a thin (1.5-mm-wide) curved spatula for transplantation. Cells (2.0×10^5 cultured epithelial cells or 1.0×10^5 freshly isolated epithelial elements) were placed into a surgically prepared pocket in the "cleared" mammary fat pad and the point of transplant marked with a fine silk suture. Mammary fat pads containing transplanted cells were removed from the host at weekly intervals over a period of 4–20 weeks after transplantation. The host's sixth gland with its intact mammary parenchyma was removed with the "cleared" fat pad. Glands were fixed, defatted, and stained with alum carmine. The stained intact mammary parenchyma and the transplant area were examined microscopically for the degree and type of mammary outgrowth.

Within 4 weeks after transplantation into the cleared fat pad, mammary epithelial cells which had been previously cultured for 14 days gave rise to normal parenchymal outgrowth (Fig. 20). This outgrowth resulting from the transplantation of individual cells was characterized by an organized branching ductal network with attached terminal end buds and alveoli. Figure 21 demonstrates that the surgical clearing technique effectively removed all parenchymal rudiments from the fat pad leaving a fine stromal background. When transplant outgrowth which had developed over an 8-week period (Fig. 22) was compared with the recipient's intact sixth mammary gland (Fig. 23), the morphological resemblance

---

FIG. 20. Transplantation site of cultured rat mammary epithelial cells showing parenchymal outgrowth after 4 weeks. Note branching ductal network with terminal end buds and alveolar-like structures. Stained with alum carmine. × 25.

FIG. 21. "Cleared" mammary fat pad area of 50-day-old Lewis female rat illustrating the effectiveness of the surgical procedure. Note absence of all parenchymal structures. Vascularization has been restored. Stained with alum carmine. × 25.

FIG. 22. Outgrowth from transplant of cultured rat mammary epithelial cells after 8 weeks of development in the "cleared" mammary fat pad. Compare the degree of development with that seen in the host's intact gland (Fig. 23). Stained with alum carmine. × 25.

FIG. 23. Intact mammary fat pad in the area of the sixth gland of a host animal. Note well-developed branching ductal network with terminal ductular units. Stained with alum carmine. × 25.

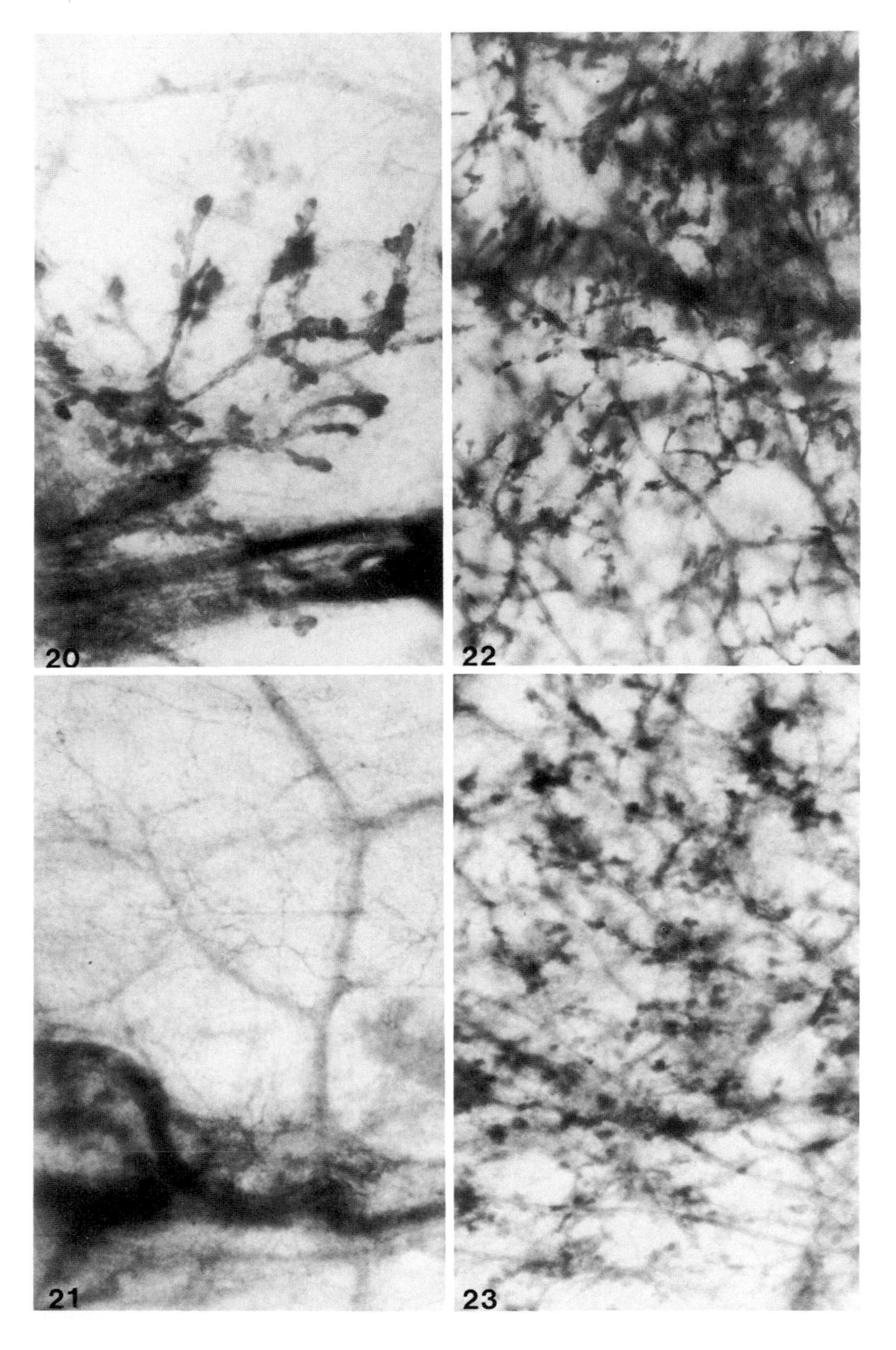
20
21
22
23

was impressive. Similar results were obtained when freshly isolated mammary epithelial elements were transplanted into the "cleared" fat pad. It is important to note that freshly isolated mammary epithelial elements obtained following ficoll gradient centrifugation consisted of cellular aggregates (hollow balls and tubes of cells) free of contaminating single cells (Fig. 6). However, after cells had been grown in culture (Fig. 7) and removed from the culture vessel, no cellular organization (hollow balls and tubes) was evident. Thus, outgrowth arising from cultured cells was the result of the cellular proliferation and structural differentiation of single cells in contrast to that arising from freshly isolated material which resulted from the proliferation of cells within an organized structure.

Transplant outgrowth never extended beyond the cleared fat pad boundaries and morphologically resembled the intact gland as the host aged. Since the transplant outgrowth underwent the same morphological changes of involution at the host gland as the animal aged, hormonal responsiveness was retained. Furthermore, if animals receiving epithelial transplants were mated, the transplant outgrowth responded to the hormonal fluctuations of pregnancy and lactation by developing an extensive lobuloalveolar morphology identical to the host's glands. Although no external openings developed for milk ejection, milk was grossly apparent in the transplant outgrowth during lactation. However, the involutionary changes caused by milk stasis resulted in a more rapid regression in the transplant outgrowth than in the host gland.

The results demonstrate that freshly isolated or cultured rat mammary epithelial cells can be successfully transplanted into the cleared mammary fat pad of an isologous host and give rise to outgrowths with typical mammary morphology. Furthermore, these cultured cells have retained their ability to respond to hormones for the morphology of their outgrowths corresponds to that of the host during periods of pregnancy, lactation, and old age involution. This system of culture-transplantation not only demonstrates the epithelial nature of the cells isolated and grown in culture but also illustrates their retention of hormonally mediated regulatory pathways for proliferation and differentiation. Furthermore, transplantation can now be used as an endpoint for *in vivo* transformation following the treatment of mammary epithelial cells with carcinogens.

Based upon these findings and the methodology developed, attempts can now be made to transplant freshly isolated or cultured normal human breast epithelial cells into the "cleared" mammary fat pad of the "nude" athymic mouse or the newly derived "nude" athymic rat. Such studies will demonstrate the epithelial nature and the retention of hormone responsiveness of human breast cells while in culture and will more critically demonstrate that the culturing of cells does not result in time in the loss of their ability to recognize and respond to signals for normal proliferation and differentiation (i.e., undergo spontaneous *in vitro* transformation). Should these experiments be successful, normal human breast epithelium can be studied in a more defined system and those factors which are thought to influence growth, differentiation, and carcinogenesis can be examined.

ACKNOWLEDGMENTS

The authors gratefully acknowledge the skillful technical assistance of Mrs. Elizabeth I. Hadaway, Mrs. Susan P. Kelley, and Mrs. M. Alfernia Dailey and thank Mrs. Florence H. Henninger for her typing of the manuscript. We are indebted to Drs. Curtis C. Harris and Benjamin F. Trump for their continued encouragement and interest in the work presented.

REFERENCES

Abraham, G. E., Maroulis, G. B., and Marshall, J. (1974). *Obstet. Gynecol.* **44,** 522-525.

Baird, D. T., and Fraser, I. S. (1974). *J. Clin. Endocrinol. Metabol.* **38,** 1009-1017.

Banerjee, S. D., Cohn, R. H., and Bernfield, M. R. (1977). *J. Cell Biol.* **73,** 445-463.

Barker, B. E., Fanger, H., and Farnes, P. (1964). *Exp. Cell Res.* **35,** 437-448.

Buehring, G. C. (1972). *J. Natl. Cancer Inst.* **49,** 1433-1434.

Buehring, G. C., and Williams, R. R. (1976). *Cancer Res.* **36,** 3742-3747.

Ceriani, R. L., Contesso, G. P., and Nataf, B. M. (1972). *Cancer Res.* **32,** 2190-2196.

Ceriani, R. L., Peterson, J. A., and Abraham, S. (1978). *In Vitro* **14,** 887-894.

DeAngelo, A. B., Stewart, M. L., and Janss, D. H. (1979). *In Vitro* **15,** 221-222 (Abstr.).

DeOme, K. B., Faulkin, L. J., Bern, H. A., and Blair, P. B. (1959). *Cancer Res.* **19,** 515-520.

Dilley, W. G. (1971). *Endocrinology* **88,** 515-517.

Elias, J. J. (1959). *Proc. Soc. Exp. Biol. Med.* **101,** 500-502.

England, P. K., Skinner, L. G., Cottrell, K. M., and Sellwood, R. A. (1974). *Br. J. Cancer* **29,** 462-469.

Esber, H. J., Payne, I. J., and Bogden, A. E. (1973). *J. Natl. Cancer Inst.* **50,** 559-562.

Flaxman, B. A. (1974). *J. Invest. Dermatol.* **63,** 48-57.

Flaxman, B. A., and Lasfargues, E. Y. (1973). *Proc. Soc. Exp. Biol. Med.* **143,** 371-374.

Foley, J. F., and Aftonomos, B. T. (1965). *J. Natl. Cancer Inst.* **34,** 217-229.

Gaffney, E. V., Polanowski, F. P., Blackburn, S. E., Lambiase, J. T., and Burke, R. E. (1976a). *Cell Diff.* **5,** 69-81.

Gaffney, E. V., Polanowski, F. P., Blackburn, S. E., and Lambiase, J. P. (1976b). *Cell Tissue Res.* **172,** 269-279.

Hallowes, R. C., Mills, R., Piggot, D., Shearer, M., Stoker, M. G. P., and Taylor-Papadimitriou, J. (1977a). *Clin. Oncol.* **3,** 81-90.

Hallowes, R. C., Rudland, P. S., Hawkins, R. A., Lewis, D. J., Bennett, D., and Durbin, H. (1977b). *Cancer Res.* **37,** 2492-2504.

Hollmann, K. H. (1974). *In* "Lactation" (B. L. Larson and V. R. Smith, eds.), Vol. I, pp. 3-95. Academic Press, New York.

Holquist, D. E., and Papanicolaou, G. N. (1956). *Ann. N.Y. Acad. Sci.* **63,** 1422-1435.

Janss, D. H., Mallan-Shibley, L. B., Ben, T. L., and Hillman, E. A. (1980). *Endocrinology* (in preparation).

Larson, B. L. (1969). *J. Dairy Sci.* **52,** 737-747.

Lasfargues, E. Y., and Moore, D. H. (1971). *In Vitro* **7,** 21-25.

Lasfargues, E. Y., and Ozzello, L. (1958). *J. Natl. Cancer Inst.* **21,** 1131.

Lockwood, D. H., Stockdale, F. E., and Topper, Y. J. (1967). *Science* **156,** 945–946.

Luft, J. H. (1961). *J. Biophys. Biochem. Cytol.* **9,** 409–414.

Lyons, M. J., and Moore, D. H. (1965). *J. Natl. Cancer Inst.* **35,** 549–565.

McDowell, E. M., and Trump, B. F. (1976). *Arch. Pathol. Lab. Med.* **100,** 405–414.

MacMahon, B., Cole, P., and Brown, J. (1973). *J. Natl. Cancer Inst.* **50,** 21–42.

McNeilly, A. S., and Chard, T. (1974). *Clin. Endocrinol.* **3,** 105–112.

Malan, L. B., and Janss, D. H. (1978). *In Vitro* **14,** 338 (Abstr.).

Malan-Shibley, L. B., DeAngelo, A. B., and Janss, D. H. (1979). *In Vitro* **15,** 222 (Abstr.).

Malan-Shibley, L. B., Janss, D. H., Kelley, S. P., and Dailey, M. A. (1980). *Endocrinology* (in preparation).

Mishell, D. R., Nakamura, R. M., Crosignani, P. G., Stone, S., Kharma, K., Nagata, Y., and Thorneycroft, I. H. (1971). *Am. J. Obstet. Gynecol.* **111,** 60–65.

Murad, T. M., and Van Haan, E. (1968). *Cancer* **21,** 1137–1149.

Orr, M. F., and McSwain, B. (1954). *Texas Rep. Biol. Med.* **12,** 916–920.

Orr, M. F., and McSwain, B. (1955). *Am. J. Pathol.* **31,** 125–141.

Ozzello, L. (1971). *In* "Pathology Annual" (S. C. Sommers, ed.), 6th Ed., pp. 1–59. Appleton, New York.

Ozzello, L. (1974). *J. Invest. Dermatol.* **63,** 19–26.

Pitelka, D. R., Kerkof, P. R., Gagne, H. J., Smith, S., and Abraham, S. (1969). *Exp. Cell Res.* **57,** 43–62.

Pretlow, T. G. (1971). *Anal. Biochem.* **41,** 248–255.

Pretlow, T. G., and Boone, C. W. (1969). *Exp. Mol. Pathol.* **11,** 139–152.

Pretlow, T. G., Dow, S. R., Murad, T. M., and Jones, J. (1974). *Am. J. Pathol.* **76,** 95–106.

Prop, F. J. A., and Wiepjes, G. J. (1973). *In* "Tissue Culture: Methods and Applications" (P. F. Kruse and M. K. Patterson, eds.), pp. 21–24. Academic Press, New York.

Reyes, F. I., Winter, J. S. D., and Faiman, C. (1977). *Am. J. Obstet. Gynecol.* **129,** 557.

Rivera, E. M., and Bern, H. A. (1961). *Endocrinology* **69,** 340–353.

Rogers, A. W. (1973). "Techniques of Autoradiography," pp. 309–310. Elsevier, Amsterdam.

Ross, G. T., Cargille, G. M., Lipsett, M. B., Rayford, P. L., Marshall, J. R., Strott, C. A., and Robard, D. (1970). *Rec. Prog. Horm. Res.* **26,** 1–62.

Russo, I. H., Ireland, M., Isenberg, W., and Russo, I. H. (1976). *In* "Proceedings, Electron Microscopy Society of America" (G. W. Bailey, ed.), Vol. XXXIV, pp. 146–147. Claitor's, Baton Rouge, Louisiana.

Sherman, B. M., and Korenman S. G. (1975). *J. Clin. Invest.* **55,** 669–706.

Stirling, J. W., and Chandler, J. A. (1976). *Virchows Arch. A. Pathol. Anat. Histol.* **372,** 205–226.

Stirling, J. W., and Chandler, J. A. (1977). *Virchows Arch. A. Pathol. Anat. Histol.* **373,** 119–132.

Stoker, M. G. P., Piggot, D., and Taylor-Papadimitriou, J. (1976). *Nature (London)* **264,** 764–767.

Tannenbaum, M., Weiss, M., and Marx, A. J. (1969). *Cancer* **23,** 958–978.

Taylor-Papadimitriou, J., Shearer, M., and Stoker, J. G. P. (1977). *Int. J. Cancer* **20,** 903–908.

Tobon, H., and Salazar, H. (1975). *J. Clin. Endocrinol. Metab.* **40,** 834–844.

Trump, B. F., Smuckler, E. A., and Benditt, E. P. (1961). *J. Ultrastruct. Res.* **5,** 343–348.

Venable, J. H., and Coggeshall, R. (1965). *J. Cell Biol.* **25,** 407–408.

Waymouth, C. (1974). *In Vitro* **10,** 97.

Wellings, S. R., and Jentoft, V. L. (1972). *J. Natl. Cancer Inst.* **49,** 329–338.

Whitescarver, J., Recher, L., Sykes, J. A., and Briggs, L. (1968). *Texas Rep. Biol. Med.* **26,** 613–628.

Williams, R. H. (1974). "Textbook of Endocrinology." Sanders, Philadelphia, Pennsylvania.

METHODS IN CELL BIOLOGY, VOLUME 21B

Chapter 6

Explant Culture: Pancreatic Islets

ARNE ANDERSSON AND CLAES HELLERSTRÖM

Department of Histology,
University of Uppsala,
Uppsala, Sweden

I. Introduction

Hormones produced by the islets of Langerhans are of crucial significance for the maintenance of glucose homeostasis. Disturbances of islet hormone biosynthesis or release will contribute to the development of either diabetes mellitus or hypoglycemic states. The mechanisms for the normal regulation of these cellular events are, however, poorly known, and there is particularly little information about the pathological changes. Since diabetes mellitus is a life-long disease of high incidence in the general population, and, in addition, leads to severe vascular complications, there is an urgent need for more research in this field.

Studies of the endocrine pancreas are hampered by its dispersion into over a million islets comprising only 1–2% of the entire gland. Furthermore, each islet is composed of several different cell types, each of which secretes a hormone of

ISBN 0-12-564140-0

specific biological significance. The clarification of the pathophysiology of this complicated organ requires the use of techniques which have been adapted for the study of the small cell aggregates represented by the islets. In this context the development of various methods for the isolation and tissue culture of islets has been of significance and several of these techniques have also been applied to culture of the human islets. In particular, the potential usefulness of the endocrine pancreas for transplantation purposes has stimulated attempts to store human islets in tissue culture over prolonged periods of time. It is the purpose of this chapter to describe the various methods for culture of the endocrine pancreas with special emphasis on those techniques which involve the use of isolated pancreatic islets.

II. Structure of the Human Endocrine Pancreas

Tissue culture of the endocrine pancreatic organ can be performed at various levels of organization of the pancreas. Thus, pieces of the intact gland can be explanted as well as isolated islets or dispersed pancreatic cells. Irrespective of which technique is employed the results reflect the histological organization of the explanted tissue. In this section we will summarize some structural features of the mammalian endocrine pancreas of special significance for the culture of this part of the gland. A comprehensive review of this subject has been published recently (Volk and Wellman, 1977).

For practical purposes the islets can be regarded as uniformly distributed within the acinar parenchyma. The total weight of the human islet organ has been estimated as 1–2 g and direct weighing of isolated human islets has shown an average dry weight of 0.8 μg per islet (Andersson *et al.*, 1976). However, this is not a true measure of the mean islet weight *in vivo* since the size distribution of the mammalian islets is not uniform, with a complete numerical predominance of small islets. However, the bulk of the total islet cell mass is made up of medium sized islets. In the isolation procedure only islets belonging to the larger size classes are obtained and this material therefore forms the basis for culture of isolated islets *in vitro*.

Although the precise cellular composition of the islet organ remains obscure, the application of immunocytochemical techniques has greatly helped to clarify this issue. At least four distinct types of islet cells have so far been well characterized both with regard to their morphology and their associated secretory products.

The predominant islet cell type is the B cell, which makes up about 60% of the human endocrine pancreas. The B cell had already been described at the beginning of this century and is by far the most well defined of all the islet cell types.

Its role as the source of insulin was recognized over four decades ago and today the knowledge of the mechanisms of insulin formation and release has greatly expanded. From a structural point of view the B cell is easily identified in the light microscope with the aid of classical histological stains or with immunocytochemical techniques (Fig. 1) and in the electron microscope by the characteristic appearance of the secretory granules (Fig. 2).

Among other islet cells the A_2 cell (or A cell) is the most well defined. This cell is the source of pancreatic glucagon, which has been suggested as contribut-

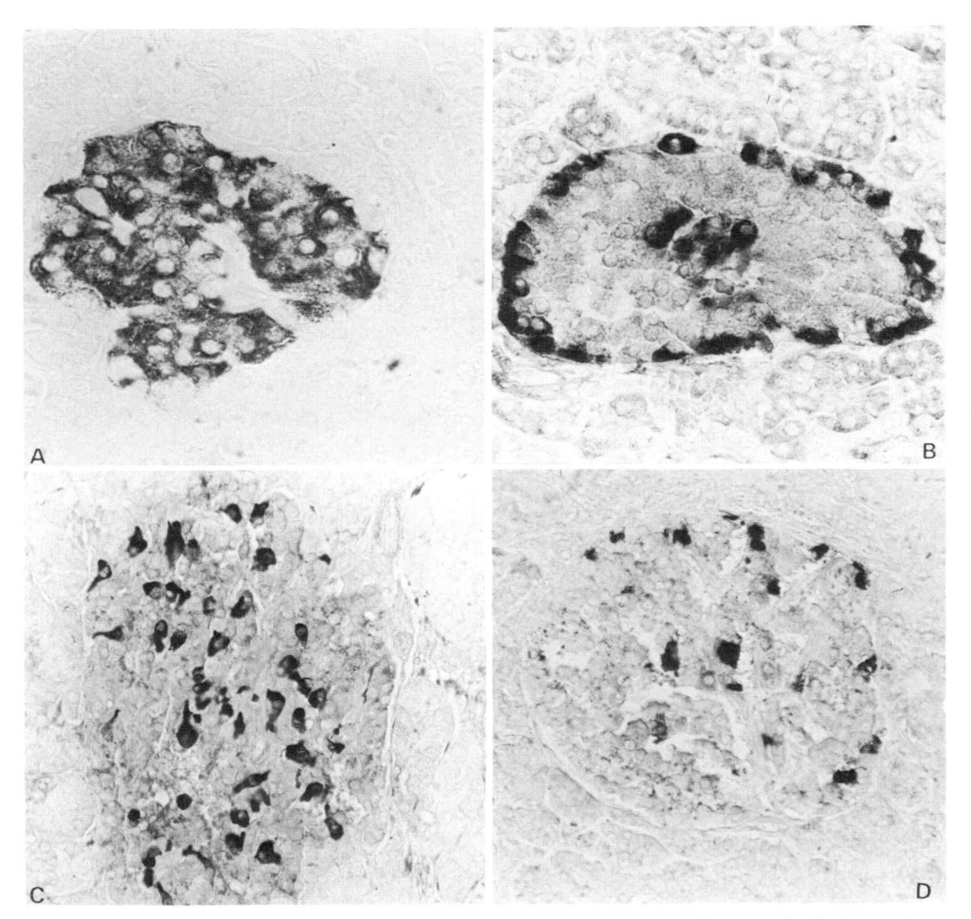

FIG. 1. Immunoperoxidase-stained sections of human pancreas showing the distribution and frequency of the various hormone-producing cells of the islets of Langerhans. Dark staining cells indicate (A) insulin-containing B cells, (B) glucagon-containing A_2 cells, (C) somatostatin-containing A_1 cells, and (D) pancreatic polypeptide containing PP cells. $\times$ 400 (A,B) and $\times$ 250 (C,D). (Stained pancreatic sections were generously placed at our disposal by Dr. L. Grimelius, Uppsala.)

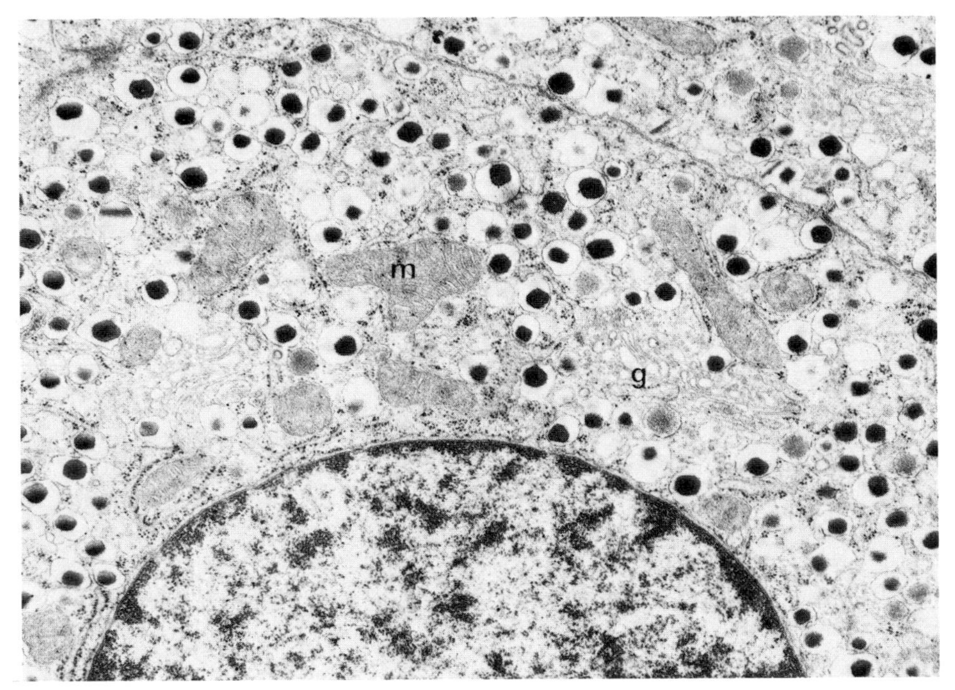

FIG. 2. Electron microscopic view of a mouse B cell in a collagenase-isolated pancreatic islet. The cell nucleus is seen in the lower part of the figure. Prominent features of the cytoplasm are the numerous secretory granules characterized by an electron-dense core surrounded by a wide membranous sac. The cytoplasm also shows sparse endoplasmic reticulum, mitochondria (m), and Golgi region (g). × 14.500. (Electron micrograph courtesy of Dr. L. A. H. Borg, Uppsala.)

ing to the development of diabetes (Unger, 1976). In some species the A_2 cell shows a positive tinctorial reaction for tryptophan, possibly reflecting the presence of this amino acid in the glucagon molecule. In the human islets these cells are typically located at the periphery and also along the islet capillaries (Fig. 1). Altogether the A_2 cells make up about 25% of the human endocrine pancreas.

The A_1 cell (or D cell) is the cell producing somatostatin or a compound with somatostatin-like immunoreactivity. While the A_1 cell usually comprises less than 10% of the islets of laboratory animals, they seem to be relatively more common in the human islets (Fig. 1). Although the precise physiological role of pancreatic somatostatin is so far obscure, it is of considerable interest that this compound is a powerful inhibitor of both insulin and glucagon secretion. This observation, together with the reported improvement of glucose homeostasis in juvenile diabetes during somatostatin infusion (Meissner *et al.*, 1975), suggests a possible glucoregulative function of somatostatin.

The fourth and most recently discovered of the islet cell types has been provisionally designated as the PP cell. As the name implies this cell is the source of pancreatic polypeptide, a protein composed of 36 amino acid residues in a straight chain. In the human islets the PP cells are relatively rare but they occur also in the acinar parenchyma (Fig. 1). Their biological significance is so far obscure but recent evidence for a marked proliferation in juvenile diabetes (Gepts *et al.*, 1977) makes them a quite interesting object of clinical studies.

Recent studies indicate an inverse relationship between the frequency of PP cells and A_2 cells in islets located in various regions of the pancreas in the rat. Thus Orci *et al.* (1977) demonstrated that islets in the duodenal part of the pancreas contain an abundance of PP cells but are virtually devoid of glucagon-producing A_2 cells. Conversely, islets in the splenic part are rich in A_2 cells and lacking in PP cells. Interestingly, the geographical distribution of these two islet varieties within the pancreas follows the vascularization of the gland in that PP-rich islets occur in the territory vascularized by the pancreatico-duodenal artery while glucagon-rich islets belong to areas supplied by the gastroduodenal and splenic arteries. These regional differences in distribution of the two cell types should be kept in mind when pancreatic cultures are employed for studies of the formation or release of glucagon or pancreatic polypeptide.

III. Review of Previous Work on Human Islet Cell Culture

The physical association between islet and acinar tissue and the complete predominance of the latter creates a practical obstacle to culture of the endocrine cells. In addition, proteolytic enzymes released from the exocrine cells may attack and degrade hormones secreted into the culture medium and also the cultured cells themselves. Therefore, there are few reports dealing with culture of the intact pancreatic gland of adult humans or experimental animals. Instead, pieces of fetal pancreas obtained before the beginning of enzyme production, isolated pancreatic islets, or suspensions of dispersed pancreatic cells have been used for culture purposes. A brief summary of previous work in this field will be given.

A. Development of Techniques for Islet Isolation

The introduction of methods for isolation of intact mammalian islets provided new possibilities for culture of the endocrine pancreas. The geometry and dimensions of the mammalian islets make these specimens an almost ideal preparation for tissue culture. The first practically useful technique for isolation of islets was reported in 1964 by Hellerström, who showed that islets from rat, mouse, and guinea pig could be microdissected free of acinar cells. It was also demonstrated that when the isolation procedure was carried out under rigorously controlled conditions the islets remained functionally and metabolically active and were thus suited for further studies in tissue culture (Andersson *et al.*, 1967). However, microdissection is time consuming and the technique was not suited to isolations of islets in excess of about 100 per experiment.

This situation changed when Moskalewski (1965) introduced a method for islet isolation by collagenase digestion of the pancreas. Subsequent modifications have now made it possible to obtain thousands of islets from laboratory animals in the same period of time previously required for collecting a hundred by microdissection. In his original publication, Moskalewski (1965) presented morphological evidence of survival and continued function of the isolated islets in tissue culture. These observations have been confirmed and extended in numerous subsequent studies and today isolated islets represent a standard preparation for the tissue culture of endocrine pancreas.

Unfortunately, the application of islet isolation procedures has met with difficulties in the case of the human pancreas. This is probably due to the more compact structure and the presence of substantial amounts of fibrous and fatty tissue within the acinar parenchyma. Useful techniques, based on the collagenase isolation procedure, have nevertheless been reported and will be described.

B. Culture of the Human Endocrine Pancreas

Current techniques for the culture of human islet cells all are based on methods which have been developed in experimental animal research. For a review of this topic the reader is referred to Andersson (1976), Lacy and Gingerich (1977), Chick *et al.* (1977), and Hellerström and Andersson (1979).

Culture of human fetal pancreas has mainly been carried out with pancreatic pieces or fragments as primary explants. Initially, these studies aimed at evaluating factors controlling differentiation of structure and function of the endocrine pancreatic cells. More recently culture of human fetal pancreas has also been applied as a means of storing the endocrine part for transplantation purposes (Agren *et al.*, 1978). Together the data so far indicate that the endocrine pancreas of human fetuses readily adapts to tissue culture conditions and maintains *in vitro* hormone biosynthesis and release for several weeks. Degeneration of acinar cells may lead to enrichment of endocrine cells and even isolated islets which may be used for extended physiological studies. These observations, together with evidence of new growth and functional differentiation during culture, make the human fetal pancreas an interesting source of tissue for clinical transplantation of B cells.

For the culture of human postnatal pancreas both pancreatic fragments, isolated islets, and dispersed pancreatic cells have been used. Difficulties in obtaining islets from the human pancreas initiated attempts to isolate such specimens by maintaining fragments of normal adult pancreas in culture with the hope that degeneration of acinar cells might lead to enrichment of surviving islets (Weber *et al.*, 1977). However, the evidence presented for islet purification by this approach remains inconclusive. In other investigations, aiming at long-term *in vitro* studies of human islet cells, small pancreatic fragments or dispersed pancreatic cells obtained from newborn babies with hypoglycemia due to nesidioblastosis were cultured over several months (Hollande *et al.*, 1976). In one of these cases insulin secretion was maintained for almost a year and cytological evidence that insulin containing B cells were present in the monolayer cultures was obtained as late as 250 days after explantation. The pancreas of newborns with nesidioblastosis therefore may be an interesting source of tissue in future attempts to establish long-term cultures of human islet cells. Adult pancreas has furthermore been utilized for monolayer cultures, which were infected with Coxsackie B3 virus (Yoon *et al.*, 1978). The demonstration of virus-containing B cells within 24 hours of infection suggests that human B cells are, indeed, susceptible to this virus strain.

The low yield of isolated islets from the adult human pancreas remains an obstacle to more extensive *in vitro* studies of islet structure and function. Nevertheless, there are data to indicate a satisfactory survival of human islets in tissue culture. Early studies by Lawson and Poutala (1974) demonstrated a continuous insulin release for several weeks after islet explantation. In more recent studies, Andersson *et al.* (1976) showed that islets isolated from the pancreas of brain-dead kidney donors maintained a significant insulin response to glucose during 1 week in culture. Exposure of islets to high glucose (16.7 m*M*) also inhibited the glucagon release whereas glucose plus theophylline acted as a stimulus. In addition, the cultured islets displayed both an active insulin biosynthesis and satisfactory ultrastructure for up to 3 weeks *in vitro*.

Altogether these data demonstrate that isolated human islets remain morphologically and functionally well preserved while maintained in tissue culture. Indeed, there are recent observations showing that with this technique human islets can survive for several months without significant decrease in the insulin response to glucose (Nielsen *et al.*, 1979). Therefore, the culture system should be suited not only for functional and morphological studies of islets *in vitro* but also for collection and storage of human islets intended for transplantation. During the culture period islets can be checked with regard to viability and microbial contamination, and, at the end of the period, they can easily be harvested. Recent observations suggest that cultured human islets also remain functionally active after transplantation (Lundgren *et al.*, 1977).

IV. Specific Requirements for Culture of Pancreatic Endocrine Cells

For reasons given above, the application of islet isolation procedures to the human pancreas has met with difficulties. Moreover, the limited availability of both adult and fetal human pancreatic material has made it necessary to perform most of the methodological experiments with explants obtained from different experimental animals.

Once isolated, the islets can be maintained in different culture systems. It is of great advantage if a technique permits the recovery of the islets at the end of the culture period by means of a nontraumatic harvesting procedure. Therefore, there are excellent possibilities for functional and structural studies of the cultured islet cells by means of methods originally developed for use with noncultured islets.

The culture method devised and employed in our laboratory involves the use of islets, which are isolated aseptically from normal mice with collagenase and then suspended in plastic petri dishes. In general, the culture medium has been medium 199 supplemented with 10–20% calf serum, penicillin (100 U/ml), streptomycin (0.1 mg/ml), and the glucose concentration has been adjusted according to the purpose of the experiment. In our initial experiments, we used petri dishes allowing cell attachment (NUNC, Roskilde, Denmark). The suspended islets gradually stuck to the bottom of the dishes and became attached by the outgrowth of a few fibroblastoid cells. At the termination of the culture period, usually after 1–3 weeks, the islets were detached either by a brief trypsinization or by careful scraping with a "rubber policeman."

The above-mentioned harvesting procedures as well as a tendency for monolayer formation of the islets after prolonged culture periods may introduce functional disturbances. We therefore examined whether the islets could be kept in free-floating culture, which was made possible either by (*a*) omitting the calf serum from the culture medium, or (*b*) culturing the islets in petri dishes which do not permit attachment of the cells.

The results of this study, the details of which have recently been published (Andersson, 1978), indicated that a substantial loss of islets occurred when the culture medium contained no serum. Some islets, presumably damaged by the collagenase digestion, also disappeared when cultured in the presence of serum. Moreover, culture in the absence of serum produced a marked diminution of glucose-stimulated insulin release (Fig. 3) and insulin biosynthesis. The latter observation correlated well with the reduced insulin content of these particular islets. It is noteworthy, however, that the presence of serum in the culture medium could partly be replaced by high concentrations of glucose.

Thus, it seems that maintaining the islets in a free-floating manner in the presence of serum provides the best conditions for islet culture in petri dishes. However, attention has also been focused on the composition of the culture

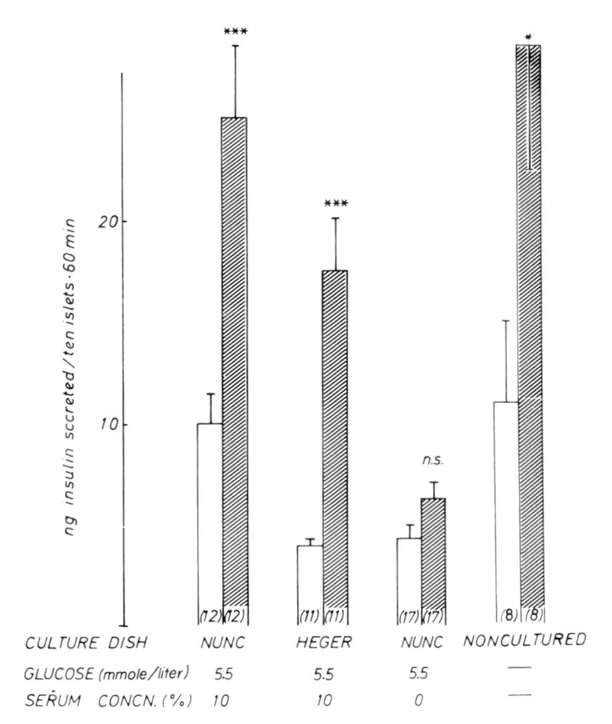

FIG. 3. Insulin release by cultured mouse islets. Islets isolated from NMRI mice were cultured for 1 week in medium 199 either free-floating in the presence of serum (HEGER dish) or in NUNC dishes permitting adherence, if serum was present in the culture medium. The insulin release from either noncultured or 1-week cultured islets was estimated in short-term incubations in 1.5 mM (open bars) or 15 mM glucose (hatched bars). Statistical significance of the difference between low and high concentrations of glucose: *, $p < 0.05$; ***, $p < 0.001$; n.s., nonsignificant.

medium. By the use of the free-floating technique, we tested the effects of different commercially available culture media on specific islet functions, during, and at the end of a 1-week-long culture period (Andersson, 1978). These studies indicated that medium RPMI-1640 was superior in preserving the glucose responsiveness of the B cells during the culture period (Fig. 4). Its ability to maintain and even enhance glucose-stimulated insulin secretion cannot, however, be ascribed only to its high content of glucose (11 mmole/liter) since adjustment of the glucose concentration of medium 199 to the same level did not enhance insulin secretion to the same degree. Another difference between media RPMI-1640 and medium 199 is the high vitamin content of RPMI-1640 and especially that of nicotinamide, the concentration of which is 40 times higher than in medium 199. Indeed, the addition of nicotinamide to medium 199, with 11 mmole/liter glucose, increased insulin secretion into the culture medium so that only small differences were observed between this particular medium and RPMI-1640. The observation that culture in the supplemented medium 199 gives

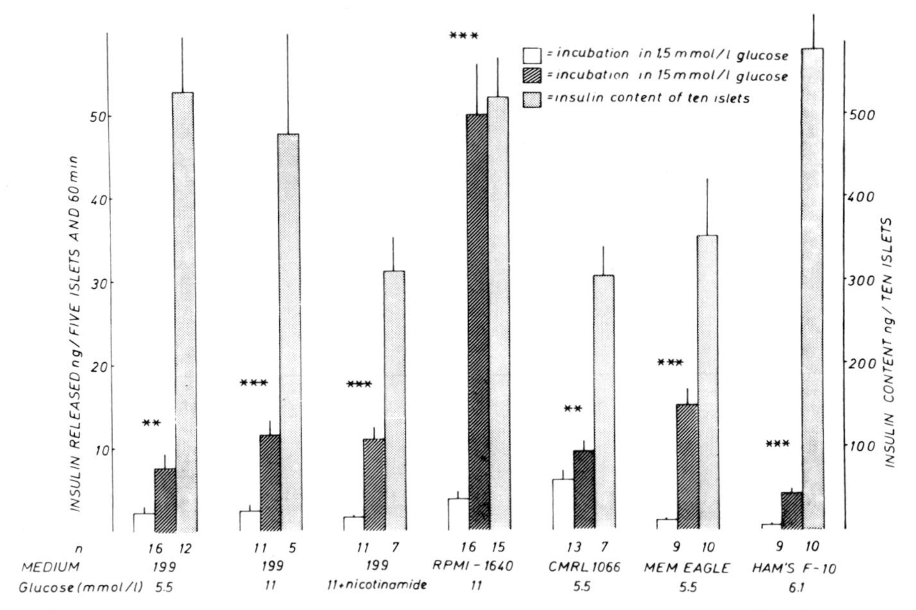

FIG. 4. Insulin release and content of cultured mouse islets. Islets were cultured free-floating for 1 week in different culture media supplemented with 10% (v/v) calf serum. The nicotinamide concentration was 8.2 μM. The insulin release was estimated in short-term incubations either in 1.5 mM (open bars) or 15 mM glucose (hatched bars). Solid bars indicate insulin content of the islets at the end of the culture period. From Andersson (1978), by permission of Springer-Verlag, Heidelberg. Statistical significance of the difference between low and high concentrations of glucose: **, $p < 0.01$; *** $p < 0.001$.

a lower insulin response to an acute glucose challenge as well as a lower islet insulin content (Fig. 4) suggests that factors other than the glucose and nicotinamide concentrations are responsible for the different suitabilities of these two media. It is of interest that the superiority of RPMI-1640 was not confined to its effects on the function of the islet B cells, since corresponding effects have also been observed on glucagon production of islets in culture. Thus, the glucagon accumulation in the culture medium was highest when RPMI-1640 was used (Fig. 5). It is worthy of note that, whereas the accumulation of glucagon generally tended to decrease with the length of the culture period, the opposite was found to be the case when the culture medium consisted of RPMI-1640.

In addition, the choice of the serum added to the medium might be of great importance. In all the studies presented we have used normal calf serum, which obviously is sufficient for maintenance of islet function *in vitro*. No comprehensive study has so far been performed to evaluate the effects of different sera on islet function. However, there is evidence to indicate that heat inactivation for 60 minutes at 56°C decreases the degradation of accumulated insulin in the culture medium (Kedinger *et al.*, 1977; Nielsen *et al.*, 1979). In spite of these reports, we found no effect of the heat-inactivated calf serum on glucose-stimulated insulin response of the cultured islets in the short-term experiments performed at the end of the culture period. Moreover, preliminary studies with culture of mouse islets in the presence of human sera have revealed that insulin degradation in the culture medium is decreased and there is some evidence suggesting a facilitated secretion of insulin, especially when sera from diabetic patients were used (Nielsen *et al.*, 1977).

Attention has also been focused on the effect of different pH values in the culture medium on the release of insulin (Brunstedt and Nielsen, 1978). Not surprisingly, the insulin release to the medium was independent of pH from 7.2 to 7.6 during the first weeks of culture. If, however, the culture period was extended for more than 2 months islets kept at pH 7.2 were found to be the best preserved, whereas culture at pH 6.8 or 7.6 was inconsistent with good survival.

In conclusion, several studies have demonstrated the importance of the composition of the culture medium for maintaining the specific functions of islet tissue in culture. It should be noted that the glucose and leucine concentrations in the medium markedly influence the structure and metabolism of islet cells in

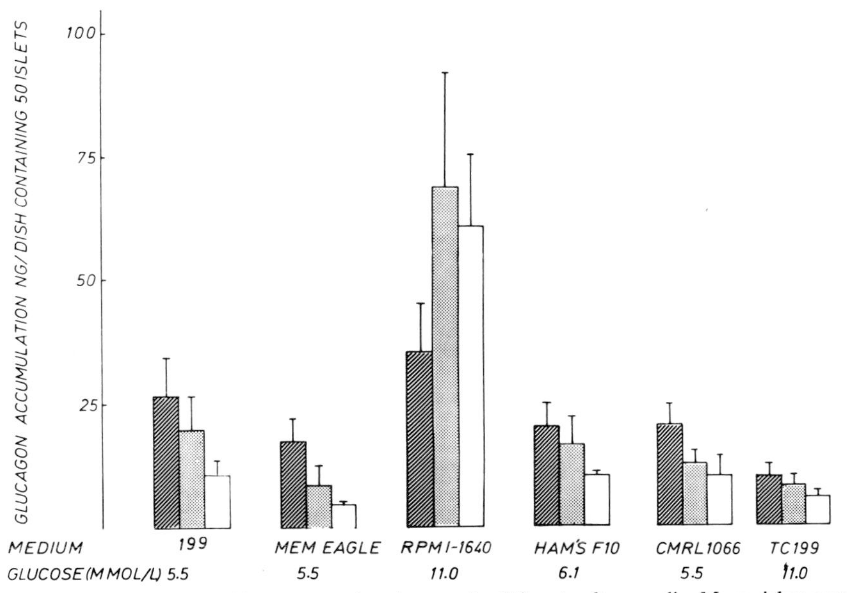

FIG. 5. Accumulation of immunoreactive glucagon in different culture media. Mouse islets were cultured free-floating in media supplemented with 10% (v/v) heat-inactivated calf serum and 100 μg/ml Aprotinin (Novo, Denmark). The medium was changed after 3 (hatched bars) or 5 (solid bars) days culture and a sample of the medium was also taken at the end of the culture period (seventh day—open bars).

culture (Andersson and Hellerström, 1972; Andersson *et al.*, 1974, 1977). For the future, it might be worthwhile to add different hormones to the culture medium, e.g., growth hormone or hydrocortisone, since preliminary studies suggest that both these hormones may affect the function of islets in culture.

V. Optimal Methodology

A. Viability Tests

Before describing the present culture techniques in more detail, some tests will be discussed that allow a structural and functional characterization of the explanted cells. The isolation procedure for islets from adult human pancreas is technically very difficult and fetal pancreatic explants contain a substantial amount of nonendocrine cells. Thus, it is important to evaluate the morphological integrity of the newly explanted cells by microscopic examinations of the donor pancreas. As regards adult pancreatic glands, which usually are obtained from brain dead kidney donors, there is a considerable difference in the ischemic time period between different specimens. The same is also true of the fetal material, for which intrauterine death may have occurred well before the delivery.

During the entire culture period the function of the explanted specimens can be examined by hormone assays of culture media samples. There are, however, many sources of error in the use of such methods. First of all, it is impossible to determine to what extent accumulated hormone in the medium has been released by an active process or has just leaked out from disintegrating islet cells. In addition, there may be a substantial degradation of the hormones in the culture medium (Andersson *et al.*, 1974). Therefore, it is suggested that tests on the viability of islet cells in culture are performed as short-term (1–3 hours) incubations according to the protocol given in Fig. 7.

After harvest, the cultured islets may be subjected to three basic tests of viability involving both structural, metabolic, and functional studies. For the metabolic experiments, we used the Cartesian diver technique to measure the respiratory rate of the cultured islets. After 1 week in culture, the rate of oxygen uptake by isolated human islets was significantly increased by addition of glucose to the diver medium (Andersson *et al.*, 1976). An alternative method is to examine the production of $^{14}CO_2$ from ^{14}C-labeled substrates, for example, [^{14}C]glucose. Such methods have been used extensively for islets isolated from both different experimental animals and man.

The most specific viability tests are those focused on hormone content, release, and biosynthesis. Such studies provide both quantitative data on the presence of endocrine cells in the explants and qualitative data on their ability to respond with an altered activity when exposed to different stimulants or inhibitors. There are several experimental systems used for these types of experiments and almost all of them are based on the availability of a sensitive radioimmunoassay. Probably, the most informative approach for evaluating the islet activity is represented by the perifusion systems, in which not only the total amount of secreted hormone can be measured but also the dynamics of the release (Andersson *et al.*, 1978).

B. Method for Culture of Adult Human Pancreatic Islets

In our studies, pancreatic glands have been obtained from brain-dead kidney donors, ranging in age from 8 to 65 years. After excision, the pancreas was transported to the laboratory in chilled (+ 5°C) Hanks' balanced salt solution (BSS) buffered with 25 m*M* HEPES and aseptic isolation of islets began 1–7 hours after circulatory arrest. The isolation procedure, essentially a modification of the conventional collagenase-digestion technique for isolation of pancreatic islets (Moskalewski, 1965), has recently been described in detail (Andersson *et al.*, 1976).

The pancreas was cut into pieces (2 × 2 cm) which were distended by injection of Hanks' BSS. Fragments were then minced, washed several times in Hanks' BSS to remove as much fat as possible, and subsequently incubated at +37°C with collagenase at a concentration of 10 mg/ml. During the approximate 30-minute digestion period, the tubes were shaken by hand and occasionally examined under the stereo microscope. When digestion was considered complete, i.e., when isolated islets could be recognized, the digest was washed three times in Hanks' BSS and the islets were collected under a stereo microscope using sterile braking pipettes. The recovered islets were placed in petri dishes containing 5 ml medium 199 supplemented with 20% calf serum, antibiotics, and 5.6 m*M* glucose and incubated in 5% CO_2 in air. Viable islets rapidly attached to the dish and the medium was changed on day 3 and then subsequently every second day. From day 3, the concentration of calf serum was reduced to 10%. After 7 or 21 days in culture, the islets were carefully scraped from the bottom of the petri dish with a thin rubber policeman and groups of islets were studied with regard to structural and functional integrity. The results of these studies have recently been published (Andersson *et al.*, 1976, 1978; Lundgren *et al.*, 1977).

All these culture experiments have been performed with tissue culture dishes and with serum supplementation of the medium, which has meant that the islets have stuck to the bottom of the dishes. Subsequent studies (Andersson, 1978), however, have indicated that these culture conditions are not optimal, and, therefore, a series of experiments has recently been undertaken in which medium RPMI-1640 was used and the islets were maintained in free-floating culture (Nielsen *et al.*, 1979). These studies revealed that adult human islets can be maintained in culture for several months with preservation of the ability to release insulin in response to glucose stimulation (Fig. 6).

C. Method for Culture of Human Fetal Pancreas

For this purpose, we have used the pancreas of human fetuses obtained at prostaglandin-induced legal abortions in weeks 12–23 of gestation. The pancreas was cut into small pieces (less than 1 mm^3), which were transferred to plastic petri dishes containing culture medium 199 supplemented with 20% calf serum. The explants were incubated as described for 6–13 days. The medium was changed after approximately 2 days of culture and then once daily. At the end of the culture period the pancreatic pieces were harvested mechanically for further examination of the structural and functional integrity of the cells.

The results suggest that there was a gradual disappearance of primitive acini, which were replaced by fibrous tissue. Simultaneously, the hormone concentrations (insulin, glucagon, somatostatin) of the explanted tissue increased between 15 and 35 times. Incubation of the cultured tissue with labeled amino acids suggested the formation of both insulin and glucagon. Moreover, short-term incubations performed with the tissue still present in the petri dishes indicated a

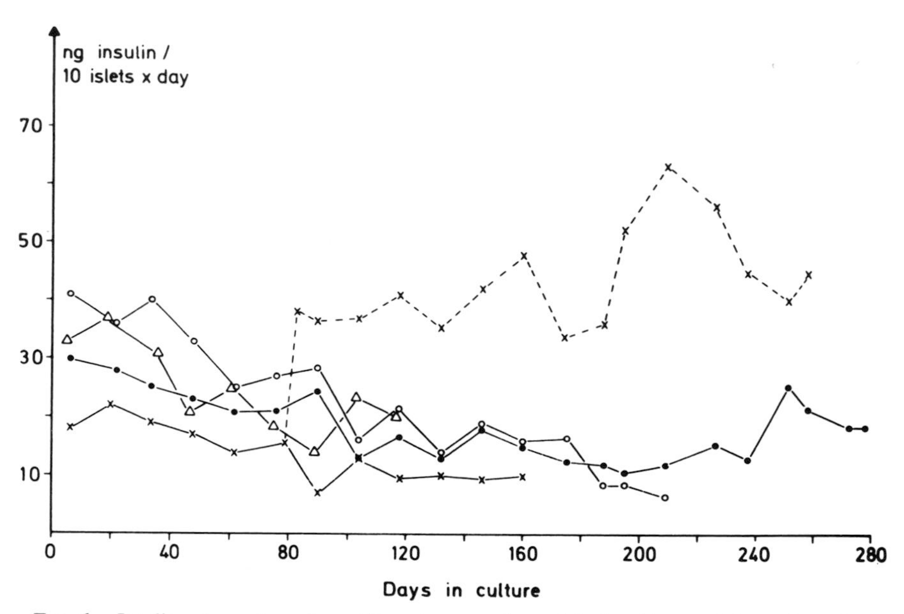

FIG. 6. Insulin release from four different preparations of isolated human islets. The culture medium was RPMI-1640 supplemented with 10% (v/v) newborn calf serum (solid lines) or 10% (v/v) normal human serum (broken line). From Nielsen *et al.* (1979) by permission of Springer-Verlag, Heidelberg.

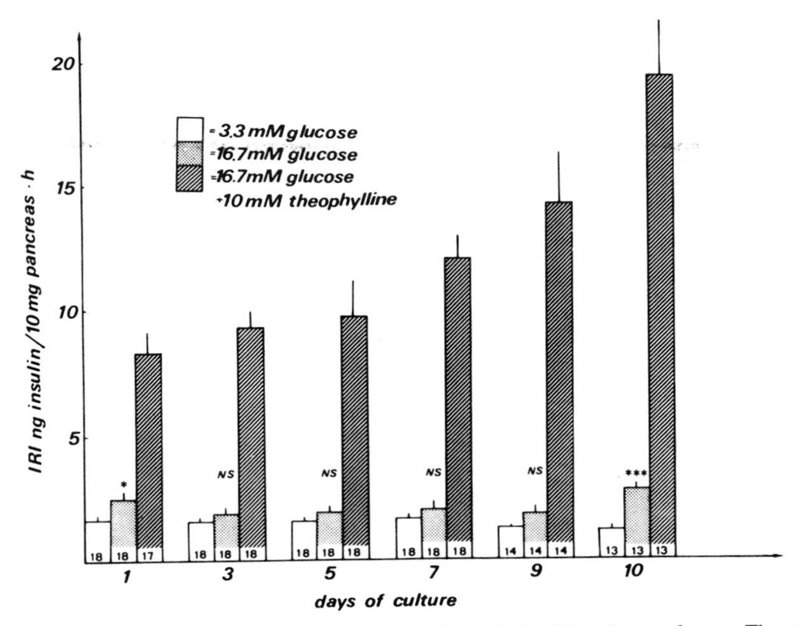

FIG. 7. Insulin secretion by cultured pancreatic pieces obtained from human fetuses. The culture medium consisted of medium 199 supplemented with 20% (v/v) calf serum. The glucose concentration of this medium was adjusted in the short-term incubation performed for estimations of insulin release. Statistical significance of the difference between low and high concentrations of glucose: *, $p<0.05$; ***, $p<0.001$; n.s., nonsignificant. (Courtesy of Dr. A. Agren.)

poor insulin response to glucose alone, whereas high glucose plus theophylline greatly stimulated the insulin release (Fig. 7).

VI. Perspectives

Although there is, to date, a rather limited knowledge of human islet culture, some trends for future applications have evolved. The general observation that islet hormone production is maintained in culture makes it likely that the system will be used for preservation of viable islets intended for transplantation. This also makes it possible to collect a large number of islets from several donors. Whether the antigenicity of the islets can be modified in culture remains to be elucidated. Furthermore, the use of cultured human islets as a research tool for more extensive studies of, for example, B-cytotropic viruses and for confirmation of previous results of animal experiments is a tempting possibility. In particular, the regulation of the human B cell replication should be a suitable topic for study. This would elucidate whether a deficient capacity for B cell regeneration could be of pathogenetic significance for the development of maturity onset diabetes. The production of human islet cell lines represents an additional avenue for research. Cloning of the human insulin gene would open fascinating perspectives both for research and production of human insulin.

Acknowledgments

The work of the authors included in this paper was supported by the Swedish Medical Research Council (16X-4527; 12X-109), the Swedish Diabetes Association, the Nordic Insulin Fund, and Stiftelsen Expressens Prenatalforskningsnämnd.

References

Agren, A., Andersson, A., Gunnarsson, R., Hellerström, C., Lindmark, G., Lundqvist, G., and Petersson, B. (1978). *Diabetologia* **14,** 213–214.

Andersson, A. (1976). *In* "Immunological Aspects of Diabetes Mellitus" (O. O. Andersen, T. Deckert and J. Nerup, eds.), pp. 283–294. *Acta Endocrinol.* (Kbh.) Suppl. 205.

Andersson, A. (1978). *Diabetologia* **14,** 397–404.

Andersson, A., and Hellerström, C. (1972). *Diabetes* **21** (Suppl. 2), 546–554.

Andersson, A., Hellerström, C., and Petersson, B. (1967). *Z. Zellforsch.* **82,** 110–117.

Andersson, A., Westman, J., and Hellerström, C. (1974). *Diabetologia* **10,** 743–753.

Andersson, A., Borg, H., Groth, C.-G., Gunnarsson, R., Hellerström, C., Lundgren, G., Westman, J., and Östman, J. (1976). *J. Clin. Invest.* **57,** 1295–1301.

Andersson, A., Nielsen, J. H., and Borg, H. (1977). *Diabetologia* **13,** 59–69.

Andersson, A., Groth, C.-G., Gunnarsson, R., Hellerström, C., Lundgren, G., Petersson, B., and Östman, J. (1978). *In* "Regulatory Mechanisms of Carbohydrate Metabolism" (V. Esmann, ed.), Vol. 42, Symp. A1, pp. 249–257. Pergamon, New York.

Brunstedt, J., and Nielsen, J. H. (1978). *Diabetologia* **15,** 181–185.

Chick, W. L., King, D. L., and Lauris, V. (1977). *In* "Pancreatic Beta Cell Culture" (E. von Wasielewski and W. L. Chick, eds.), pp. 85–91. Excerpta Medica, Amsterdam.

Gepts, W., De May, J., and Marichal-Pipeleers, M. (1977). *Diabetologia* **13,** 27–34.

Hellerström, C. (1964). *Acta Endocrinol.* (Kbh.) **35,** 122–132.

Hellerström, C., and Andersson, A. (1979). *In* "Carbohydrate Metabolism and Its Disorders" (D. F. Steiner, P. J. Randle and W. J. Whelan, eds.), Vol. III. Academic Press, New York (in press).

Hollande, E., Giron, B., Lehy, T., Accary, J. D., and Roze, D. (1976). *Gastroenterology* **71,** 255–262.

Kedinger, M., Moody, A. J., Launay, J. V., and Haffen, K. (1977). *Experientia* **33,** 972–973.

Lacy, P. E., and Gingerich, R. L. (1977). *In* "Pancreatic Beta Cell Culture" (E. von Wasielewski and W. L. Chick, eds.), pp. 37–45. Excerpta Medica, Amsterdam.

Lawson, R. K., and Poutala, S. D. (1974). *Surg. Forum* **25,** 377–379.

Lundgren, G., Andersson, A., Borg, H., Buschard, K., Groth, C.-G., Gunnarsson, R., Hellerström, C., Petersson, B., and Östman, J. (1977). *Transpl. Proc.* **9,** 237–240.

Meissner, C., Thum, C., Beischer, W., Winteler, G., Schröder, K. E., and Pfeiffer, E. F. (1975). *Diabetes* **24,** 988–996.

Moskalewski, S. (1965). *Gen. Comp. Endocrinol.* **5,** 342–353.

Nielsen, J. H., Eff, C., Deckert, T., Jensen, D., Brunstedt, K., and Andersson, A. (1977). *Diabetologia* **13,** 421.

Nielsen, J. H., Brunstedt, J., Andersson, A., and Frimodt-Møller, C. (1979). *Diabetologia* **16,** 97-100.

Orci, L., Amherdt, M., Baetens, D., Malaisse-Lagae, F., Ravazzola, M., and Rufener, C. (1977). *In* "Pancreatic Beta Cell Culture" (E. von Wasielewski and W. L. Chick, eds.), pp. 9-22. Excerpta Medica, Amsterdam.

Unger, R. H. (1976). *In* "The Year in Metabolism" (N. Freinkel, ed.), pp. 73-102. Plenum, New York.

Volk, B. W., and Wellman, K. F. (1977). "The Diabetic Pancreas." Plenum, New York.

Weber, C. J., Hardy, M. A., Lerner, R. L., and Reemtsma, K. (1977). *Surgery* **81,** 270-273.

Yoon, I. W., Onodera, T., Jensen, A. B., and Notkins, A. L. (1978). *Diabetes* **27,** 778-781.

Chapter 7

Primary Cultures of Human Prostate

L. M. FRANKS

Department of Cellular Pathology,
Imperial Cancer Research Fund,
London, England

I. Introduction

There have been many reports on the culture of human prostatic tissue since the first report by Burrows *et al.* (1917). Most are "one-off" jobs, i.e., the authors publish a single paper and then apparently do not continue the work. Many fail to give all the relevant details of techniques used, and it is often difficult to establish the exact numbers of specimens and cultures or even the precise source. The vast majority have not cultured normal tissues but have used surgical specimens from patients with benign prostatic hyperplasia or cancer. Some of those who have had access to normal material do not mention differences in their results when culturing normal or abnormal tissues but group all together. The exact source and nature of the material is often not given and few

ISBN 0-12-564140-0

characterize the cells obtained. Fortunately, there are some exceptions to the general rule and some general points which can be learned. First, normal human prostate in a viable state is very difficult to obtain. Second, a cellular outgrowth from prostatic tissue can be obtained using a wide range of culture conditions and medium, but there seems to be no immediately obvious differences between the outgrowth from normal, benign, hyperplastic, or tumor tissue. Last, there is no method for absolute identification of different cell types. These are problems which are not unique to the prostate.

In this chapter I shall assess the methods and media used and discuss some of the problems. Three main techniques are available: outgrowth culture, epithelial suspension culture, and organ culture. Since there appears to be little to be learned from a detailed review of all the published work, I shall concentrate on those using normal tissue or those with techniques of special interest.

II. Outgrowth Culture

A. Methods

I have found eight published reports and three preliminary communications in *Prostatic Cancer News Letter* on the culture of normal prostate. These are listed in Table I together with the methods used. All date from 1975 onward. Five specimens were from transplant donors (Webber, 1975, 1976), one aged 6 years (ages of the others are not given), and one was from a neonate with respiratory failure (Lechner *et al.*, 1978). These can be regarded as normal. The others, with the exception of a child dying with septicemia, either give no details or are from cancer patients. Seven methods have been used:

1. Standard explant culture.
2. Maitland cultures (Maitland and Maitland, 1928) and rocker cultures (Trump, 1977; Sanefuji *et al.*, 1978).
3. Suspension method of Stonington and Hemmingsen (1971).
4. Spillage method of Lasfargues and Ozzelo (1958) and Lechner *et al.* (1978).
5. Enzymatic dissociation (Kaighn and Babcock, 1975).
6. Thin layer steady-state method (McLimans, 1978a,b).
7. Microcarrier spinfilter suspension system (McLimans *et al.*, 1977).

Explant cultures have been used by many workers using abnormal tissues, but there are only brief reports on normal tissue from two groups in this series (Lechner *et al.*, 1978; McLimans, 1978a). The Maitland method, used by Geder

TABLE I

CULTURE OF HUMAN PROSTATE

Tissue	Source	Method	Reference
6-year-old	Transplant donor	Suspension	Webber (1975)
4 normal	Transplant donor	Suspension	Webber (1976)
Normal	Surgical (?cancer bladder)	Enzyme dissociation	Kaighn and Babcock (1975)
3-year-old	*H. influenza* Septicemia	Suspension	Rapp *et al.* (1975)
18-year-old	No details given	No details given	
Normal 30 cultures (? patients)	Unspecified	Maitland after collagenase	Geder *et al.* (1977)
Normal (numbers not given)	Autopsy	Rocker (? Maitland)	Trump (1977, 1978) Sanefuji *et al.* (1978)
Normal	No details given	Microcarrier spin filter suspension	McLimans *et al.* (1977)
Normal	Baboon		
9 Normal	Autopsy (cancer patients)	(i) Thin layer steady state	McLimans *et al.* (1977)
6 Normal	Surgical (cancer bladder)	(ii) Explant	
Normal	Baboon		
Neonatal	Respiratory failure	(i) Spillage (ii) Explant	Lechner *et al.* (1978)

et al. (1977), and the method used by Trump (1977) and Sanefugi *et al.* (1978) is basically a suspension system in which culture flasks are kept on a rocker. In short-term cultures the explants do not adhere to the containers, but in long-term cultures attachment takes place and outgrowths develop. In the suspension method (Stonington and Hemmingsen, 1971; Webber *et al.*, 1974), tissue fragments (1 mm^3) were explanted in medium in either microtest plates or plastic petri dishes, which were stirred (not shaken—J. Bond, personal communication) once daily to prevent attachment. The surfaces of the explants became covered with cuboidal epithelium and after about 7 days explants were transferred to plastic petri dishes and medium was added so that the explant was not completely submerged. Epithelial cells then grew out from the surface layer. The original paper gives 9–15 days as the optimal time to keep tissues in suspension. The spillage method was used by Lechner *et al.* (1978) and compared with explant cultures. The tissue was minced in cold medium, transferred in the medium to a conical centrifuge tube, and allowed to stand in an ice bath for 5 minutes. The supernatant containing small clusters of cells dislodged during the mincing was then explanted into plastic petri dishes with fresh growth medium. In the enzymatic dissociation method (Kaighn and Babcock, 1975) the minced tissue was washed three times in a 50-ml Erlenmeyer flask in Moscona saline with 1% chicken serum, then treated with CTC (0.1% collagenase, 0.1% trypsin, and 1% chicken serum in Moscona saline) and gently shaken in a waterbath at 37°C. After 20 minutes the supernatant was removed and fresh CTC was added. This process was carried out five times in all. The supernatants were pooled and centrifuged at low speed. The cell pellet was suspended in fresh medium and clonal cultures were established, using 6 ml of suspension in 100-mm plastic petri dishes. The thin layer steady-state method uses a somewhat complicated system described in detail by McLimans *et al.* (see McLimans, 1978b, for references). The cultures are kept in T 30 flasks held at an angle of 25° from the horizontal and rocked through 50° every 3 minutes so that a thin film of nutrient medium forms over the surface of the cultures at regular intervals. Since the cells are covered by a thin layer of medium, gas exchange is thought to be improved. There is also a perfusion system for the medium which maintains constant volume, pH, and gas atmosphere. The microcarrier spin filter suspension system was also established by McLimans *et al.* (1977). In this system a tissue mince or trypsinized cell suspensions are grown on gently stirred or stationary Sephadex beads in a fermentor flask with a medium replenishment system similar to that in the thin layer steady-state system. There are several modifications to the basic methods, e.g., coating of culture surface in any of the systems with collagen, variation in enzyme treatment in tissue preparation, selective trypsinization of mixed cultures to remove fibroblasts, etc.

B. Results of Explant Cultures

All explant methods seem to produce an outgrowth of cells, some epithelial, some mesenchymal. The mesenchymal cells, the so-called fibroblasts seen in culture, are almost certainly derived from vasoformative mesenchyme (Franks and Wilson, 1970; Franks and Cooper, 1972). Each group of workers seems to suggest that the method they describe gives a greater yield of epithelium than earlier systems, although there are few direct comparisons (see, for example, McLimans, 1978a) and the continuing series of reports on "new" different systems suggests that existing methods are not entirely satisfactory. The series described by Rapp *et al.* (1975) and Geder *et al.* (1977) seemed to give rise to mainly mesenchymal outgrowths. None of these authors describe any distinctive features of the epithelial cells derived from normal tissues, as compared to benign or tumor tissues. Webber *et al.* (1974) and Trump (1977) describe the reepithelialization of the surface of the explants by cells growing out of the cut edges of glands. Some of these cells secreted neutral and sialomucins (Trump, 1977). Ultrastructurally (Webber, 1975) the cells had some epithelial markers, including junctional complexes and basal lamina (Trump, 1977). The cells were initially columnar, but with time became cuboidal and sometimes multilayered and resembled squamous cells. Trump (1978) noted that the basal cells were more resistant to degenerative changes. Webber (1976) described cytopathic effects in the outgrowths in cultures from 4 normal, 26 benign, and 3 carcinoma cases. The effects included polykaryocytosis (multinucleated cells) first seen at 2–3 days, nuclear inclusions, and rounding up and clumping of cells. Ultrastructural changes said to resemble those induced by herpes virus, especially HSV 2, were also described. However, these changes are not generally accepted as being of viral origin. For example, some of the nuclear alterations could be regarded as nuclear bodies which are commonly found in cells *in vitro,* particularly if antibiotics are present in the medium (see Franks and Wilson, 1977; Bouteille *et al.*, 1974). Although virus-like particles have been described in human prostatic carcinoma and benign hyperplastic tissues (see for example Ohtsuki *et al.*, 1976), these differ in structure and distribution from those described by Webber. The Ohtsuki group failed to find virus-like particles in 15 different human prostate tissue cultures. Although Trump (1978) noted the presence of basal cells in explants, most groups, with the exception of Lechner *et al.* (1978), have described only one type of epithelial cell in the outgrowths. Lechner *et al.* (1978) described two distinct types of epithelial cells. One cell type, NP2-s, was composed of small cuboidal cells growing in a closely packed cobblestone pattern. Ultrastructurally they resembled prostatic epithelium. These cells were found only in "spillage" cultures. NP2-e cells were larger, well spread cells with ruffled membranes. These were found in explant cultures only but there was no ultrastructural confirmation of their epithelial nature. Both cell types were karyotypically normal and did not grow in agar.

The cells in the outgrowths in all culture systems survived for variable periods, up to 14 weeks or more, but even under apparently identical conditions some groups of cells survive for long periods while others, perhaps in the same dish, die. This is, of course, a common experience in cell culture, but it does suggest that there are differences in the cells not related to morphology. I shall discuss this later. It is difficult to compare cell survival times using the different methods of culture, but those cells which do survive seem to behave similarly in all types of culture. McLimans (1978a) compared steady-state with static cultures, and autopsy material with surgical material. Twenty-nine of forty-one (71%) steady-state cultures gave good growth as compared to 2 of 23 (9%) static cultures. Twenty-three of twenty-nine (79%) cultures of autopsy specimens gave good growth as compared with 6 of 12 (50%) cultures of surgical specimens. There was no difference in survival times of those cultures which grew 60–70 days.

C. Factors Influencing the Outgrowth of Cells: The Effects of Medium, Serum, and Hormones

The influence of the nonserum component of medium on cell growth is probably not vital since outgrowth occurred in cultures using many different media, although conditions may not have been optimal. Kaighn (1977) established optimal levels of cysteine and glutamine for his prostate cultures, in medium PFHR-1 (Lechner *et al.*, 1978). McLimans *et al.* (1977) used McCoy's medium 5-A, with up to 400 mg/100 ml of glucose, supplemented with hydrocortisone (1 μg/ml), insulin (5 μg/ml), and, for some experiments, testosterone (0.1 μg/ml) and zinc (0.1 mg/ml). He also recommends strict control of pH (optimum 7.35), p_2CO 5% (2% for baboons), osmolarity 3.10–3.15 mOsm, and medium feed rates of 1–10% culture medium per hour as determined by alterations in pH. McLimans (personal communication) also points out that other constituents not usually found in the cellular environment may be present in media. For example, he finds that ammonia may be present in concentrations much greater than that found in human serum in severe liver disease. The influence of factors such as these are rarely considered. Logically, it seems likely that detailed alterations and controls of this sort would produce more optimal conditions but, in the present state of knowledge of defined media, a more important role is played by the serum. McLimans *et al.* (1977) supplements his medium with 10% horse serum. Webber *et al.* (1974) report that 20% fetal bovine serum gave the best production of epithelium in the early stages of culture, however, after attachment, 10 to 20% of horse serum favored the outgrowth of epithelium. Kaighn (1977) showed that growth and cloning efficiency were critically dependent on particular batches of serum and that these parameters were influenced by a factor in dialyzed serum. They recommended 10–17% of fetal bovine serum from a selected batch. Of course, the problem of serum is common to all types of cell and tissue culture.

The effects of hormones on survival or normal prostate cultures has received little attention, although Rohl (1959) reported a response of some cells from hyperplastic and tumor tissues and there are many reports on the response of organ cultures. The most detailed study is that of Schroeder and Mackensen (1974) who extracted steroids from the serum they used and found that this, or the addition of androgens, did not influence the growth of cells from hyperplastic or tumor tissues. Insulin and hydrocortisone have been used by McLimans *et al.* (1977) but as part of a general medium rather than for its specific effects on the prostate [see Hodges (1976) for general discussion]. Interestingly enough, the use of medium containing insulin (Trowell's T8 medium, 1959) did reveal an unsuspected hyperplastic response of normal mouse prostate to insulin (Franks, 1959, 1961). This will be discussed later.

The gas phase in most culture systems has been either 5 or 10% CO_2 in air, but Trump (1977) used 5% CO_2, 45% O_2, and 55% N_2 in his cultures. The standard range of antibiotics was used in most systems but McLimans (1977) recommends gentamicin.

D. Characterization of Cells in the Outgrowth

The absolute identification of cells in tissue culture is a well-known problem which I have discussed in considerable detail (Franks and Wilson, 1977). First, we have to establish that the cells are epithelial and second that they are the organ-specific cells we are interested in. In culture, hexagonal cells with well-defined margins growing in closely packed sheets are usually described as epithelial, and spindle-shaped cells growing in bundles and forming a meshwork are described as fibroblastic. The early cell culturists were aware that the terms were inappropriate since a wide range of variation in appearances occurs depending on medium, substrate, cell number, etc. Light microscope morphology gives a general guide although even apparently typical epithelial cells, e.g., in cultures of colon, have been shown to be neural in origin (Franks *et al.,* 1978).

Ultrastructure provides some help in cell identification since there are a few specific epithelial markers, but even here there are problems (see Franks and Wilson, 1977, for full discussion). The adaptation of cells to growth *in vitro* requires a modification of cell function to allow survival under conditions that differ greatly from those *in vivo* and, in general, this involves a loss of organized structure to a greater or lesser degree, usually accompanied by an increase in cell mobility. The nutrient and gas exchange systems are also less closely regulated than *in vivo*. Cells that survive *in vitro* are therefore required to adopt certain common metabolic and functional patterns which differ from those found *in vivo*. These are mirrored by structural changes and, since the tissue culture conditions are usually standardized, the cells, whatever their origin, adopt a standard undifferentiated pattern. Thus, most tissue culture cells are similar in surface structure and mitochondrial pattern, and show an increase in pinocytosis and phagocytosis and an alteration in the distribution of intracellular filaments associated with attachment and cell movement. The majority of cells seen in sections from cultures have this undifferentiated pattern and, regardless of their origin, cannot easily be distinguished from one another. Because of the physical requirement that the cells grow in flat sheets, even differentiated characters that may be retained by some cells, e.g., specialized junctional complexes or secretory products, may show considerable modification. Since many of the specialized structures are limited to small areas of the cell surface, the chance of finding such specialized features is likely to be small unless there is deliberate selection. Therefore, in mixed cultures it is not possible to identify the majority of cells with any certainty. Finally, cells in culture are usually selected by their ability to proliferate *in vitro*. Consequently, not only are mitotic cells common, but cells in an active growth cycle are less likely to demonstrate differentiated characters.

A more unfortunate feature is that some appearances—particularly of cell contacts—may be misinterpreted. The spot desmosome is a characteristic of epithelial cells and its distinguishing features are the central lamella and the presence of 10-nm filaments. Intermediate junctions are present in many cells *in vitro,* epithelial and mesenchymal. These junctions never have a central lamella and are associated with 7.5-nm filaments but there are many reports in the literature which show intermediate junctions, described as desmosomes, used to identify the cells as epithelial.

Having identified the cells as epithelial it is necessary to demonstrate a prostate-specific marker. McLimans *et al.* (1977) list various enzymes, aminopeptidases, β-glucuronidase, acid phosphatase, 5α-reductase, zinc, citric acid, and sulfated sialomucins, none of which are absolutely specific for prostate. Many workers have used a histochemical method for acid phosphatase as a marker but Kaighn and Babcock (1975) have shown conclusively that many non-prostatic cell lines had abundant tartrate-inhibited acid phosphatase activity. This is not altogether surprising since it seems likely that prostate-specific secretory acid phosphatase is present as a cytoplasmic nonlysosomal enzyme which is likely to be lost using the standard methods of fixation.

Major functional markers would be the production of specific products, the presence of hormone receptors, and the response to hormones. There is no detailed information on any of these markers in normal human prostate cultures although work is in progress (e.g., Lechner *et al.,* 1978). An organ-specific immunological marker such as one we have described in the bladder (Nathrath *et al.,* 1979) or Ceriani *et al.* (1978) described in the breast might be worth looking for in the prostate. There may be a response to hormones but the results are not uniform. In normal mouse cultures we have shown (Franks and Barton, 1960) that cells in the main mass of an explant showed degenerative changes which could be prevented by the addition of testosterone; i.e., they were normally responsive, whereas cells in the outgrowth were grossly abnormal but showed no degenerative changes and did not secrete when treated with testosterone. The significance of this finding will be discussed later.

E. What Are the Outgrowth Cells?

There is little direct evidence from human prostate cultures, but studies on cultures of other mammalian tissues have answered this question. Although most parenchymal cells die *in vitro* within the first few days, some will survive for several months. However, these are not necessarily the original differentiated cells. The outgrowth cells described by Franks and Barton (1960) in mouse prostate cultures were presumably relatively undifferentiated stem cells. Nearly

60 years ago Champy (1920) described loss of differentiation in cultures of guinea pig prostate and repopulation by undifferentiated cells, as shown in the following (translated):

> Fragments of guinea-pig prostate (presumably inner lobe) coagulate seminal vesicle fluid, even after washing in saline or storage on ice for 4 days. After culture in plasma (from the same animal) the reaction becomes weaker with time. It is present for the first two days but has disappeared on the third and fourth days. Histological examination shows that the epithelium has survived and multiplied but the characteristic globular secretion is lost. Shortly after the culture is started the luminal portion of the cells breaks off and rapidly disappears. Champy suggests that this area is rich in enzymes and after loss of this preformed material the cells are unable to secrete more i.e. they have lost their physiological differentiation.

In the adult mouse salivary gland (Wigley and Franks, 1976) acinar cells are all degenerate within 24–48 hours and the cultures are repopulated from a stem cell population, in this case granular tubular cells. These cells retain specialized granular duct functions such as protease and amylase production for about 14 days, although the cells will survive for over 100 days. Similar changes take place in adult colon cultures (Defries and Franks, 1977) and skin (Rowden *et al.*, 1975). In the bladder the outgrowth from the basal stem cells can be clearly seen (Summerhayes and Franks, 1979). In this case the cells in the outgrowth resemble the partly differentiated intermediate cells in the bladder. They fail to differentiate completely and never form the characteristic asymmetric luminal bladder cell membrane. There seems to be little doubt that surviving epithelial cells in outgrowth cultures are derived from stem cells and are not fully differentiated. The basal cells described by Trump (1977) in culture of human prostate may be the stem cell source of surviving epithelial cells. These cells have been described in human (Mao and Angrist, 1966) and rodent prostate (Rowlatt and Franks, 1964), but their exact function is not known.

F. Conclusions—Outgrowth Culture

The conclusions to be drawn are that many epithelial cells can be maintained in primary culture for a variable period of time. In most cases, the differentiated cells present in the original explants die and the cultures are repopulated by cells derived from a stem cell population. These cells may retain some but not all their differentiated characters. The ultrastructure of the cells may resemble that of the organ of origin, although they may not respond to normal stimuli. Others may lose their normal ultrastructure but still retain a capacity to respond to stimuli, e.g., specific trophic hormones.

The failure of differentiated epithelial cells to survive *in vitro* may be due to cell damage during preparation of the cultures, to the inability of normal cells to adapt metabolically to *in vitro* conditions, to deficiencies in the medium, to a requirement for a stromal product, or to any combination of these factors. Differentiated tumor cells that can grow *in vitro* may be more resistant to damage, more capable of metabolic adaptation, or more able to synthesize essential nutrients. The problem of cell damage is well known (see, e.g., Franks *et al.,* 1970) and need not be discussed further.

Another explanation for failure to survive and proliferate may be loss of stroma. The importance of the stroma in embryonic growth and development has been recognized for many years (Grobstein, 1964). Little is known about the mutual interdependence of the stroma and epithelium in the adult, but there is some evidence to suggest that both epithelium and stroma are necessary for normal growth and function in the cornea (Herrman, 1960), breast (Lasfargues, 1957), and mouse prostate (Franks and Barton, 1960; Franks, 1963). Autoradiographic evidence suggests that RNA synthesis in the prostate can proceed in the absence of stroma, but that DNA synthesis cannot (Franks *et al.,* 1970).

A final point is that many adult epithelial cells may have a finite lifespan and may be able to go through only a small number of division cycles. The number of stem cells capable of continuous division in each culture may be very small. Although these conclusions are based on a wider study on the origin of epithelial cells in culutre (Franks and Wilson, 1977) they provide an accurate summary of the situation for the human prostate.

III. Epithelial Suspension Culture

One possible method for obtaining pure cultures of epithelial cells is to separate the cells from mesenchymal contaminants before culture. There are no reports on this technique using normal human tissue but there are three using benign hyperplastic or tumor tissue; one group, my own (Franks *et al.,* 1970), used a mechanical press and the others (Helms *et al.,* 1976; Pretlow *et al.,* 1977) used enzyme dissociation followed by separation of cells on a gradient.

The mechanical system produces a good yield of cells but, although they incorporated amino acids and uridine, they did not incorporate thymidine, nor did they grow in culture. Interestingly, the cells separated from the stroma at the basal cell layer, so that the basal cells, the presumed stem cells in the prostate, were mostly absent from the suspension.

The tissue for the preparation of cell suspensions was selected from epithelial nodules surgically removed from patients with benign prostatic hyperplasia. Cells can be separated from some tissues by squeezing the tissue between two glass slides held between finger and thumb. The method is simple but the yield is low and sterilization of some of the equipment used is inconvenient. A number of mechanical variants were therefore designed to avoid some of these drawbacks.

The basic principle in all models is that constant pressure without torsion is applied to the tissue, which lies between two smooth glass or stainless-steel plates. The diameter of the tissue should be about one-third of the diameter of the plate and its thickness about one-third of its greatest diameter, but the exact size is not critical. Fine radial incisions are made in the tissue to open up as many epithelial tubules as possible. The incized tissue is placed between the plates and immersed in balanced salt solution or a tissue culture medium in a container which should be wide enough to allow a good clearance between its walls and the upper plate. For metal presses 250-ml Pyrex centrifuge bottles or beer bottles with the tops cut off are used. A preliminary study of locally available brands is recommended to decide on the most suitable container. Two different models were used. In the metal press the upper plate moves on three guide rods which pass through holes in the outer edges. Pressure is applied by a screw, the lower end of which is unthreaded and rotates freely in a cavity in the top plate. The screw is retained in this plate by a removable split collar which fits loosely into a horizontal groove in the unthreaded portion of the screw. When only small pieces of tissue are available glass presses are more conveniently used. These are made by fusing an optically flat glass disc about 1.5 cm in diameter to the bottom of a piece of stout walled glass tubing of slightly larger diameter, which terminates in a standard B 23 ground glass female joint. The overall length is about 12 cm. The dimensions are not critical. A bulb is blown near the end of the tubing to act as a reservoir and above this a length of glass tubing about 0.5 cm in diameter, terminating in a standard B19 male joint is inserted. Fluid or cell suspensions are added or removed through this side arm. The upper plate is formed from stout walled glass tubing, the end of which is flattened and sealed. This moves freely up and down through a stopper of Teflon or solid nylon, but the fit should be tight enough so that the rod cannot twist easily. In use, the incised tissue is placed on the bottom plate, the stopper and glass rod are inserted, and the tissue is held lightly in place. Balanced salt solution or tissue culture medium is inserted through the side arm, up to the level of the bulb, and pressure is then applied by the rod.

The epithelial cells are torn off their basement membrane and emerge through the incisions in the piece of tissue. Once they emerge they are protected from further pressure from the plates by the residual tissue and if the container is wide enough there is much less chance of torsional damage to the cells by fluid streaming between the press and the container wall, as occurs in conventional tissue presses. The cells can be concentrated by sedimentation or centrifuging.

The techniques for density separation have been discussed by Pretlow *et al.* (1975). They recommend that tissues should be put into ice-cold culture medium containing 10% fetal bovine serum as soon as possible after removal. Surgically removed tissue gave a better yield of cells than did autopsy material. The prostates were minced in cold medium and digested in successive changes of enzyme solution. For hamster tissue 0.25% trypsin gave the best results, but for human tissue (benign hyperplasia and tumor) 0.1% Pronase was used. The fragments (human) were digested for 14 20-minute periods in successive changes of enzyme solution. The cells in the solutions from each of the 20-minute digestions were decanted and cooled in an ice bath for 5 minutes. The suspensions were then centrifuged at 97 *g* for 7.5 minutes to sediment the cells. The cells were resuspended in 5 volumes of medium containing 10% fetal bovine serum and kept in an ice bath until the tissue was exhaustively digested. The first four digestions were discarded because they contained many red blood cells and cells that did not exclude trypan blue. The suspensions of cells from the fifth through the fourteenth digestions were filtered through a single layer of Nitex (Tobler, Ernst and Traber, Inc., Elmsford, N.Y.) having a pore diameter of 100 μm. The cell suspension was then purified by velocity sedimentation in an isokinetic gradient (see Pretlow, 1971, for details). Over 90% of the cells excluded trypan blue and could be grown in short-term culture (further details not given).

The mechanical separation method will produce a good yield of differentiated epithelial cells but they are unlikely to grow in culture. The enzyme dissociation method is likely to produce a mixed population of cells, since the cells are separated by density only.

IV. Organ Culture

There appear to be no reports of organ culture of normal human prostate, although there are many using benign hyperplastic or tumor tissue. There are, of course, many reports on organ culture of rodent prostates, mainly based on the pioneer work of Lasnitzki (1974, for review). Most workers have used a modified Trowell technique (1959), in which tissues, on a piece of inert substrate (e.g., Millipore filter, rayon, or nylon mesh) are supported on a perforated metal stand so that the cultures are at the gas–liquid interface. There are modifications in which the substrate is on agar gel or collagen. Earlier workers used plasma clot cultures and others (e.g., McRae *et al.*, 1973) have used Leighton tubes or the steady-state system of McLimans (1978a,b). A range of media, sera, and gas phase similar to those used in cell cultures has also been used for organ culture. The general conclusions from organ cultures of human tissues is that abnormal tissues can be maintained relatively easily, even in androgen-free medium. In addition, these tissues will respond to steroids in the expected way (e.g., Lasnitzki *et al.*, 1975) but to a limited extent. The normal rodent prostate responds to steroids much more actively both morphologically and metabolically (e.g., Lasnitzki *et al.*, 1974), but a review of this work is outside the scope of this chapter. The relative lack of response in the abnormal human tissues probably reflects the intrinsically reduced sensitivity of the cells. It may also reflect alterations in the cells as a consequence of the culture conditions and preparation, since metaplastic changes, squamous or mucous, are common in the cultures (e.g., Franks, 1975: Trump, 1977). From work on the rodent prostate some of these changes may be a regenerative response to sublethal trauma (Franks, 1959) or due to deficiencies in the medium, e.g., vitamin A (Lasnitzki, 1955). Another point to be learned from rodent cultures is that a striking response to hormones not apparent *in vivo* may be found *in vitro*. Thus I found that there was a striking hyperplastic response of normal mouse prostate to insulin, a response which was inhibited by normal serum or by transplantation to a normal host, i.e., inhibition by the normal endocrine environment (Franks, 1959, 1961).

The preferred method for organ culture would seem to be the modified Trowell method. The general methodology is described in detail by Hodges (1976). This extensive review also considers alternative methods and discusses media and gas requirements. We find that the size of the explant is of less importance than the thickness, which ideally should be less than 0.1 cm. Whole mouse prostates can be cultured in this way as can slices of human prostatic tissue 1.5 cm^2 or more. The great virtue of using large pieces is that it reduces the areas at the cut edges where post-traumatic regenerative changes take place. These regenerating cells seem not to respond normally to physiological stimuli. We use a simple slicer (Figs. 1 and 2) made from a medical flat prescription bottle, with a 3 × 1 in. glass slide stuck to the bottom of its narrow edge. A piece of filter paper (3 × 1

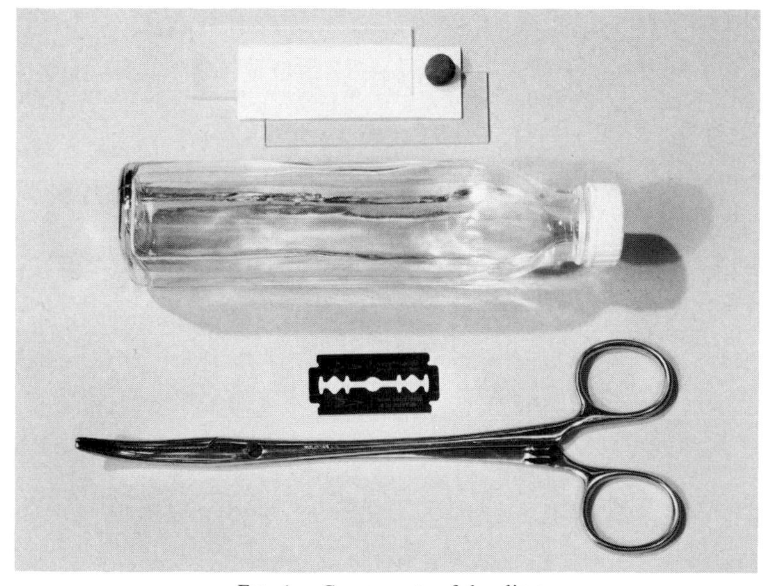

FIG. 1. Components of the slicer.

FIG. 2. The slicer in use. For some tissues it is more convenient to have the slide attached nearer the bottom of the bottle.

in.) is fixed to the slide with a drop of saline. The tissue to be sliced is placed on the end of the slide and a second frosted glass slide is placed on top and held firmly at the other end to maintain the tissue in position. The slice is cut using a razor blade clamped in a pair of artery forceps. If a cold surface is required, the bottle is partly filled with water and put in the freezing compartment of a refrigerator, after sterilization.

V. Conclusions

The problems involved in culture of normal human prostate are the same as those using any other tissue. The main lesson to be learned is that the method selected must depend on the question asked. Normal human prostate can be maintained in cell culture but it seems almost certain that the epithelial cells which survive are stem cells which are not fully differentiated and may not respond to normal stimuli. These cells may be valuable for studies on growth. The general problem of maintenance of differentiated cells *in vitro* is not unique to the normal human prostate but also applies to the mouse, rat, and guinea pig prostate, and to most other organs and species (see Wigley, 1975, for extensive discussion and references). For small scale studies almost any standard explant method seems to serve. The spillage method used by Lechner *et al.* (1978) seems to produce clones of cells but this introduces a major element of selection. For large scale work the systems devised by McLimans, steady state or microcarrier spin filter suspension, are the only methods available. Both seem to involve a considerable capital outlay and much expertise. You would really need to have a good reason—and funds—before embarking on these systems. Again, they are likely to produce stem cells.

For experiments using fully differentiated cells mechanical separation is a possible method but I have not used this on normal human prostate. These cells should be of use for immediate biochemical or physiological experiments but they will not grow since the method seems to select against stem cells. The alternative method is the enzyme separation method, which is likely to produce a mixed population of differentiated and stem cells and perhaps mesenchymal cells. The method has not been used extensively on normal tissues and it would be necessary to establish optimal conditions for survival of specific cell types.

The organ culture system, preferably using large, thin slices, and a modified Trowell technique, seems to provide the best method for physiological experiments, but there are basic problems involved in the assessment of results. I have discussed these in detail elsewhere (Franks, 1976).

There seems to be little doubt that normal prostate cells can be maintained in culture for only a very short time whatever culture technique is used. Until we can produce culture conditions which are as well defined and controlled as they are *in vitro* this is likely to remain so. McLimans (personal communication) provides a fitting final comment:

> Such (culture) conditions vary so much from laboratory to the other. Further, our understanding of the physiologic parameters required is so meager, the nature of the cell that should be evaluated so questionable, the required cell population level so ill-defined, that it becomes almost an impossible task.
>
> It seems to me that the greatest threat in interpretation, is the fact that much of the literature deals with a particular condition that was set up at the time the culture was initiated or a media change made, with no further effort made to control the devleoping transient states that occur prior to the next media change. At some point in the transient state curve optimal conditions were encoutnered—perhaps not at all at the present level of that parameter. Just what value experimental observations of this type are, I am not sure.
>
> To continue, there are the many observations that relate to what is considered optimal conditions for a particular *cell line*. It seems to me that these reports relate to a cell that has adapted to an artificial environment—artificial in terms of the physiology of the tissue of origin and its optimum conditions. As such, the determination of optimal conditions become somewhat an exercise in methodology or "bibliography" with limited critical application to man or his pathology. Further little account seems to be taken of cell density which may be the most important parameter of all.

References

Bouteille, M., Laval, M., and Dupuy-Coin, A. M. (1974). *In* "The Cell Nucleus: Nuclearstructure" (H. Busch, ed.), Vol. I, pp. 5–64. Academic Press, New York.

Burrows, M. T., Burns, J. E., and Suzuki, Y. (1917). *J. Urol.* **1,** 3–15.
Ceriani, R. L., Peterson, J. A., and Abraham, S. (1978). *J. Natl. Cancer Inst.* **61,** 747–751.
Champy, C. (1920). *C. R. Soc. Biol.* **83,** 842.
Defries, E. A., and Franks, L. M. (1977). *J. Natl. Cancer Inst.* **58,** 1323–1328.
Franks, L. M. (1959). *Br. J. Cancer* **13,** 59–68.
Franks, L. M. (1961). *Exp. Cell Res.* **22,** 56–72.
Franks, L. M. (1963). *Natl. Cancer Inst. Monogr.* **11,** 83–94.
Franks, L. M. (1975). *In* "Benign Prostatic Hyperplasia" (J. Grayhack, J. Wilson and M. Scherbenske, eds.), pp. 244–248, HEW, Washington, D.C.
Franks, L. M. (1976). *In* "Organ Culture and Biomedical Research" (M. Balls and M. Monnickendam, eds.), pp. 549–556. Cambridge Univ. Press, Cambridge.
Franks, L. M., and Barton, A. A. (1969). *Exp. Cell Res.* **19,** 35–50.
Franks, L. M., and Cooper, R. W. (1972). *Int. J. Cancer* **9,** 19–29.
Franks, L. M., and Wilson, P. (1970). *Eur. J. Cancer* **6,** 517–523.
Franks, L. M., and Wilson, P. (1977). *Int. Rev. Cytol.* **48,** 55–139.
Franks, L. M., Riddle, P. N., Carbonell, A. W., and Gey, G. O. (1970). *J. Pathol.* **100,** 113–119.
Franks, L. M., Hamilton, E., and Hemmings, V. J. (1978). *J. Pathol.* **124,** 19–22.
Geder, L., Sanford, E. J., Rohner, R. J., and Rapp, F. (1977). *Cancer Treat. Rep.* **61,** 139–146.
Grobstein, C. (1964). *Science* **143,** 643–650.
Helms, S. R., Pretlow, T. G., II, Bueschen, A. J., Lloyd, K. L., and Murad, T. K. (1976). *Cancer Res.* **36,** 481–486.
Herrman, H. (1960). *Science* 132–134.
Hodges, G. M. (1976). *In* "Organ Culture in Biomedical Research" (M. Balls and M. Monnickendam, eds.), pp. 15–59. Cambridge Univ. Press, Cambridge.
Kaighn, M. E. (1977). *Cancer Treat. Rep.* **61,** 147–151.
Kaighn, M. E., and Babcock, M. S. (1975). *Cancer Chemother. Rep.* **59,** 59–63.
Lasfargues, E. Y. (1957). *Exp. Cell Res.* **13,** 553.
Lasfargues, E. Y., and Ozzelo, L. (1958). *J. Natl. Cancer Inst.* **21,** 249–272.
Lasnitzki, I. (1955). *Br. J. Cancer* **9,** 434–441.
Lasnitzki, I. (1974). *In* "Structure and Function in Mammals" (D. Brandes, ed.), pp. 348–364. Academic Press, New York.
Lasnitzki, I., Whitaker, R. H., and Withycombe, J. F. R. (1975). *Br. J. Cancer* **32,** 168–178.
Lechner, J. F., Narayan, S. K., Ohnuki, Y., Babcock, M. S., Jones, L. W., and Kaighn, M. E. (1978). *J. Natl. Cancer Inst.* **60,** 794–801.
McLimans, W. F. (1978a). *Prosta. Cancer Newslett.* **5,** 3–4.
McLimans, W. F. (1978b). *Methods Enzymol.* **58,** 194–211.
McLimans, W. F., Kwasniewski, B., Robinson, F. O., Chu, T., Ming, Sufrin, G., and Gailani, S. (1977). *Cancer Treat. Rep.* **61,** 161–165.
McRae, C. U., Ghanadian, R., Fotherby, K., and Chisholm, G. D. (1973). *Br. J. Urol.* **45,** 156–162.
Maitland, H. B., and Maitland, M. C. (1928). *Lancet* **215,** 596–597.
Mao, P., and Angrist, A. (1966). *Lab. Invest.* **15,** 1768–1782.
Nathrath, W. B. J., Detheridge, F., and Franks, L. M. (1979). *J. Natl. Cancer Inst.* **63,** 1323–1330.
Ohtsuki, Y., Seman, G., Maruyama, K., Bowen, J. M., Johnson, D. E., and Dmochowski, L. (1976). *Cancer* **37,** 2295–2305.
Pretlow, T. G., II. (1971). *Anal. Biochem.* **41,** 248–255.
Pretlow, T. G., II, Weir, E. E., and Zettergren, J. G. (1975). *In* "International Review of Experimental Pathology" (G. W. Richter and M. A. Epstein, eds.), pp. 91–204. Academic Press, New York.
Pretlow, T. G., II, Brattain, M. G., and Kreisberg, J. I. (1977). *Cancer Treat. Rep.* **61,** 157–160.

Rapp, F., Geder, L., Murasko, D., Lausch, R., Ladda, R., Huang, E-S., and Webber, M. M. (1975). *J. Virol.* **16,** 982–990.

Rohl, L. (1959). *Acta Chir. Scand.* (Suppl.) 240.

Rowden, G., Lewis, M. G., Sheikh, K. M., and Summerlin, W. T. (1975). *J. Pathol.* **117,** 139–149.

Rowlatt, C., and Franks, L. M. (1964). *Nature (London)* **202,** 707–708.

Sanefuji, H., Heatfield, B. M., and Trump, B. F. (1978). *Tissue Culture Assoc. Manual* **4,** 855–856.

Schroeder, F. H., and Mackensen, S. J. (1974). *Invest Urol.* **12,** 176–181.

Stonington, O. G., and Hemmingsen, H. (1971). *J. Urol.* **106,** 393–400.

Summerhayes, I., and Franks, L. M. (1979). *J. Natl. Cancer Inst.* **62,** 1017–1023.

Trowell, O. A. (1959). *Exp. Cell Res.* **16,** 118–145.

Trump, B. F. (1977). *Prostat. Cancer Newslett.* **4,** 2–3.

Trump, B. F. (1978). *Prostat. Cancer Newslett.* **5,** 5–6.

Webber, M. M. (1975). *J. Ultrastruct. Res.* **50,** 89–102.

Webber, M. M. (1976). *Invest. Urol.* **13,** 259–270.

Webber, M. M., Stonington, O. G., and Poche, P. A. (1974). *In Vitro* **10,** 196–205.

Wigley, C. B. (1975). *Differentiation* **4,** 25–55.

Wigley, C. B., and Franks, L. M. (1976). *J. Cell Sci.* **20,** 149–165.

METHODS IN CELL BIOLOGY, VOLUME 21B

Chapter 8

Long-Term Explant Culture of Normal Human Prostate[1,2]

BARRY M. HEATFIELD, HAYATO SANEFUJI, AND BENJAMIN F. TRUMP

Department of Pathology,
University of Maryland School of Medicine,
Baltimore, Maryland

---

[1]This is contribution #743 from the Cellular Pathobiology Laboratory.
[2]This work was supported by NCI Grant, CA 15798.

ISBN 0-12-564140-0

I. Introduction

Despite inherent difficulties, the need for models of neoplasia based on human tissues has recently been emphasized (Harris *et al.*, 1978). Development of such models using human tissues is of special importance where animal models have proven unsatisfactory, as is presently the case for human prostatic adenocarcinoma (Franks, 1977; Handelsman, 1977; Merchant, 1977). Such investigation of human tissues must, of necessity, be conducted *in vitro*. To date, there have been few successful attempts to culture human prostate for prolonged periods, and no detailed baseline studies have been performed with explant cultures of this tissue. In previous work, human prostate was obtained surgically either from patients with benign prostatic hypertrophy (BPH) (Franks *et al.*, 1970; Schrodt and Foreman, 1971; Stonington and Hemmingsen, 1971; McMahon and Thomas, 1973; McRae *et al.*, 1973; Harbitz *et al.*, 1974; Lasnitzki *et al.*, 1975; Dilley and Birkhoff, 1977) or at the time of cystectomy (Noyes, 1975). BPH is a disease of clinical significance, which frequently occurs in men over the age of 50, and although cytologic, ultrastructural, and other features do not appear to differ significantly from "normal" prostate of young adults, nevertheless, BPH is not considered "normal" prostate (Franks, 1977). Similarly, prostate obtained at the time of cystectomy, usually from older patients, may already be hyperplastic without clinical symptoms. Moreover, a further consideration in the use of BPH as a model of normal prostate relates to the occasional presence of neoplastic foci in pathologic tissue samples, which renders such tissue of little value for studies on *in vitro* carcinogenesis.

Summarized in this chapter are recent detailed findings of baseline studies on morphologic and histochemical alterations during long-term explant culture (24 weeks) of normal human prostate obtained at immediate autopsy of young adults. These findings demonstrate survival, mitosis, migration, and differentiation of prostatic basal cells *in vitro* (Heatfield *et al.*, 1980; Sanefuji *et al.*, 1980).

II. Methods and Materials

A. Tissues

Normal prostate tissue was obtained from young adult males 16 to 31 years old during immediate autopsy following accidental death, usually as a consequence of severe head trauma (Trump *et al.*, 1975). Tissues were rapidly dissected using sterile technique, placed in plastic cups (Falcon, Oxnard, CA) containing ice-chilled Leibovitz medium L-15 [Grand Island Biological Co. (GIBCO), Grand Island, NY] supplemented with 300 U of penicillin (GIBCO), 300 μg of streptomycin (GIBCO), and 40 μg of gentamicin (Schering Pharmaceutical, Kenilworth, NJ) per ml, and transported in ice to the culture laboratory (Sanefuji *et al.*, 1978).

B. Culture Medium

Prior to culture, tissues were placed in a laminar flow hood (Bellco, Vineland, NJ), and sliced 1 mm in thickness with razor blades on a dental wax plate (Polysciences, Warrington, PA) previously wiped with 70% ethanol. Slices were then placed in L-15 medium in petri dishes (100 × 20 mm) using aseptic surgical technique. They were then trimmed into small 1 × 1 × 10-mm pieces with a scalpel, gently washed in three changes of L-15, and transferred to petri dishes (60 × 15 mm; 5–10 pieces per dish). Added to each dish were 3 ml of CMRL 1066 medium (GIBCO) supplemented with 0.1 μg/ml hydrocortisone hemisuccinate (Schwarz/Mann, Rockville, MD), 1 μg/ml bovine recrystallized insulin (Schwarz/Mann), 2 m*M* L-glutamine (GIBCO), 5% heat-inactivated fetal bovine serum (GIBCO), and the same concentrations of antibiotics present in L-15 described above.

C. Incubation

Culture dishes were placed in a controlled atmosphere culture chamber (Bellco) and gassed with a mixture consisting of 45% O_2, 50% N_2, and 5% CO_2 for 5 minutes. The chamber was then positioned on a rocker platform (Bellco) in an incubator (Forma Scientific, Marietta, OH) at 37°C, and rocked at 10 cycles/minute, permitting exposure of explant surfaces to the gas phase approximately 50% of the rocking cycle. The culture medium and atmosphere were replaced three times each week.

D. Sampling and Fixation

Explants were sampled at intervals during culture, including 0-time culture, up to a maximum of 24 weeks. For correlative light microscopy (LM), histochemistry, and transmission and scanning electron microscopy (TEM, SEM), explants were fixed in mixed aldehydes (4% formaldehyde/1% glutaraldehyde) in 200 mOsm phosphate buffer (McDowell and Trump, 1976).

E. Light Microscopy

For light microscopy (LM), fixed tissues were dehydrated through graded alcohols and embedded in paraffin employing conventional techniques. Histologic sections were cut 5 μm in thickness and routinely stained with hematoxylin and eosin (H&E).

F. Autoradiography

For detection of mitotic cells, explants were pulse-labeled for 6 hours with [^{3}H]thymidine (specific activity, 6.7 Ci/m*M*, New England Nuclear, Boston, MA) at an activity of 25 μCi/ml after culture for 2 days. Some explants were then fixed (0-time incubation); others were washed in fresh unlabeled medium, cultured for an additional period of 3 (5 days culture) or 7 (9 days culture) days postincubation, and then fixed to follow the fate of labeled cells. Fixed tissues were embedded in paraffin as above. Histologic sections 3–4 μm in thickness were deparaffinized, placed in two changes of absolute ethanol, air dried, and coated under safelight conditions with NTB-2 nuclear track emulsion (Eastman Kodak, Rochester, NY) diluted 1:2 with distilled water. Coated slides were then dessicated, exposed at 4°C for 1 or more weeks in the dark, developed with Dektol (Kodak), and fixed with acid fixer (Kodak). The sections were then stained with H&E.

G. Histochemistry

For the study of neutral and acidic mucosubstances, serial histologic sections were stained with periodic acid–Schiff (PAS) with and without diastase digestion, alcian blue (pH 2.5)/PAS sequence (AB/PAS), aldehyde fuchsin (pH 1.0)/alcian blue (pH 2.5) sequence (AF/AB), and mucicarmine. To block free aldehyde groups contributed by the glutaraldehyde component of the fixative, which otherwise would react with the Schiff reagent giving a false positive, tissue sections were treated with dimedone (Pearse, 1968).

H. Electron Microscopy

For transmission electron microscopy (TEM), fixed tissues were washed three times in 0.2 *M* sodium cacodylate buffer containing 7% sucrose, then postfixed in 1% OsO_4. Washed tissues were stained with 1% uranyl acetate *en bloc*, dehydrated through a series of graded ethanols and propylene oxide, and embedded in Epon 812. For light microscopy and selection of suitable areas for ultrathin sectioning, semithin sections 1 μm in thickness were stained with toluidine blue (Trump *et al.*, 1961). Ultrathin sections were stained with uranyl magnesium acetate and lead citrate, and examined in a JEOL 100B transmission electron microscope.

For scanning electron microscopy (SEM), fixed tissues were washed three times in 0.2 *M* cacodylate buffer, postfixed in 2% OsO_4 for 1 to 2 hours, washed 5–10 times in 0.2 *M* cacodylate buffer, then transferred to saturated thiocarbohydrazide (TCH) (Kelley *et al.*, 1973) in distilled water for 20 minutes at room temperature. TCH-treated specimens were washed 5–10 times in distilled water, placed in OsO_4 for 30 minutes to 1 hour, dehydrated through graded ethanols, and critical point dried (Model E 3000, Polysciences) with liquid CO_2. Dried specimens were mounted on aluminum stubs (EJ Fjeld, Maynard, MA) with silver conductive paint (ACME Chemicals and Insulation, New Haven, CN), sputter-coated (Technics, Alexandria, VA) with gold-palladium for 5 minutes, and examined with an AMR-1000 scanning electron microscope.

III. Results

A. Tissue Fixation

Marked shrinkage of normal human prostate occurred during fixation. This was manifested by bulging and distortion of the normal glandular architecture at cut surfaces as seen by SEM at 0-time culture. However, if initial fixation for several hours was followed by additional trimming to expose glands within tissue pieces, then further fixation of retrimmed pieces, preparations with a remarkable resemblance to paraffin sections, but seen in three-dimensional view by SEM, could be obtained. At subsequent sampling intervals, such retrimming also facilitated study of changes in glandular epithelium within explants.

B. Zerotime Culture

LM of paraffin or Epon-embedded tissue revealed numerous tubuloalveolar glands lined by epithelium and surrounded by dense fibromuscular stroma (Fig. 1). Sometimes present within the gland lumens were small, eosinophilic corpora amylacea with a lamellar structure. Two types of epithelial cells were present lining acinar spaces.

1. Secretory Cells

By LM, this cell type was columnar to cuboidal in profile (Fig. 1). In the basal portion of secretory cells was a large rounded nucleus and a single nucleolus. The apical portion of these cells appeared foamy. By histochemistry, small diastase resistant, PAS-positive granules of variable size were seen within the cytoplasm

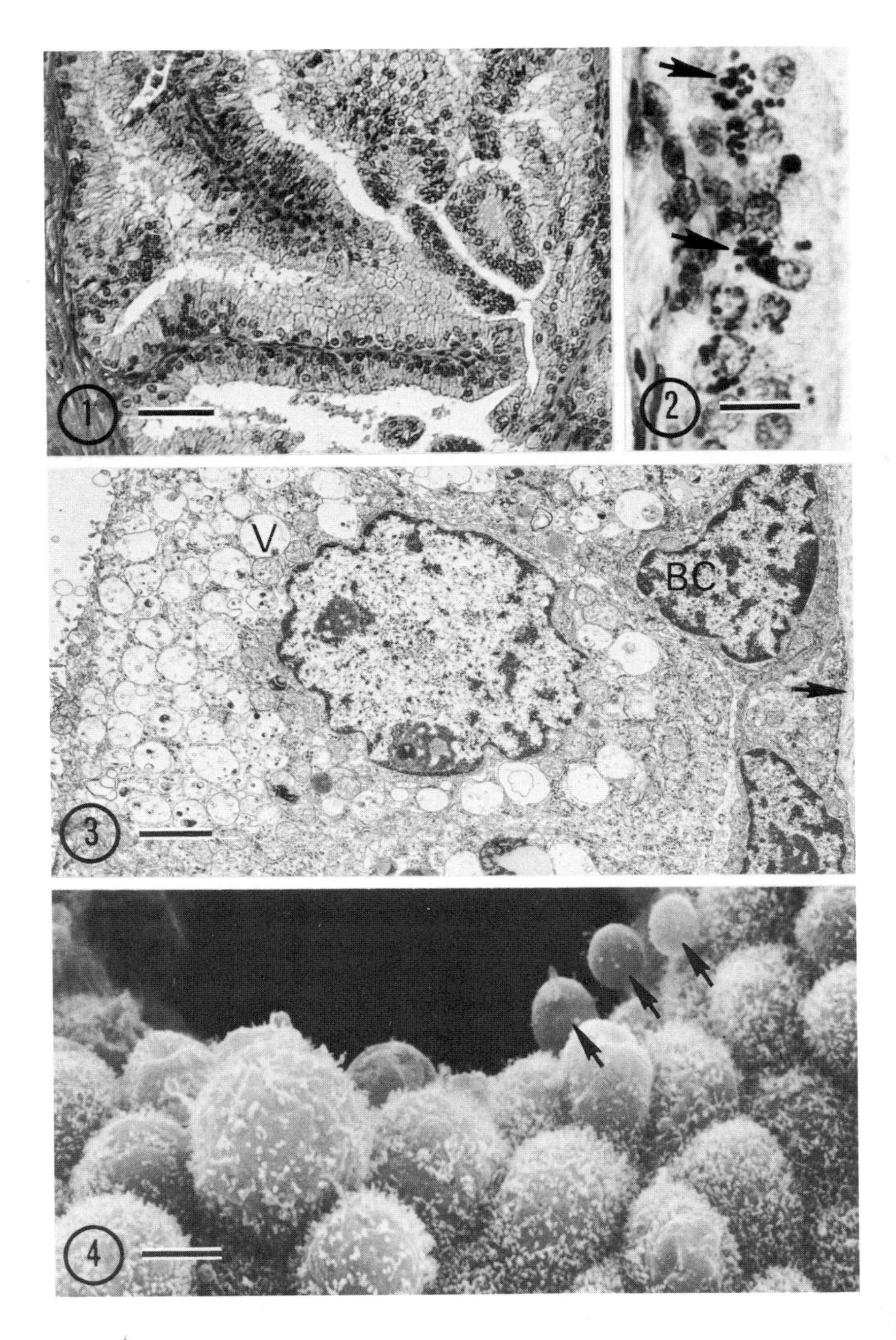
1
2
V
BC
3
4

of these cells, or occurred individually or as aggregates within the acinar lumen (Fig. 2). Secretory cells and intraluminal material failed to stain for acidic components with alcian blue (pH 2.5) or aldehyde fuchsin (pH 1.0).

By TEM, secretory cells extended from the basal lamina to the acinar lumen (Fig. 3). These cells were polarized with respect to the distribution of membrane specializations and organelles, especially secretory vacuoles, which were particularly abundant within the apical cell cytoplasm. Secretory vacuoles were devoid of contents, except for occasional heterogeneous inclusions such as irregular dense bodies, small vesicles, and occasional membrane whorls. A variety of other organelles also were present, including a well-developed supranuclear Golgi complex. Mitochondria, sparse profiles of rough endoplasmic reticulum (RER), an occasional lipid droplet, and free polysomes were distributed within the basal portion of secretory cells, but sometimes were seen in other regions of the cell as well. Lysosomes were scattered within apical and basal zones of secretory cells, similar to that of PAS-positive granules mentioned previously. Membrane specializations also were polarized in distribution. Microvilli were present on apical cell surfaces, while junctional complexes and microplicate folds of plasma membrane were seen along lateral cell surfaces. Blebbing, disruption, and even loss of the apical plasma membrane also were observed.

By SEM, folds and papillary projections of the walls of glandular structures were covered by compact masses of secretory cells and were readily distinguished from surrounding stroma. The apical contour of secretory cells was typically convex, but varied from flattened to hemispherical, and smooth to wrinkled (Fig. 4). Microvilli were especially numerous around the periphery of each cell, but also were scattered over other portions. Pores or holes in the apical plasma membrane were common, as were blebs of various size, which sometimes involved the entire apical surface. When cut longitudinally, nuclei, secretory vacuoles, and other organelles of secretory cells were revealed.

2. Basal Cells

By LM, this cell type was triangular or flattened along the basement membrane and appeared nestled among basal profiles of secretory cells. Hyper-

---

Fig. 1. LM of normal human prostate, 0-time culture. Fibromuscular stroma surrounds glands consisting of papillary infoldings covered by secretory epithelium. H&E. Bar 75 μm.

Fig. 2. LM of normal prostate, 0-time culture. Neutral mucosubstances, present in cytoplasm of secretory cells, appear as small, dark droplets (arrows). PAS. Bar = 20 μm.

Fig. 3. TEM of normal prostate, 0-time culture. Glandular epithelium consists of secretory cells and basal cells. Secretory cells are characterized by numerous vacuoles (V) in the apical cytoplasm; basal cells (BC) lack secretory vacuoles, and are oriented along the basal lamina (arrow). Bar = 2 μm.

Fig. 4. SEM of normal prostate, 0-time culture. Convex apical surfaces of secretory cells possess short microvilli and occasional secretory blebs (arrows). Bar = 4 μm.

chromatic nuclei with inconspicuous nucleoli were spindle-like or triangular in shape. These cells possessed little cytoplasm, and no discernible mucosubstances.

By TEM, basal cells differed significantly from their secretory counterparts. Secretory vacuoles were absent, as was polarization in the distribution of intracellular organelles (Fig. 3). Junctional complexes and occasional microplicate folds of the plasma membrane were seen; microvilli were absent. Except for free polysomes, organelles were sparse; the Golgi complex was poorly developed. Appearing at the junction of the epithelium and stroma was a thin amorphous basal lamina.

C. One-Week Culture

Within the first 2–3 days of culture, widespread degeneration and necrosis of secretory cells occurred in glands within explants, and in those exposed at explant surfaces as seen by LM. Lumens of these glands were filled with eosinophilic debris. The extent of degeneration varied with time and from acinus to acinus, but was most striking deep within explants. Lining glands denuded of secretory cells were other epithelial cells with a histologic location similar to that of basal cells. With increasing periods of time up to 1 week, these lining cells did not degenerate and slough into glandular lumens but appeared to repopulate acinar structures forming a conspicuous epithelium of polygonal cells separated by widened intercellular spaces. Continuous with epithelium lining acinar spaces and ducts within glands was a sheet of similar epithelial cells covering cut surfaces of explants. A single layer of migratory cells was seen advancing over bare areas; elsewhere the epithelial layer was 2–3 cells in thickness, appearing multilayered. Cells comprising this epithelium varied in profile from columnar to polygonal, or appeared flattened along the basement membrane. In spite of conspicuously widened intercellular spaces, points of attachment were apparent between cells, especially along the outer epithelial surface.

Following the initial period of exposure to [^{3}H]thymidine, cells with labeled nuclei were detected by autoradiography at explant surfaces and among cells lining acinar spaces within explants. In most instances, labeled cells in acini were present at sites corresponding histologically to the location of basal cells. Se-

---

FIG. 5. Autoradiogram of normal prostate explant, 0-time labeling with [^{3}H]thymidine (2 days culture). Acinus with intact secretory cells, but lacking cells with labeled nuclei, is shown. H&E Bar = 20 μm.

FIG. 6. As in Fig. 5. Acinus contains some necrotic debris (arrow). Nuclei of several cells in a basal location are labeled (arrow heads). H&E. Bar = 20 μm.

FIG. 7. As in Fig. 5, 3 days after labeling (5 days culture). Surface of explant is covered by a multilayered epithelium. Nuclei of many of the cells are labeled (arrows). H&E. Bar = 20 μm.

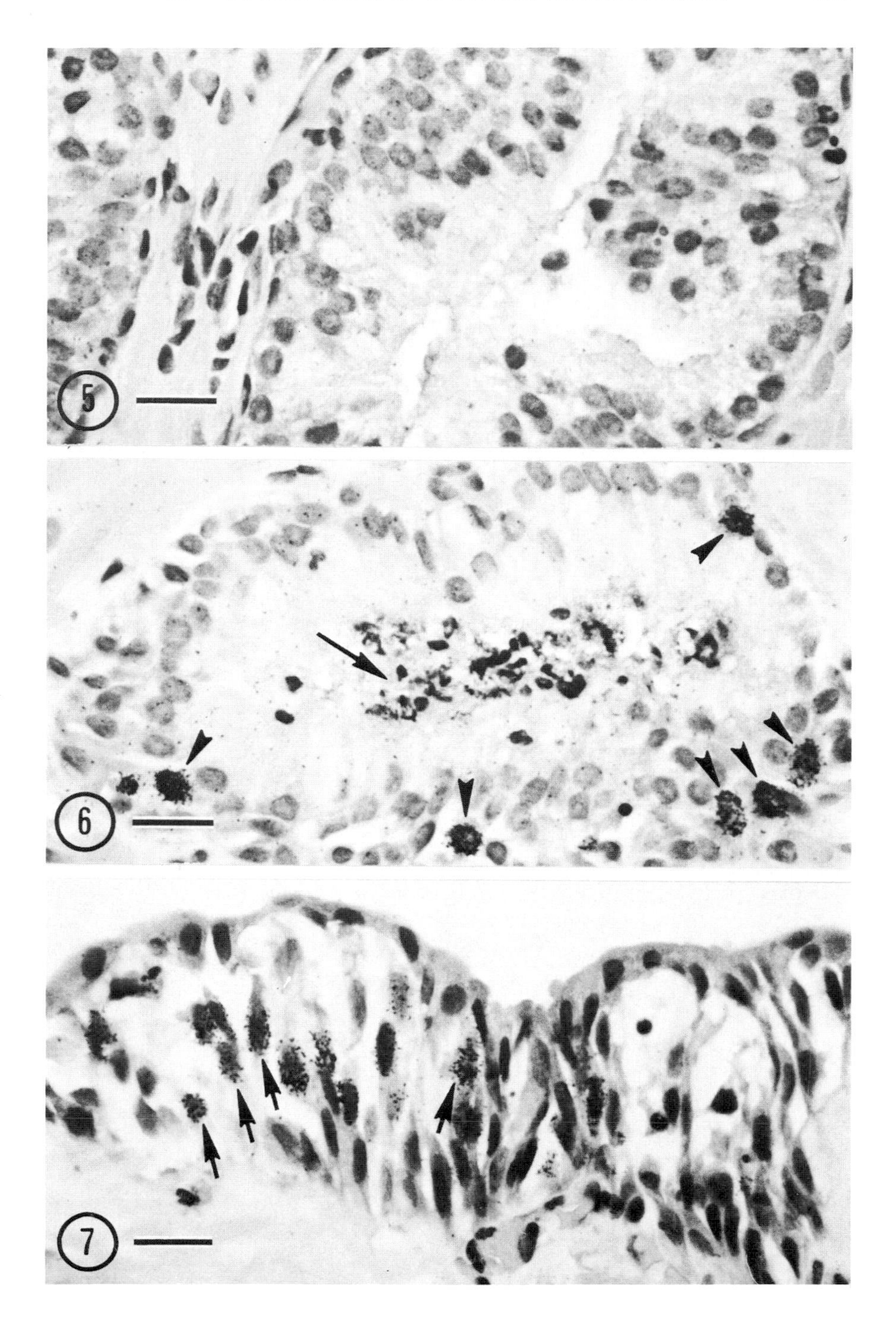
5
6
7

cretory cells were unlabeled. There appeared to be a direct correlation between the extent of degeneration and necrosis of secretory cells and labeling of basal cells. In intact acini with no detectable degeneration or necrosis of secretory cells, few basal cells were labeled (Fig. 5). However, in acini characterized by degeneration and necrosis of secretory cells, or absence of secretory cells, numerous labeled basal cells were seen (Fig. 6). Labeling of surface cells covering portions of explant surfaces was common, whether the epithelial sheet was of single or multilayered construction. More labeled cells were seen in surface epithelium adjacent to sites where glands or ducts opened onto the former cut surfaces, although larger numbers of cells also were present at these sites.

At 3 days following incubation in label (5 days culture), labeled cells were seen within epithelium which completely covered the explant as well as that lining ducts and acinar spaces within explants. Again, labeled cells of the surface epithelium were most numerous in the vicinity of glands or ducts. Labeled cells sometimes occurred individually or in clusters. In one explant numerous tall columnar epithelial cells present on the surface at one locus were labeled (Fig. 7). In other regions, few surface cells were labeled. The number of labeled cells seemed to increase and the labeling intensity was less compared to 0-time incubation, suggesting division of previously labeled cells and dilution of label. Within the explant interior, label was not seen within cell debris, nor in acinar spaces. Here, too, labeling intensity was less than at 0-time. In general, the majority of cells appeared unlabeled at this time interval.

At 7 days following incubation in label (9 days culture), labeled material was seen associated with eosinophilic cell debris in some acini deep within the explant, where even death of basal cells eventually occurred. In many instances, label was associated with pyknotic nuclei free in the lumen. However, relatively large numbers of labeled cells were distributed in a pattern similar to that seen 3 days after incubation in the label, as described previously.

Apical portions, particularly of cells extending to the outer surface of the multilayered epithelium surrounding the explant and lining ducts and acinar spaces within explants, possessed numerous PAS-positive diastase-resistant granules. At the cell apex, granules also stained with AB (pH 2.5) (see Fig. 11) but not with AF (pH 1.0). More proximally situated granules typically stained only with PAS. AB-positive mucosubstances also stained with mucicarmine. Surfaces of these epithelial cells showed affinity for both AB (pH 2.5) and mucicarmine.

TEM confirmed LM observations of striking degenerative changes in normal prostatic secretory cells lining ducts and glands within explants (Fig. 8). Degenerative cytoplasmic changes included appearance of numerous phagosomes, membrane disruption, nuclear pyknosis, and organelle dissolution. Sloughing of these cells resulted in accumulation of necrotic debris within acinar lumens. Of interest, epithelial cells with ultrastructural features identical to basal cells, i.e., cells oriented along the basal lamina and lacking secretory vacuoles, remained viable. These cells presumably served as precursor cells, incorporated labeled thymidine as demonstrated by autoradiographic studies described previously, divided, repopulated ducts and acinar spaces denuded of secretory cells, and migrated onto explant surfaces forming a differentiated, mucus-secreting epithelium. Cells comprising this epithelium possessed a number of epithelial characteristics including microvilli on apical surfaces, junctional complexes, and basal lamina. Epithelial cells were columnar to polygonal, and either extended from the basal lamina to the outer surface, were intermediate in position, or were somewhat flattened along the basal lamina. Widened intercellular spaces were penetrated by microplicate folds of plasma membrane which were free or interdigitated with those of adjacent cells. Tight junctions connected apical portions of adjacent cells at the apical surface; small desmosomes were present elsewhere. Numerous microvilli covered apical surfaces of outermost cells. Especially abundant in apical portions of these well-differentiated epithelial cells were vacuoles enclosing mucus-like material (Fig. 9), which probably accounted for histochemical staining for mucosubstances described previously. Within epithelial cells were nuclei with prominent nucleoli, one or more well-developed Golgi complexes, mitochondria, RER, lysosomes, sometimes numerous myelin figures, occasional lipid droplets, and other organelles.

SEM of epithelial cells revealed flattened to markedly convex apical surfaces covered with variable numbers of short microvilli (Fig. 10). In surface view adjacent polygonal borders of these cells were tightly adherent. Where exposed, lateral cell surfaces were characterized by numerous microplicate folds. In contrast, migratory cells covering bare areas of explant surfaces were squamous-like, with scalloped overlapping borders.

D. Two-Week Culture

Nearly identical features characterizing the epithelium of explants at early intervals were seen by LM after culture for 2 weeks. These features included a multilayered arrangement of polygonal epithelial cells, both on explant surfaces and lining glands and ducts within explants. Acinar lumens contained abundant cell debris. Occasional mitotic figures were seen among epithelial cells lining acinar spaces and covering explant surfaces.

Marked synthesis and secretion of mucosubstances continued to characterize these well-differentiated epithelial cells. Material with affinity for PAS/AB (pH 2.5) (Fig. 11), and mucicarmine also was present in acinar lumens together with necrotic debris.

By TEM, few ultrastructural changes in epithelial cells were noted compared to earlier time periods. Fewer columnar cells were seen, and microvilli appeared shorter and more sparsely distributed on apical cell surfaces. In epithelial cells

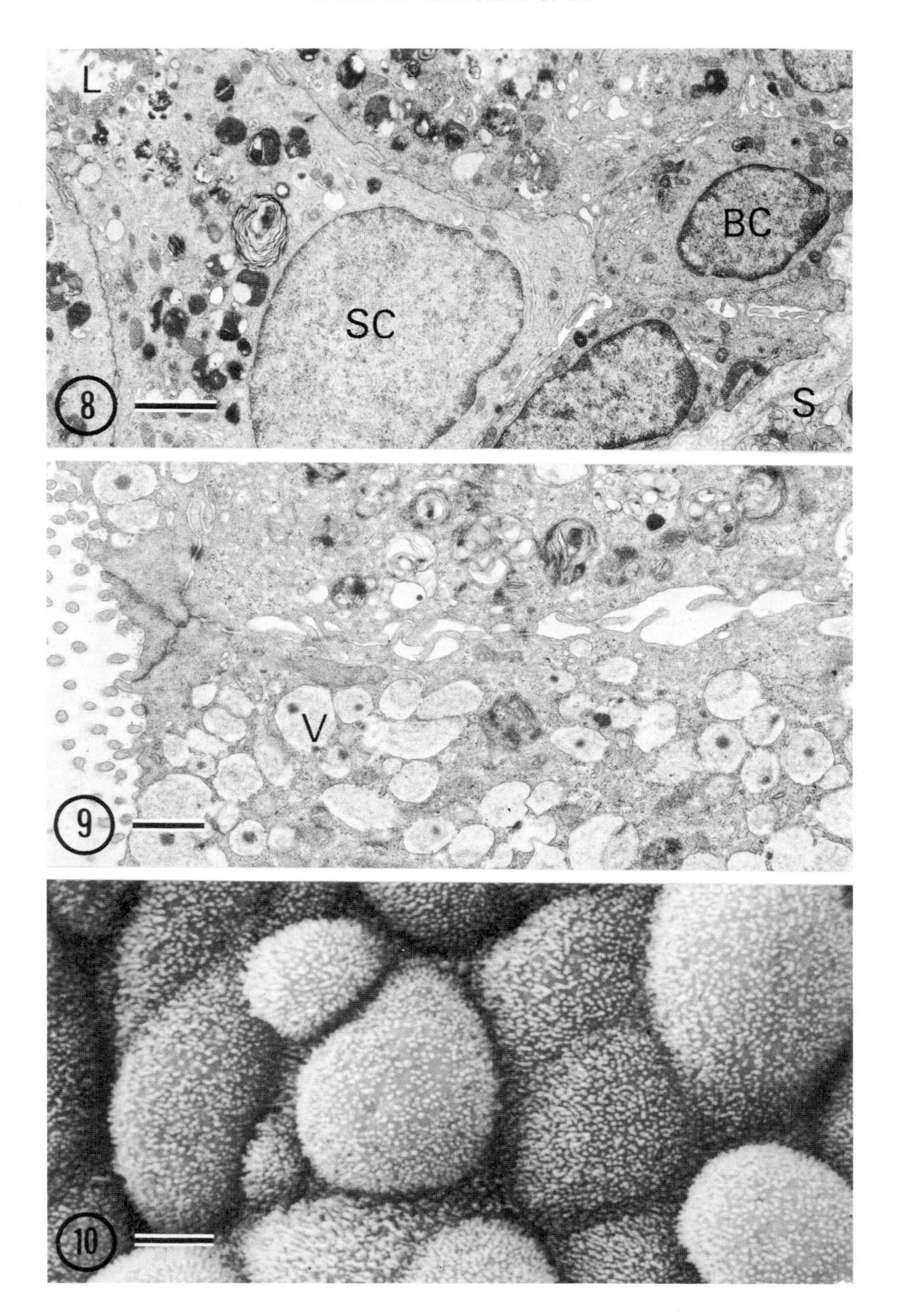
L
SC
BC
S
8
V
9
10

within explants, phagosomes, myelin figures, and lipid droplets were more frequently seen than in cells on explant surfaces.

Epithelial cells again possessed polygonal surface profiles, and varied from flattened to convex, as noted previously, by SEM. Convex cell surfaces possessed somewhat longer microvilli than flattened cells. Epithelial cells with similar surface features characterized acini within explants.

E. Three- to Twenty-four-Week Culture

As seen by LM throughout this interval, the multilayered epithelium covering explant surfaces and ducts, and lining debris-clogged acinar spaces within explants exhibited few histologic changes from earlier sampling periods (Fig. 12). Mitotic figures were occasionally seen.

In contrast, histochemical analyses revealed a decline in synthesis of mucosubstances. Fewer apical granules were seen in surface or acinar epithelial cells, although qualitatively, staining was similar to earlier time intervals. Cell surfaces continued to be AB- (pH 2.5) and mucicarmine-positive.

During this period, ultrastructural features of explant surface cells and acinar cells within explants revealed by TEM were consistent with previous observations (Fig. 13). Terminal bars were present, and numerous small vacuoles enclosing mucus-like material were still seen in some cells at 7 weeks, and at later intervals. At 24 weeks, calcification of residual cell debris was noted.

By SEM topographic features of apical surfaces of epithelial cells at explant surfaces or lining acinar structures within explants were again similar to earlier observations. Throughout this period flattened to convex apical cell surfaces were seen, although the former predominated (Fig. 14). Occasional blebs of the apical plasma membrane were present. Cell borders were tightly adherent as before; terminal bars were conspicuous. Numerous, stubby microvilli characterized apical cell surfaces, while microplicate folds were present on lateral cell surfaces and on epithelial cells oriented along the basal lamina.

FIG. 8. TEM of normal prostate, 3 days culture. Degenerating secretory cells (SC) and intact basal cells (BC) are seen adjacent to stroma (S). In apical portions of secretory cells, near the acinar lumen (L), are numerous phagosomes, some with myelin-like material; loss of organelles and membrane rupture also are seen. Basal cells possess only a few phagosomes; other organelles are intact. Bar = 2 μm.

FIG. 9. TEM of normal prostate, 1 week culture. Apical portions of columnar cells of the epithelium on explant surfaces are illustrated. Cell surfaces possess numerous microvilli; apical cytoplasm is packed with vacuoles (V) enclosing flocculent mucus-like material with an eccentric dense core. Junctional complexes, microplicate folds of lateral plasma membrane, and widened intercellular spaces characterize epithelial cells. Numerous phagosomes are present in one of the cells (upper right). Bar = 1 μm.

FIG. 10. SEM of normal prostate, 4 days culture. Markedly convex apical surfaces are covered by short microvilli. Bar = 2 μm.

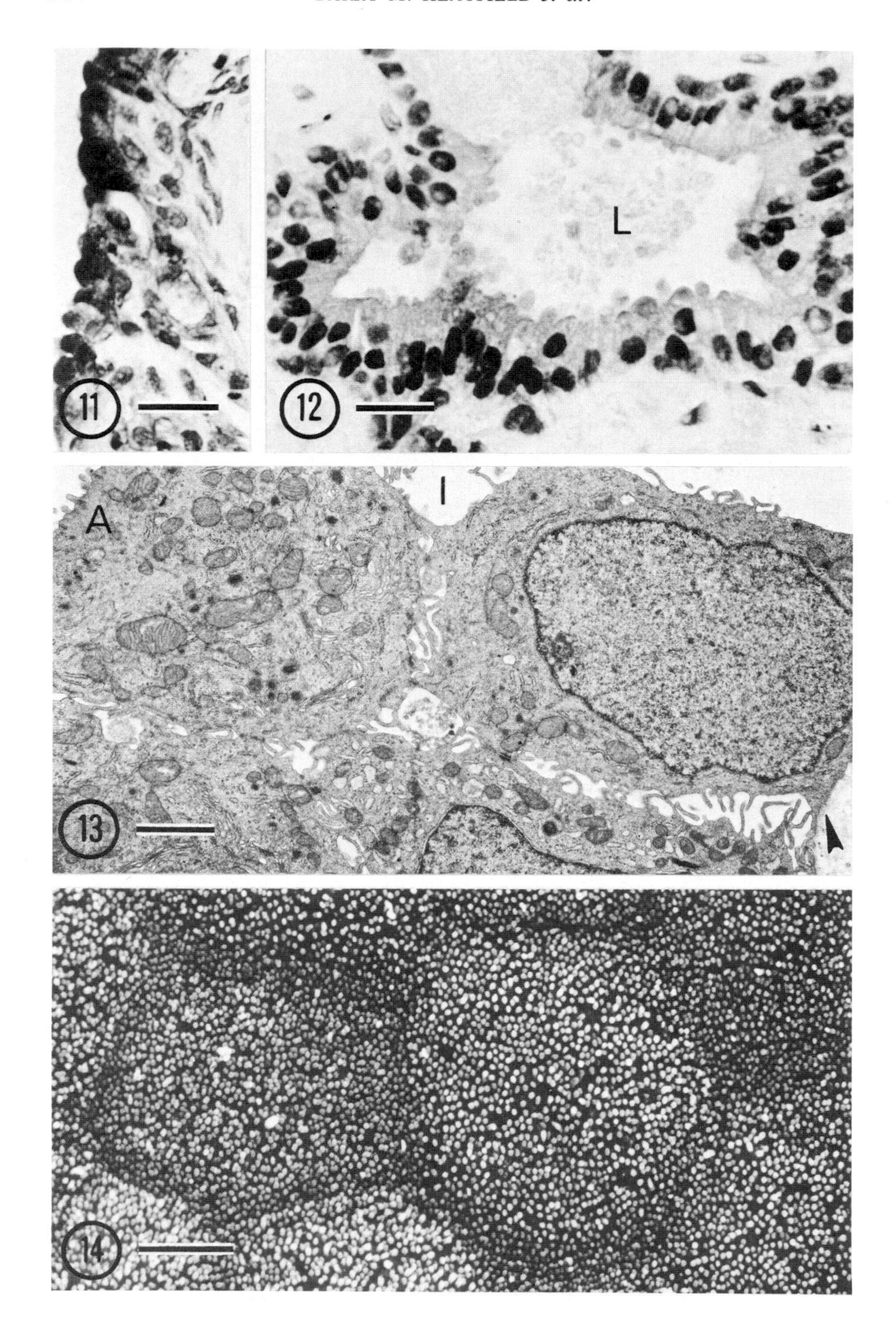
11
12
L
I
A
13
14

IV. Discussion

A. Sources of Tissue

Most often, human prostate tissue is removed surgically because of clinically significant BPH. Rarely, prostatic tissue may be sampled at the time of cystectomy (removal of bladder). Obstruction of the posterior urethra due to BPH requires surgical intervention. Relief of urethral compression is usually achieved by transurethral resection, involving either use of a heated platinum wire loop or "cold punch" technique to excavate excessive tissue surrounding the urethra. The former technique damages the outer 1 mm or more of the tissue chip removed, which sometimes results in poor viability *in vitro*. If hyperplasia involves primarily the fibromuscular component, rather than the epithelial component as sometimes happens, chips may yield few or no viable epithelial cells. Concern for the possible presence of latent carcinoma may dictate processing of many or most of the chips for diagnostic purposes, leaving little for experimental use. Abundant and viable tissue may be obtained by total prostatectomy for BPH. However, this surgical approach is infrequently utilized. Biopsy samples taken at the time of preliminary diagnosis may be damaged, usually are of very small dimensions and insufficient to provide tissue for both diagnostic requirements and *in vitro* study. Even if viable tissue can be obtained from BPH, it is not considered normal prostate (Franks, 1977), and, despite its availability, is not suitable for *in vitro* studies where responses of normal tissues are required.

The only approach currently available to obtain abundant and viable normal prostate for *in vitro* studies is that based on the immediate autopsy (Trump *et al.*, 1975), utilized in the work summarized here. This approach is made possible through cooperation with the Shock Trauma Unit, Department of Emergency Medical Services, Maryland Institute for Emergency Medicine, Baltimore, Maryland. Within a few minutes of somatic death, rapid sampling of tissues is begun. Prostate tissue is quickly placed in chilled transport medium, further

---

FIG. 11. LM of normal prostate, 2 week culture. Darkly stained neutral and acidic mucosubstances are present in apical cytoplasm of outer-most cells of epithelium lining a duct near the adjacent cut surface of the explant. AB (pH 2.5)/PAS. Bar = 20 μm.

FIG. 12. LM of normal prostate, 5 week culture. Acinus within an explant is lined by an epithelium similar to that covering explant surfaces. The lumen (L) contains heterogeneous necrotic debris. H&E. Bar = 20 μm.

FIG. 13. TEM of normal prostate, 9 week culture. View of the epithelium on explant surfaces. Epithelial characteristics include microvilli at apical surfaces (A), junctional complexes, and basal lamina (arrow head). Also present are lateral microplicate folds in widened intercellular spaces (I), large nuclei, numerous mitochondria, and multiple Golgi complexes. Bar = 2 μm.

FIG. 14. SEM of normal prostate, 24 week culture. Rounded apical surfaces covered with short microvilli, and similar to earlier periods of culture, are shown. Bar = 2 μm.

trimmed, and placed in culture as described previously. Viability is excellent, contamination of cultured tissues is infrequent, and long-term *in vitro* maintenance of epithelial cells is possible. Since immediate autopsy can be performed on young adults, in whom BPH or latent adenocarcinoma is unlikely, *normal* prostate tissue can be obtained. Since virtually the entire prostate is available for study, numerous explants can be prepared enabling long-term controlled experiments *in vitro,* with subsequent analyses by multiple techniques including correlative LM, histochemistry, TEM, SEM, and biochemistry.

B. Explant Culture Techniques

In most *in vitro* studies of human prostate, tissue (BPH) is collected, transported in chilled balanced salt solution, trimmed into small pieces or slices several millimeters in maximum thickness, and incubated in a variety of media, such as Eagle's MEM, McCoy's 5a, Parker's 199, Trowell's T8, or Waymouth's MB 752/1, to which has been added 5–20% fetal bovine serum and antibiotics (Schrodt and Foreman, 1971; McMahon and Thomas, 1973; Harbitz, 1973; McRae *et al.,* 1973; Lasnitzki *et al.,* 1975; Noyes, 1975; Dilley and Birkhoff, 1977). Explants are placed on thin slabs of agar, lens paper, or gelfoam sponge bathed in aqueous medium in petri dishes or other suitable container. Incubation is commonly conducted at 37°C in an atmosphere of 5% CO_2 and 95% O_2. Medium and atmosphere are usually replaced every 2–4 days.

In our laboratory, we have successfully maintained explants of human prostate for periods up to 24 weeks. We attribute these results to a number of features of our culture technique. Those which seem particularly advantageous include rapid collection of tissue followed by chilling in transport medium. To minimize bacterial contamination, both transport medium and incubation medium contain antibiotics, and tissues are trimmed using aseptic surgical technique and a laminar flow hood. Thin pieces of tissue are used. One obvious variable in our technique which differs from that of other studies relates to composition of the gas phase. Whereas most investigators have used a mixture of 5% CO_2 and 95% O_2, we employ a mixture of 5% CO_2, 45% O_2, and 50% N_2. High concentrations of O_2 may be toxic to prostate epithelial cells. In addition, rocking cultures results in circulation of medium and exposure of explant surfaces, facilitating rapid exchange of gases.

C. Viability and Maintenance of Epithelial Cells *in Vitro*

In previous *in vitro* studies with explants of human prostate, viable epithelium was maintained from periods of only a few days to less than 2 weeks (Schrodt and Foreman, 1971; McMahon and Thomas, 1973; Harbitz, 1973; McRae *et al.,* 1973; Harbitz *et al.,* 1974; Lasnitzki *et al.,* 1975), to 3 weeks (Stonington and

Hemmingsen, 1971), 6 weeks (Noyes, 1975), and as long as 48 months (Franks *et al.*, 1970), although details are lacking. In our studies we have maintained normal prostate epithelial cells (basal cell derivatives) in explant culture up to 24 weeks.

One of the most striking changes which takes place in prostatic epithelium during early periods *in vitro* is alteration in secretory cells, as discussed in more detail later. Testosterone and dihydrotestosterone have been added to culture media in an attempt to maintain differentiation of secretory epithelial cells during explant culture (Schrodt and Foreman, 1971; McMahon and Thomas, 1973; McRae *et al.*, 1973; Harbitz *et al.*, 1974; Dilley and Birkhoff, 1977) but without success. Addition of estradiol also was not accompanied by any histologic changes in secretory cells (Harbitz *et al.*, 1974; Dilley and Birkhoff, 1977). However, Lasnitzki *et al.* (1975) reported maintenance of differentiated secretory epithelial cells as long as 6 days in the presence of testosterone or dihydrotestosterone. Over the same interval, estradiol induced degenerative changes in secretory cells. Effects on maintenance of differentiation of secretory epithelial cells by other agents such as synthetic retinoids (vitamin A) await to be investigated.

D. Response of Epithelial Cells to Culture Conditions

Prior to culture (0-time) morphologic observations on human prostate epithelium in our studies and those of other investigators (Franks *et al.*, 1970; Schrodt and Foreman, 1971; McMahon and Thomas, 1973; Harbitz, 1973; McRae *et al.*, 1973; Harbitz *et al.*, 1974; Lasnitziki *et al.*, 1975; Noyes, 1975; Dilley and Birkhoff, 1977) are consistent with detailed ultrastructural reports (Brandes *et al.*, 1964; Mao and Angrist, 1966; Fisher and Sieracki, 1970). Epithelium of normal prostate or BPH consists of two cell types: (*a*) columnar to cuboidal secretory cells and (*b*) nonsecretory basal cells. Secretory cells extend from the basal lamina to the gland lumen and show polarization in the distribution of cellular organelles, particularly secretory vacuoles, which pack the apical cytoplasm. In contrast, basal cells are oriented along the basal lamina and do not extend to the luminal surface. Ultrastructurally, basal cells appear functionally undifferentiated, lacking secretory vacuoles and possessing a poorly developed Golgi complex. Based on apical surface features seen by SEM, several categories of secretory epithelial cells have been described (Stone *et al.*, 1977; Mickey *et al.*, 1977). Cells with flat surfaces are thought to be cuboidal and relatively inactive, while those with convex surfaces are active columnar cells (Stone *et al.*, 1976). Among secretory cells are: (*a*) microvillus cells, possessing numerous microvilli (90% or more of cells present), (*b*) bare cells lacking membrane specializations, and (*c*) ruffled cells, characterized by numerous microplicate folds (Stone *et al.*, 1976, 1977; Mickey *et al.*, 1977). In our studies, few ruffled cells were identified in normal human prostate, while microvillus cells, bare cells, and other cells intermediate between these two extremes, were common.

The principal histologic, histochemical, and fine structural changes in normal human prostate during long-term explant culture based on our work can be summarized as follows. During early periods *in vitro,* degeneration and necrosis of secretory epithelial cells in glandular acini within explants takes place. In these acini, cells oriented along the basal lamina and with the ultrastructural characteristics of basal cells remain viable during degeneration and sloughing of secretory cells. Following sloughing of secretory cells, and accumulation of necrotic debris in gland lumens, ducts and acinar structures become repopulated by cells with epithelial features, including microvilli, junctional complexes, multiple Golgi complexes, and basal lamina. Division and subsequent migration of these cells, concomitant with loss of secretory cells, is indicated by the presence of mitotic figures and patterns of incorporation of [^{3}H]thymidine and distribution of labeled cells in pulse-labeling experiments. Epithelial cells also colonize bare explant surfaces, and form a multilayered, well-differentiated epithelium of columnar or polygonal basal cell derivatives continuous with that lining ducts and glands within explants. Basal cell derivatives are similar in fine structure to those covering explant surfaces. In contrast to normal secretory cells, numerous apical vacuoles in basal cell derivatives at explant surfaces and within explants contain mucosubstances stainable histochemically for neutral and acidic components. Initially, synthesis and secretion of this material are marked, but gradually decline with increasing periods of culture. Fewer columnar cells are seen at later intervals, and convex apical surfaces tend to flatten.

Since it has been possible to follow the sequence of events in detail during the earliest periods of culture in our studies, it can be concluded that normal secretory cells of human prostate do not survive explant culture for extended periods, but that epithelial cells with characteristics similar to basal cells do survive, proliferate, migrate, and differentiate into mucus-producing cells.

In most previous studies where degeneration and necrosis of epithelial cells were noted, no clear distinction was made between secretory cells and basal cells (Schrodt and Foreman, 1971; McMahon and Thomas, 1973; Lasnitzki *et al.,* 1975; Noyes, 1975; Dilley and Birkhoff, 1977). Maintenance of secretory cells with a columnar profile for as long as 10 days has been reported (McMahon and Thomas, 1973), although reduction in secretory vacuoles and cell height was also seen after only a few days *in vitro* (Schrodt and Foreman, 1971; Dilley and Birkhoff, 1977). Squamous changes have also been noted during early culture periods (Schrodt and Foreman, 1971; McMahon and Thomas, 1973; Lasnitzki *et al.,* 1975; Noyes, 1975; Dilley and Birkhoff, 1977). The description by Lasnitzki *et al.* (1975) of epithelium of cultured explants as consisting of a layer of cells interconnected by tonofibrils sandwiched between an outer layer of cuboidal cells and an inner layer of basal-like cells is similar to that seen in our studies. Schrodt and Foreman (1971) examined the ultrastructure of secretory epithelial cells of BPH up to 9 days *in vitro*. A reduction in cell height and numbers of secretory vacuoles was noted up to 6 days; degeneration and necrosis of secretory cells were common thereafter. Squamous-like cells with tonofilaments similar to those seen in our studies also were noted. Schrodt and Foreman (1971) considered the possibility that squamous-like epithelial cells seen in the older cultures represented altered basal cells of prostatic acini. With our work as a background, it seems most unlikely that cells surviving longer periods of explant culture and seen by Franks *et al.* (1970), Stonington and Hemmingsen (1971), and Noyes (1975) were viable secretory cells. It is more probable that these cells were derived from basal cells.

With the exception of one report (Lasnitzki *et al.*, 1975), addition of steroid hormones to explant cultures of human prostate (BPH), as mentioned previously, did not result in maintenance of differentiation of secretory cells for periods longer than that of controls (Schrodt and Foreman, 1971; McMahon and Thomas, 1973; McRae *et al.*, 1973; Harbitz *et al.*, 1974; Dilley and Birkhoff, 1977).

E. Proliferation of Epithelial Cells *in Vitro*

In pulse-labeling autoradiographic studies reported here, mitotic epithelial cells of normal human prostate explants labeled at 2 days culture incorporated [^{3}H]thymidine. Labeled epithelial cells corresponded in histologic location to basal cells; secretory cells did not incorporate label. Incorporation of label by basal cells occurred concomitantly with degeneration of secretory cells. Occasional migratory epithelial cells already on explant surfaces also were labeled at this early period. Incorporation of label was followed by mitosis and migration of labeled cells onto explant surfaces from ducts or acini within explants at 3 and 7 days postlabeling. In histologic sections, mitotic figures also were seen, supporting autoradiographic findings.

These observations augment and are consistent with other studies, which similarly have reported both mitotic figures (McMahon and Thomas, 1973; McRae *et al.*, 1973; Dilley and Birkhoff, 1977) and autoradiographic detection of [^{3}H]thymidine incorporation during explant culture of BPH (McMahon and Thomas, 1973; Harbitz *et al.*, 1974; Dermer, 1978). After 3–4 days, mitotic activity appeared to peak (McMahon and Thomas, 1973) and was not enhanced by testosterone, dihydrotestosterone, or estradiol (Harbitz *et al.*, 1974). On the other hand, Lasnitzki *et al.* (1975) observed stimulation of mitosis by testosterone and dihydrotestosterone but not estradiol. Mitosis of cells within acini (McMahon and Thomas, 1973; Harbitz *et al.*, 1974; Lasnitzki *et al.*, 1975; Dilley and Birkhoff, 1977; Dermer, 1978), and migration from exposed ducts and acini onto surfaces of explants (Stonington and Hemmingsen, 1971; McMahon and Thomas, 1973; McRae *et al.*, 1973; Dilley and Birkhoff, 1977) and culture dishes (McMahon and Thomas, 1973; McRae *et al.*, 1973; Franks *et al.*, 1970), have been reported previously.

F. Identification of Epithelial Cell Types *in Vitro*

Only two types of epithelial cells have been described in human prostate. The ease with which morphologic distinction can be made between secretory cells with secretory vacuoles and basal cells without secretory vacuoles enables identification of these two cell types in explant culture, even though occasional transitional forms possessing only a few secretory vacuoles have been reported (Mao and Angrist, 1966; Fisher and Sieracki, 1970). During early periods of culture, secretory cells degenerate, while other epithelial cells, which are identical to basal cells in location within the acinar epithelium, and in ultrastructural features, survive and incorporate [^{3}H]thymidine. When mechanically separated from the stroma, however, prostatic epithelium failed to survive and incorporate [^{3}H]thymidine *in vitro* (Franks *et al.*, 1970). Basal cells were clearly disrupted during mechanical separation as seen by TEM, which probably accounts for lack of subsequent growth *in vitro*. Nevertheless, maintenance of normal stromal–epithelial contact may also be important in stimulating growth during culture, since intact acini isolated from stroma by enzymatic digestion also failed to grow (Stone *et al.*, 1976). That cells comprising the well-differentiated epithelium on explant surfaces are derivatives of acinar basal cells is suggested by continuity of epithelium between duct, glands, and surfaces of explants, similarity in histochemical staining characteristics of cytoplasmic mucosubstances, similarities in ultrastructure, and patterns of cell division and migration from exposed ducts and glands onto cut surfaces of explants.

In previous reports, the origin of proliferating epithelial cells during explant culture was attributed to acinar epithelial cells, without distinction as to cell type (Stonington and Hemmingsen, 1971; Harbitz *et al.*, 1974; Lasnitzki *et al.*, 1975; Dilley and Birkhoff, 1977), or to the outer or basal layer of epithelial cells lining the acinar space (McMahon and Thomas, 1973; Dermer, 1978).

V. Perspectives

A. Current Applications and Findings

Little definitive information is available on the role of basal cells in human prostate. Basal cells are still sometimes referred to as myoepithelial cells (Tannenbaum, 1977), although they do not possess numerous myofibrils characteristic of smooth muscle cells present in the surrounding stroma (Brandes *et al.*,

1964; Fisher and Sieracki, 1970). Basal cells more frequently are regarded as "reserve" cells (Brandes *et al.*, 1964; Mao and Angrist, 1966; Fisher and Sieracki, 1970; Dermer, 1978), implying a capacity to differentiate into secretory cells under appropriate stimulation. Basal cells are capable of proliferating as discussed previously and may be a source of secretory cells, as evidenced by incorporation of [^{3}H]thymidine by epithelial cells of BPH *in vitro* (Dermer, 1978). Our studies so far have demonstrated the capacity of basal cells to divide, migrate, and differentiate into a mucus-secreting epithelium. The factor(s) stimulating these events is not yet known. However, incorporation of [^{3}H]thymidine by basal cells as described here occurred concomitantly with degeneration of secretory cells within a given acinus during early culture of explants, suggesting stimulation due to loss of cell–cell contact and/or release of substances by degenerating secretory cells.

The possibility that basal cells play an important role in the histogenesis of prostatic carcinoma has not received much attention in the past. Carcinoma of the prostate may develop from secretory cells following senile atrophy, rather than directly from hyperplastic acini of BPH (Rich, 1935; Moore, 1943; Andrews, 1949; Franks, 1954; Liavag, 1968). However, a statistical association between prostate carcinoma and BPH has been demonstrated by Edwards *et al.* (1953). More recently, an origin from active or only moderately atrophic glands, rather than postatrophic hyperplastic glands, has been postulated by McNeal (1969). His hypothesis involves either (*a*) progressive anaplastic changes in hyperplastic foci of active duct epithelium to poorly differentiated carcinoma *in situ,* followed by invasion, or (*b*) gradual transition of newly formed acini, originating from ductal hyperplastic foci, to moderately well-differentiated carcinoma showing gland formation.

The results of our studies strongly imply an important role of basal cells in histogenesis of prostatic adenocarcinoma, since they are capable of proliferation and differentiation *in vitro*. Consistent with this hypothesis is the finding that, although not histochemically demonstrable in normal human prostate, mucosubstances frequently are detectable in prostatic adenocarcinoma (Franks *et al.*, 1964; Hukill and Vidone, 1967) and possess staining affinities (neutral and acidic components) similar to those of mucosubstances synthesized and secreted by basal cell derivatives during explant culture, as described previously. Moreover, we recently obtained morphologic evidence of premalignant and malignant changes in basal cell derivatives after exposure of explants of normal human prostate to the direct-acting chemical carcinogen *N*-methyl-*N'*-nitro-*N*-nitrosoguanidine (MNNG), for 4 weeks *in vitro* (Sanefuji *et al.*, 1979). Morphologic changes compared to controls included loss of tight cell–cell contact and organizational pattern accompanying reduction or loss of junctional complexes, shift in profile from columnar or polygonal to squamous-like cells lacking microvilli, and loss of basal lamina. Carcinogen-treated cells also appeared anaplastic, with pleomorphic nuclei and prominent nucleoli. Local invasion of adjacent stroma also was suggested by morphologic observations. Of significance, morphologic features of these epithelial cells also were characteristic of neoplastic cells of well- to moderately differentiated prostatic adenocarcinoma maintained *in vitro* in parallel experiments (Heatfield *et al.*, 1979).

B. Future Prospects

The present chapter summarizes methods for collection of normal human prostate, techniques for long-term explant culture of prostatic basal cells, and detailed morphologic responses of these cells to culture conditions in baseline histochemical, autoradiographic, and ultrastructural studies. Based on this information, a number of problem areas have emerged which, hopefully, will stimulate interest and further investigation.

For example, rapid degeneration and necrosis of secretory cells *in vitro,* which we and others have observed, raise questions regarding the basis of maintenance of secretory cells *in vivo* and pose a challenge to experimentally alter this seemingly inevitable course of destruction. It is well known that maintenance of differentiation of prostatic secretory epithelial cells *in vivo* is androgen-dependent. However, the effect on viability and/or maintenance of differentiation of prostatic secretory cells by addition of androgens to culture media remains controversial, and reexamination of this problem appears justified. Addition of steroid hormones, synthetic retinoids, etc., to culture media should be pursued in an attempt to define the basis of secretory cell death *in vitro*. Such an approach also may give valuable insight into the pathophysiology of prostatic epithelium.

Isolation of prostatic basal cells as a consequence of degeneration and necrosis of secretory cells during explant culture will enable detailed study of the role of basal cells in normal human prostate, through investigation of their responses to a variety of stimuli, including steroid hormones, retinoids, and other agents.

It should also be possible to study the role of stromal–epithelial interactions in normal growth and differentiation of basal cells of human prostate by examining the effects of separation of basal cells from stroma by enzymatic digestion and *in vitro* culture with and without an artificial collagen substratum.

Although not summarized here, we have found that, in general, the response of explants of BPH to culture conditions was identical to that of normal prostate of young adults (Heatfield *et al.*, 1980; Sanefuji *et al.*, 1980). Assuming that prostatic basal cells are the stem cell population for replacement of secretory cells, it may be possible to detect subtle differences between the basal cell populations of normal prostate and BPH *in vitro,* relating to growth kinetics, binding and metabolism of hormones, etc., which could lead to new avenues of therapy for BPH.

Further analysis of the neoplastic response to carcinogens, described previously, should involve xenotransplantation of carcinogen-exposed explants into an immunologically deficient animal model, such as the athymic nude mouse, to test tumorigenicity. Also of interest are subsequent effects of removal of the carcinogen, and alteration in the course of carcinogenesis *in vitro* by steroid hormones, retinoids, and other modifying factors.

If a transplantable human prostatic tumor model is developed based on carcinogenesis of prostatic explants *in vitro,* then a whole host of other applications will be possible, similar to those proposed by Merchant (1977), based on development of an *in vivo* model. For example, studies on the etiology of human adenocarcinoma, mechanisms of carcinogenesis, screening of antitumor agents, hormone metabolism, etc., could be conducted *in vitro;* tumor progression, metastasis, hormone metabolism, tumor genetics, immunotherapy, chemotherapy, hormone therapy, radiation therapy, and studies of morphologic and biochemical differences between normal and malignant prostatic epithelial cells could be attempted *in vivo*.

REFERENCES

Andrews, G. S. (1949). *J. Clin. Pathol.* **2,** 197–208.

Brandes, D., Kirchheim, D., and Scott, W. W. (1964). *Lab. Invest.* **13,** 1541–1560.

Dermer, G. B. (1978). *Cancer* **41,** 1857–1862.

Dilley, W. G., and Birkhoff, J. D. (1977). *Invest. Urol.* **15,** 83–86.

Edwards, C. N., Steinthorsson, E., and Nicholson, D. (1953). *Cancer* **6,** 531–554.

Fisher, E. R., and Sieracki, J. C. (1970). *In* "Pathology Annual" (S. C. Sommers, ed.), Vol. 5, pp. 1–26. Appleton, New York.

Franks, L. M. (1954). *J. Pathol. Bacteriol.* **68,** 617–621.

Franks, L. M. (1977). *Urol. Res.* **5,** 159–162.

Franks, L. M., O'Shea, J. D., and Thomson, A. E. R. (1964). *Cancer* **17,** 983–991.

Franks, L. M., Riddle, P. N., Carbonell, A. W., and Gey, G. O. (1970). *J. Pathol.* **100,** 113–119.

Handelsman, H. (1977). *Oncology* **34,** 96–99.

Harbitz, T. B. (1973). *Scand. J. Urol. Nephrol.* **7,** 6–13.

Harbitz, T. B., Falkanger, B., and Sander, S. (1974). *Acta Pathol. Microbiol. Scand. A.* (Suppl.) **248,** 89–93.

Harris, C. C., Saffiotti, U., and Trump, B. F. (1978). *Cancer Res.* **38,** 474–475.

Heatfield, B. M., Sanefuji, H., and Trump, B. F. (1979). "Scanning Electron Microscopy/1979/III" SEM, AMF O'Hare, Illinois. 645–655.

Heatfield, B. M., Sanefuji, H., and Trump, B. F. (1980). (Submitted).

Hukill, P. B., and Vidone, R. A. (1967). *Lab. Invest.* **16,** 395–406.

Kelley, R. O., Dekker, R. A. F., and Bluemink, J. G. (1973). *J. Ultrastruct. Res.* **45,** 254–258.

Lasnitzki, I., Whitaker, R. H., and Withycombe, J. F. R. (1975). *Br. J. Cancer* **32,** 168–178.

Liavag, I. (1968). *Acta Pathol. Microbiol. Scand.* **73,** 338–350.

Mao, P., and Angrist, A. (1966). *Lab. Invest.* **15,** 1768–1782.

McDowell, E. M., and Trump, B. F. (1976). *Arch. Pathol. Lab. Med.* **100,** 405–414.

McMahon, M. J., and Thomas, G. H. (1973). *Br. J. Cancer* **27,** 323–335.

McNeal, J. E. (1969). *Cancer* **23,** 24–34.

McRae, C. U., Ghanadian, R., Fotherby, K., and Chisholm, G. D. (1973). *Br. J. Urol.* **45,** 156–162.
Merchant, D. J. (1977). *Oncology* **34,** 100–101.
Mickey, D. D., Stone, K. R., Stone, M. P., and Paulson, D. F. (1977). *Cancer Treat. Rep.* **61,** 133–138.
Moore, R. A. (1943). *J. Urol.* **50,** 680–710.
Noyes, W. F. (1975). *Cancer Chemother. Rep.* **59,** 67–71.
Pearse, A. G. E. (1968). "Histochemistry: Theoretical and Applied." Little, Brown, Boston.
Rich, A. R. (1935). *J. Urol.* **33,** 215–223.
Sanefuji, H., Heatfield, B. M., and Trump, B. F. (1978). *Tissue Cult. Assoc. Man.* **4,** 855–856.
Sanefuji, H., Heatfield, B. M., and Trump, B. F. (1979). "Scanning Electron Microscopy/1979/III" SEM, AMF O'Hare, Illinois. 657–663.
Sanefuji, H., Heatfield, B. M., Trump, B. F., and Young, J. D. (1980). (Submitted).
Schrodt, G. R., and Foreman, C. D. (1971). *Invest. Urol.* **9,** 85–94.
Stone, K. R., Stone, M. P., and Paulson, D. F. (1976). *Invest. Urol.* **14,** 79–82.
Stone, M. P., Stone, K. R., Ingram, P., Mickey, D. D., and Paulson, D. F. (1977). *Urol. Res.* **5,** 185–200.
Stonington, O. G., and Hemmingsen, H. (1971). *J. Urol.* **106,** 393–400.
Tannenbaum, M. (1977). *In* "Urologic Pathology: The Prostate" (M. Tannenbaum, ed.), pp. 303–397. Lea & Febiger, Philadelphia, Pennsylvania.
Trump, B. F., Smuckler, E. A., and Benditt, E. P. (1961). *J. Ultrastruct. Res.* **5,** 343–348.
Trump, B. F., Valigorsky, J. M., Jones, R. T., Mergner, W. J., Garcia, J. H., and Cowley, R. A. (1975). *Hum. Pathol.* **6,** 499–516.

Chapter 9

Normal Human Prostate Epithelial Cell Cultures[1]

JOHN F. LECHNER, MERRILL S. BABCOCK, MAUREEN MARNELL, K. SHANKAR NARAYAN, AND M. EDWARD KAIGHN

Pasadena Foundation for Medical Research, Pasadena, California

---

[1]Supported by Public Health Service (PHS) grant R26-CA19826-01 from the National Cancer Institute through the National Prostatic Cancer Project, and PHS contract N01-CP-65850 from the Division of Cancer Cause and Prevention, National Cancer Institute (NCI). We thank K. Gass, B. Serar, and T. Smith for technical assistance.

ISBN 0-12-564140-0

I. Introduction

For a number of years it has been possible to prepare primary explant cultures from the normal as well as the hyperplastic and neoplastic human prostate (Burrows *et al.*, 1917; Franks, 1977). The difficulty arose when subcultures were attempted (Webber, 1974; Stone *et al.*, 1976). Recent developments have altered this situation. In 1976, we isolated the first metastatic prostatic adenocarcinoma using a collagenase–trypsin (CTC) mixture (Kaighn *et al.*, 1978, 1979). This method had been used previously with many types of differentiated cells (Kaighn, 1974; Kaighn and Babcock, 1975). Efforts to apply CTC dissociation to the normal prostate, either as primary explants or fresh tissue, were uniformly unsuccessful. When mixtures of other enzymes also failed, a two-step nonenzymatic procedure was developed and successfully applied. Details of the development of this procedure are presented here.

Both transmission (TEM) and scanning electron microscopy (SEM) have been used to characterize the normal prostatic epithelial cells. We have also found them to have the normal human male karyotype by chromosome banding (Lechner *et al.*, 1978).

In addition to the problem of cell dispersion, effective culture media for normal cells have not been available (Ham, 1974). Recently, a concerted effort has been made to remedy this situation (Ham and McKeehan, 1978). Our laboratory, in consultation with Dr. Ham, has been working to define the medium for normal epithelial cells. Both laboratories have independently applied the principles of enzyme kinetics to analysis of growth requirements (McKeehan and Ham, 1978; Lechner and Kaighn, 1979a). Use of this approach reduced the serum requirement of normal prostatic epithelial cells from 5.7 to 0.2% for half-maximal growth. Synergistic serum-sparing was brought about by hydrocortisone in combination with peptide growth factors, either epidermal growth factor (EGF) or fibroblast growth factor (FGF). In addition to its serum-sparing activity, EGF significantly extended the culture lifespan and enhanced migration of prostatic epithelial cells.

II. Materials and Methods

These methods, which have evolved by continually modifying our techniques, are the procedures that we currently use.

A. Nutrient Medium

The nutrient medium, PFMR-4, was developed by modifying F12K (Kaighn, 1973). PFMR-4 differs from F12K in the following components: sodium bicarbonate and sodium chloride were reduced from 3×10^{-2} to 1.4×10^{-2} *M* and 1.3×10^{-1} to 1.0×10^{-1} *M*, respectively; HEPES (*N*-2-hydroxyethylpiperazine-*N*'-2-ethanesulfonic acid) buffer, 3×10^{-2} *M* was included; cysteine was replaced by cystine, 1.5×20^{-4} *M;* trace elements (McKeehan and Ham, 1976) were included; the osmolality was reduced from 320 to 280 mOsm/kg.

PFMR-4 was compounded by mixing 1 volume of nutrient mixture concentrate (NMC) with 3 volumes of buffered salts diluent (BSD). The NMC was prepared by mixing 11 stock solutions. Each NMC stock was made as described by Ham (1972) for use in preparation of nutrient mixture, F12. The composition and volume of each stock in 1 liter of NMC are listed in Table I. These stock solutions are stable for at least 1 year when stored at −70°C. The components of BSD are shown in Table II.

Growth medium was prepared by adding fetal bovine serum (FBS) to PFMR-4. Unless stated otherwise, growth medium contained 7% FBS (2.73 mg/ml FBSP). Fetal bovine serum protein (FBSP) concentrations were measured by the Folin technique (Lowry *et al.*, 1951) with bovine serum albumin as reference.

Stock solutions of purified, lyophilized epidermal growth factor (EGF) and fibroblast growth factor (FGF) (Collaborative Research, Inc., Waltham, MA) were prepared in growth medium. Hydrocortisone (HC) (4-pregnen-11β,17α,21-triol-3,20-dione, Steraloids, Inc., Wilton, NH) was dissolved in absolute ethanol. The final ethanol concentration in the growth medium did not exceed 0.1%. Except for HC, all solutions used to culture cells were sterilized through a series of progressively smaller membrane filters to a final pore size of 0.22 nm. Before use, the assembled filter units were washed with distilled H_2O (80°C) to remove detergent (Ham, 1972).

B. Primary Cultures

The prostate was removed within 30 hours of death from neonates and infants who had died due to respiratory-related causes. The tissue was washed and then minced with scissors in cold PFMR-4 containing 12% FBS, penicillin (100 units/ml), kanamycin (100 μg/ml), and 1% PVP (polyvinylpyrrolidone, Calbiochem #5295). The tissue fragments and accompanying media were transferred to a conical centrifuge tube and allowed to stand for 5 minutes in an ice bath. Small clusters of cells which were dislodged or "spilled" (Lasfargues and Ozzelo, 1958) during mincing were separated from the tissue fragments by this step. The spilled cells in the supernatant were cultured in 100-mm petri dishes (Corning) in 5 ml of growth medium. The minced fragments were explanted in 100-mm dishes containing 2 ml of growth medium. After the fragments had attached (1–2 days incubation) the medium volume was increased to 8 ml. Both "spill" and explant cultures received fresh medium at 3–4 day intervals. Foreskin fibroblast cultures were isolated by the explant procedure from each neonate.

TABLE I

NUTRIENT MIXTURE CONCENTRATE (NMC)

Stock (amount used/liter)[a]	Component	Amount in stock (gm/liter)	Amount in PFMR-4 (moles/liter)[b]
1a (80 ml)[c,d]	L-Arginine·HCl	21.070	2.0 E-3
	Choline chloride	0.690	1.0 E-4
	L-Histidine·HCl·H_2O	2.100	2.0 E-4
	L-Isoleucine	0.390	6.0 E-5
	L-Leucine	1.310	2.0 E-4
	L-Lysine·HCl	3.650	4.0 E-4
	L-Methionine	0.448	6.0 E-5
1b (80 ml)[c,d,e]	L-Phenylalanine	0.496	6.0 E-5
	L-Serine	1.050	2.0 E-4
	L-Threonine	1.190	2.0 E-4
	L-Tryptophan	0.204	2.0 E-5
	L-Tyrosine	0.544	6.0 E-5
	L-Valine	1.170	2.0 E-4
2 (20 ml)[c]	Biotin	0.013	3.0 E-7
	Calcium pantothenate	0.095	1.0 E-6
	Niacinamide	0.007	3.0 E-7
	Pyridoxine·HCl	0.012	3.0 E-7
	Thiamin·HCl	0.067	1.0 E-6
3 (20 ml)[c,f,g]	Folic acid	0.264	3.0 E-6
	$Na_2HPO_4 \cdot 7H_2O$	13.400	8.1 E-4
4 (10 ml)[h]	$FeSO_4 \cdot 7H_2O$	0.320	3.0 E-6
5 (0.67 ml)	Phenol red	5.000	5.9 E-6
6 (40 ml)[c,g]	L-Glutamine	29.200	2.0 E-3
	Riboflavin	0.004	1.0 E-7
	Sodium pyruvate	22.000	2.0 E-3
7 (40 ml)[i]	L-Cystine	3.600	1.5 E-4
8 (40 ml)[c]	L-Asparagine	3.000	2.0 E-4
	L-Proline	7.000	6.0 E-4
	Putrescine·2HCl	0.032	2.0 E-6
	Vitamin B_{12}	0.136	1.0 E-6
9 (40 ml)[j]	L-Alanine	1.800	2.0 E-4
	L-Aspartic acid	2.600	2.0 E-4

(continued)

The plates were incubated at 36.5°C in a Forma incubator (Model #3025) equipped to maintain 99% relative humidity, and to control the CO_2 level at 2%.

C. Subcultures

The nonenzymatic "K-Passing" method was used to dissociate primary epithelial cultures and for all subcultures through the fourth passage. First, the culture was washed twice with HBS, then incubated for 15 minutes at 37°C in

TABLE I (*continued*)

Stock (amount used/liter)[a]	Component	Amount in stock (gm/liter)	Amount in PFMR-4 (moles/liter)[b]
	L-Glutamic acid	2.900	2.0 E-4
	Glycine	1.500	2.0 E-4
10 (40 ml)[c,k,l,m]	Hypoxanthine	0.400	3.0 E-5
	myo-Inositol	1.800	1.0 E-4
	Lipoic acid	0.021	1.0 E-6
	Thymidine	0.073	3.0 E-6
	$CuSO_4 \cdot 5H_2O$	0.0002	1.0 E-8
11 (2 ml)	$ZnSO_4 \cdot 7H_2O$	0.288	5.0 E-7
No stock[n]	NaCl	7.600	1.0 E-1
Adjust pH to 7.4 with 1*N*	NaOH		

[a] Add stocks to 500 ml distilled H_2O. Bring final volume of NMC to 1 liter after all components have been added.

[b] The letter E refers to the exponent of power of 10. Thus for example, 2.0 E-2 means 2.0×10^{-2}.

[c] Dissolve in order listed in 700 ml distilled H_2O, then bring to volume. Never add second compound until first is completely dissolved.

[d] Freshly thawed stock contains precipitate which dissolves on warming.

[e] Phenylalanine and tryptophan are each dissolved in 100 ml hot distilled H_2O containing one drop of concentrated HCl. Add to stock after cooling.

[f] Folic acid will not dissolve unless $Na_2HPO_4 \cdot 7H_2O$ is first dissolved.

[g] Avoid prolonged exposure to light.

[h] Dissolve $FeSO_4 \cdot 7H_2O$ by adding 0.5 ml of concentrated HCl.

[i] Dissolve cystine by adding 0.5 ml concentrated HCl.

[j] Aspartic and glutamic acids are added to 900 ml distilled H_2O containing 1 ml of phenol red (stock 5). While stirring, NaOH (1.0 *N*) is added dropwise to maintain neutrality (orange) as the aspartic and glutamic acids dissolve. Alanine and glycine are then dissolved and the volume is adjusted to 1 liter.

[k] Hypoxanthine is dissolved in 100 ml of boiling distilled H_2O.

[l] Lipoic acid is dissolved in a few drops of 1.0 *N* NaOH, diluted with distilled H_2O, then added to stock.

[m] Make up 2.0 mg/100 ml solution of $CuSO_4 \cdot 5H_2O$; sterilize, then add 10 ml/liter stock 10.

[n] Solid NaCl is added directly to NMC prior to adjusting overall volume to 1 liter.

Solution A. Solution A was then replaced with Solution B. After 15 minutes at 37°C, the cells were dislodged from the plates by gentle pipetting, sedimented by centrifugation at 125 *g*, and suspended in growth medium. The first subculture was seeded at 4×10^5 cells per collagen coated dish (Hauschka and Konigsberg, 1966). Collagen was omitted in all subsequent passages.

Solution A contains (in grams/liter): KCl, 8.4; NaCl, 7.1; $NaH_2PO_4 \cdot 7H_2O$, 1.9; KH_2PO_4, 0.005; $KHCO_3$, 1.0; PVP, 10; dialyzed FBSP (McKeehan and Ham, 1978), 9.75; glucose, 3.4; and HEPES, 4.8. Solution B contains (in grams/liter):

TABLE II

BALANCED SALTS DILUENT (BSD)

Stock (amount used/liter)[a]	Component	Amount in stock (gm/liter)	Amount in PFMR-4 (moles/liter)[b]
V (50 ml)[c]	NaCl	105.000	1.0 E-1
	KCl	7.600	3.8 E-3
	$Na_2HPO_4 \cdot 7H_2O$	4.000	8.1 E-4
	KH_2PO_4	1.580	4.3 E-4
VI (50 ml)[c]	$MgSO_4 \cdot 7H_2O$	1.050	1.6 E-4
	$MgCl_2 \cdot 6H_2O$	2.825	5.2 E-4
	$CaCl_2 \cdot 2H_2O$	3.600	9.2 E-4
No stock[d]	Glucose	1.260	7.0 E-3
Phenol red (0.33 ml)	Phenol red	5.000	5.9 E-6
No stock[d]	HEPES	9.520	3.0 E-2
pH adjusted to 7.40 with 1.0 *N* NaOH			
No stock[d]	$NaHCO_3$	1.600	1.4 E-2
Trace elements (13.3 ml)[e]	H_2SeO_3	0.3869 mg/liter	3.0 E-8
	$MnCl_2 \cdot 4H_2O$	0.0198 mg/liter	1.0 E-9
	$Na_2SiO_3 \cdot 9H_2O$	14.2100 mg/liter	5.0 E-7
	$(NH_4)_6Mo_7O_{24} \cdot 4H_2O$	0.1236 mg/liter	1.0 E-9
	NH_4VO_3	0.0585 mg/liter	5.0 E-9
	$NiSO_4 \cdot 6H_2O$	0.0131 mg/liter	5.0 E-10
	$SnCl_2 \cdot 2H_2O$	0.0113 mg/liter	5.0 E-10

[a] Add stocks to 500 ml distilled H_2O. Bring final volume of BSD to 1 liter after all components have been added.

[b] The letter E refers to the exponent of power of 10. Thus, 1.0 E-1 means 1.0×10^{-1}.

[c] Dissolve in order listed in 700 ml distilled H_2O, then bring volume to 1 liter.

[d] Powdered component added directly.

[e] Concentrated solutions (at 1.0 E-4) are used in compounding the trace element stock. $SnCl_2 \cdot 2H_2O$ stock is prepared in 2.0 E-2 *M* HCl.

KCl, 0.2; NaCl, 7.8; $NaH_2PO_4 \cdot 7H_2O$, 1.9; PVP, 10; glucose, 1.7; EGTA (ethylenebis(oxyethylnitrilo)tetracetic acid), 0.2; and HEPES, 4.8. The pH of each of the above solutions was adjusted to 7.4 with NaOH.

At the fourth passage, either trypsin or urea solutions were used to dissociate the cells. For both of these methods, the monolayers were first washed with HEPES-buffered saline (HBS), then incubated for 3–5 minutes in either trypsin (PET) or urea/B solutions. HBS contains (in grams/liter): NaCl, 7.1; KCl, 0.2; $NaH_2PO_4 \cdot 7H_2O$, 1.9; glucose, 1.7; and HEPES, 4.8; pH 7.4. PET consists of 1% PVP, 0.004% EGTA, and 0.02% trypsin in HBS. Urea/B consists of 0.5 *M* urea in "K-Pass" Solution B. For both methods, the dissociated cells were dislodged from the plate by gentle pipetting, washed by centrifugation (125 *g*, 5 minutes), and plated at 2×10^5 cells per dish.

D. Cryopreservation

Cultures were preserved in liquid nitrogen after 14, 17, and 20 population doublings (passages 3, 4, and 5). This permitted repeated experiments to be carried out with the cells at similar population doubling levels. For storage in liquid nitrogen, the cells were suspended in PFMR-4 medium, supplemented with 20% FBS and 7.5% dimethylsulfoxide. One to three million cells were dispensed into 1.2-ml borosilicate ampoules, sealed, and cooled to −70°C in a Linde Biological Freezer at the rate of −1°C/minute. The ampoules were then transferred directly to a liquid nitrogen refrigerator.

E. Soft Agar

A modification of the agar suspension culture procedure of Macpherson and Montagnier (1964) was used to evaluate anchorage-independent growth. Complete agar medium (CAM) was prepared by mixing 39 volumes of nonautoclavable agar medium (NAAM) with 61 volumes of autoclavable agar medium (AAM). One liter of NAAM was prepared by mixing: NMC, 581 ml; FBS, 179 ml; BSD stock V, 87 ml; glucose, 2.1 gm; HEPES, 18.3 gm: $NaHCO_3$, 3.1 gm in water to volume. The pH was adjusted to 7.4 with 1 *N* NaOH. One liter of AAM was prepared by mixing: BSD stock VI, 55.8 ml; 0.5% phenol red, 0.5 ml; Bacto-Agar (Difco), 4.9 gm; and water to volume. Plates were prepared 1 day in advance with a 1.5 ml CAM bottom layer. The cells were then inoculated in a 3-ml top layer. Melted AAM and NAAM were both brought to 37°C prior to mixing. Mixing, inoculating, and dispensing were all done within 10 minutes. The plates were incubated for 3 weeks before scoring for colonies.

F. Scanning Electron Microscopy (SEM)

Cultures were fixed and processed *in situ* for electron microscopy as described previously (Lechner *et al.*, 1978; Kaighn *et al.*, 1980). Cultures grown on 12-mm glass coverslips were first rinsed with HBS and then fixed at 4°C for 30–60 minutes in a mixture of 2% glutaraldehyde and 1% formaldehyde in 0.1 *M* sodium cacodylate buffer, pH 7.4, containing 0.10% calcium chloride. The samples were rinsed in cacodylate buffer containing 0.25 *M* sucrose and further fixed with 2% osmium tetroxide in 0.1 *M* cacodylate buffer, pH 7.4, for 1–2 hours at room temperature. Following fixation, the specimens were dehydrated in a graded series of ethanol and critical point-dried from CO_2. The specimens were sputter coated with gold–palladium prior to examination in the SEM.

G. Clonal Growth Assay

Frozen ampoules of cells were thawed at 37°C and cultured in growth medium for 1 day prior to an experiment. The cells were then dissociated with PET containing DNase I (Worthington), 0.125 μg/ml (Kaighn, 1973), then washed with HBS by gentle centrifugation before clonal plating. Four replicate petri dishes (Lux, 60 mm) per variable were used. Each dish received 200 cells in a final volume of 4 ml of medium. After 6–8 days growth, the clones were fixed in 10% formalin and stained with 0.25% crystal violet.

H. Clonal Growth Rate

The growth rate (R) was defined as the average number of population doublings per day. To measure this, the enlarged image of a clone was projected onto a screen through a prism attached to the ocular of an inverted microscope. A total of 16 clones (4 per culture dish) selected randomly were counted manually. The average number of cells per clone was corrected for the background growth which occurred in the absence of serum (3 cells per clone). The value was then converted to population doublings by dividing $\log_{10}$ (cells per clone) by $\log_{10}2$. R was obtained by dividing this value by the number of days of growth.

I. Growth Rate Parameters

The enzyme kinetic theory and formulations that Ellem and Gierthy (1977) applied to the measurement of DNA synthesis were adopted to relate R to mitogen concentration. Clonal growth rate (R) and the level of mitogen were defined as the velocity and substrate parameters, respectively. The influence of substrate concentration on R was measured in dose–response experiments. The data were analyzed by the Lineweaver–Burk method in which the reciprocal of R was plotted against the reciprocal of substrate concentration. The theoretical maximal growth rate, R^{T}_{MAX} was defined as the reciprocal of the Y intercept. The substrate concentration at which half-maximal growth occurred, $K^{mitogen}_{m}$, was the negative reciprocal of the X intercept.

Each dose–response experiment was analyzed by linear regression analyses according to formulas of Ellem and Gierthy (1977). Student's t test was used to evaluate the significance of the difference between R^{T}_{MAX} or $K^{mitogen}_{m}$ values derived from different experimental groups. All calculations were made with a programmable pocket calculator.

J. Culture Longevity Measurement

A modified clonal growth assay was used. This technique has several advantages. Cells growing at clonal density do not alter the composition of the medium significantly (McKeehan and Ham, 1978). Thus, the concentration of growth factors did not change appreciably during the experiment. The number of cells needed to initiate the experiment is less than 2500, and numerous replicate experiments can be conducted with the same culture. This is an important consideration, since normal cells have a finite population doubling potential.

Initially, eight sister clonal plates were inoculated. After incubation, four were stained and both the clonal plating efficiency and the average number of population doublings were determined. A subsequent series of eight clonal plates was then established using the pooled cells dissociated from the remaining unstained sister plates. This procedure was continued until cells no longer formed colonies.

III. Results

A. Primary Cultures

Two years ago, we obtained the prostate of a neonate (NP-1) who had died from respiratory-distress syndrome. The prostate was washed with PFMR-4, then minced to 1–2 mm^3 fragments. The fragments were allowed to sediment in a conical centrifuge tube. The "spilled" cells (supernatant) and fragments (sediment) were treated separately. The supernatant was divided and used to inoculate two 100-mm petri dishes. The fragments were divided into four equal portions; each was digested by a different enzyme preparation.

The enzyme preparations used were: A, 0.15% Pronase in HBS; B, CTC (0.1% collagenase, 0.1% trypsin, 1% chicken serum in HBS); C, 0.1% collagenase, 0.1% elastase, 0.1% hyaluronidase in HBS; and D, 0.1% trypsin, 0.02% EDTA in HBS. In all cases the fragments were incubated with 10 volumes excess of enzyme solution for 20 minutes at 37°C. The cells were allowed to sediment at 1 *g* for 5 minutes. The supernatant containing free cells was removed, washed twice with FBS-supplemented medium, and plated. Ten volumes of fresh enzyme solution was then added to the fragments. A total of three progressive dissociations were done on each portion of tissue. The results were as follows: Pronase (A) was toxic; only a few fibroblasts survived on the first dissociation plate. CTC (B) treatment yielded a few small aggregates of epithelial cells interspersed between a heavy monolayer of fibroblastic cells. The mixture of enzymes (C) resulted in fibroblastic cells only. No cells survived exposure to trypsin/EDTA (D).

The "spill" cultures differed significantly from those treated with enzymes. After 1 week of incubation, each plate contained roughly 500 epithelial colonies which varied in size from 10 to more than 100 cells per colony (Fig. 1A). There were no visible fibroblastic colonies.

Primary cultures have been established from subsequently acquired neonatal prostates using the "spill" and explant techniques. In addition, cold collagenase digestions of mince fragments (0.3% collagenase in growth medium at 4°C for 5–16 hours) have been tried. Although this procedure has been useful in estab-

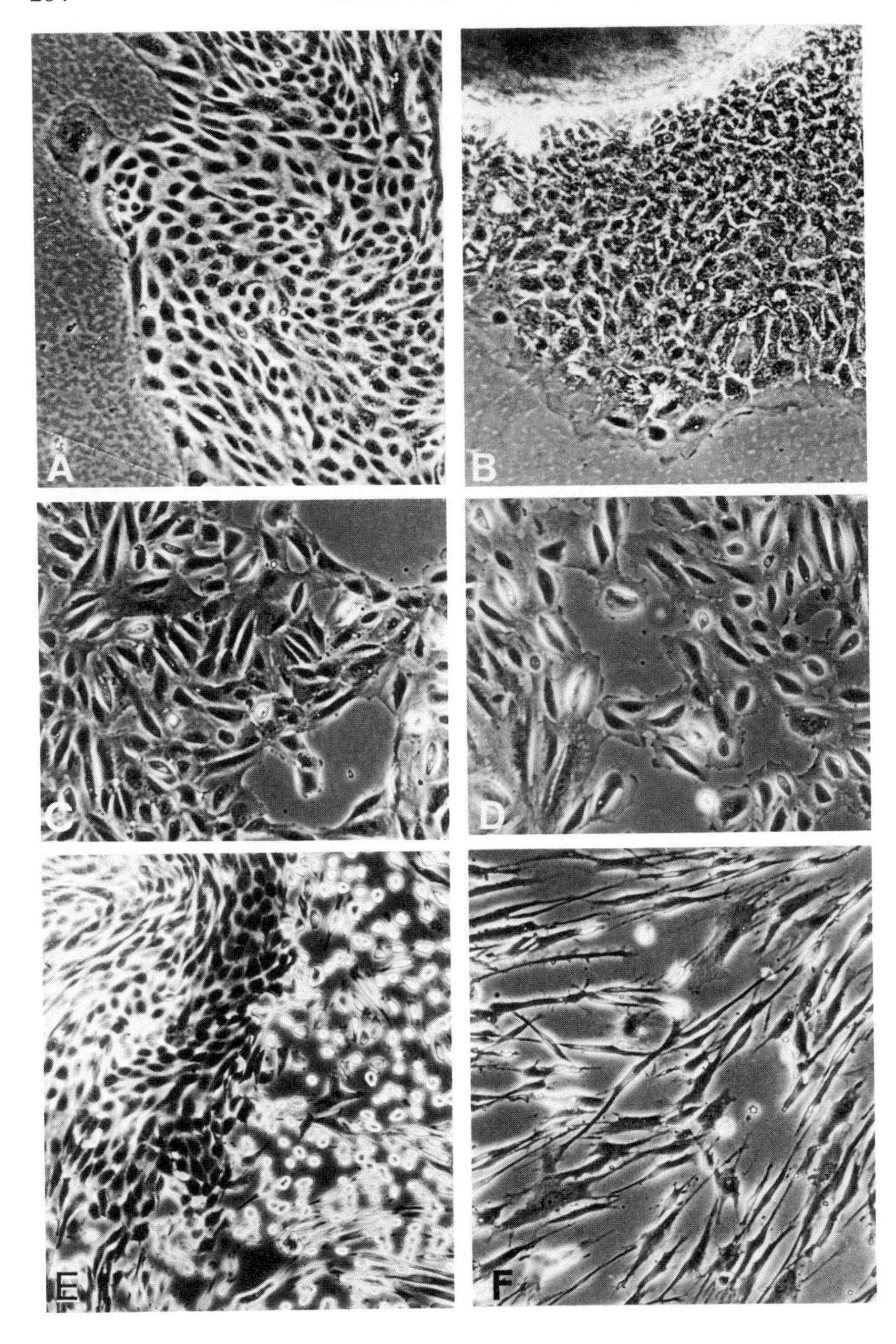
A
B
C
D
E
F

lishing primary cultures from prostatic adenocarcinoma tissue, excessive fibroblastic contamination precluded its use with normal epithelial tissue.

Replicative epithelial cultures have also been developed from explanted mince fragments. These cells rapidly migrated from the explant fragment and had a different appearance than the "spilled" cells (Fig. 1B). The explants were subcultured before fibroblasts migrated from the fragments using the nonenzymatic procedure (see following). By the fourth passage, both explant (e) and "spill" (s) cultures were similar in appearance (Fig. 1C and D).

B. Subculture of Primary Cultures

The two primary "spill" cultures from the first neonatal prostate were used to evaluate subculture methods. Since exposure to enzymes during dissociation resulted in predominately fibroblastic or nonviable cultures, this approach was avoided. First, simple scraping with a rubber policeman was tried. The cells of one semiconfluent plate were scraped from the dish and transferred to six subculture plates. Although these cultures eventually reached confluency, more than 95% of the cells were killed.

Secondary cultures were subcultured both by scraping and by exposure to various formulations containing a high concentration (0.15 *M*) of K^+. The rationale for using high K^+ to dissociate primary cultures was based on our unreported observation that leukemia cells reversibly expanded in Na^+-free medium containing 0.15 *M* K^+. In addition, it was known that high levels of extracellular K^+ rapidly affect the membrane potential of cultured cells (Orr *et al.*, 1972). Therefore, we suspected elevated K^+ levels might cause expansion of epithelial cells by alteration of their membranes to loosen the cell-to-cell contacts.

The effectiveness of each of these variations to dissociate and free the cells from the dish with minimal damage was assessed by observing their rate of reattachment. None of the formulations was superior to simple scraping. Most high K^+ formulations quickly induced the cells to round-up and separate from one another. Prolonged incubation was required to release them from the plate. Reattachment was invariably poor.

Initially, the best formulation consisted of PFMR-4 medium in which Na^+ was

---

FIG. 1. (A) Primary "spill" colony of prostatic epithelial cells after 1 week incubation. (B) Primary explant outgrowth of prostatic epithelial cells after 1 week incubation. (C) Phase micrograph of NP-2s culture at fourth passage. Replicative culture originated from primary "spill" cultures. (D) Phase micrograph of NP-2e culture at fourth passage. Replicative culture originated from primary explant cultures. (E) Phase micrograph of NP-2s culture after 10 minutes exposure to "K-Pass" Solution A. (F) Phase micrograph of fibroblastic skin culture (NF2) after 10 minutes exposure to "K-Pass" Solution A. Magnification A–F, ×120.

replaced by K^+. The culture (NP-1) was washed several times with HBS, then twice with Na^+-free PFMR-4 containing 0.15 *M* K^+. This "K-passing" medium was supplemented with 2.5% dialyzed FBS. After 15 minutes incubation at 37°C, the dish was washed with fresh "K-passing" medium. The cells rounded-up and some had detached after an additional 45 minutes at 37°C. At this point growth medium (PFMR-4 with 10 FBS) was added. The cells were freed from the plate by gentle pipetting, washed, and replated. Although viability was poor, successful passage of epithelial cells was accomplished.

When a second neonatal specimen became available, the method was refined. The following two-step procedure was developed. First, the culture was washed twice, then incubated for 15 minutes at 37°C in K^+ solution. K^+ solution was then replaced with a second solution containing enzymes or chelating agents. When the Na^+ in the HBS formulation was replaced by K^+, the cells rounded-up within 15 minutes but failed to detach from the dish. Sequential treatments with trypsin (0.1% in HBS) or EDTA (0.02% in HBS) dispersed the K^+-treated cells, but cell viability was less than 10%. However, 50% of the K^+-treated cells which were dislodged with EGTA (0.02% in HBS) were viable.

High osmolality had been reported as an effective means to dissociate liver tissue for primary cell culture (McLimans, 1969). We found that when the osmolality of the high K^+ medium was raised to 514 mOsm/kg (Solution A, see Section II) the cells rapidly separated from their neighbors (Fig. 1E). These rounded cells were freed from the plate by subsequent incubation with EGTA (Solution B, see Section II). This procedure yielded epithelial cells with a viability greater than 90%. Fibroblastic cells can also be subcultured with the "K-Pass" procedure. However, this cell type does not round-up in Solution A (Fig. 1F).

NP-2s cultures exposed to the "K-Passing" solutions have been examined in the scanning electron microscope in order to assess the effects of these solutions on cell-to-cell associations and on attachment of cells to the culture dish surface. After 10–30 seconds in Solution A, the cells showed perturbations over the nuclear and perinuclear regions (Fig. 2A). During the course of the next 2–5 minutes a majority of the cells showed a pronounced rolling of the cytoplasm around the nucleus (Fig. 2B). Extensive cytoplasmic projections still anchored the cells to the culture dish surface. Cell-to-cell connections were still apparent after 10 minutes incubation (Fig. 3A). However, prolonged incubation in Solution A resulted in the rounding-up of most cells.

The replicative cultures, which originated from both "spill" and explant pri-

FIG. 2. (A) NP-2s, second passage fixed 20 seconds after exposure to "K-Pass" Solution A. Note nuclear swelling. ×400. (B) NP-2s, second passage after 2 minutes in "K-Pass" Solution A. The cells are in varying stages of rounding and detachment. Note presence of numerous cell attachments to culture surface. ×650.

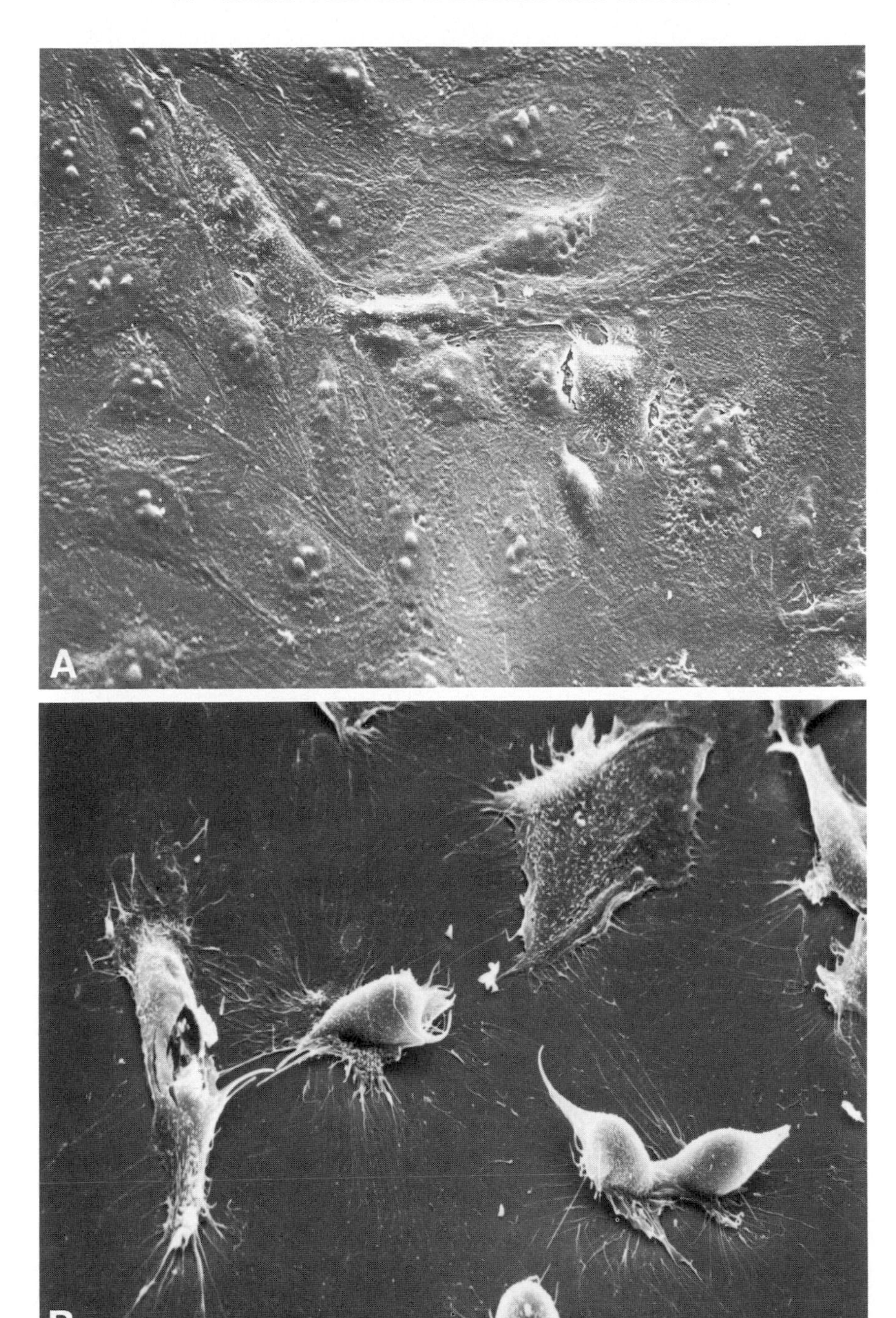
A
B

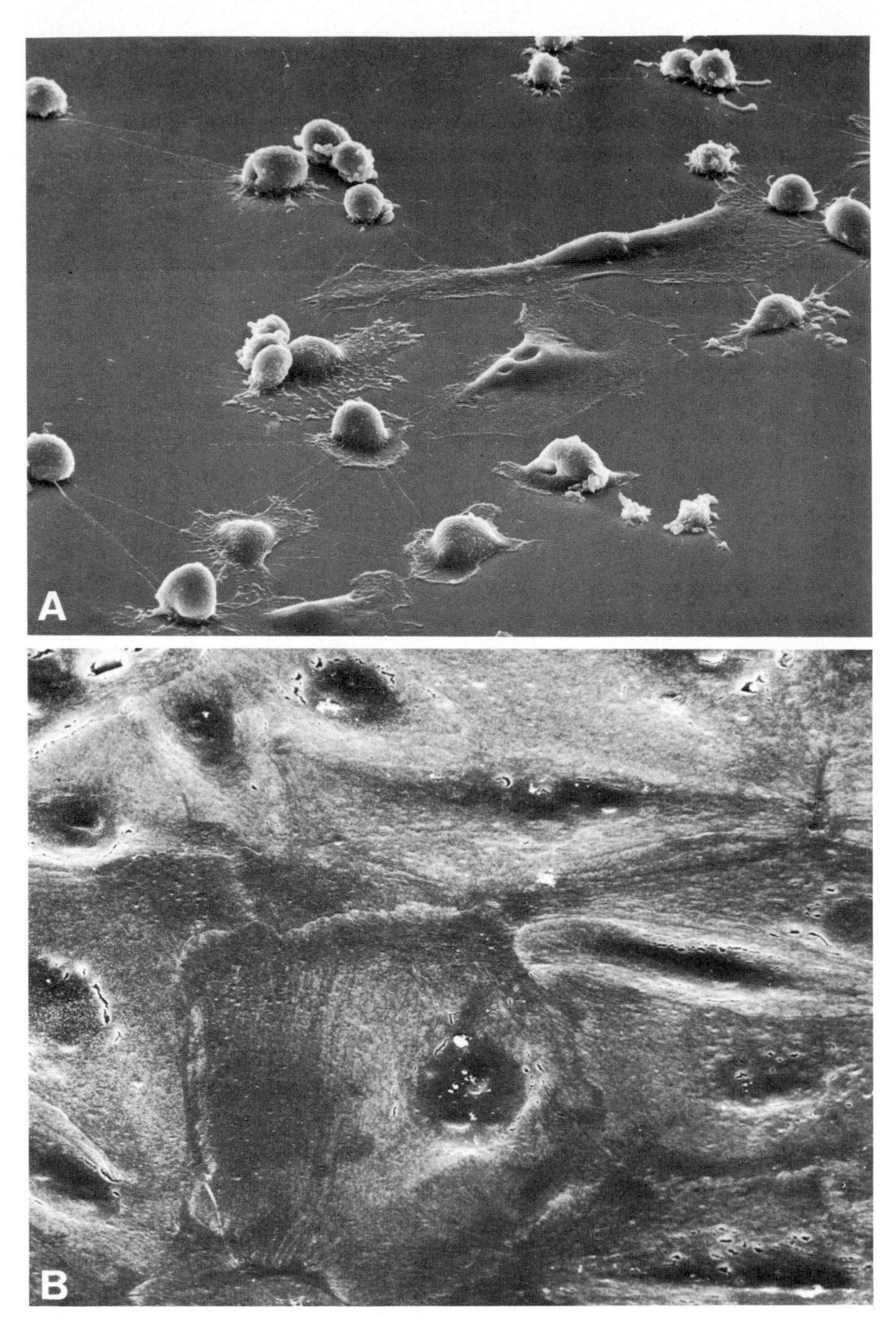

FIG. 3. (A) NP-2s, second passage after 10 minutes in "K-Pass" Solution A showing most cells completely rounded. ×650. (B) Control NP-2s, fourth passage. Note epithelioid shape of cells and close intercellular junctions. The cells exhibit no surface modifications. ×750.

mary cultures, were normal and epithelial by the criteria of ultrastructure, karyotype, and inability to grow in soft agar (Lechner *et al.,* 1978). Transmission electron microscopy at both the primary and fourth passages revealed cells with features typical of normal glandular epithelium. These cultures retained the normal human male karyotype at least through the ninth passage (25 population doublings). They did not grow in soft agar at the fourth passage. The surface of cultures at the fourth to sixth passages was smooth when examined by SEM (Fig. 3B). This is a usual characteristic of normal nontumorigenic cells.

C. Kinetic Analysis of Growth Factor Requirements

Selection of a superior lot of FBS was an important step toward developing these replicative normal epithelial cultures (Lechner *et al.,* 1978). This lot of FBS had been selected by clonal growth and plating efficiency criteria (Kaighn, 1974) using a fastidious, benign human prostatic epithelial culture (Kaighn and Babcock, 1975). Subsequently, normal human prostatic epithelial cells (NP-2s) were used to select a new lot of FBS. The test results with 23 different serum samples from 7 suppliers (Fig. 4A) revealed substantial differences in growth-promoting activities. Only two of the six lots which supported rapid growth of NP-2s promoted excellent growth of the genetically identical fibroblast culture, NF2. On the other hand, most lots which supported excellent growth of NF2 were unsuitable for NP-2s.

We sought to circumvent the problem of serum variability by supplementing serum-containing medium with several putative mitogenic factors. In preparation for these experiments, we determined the FBSP requirements for both NP-2s and NF2 by clonal assay dose–response experiments (Lechner and Kaighn, 1979a). The rate of cell multiplication was found to be proportional to FBSP concentration in a manner analogous to the effect of substrate concentration on enzymatic reaction rates (Fig. 5). The growth rate increased with serum concentration until a maximal rate was approached.

Since growth rate (velocity) was dependent upon serum (substrate) concentration until the maximal rate was approached, we utilized the principles of Lineweaver–Burk analysis (Ellem and Gierthy, 1977) to determine the $R_{\mathrm{MAX}}^{\mathrm{T\text{-}FBSP}}$ (Y axis) and the concentration of serum required for half-maximal growth, $K_{\mathrm{m}}^{\mathrm{FBSP}}$ (X axis). Using this approach, the levels of FBSP required for the normal epithelial and fibroblastic cells to multiply at half-maximal rates were found to be 2230 μg FBSP/ml (5.7% FBS) and 1480 μg FBSP/ml (3.8% FBS), respectively (Table III).

Serum was then adjusted to a suboptimal level to permit further increase in growth rate by the mitogen. Two growth factors (EGF and FGF) significantly increased the plating efficiency as well as the number of normal epithelial cells per clone. They also increased the number of fibroblastic cells per clone but did not significantly affect plating efficiency (Table IV). The relative potencies of EGF and FGF were assayed by titration. Regression analyses of the Lineweaver–Burk plots of these data showed that EGF was 12 times as potent a mitogen as FGF for NP-2s (K_m^{EGF} = 40 p*M*; K_m^{FGF} = 500 m*M*). In contrast, EGF and FGF were equally potent mitogens for NF2 (K_m^{EGF} = 86 p*M*; K_m^{FGF} = 127 p*M*, $p > 0.90$).

The extent to which each of these factors (5 ng/ml EGF and 30 ng/ml FGF) could substitute for serum was determined by varying the serum concentration (Fig. 6). Regression analyses of Lineweaver–Burk plots (Fig. 7) showed that both factors significantly reduced the FBSP requirements of NP-2s and NF2 (Table III). Since these data showed that EGF decreased the K_m^{EGF}, we determined whether FBSP changed the K_m^{EGF}. FBSP did not significantly affect the K_m^{EGF} (Table V).

The action of hydrocortisone (HC) was investigated since it has been reported to stimulate growth in a number of cell systems (Gospodarowicz *et al.*, 1978). NP-2s and NF2 were cloned in media containing a fixed level of FBSP and graded concentrations of HC. Hydrocortisone alone did not significantly increase the growth rates of either normal cell line. However, it did potentiate the activity of EGF and FGF (Fig. 8). Whereas EGF or FGF alone increased the growth rate 50%, the combination of either growth factor with HC doubled the growth rate of both cell lines.

The K_m^{EGF} of the normal epithelial cells was measured in both the presence and absence of 1×10^{-6} *M* HC. The values in the presence (29.5 p*M*) and the absence (40.0 p*M*) of HC were not significantly different ($p > 0.90$).

The combination of either EGF or FGF with HC synergistically reduced the K_m^{EGF} for NP-2s and NF2 cells 25-fold (Table III). Growth rates in media supplemented with no more than 200 μg FBSP/ml plus 1×10^{-6} *M* HC and either 5 ng/ml EGF or 30 ng/ml FGF were not significantly different from those observed in media containing 5850 μg FBSP/ml (15% serum) alone (Fig. 9; compare with Fig. 5).

D. Effect of EGF on Culture Longevity

The effect of EGF on culture longevity was investigated because it had been shown that this factor increases the culture lifespan of human skin keratinocytes (Rheinwald and Green, 1977). The epithelial and fibroblastic cultures responded

FIG. 4. (A) Comparative growth-promoting activity of 23 different lots of FBS on NP-2s. Each plate represents the average result of three replicates, each supplemented with 10% FBS. The plates were stained after 7 days incubation. (B) Appearance of NP-2s (fourth passage) after 5 days in PFMR-4 containing 1 mg/ml FBSP and 5 ng/ml EGF. The cell surfaces have many microvilli particularly over the nuclear region. Note the relatively loose attachment of the cells to the growing surface. ×1150.

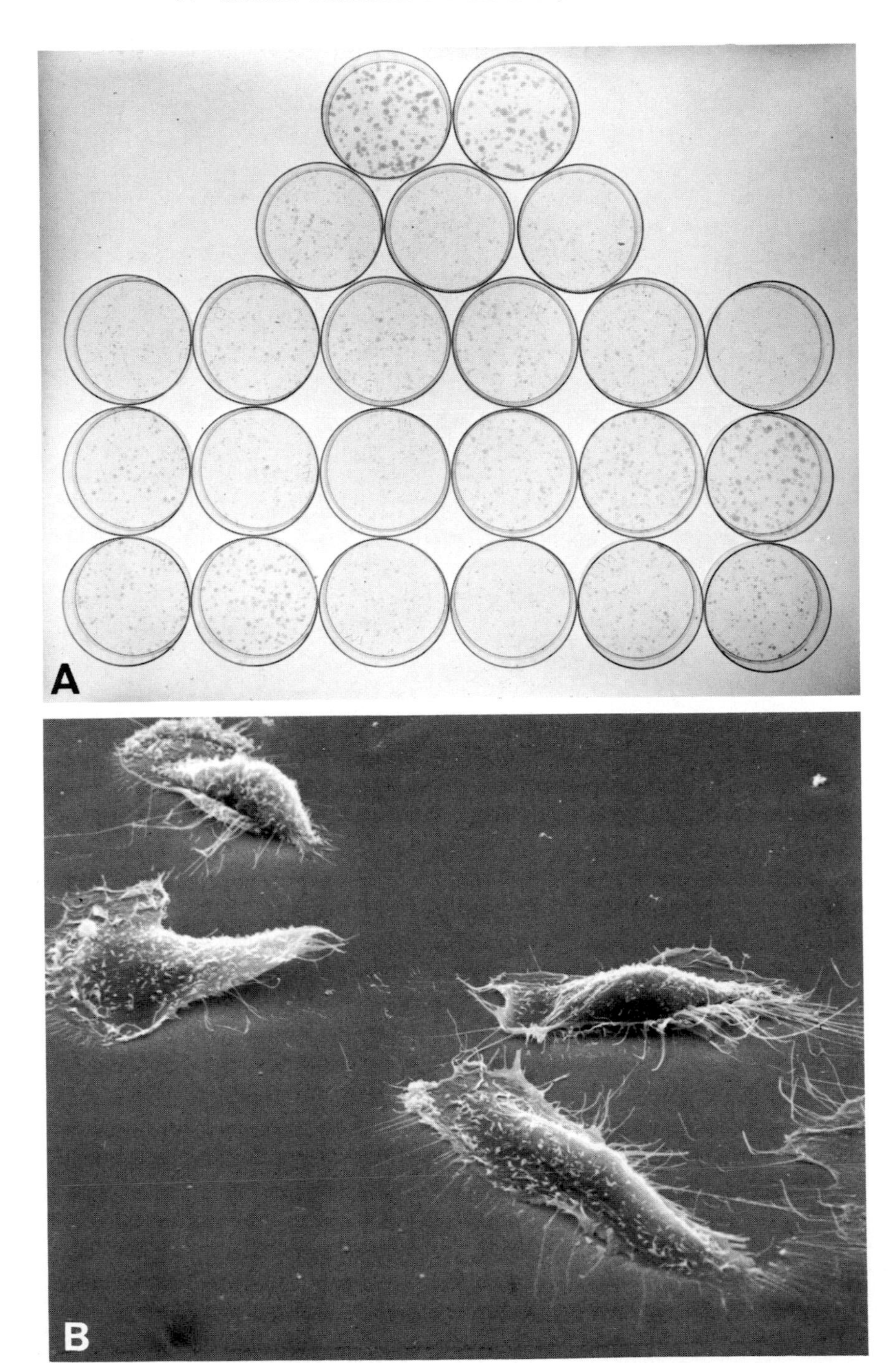
A
B

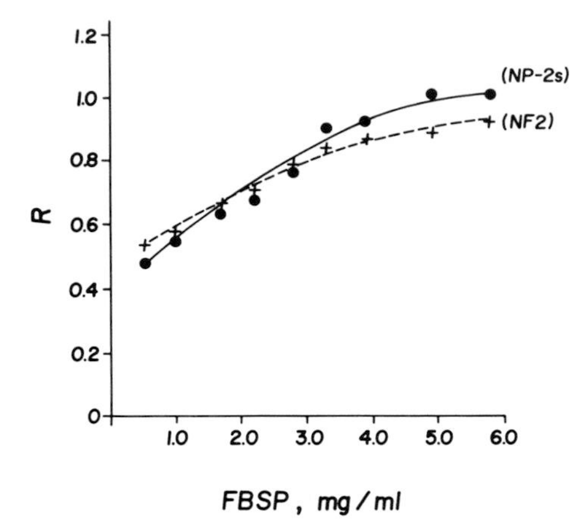

FIG. 5. Effect of FBSP on clonal growth rates. Cells were plated at 300 cells per dish in PFMR-4 medium containing increasing concentrations of FBSP. After 7 days incubation, the plates were stained and the growth rate (*R*) was determined as described in Section II. (●———●), NP-2s; (+———+), NF2.

TABLE III

SYNERGISTIC REDUCTION OF FBSP REQUIREMENTS OF NORMAL CELLS

	K_m^{FBSP} [a]	*p* value[b]	$R_{MAX}^{T\text{-}FBSP}$ [c]	*p* value[d]
NP-2s				
FBSP	2230 (987)	—	1.36 (0.13)	—
FBSP + EGF	627 (107)	<0.01	1.27 (0.13)	>0.90
FBSP + FGF	467 (60)	<0.01	1.18 (0.27)	>0.90
FBSP + EGF + HC	84 (20)	<0.01	1.12 (0.24)	>0.90
FBSP + FGF + HC	78 (30)	<0.01	1.19 (0.17)	>0.60
NF2				
FBSP	1480 (730)	—	1.26 (0.12)	—
FBSP + EGF	490 (162)	<0.01	1.22 (0.10)	>0.90
FBSP + FGF	350 (83)	<0.01	1.15 (0.15)	>0.90
FBSP + EGF + HC	40 (19)	<0.01	1.11 (0.09)	>0.90
FBSP + FGF + HC	34 (17)	<0.01	1.19 (0.17)	>0.90

[a] K_m^{FBSP} = μg/ml ($\pm$ standard error, μg/ml).

[b] *p* value (Student's *t* test) comparing K_m^{FBSP} value with factors to K_m^{FBSP} value without factors.

[c] $R_{MAX}^{T\text{-}FBSP}$ = theoretical maximal growth rate. Values derived by extrapolation using least-squares regression analyses ($\pm$ standard error).

[d] *p* value (Student's *t* test) comparing $R_{MAX}^{T\text{-}FBSP}$ value with factors to $R_{MAX}^{T\text{-}FBSP}$ value without factors.

TABLE IV

EFFECT OF EGF AND FGF ON CLONE SIZE

Cell line[a]	FBSP	Additions	Cell/clone[b]	p values[c]	PE[d]
NP-2s	(a) 2700	None	31 ± 18		5
	(b) 2700	EGF[e]	168 ± 79	b > a; 0.005	25
	(c) 2700	FGF[f]	119 ± 44	c > a; 0.005	26
NF2	(a) 2700	None	32 ± 13		28
	(b) 2700	EGF	73 ± 47	b > a; 0.005	29
	(c) 2700	FGF	101 ± 55	c > a; 0.005	33

[a] FBSP, μg/ml.
[b] Average of 16 clones ± standard error of the mean. Cells incubated for 7 days.
[c] From Student's t test.
[d] Plating efficiency (%) = number of clones/number of cells inoculated.
[e] EGF, 5 ng/ml.
[f] FGF, 30 ng/ml.

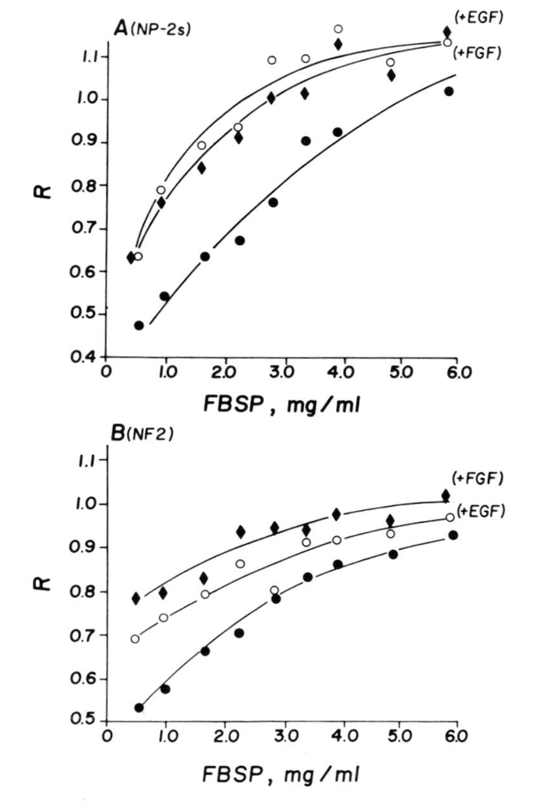

FIG. 6. Serum sparing by growth factors. Serum dose-response in the absence of factors (●——●), in the presence of 5 ng/ml EGF (○——○), and in the presence of 30 ng/ml FGF (◆——◆). Graph A, NP-2s; Graph B, NF2.

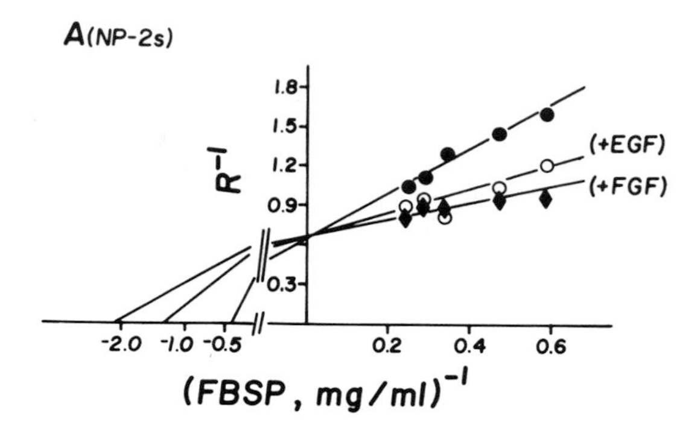

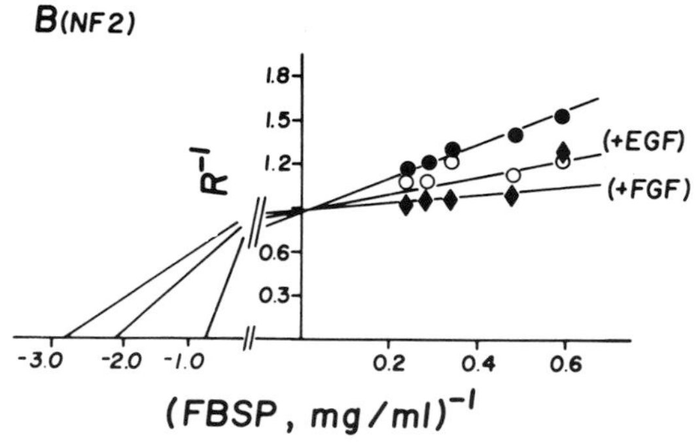

FIG. 7. Lineweaver–Burk plots of FBSP dose-response data (R^{-1} vs $FBSP^{-1}$). $R_{MAX}^{T\text{-}FBSP}$ and K_m^{FBSP} values were ascertained by linear least-squares regression to the Y and X intercepts, respectively. (A) NP-2s; (B) NF2. (●——●), FBSP without factors; (○——○), FBSP plus EGF; (◆——◆), FBSP plus FGF.

TABLE V

EFFECT OF FBSP ON K_m^{EGF}

μg FBSP/ml	K_m^{EGF} [a]	K_m^{EGF} [b]
1000	0.240 (0.093)	40
2000	0.290 (0.050)	48
2500	0.170 (0.030)	28
3000	0.260 (0.030)	43

[a] K_m^{EGF}, ng/ml (± standard error).

[b] K_m^{EGF}, pM.

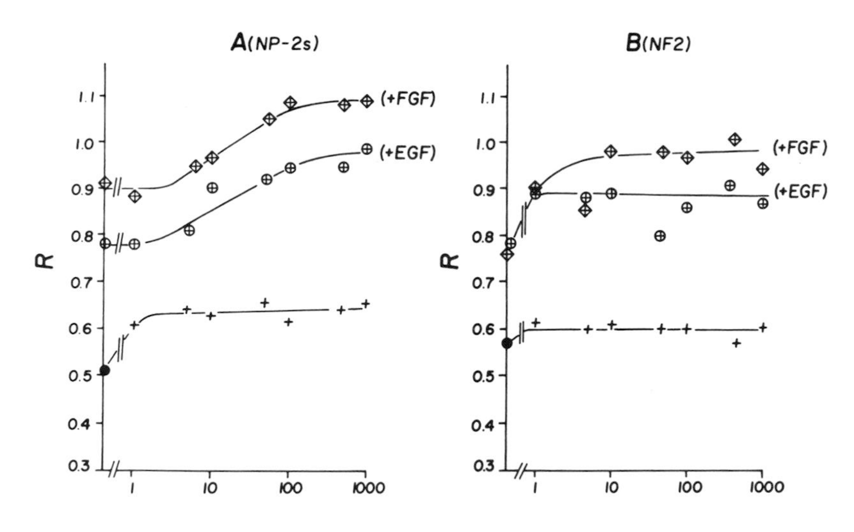

FIG. 8. Potentiation of growth factor activity by hydrocortisone. Graph A, NP-2s; graph B, NF2. Cells were cloned in PFMR-4 containing 1000 μg FBSP/ml (+——+) and either 5 ng/ml EGF (⊕——⊕) or 30 ng/ml FGF (◇——◇) supplemented with increasing concentrations of hydrocortisone (HC).

differently (Table VI). EGF improved both the clonal plating efficiency and growth rate of NP-2s and extended its culture lifespan by at least 10 population doublings (30%). In EGF-supplemented medium, the initial population doubling time (PDT) was 21 hours. This increased to 35 hours as the cells aged. Without

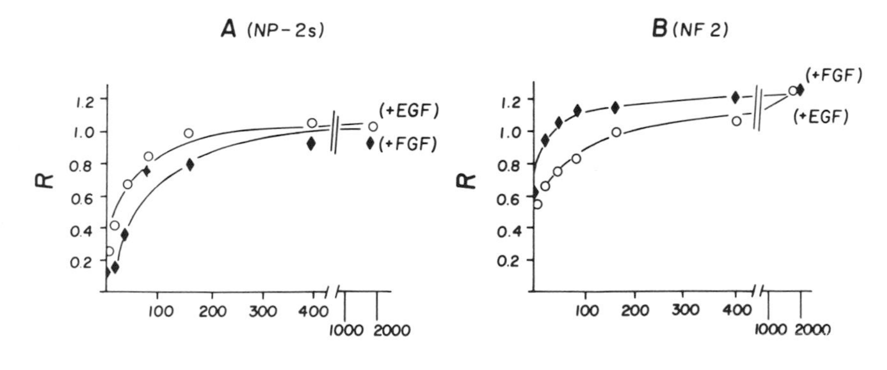

FIG. 9. Synergistic serum sparing by hydrocortisone (1×10^{-6} moles/liter) in combination with either EGF (○——○), 5 ng/ml, or FGF, (◆——◆), 30 ng/ml. Graph A, NP-2s; graph B, NF2. The cells were cloned in PFMR-4 containing growth factor and hormone supplemented with increasing levels of FBSP.

TABLE VI

CLONAL ANALYSIS OF LONGEVITY

Clonal passage	NP-2s				NF2			
	PE[a]	PD[b]	PDT[c]	CPD[d]	PE	PD	PDT	CPD
1. E_1[e]	5.3	11.3	21.1	26.3	32	9.04	26.5	33
C_1	5.0	8.12	29.5	23.1	38	8.16	29.4	32.2
2. E_1-E_2	25.0	9.01	21.3	35.3	18.0	8.6	27.9	41.6
E_1-C	25.3	7.04	27.3	33.3	ND[f]	—	—	—
C_1-E	11.0	8.72	22.0	21.8	ND	—	—	—
C_1-C_2	6.4	6.91	27.8	30.0	25.0	7.06	33.8	39.0
3. E_2-E_3	15	8.46	36.8	43.8	22.0	6.96	34.3	48.6
C_2-C_3	4.7	4.95	63.0	34.9	16.0	5.85	41.4	44.8
C_2-E	5.7	8.57	36.4	40.4	ND	—	—	—
4. E_3-E_4	2.0	5.09	33.0	48.9	14.0	10.46	34.3	59.1
C_3-C_4	0.07	2.4	70.0	37.3	21.0	10.71	33.8	55.5
5. E_4-E_5	ND	—	—	—	8.6	5.50	57.1	64.6
C_4-C_5	ND	—	—	—	22	5.53	57.1	61.0
6. E_5-E_6	ND	—	—	—	1	4.81	70.6	69.4
C_5-C_6	ND	—	—	—	5	5.12	66.7	66.1

[a] Plating efficiency (%).
[b] PD, average number of population doublings per clone.
[c] PDT, population doubling time in hours.
[d] CPD, cumulative *in vitro* population doublings.
[e] Letters denote clonal history. For example C_2-E means that the cells were first cloned in control C medium, subsequently subcloned (second passage) into control C medium then subcloned (third passage) into EGF supplemented medium.
[f] ND, not done.

EGF, the PDT increased from 30 to 70 hours. EGF was an effective mitogen even when the cells were first exposed to the growth factor after 30 population doublings. At this time, PDT was decreased from 63 to 36 hours.

Clonal plating efficiency of the fibroblast line, NF2, was somewhat depressed (38 to 32%) by supplementing the medium with EGF. However, it was mitogenic for approximately 45 population doublings. Thereafter, its effectiveness was marginal. EGF did not prolong the lifespan of this line. As was observed with the epithelial cells, the PDT increased from 28 to 170 hours as the cells aged, Hydrocortisone plus EGF extended the division potential of the epithelial cells to a greater extent than did EGF alone. An additional five to ten population doublings (data not shown) were obtained when HC was continually present. Hydrocortisone (± EGF) did not alter the lifespan of the fibroblasts.

E. Influence of EGF on Epithelial Cell Migration

While counting colonies, it was observed that EGF-treated cells formed spreading colonies with intercellular distances three to four times those of comparable controls. By counting the number of cells per unit area of similar sized colonies, we found that EGF-grown colonies had an average of 41 cells/mm^2, whereas control colonies averaged 60 cells/mm^2.

The surface structure of NP-2s cultures treated with EGF for both brief (5 to 120 minute) and extended (up to several days) duration has been studied by SEM. EGF-treated cells were generally smaller and displayed a significant swelling over the nuclear region. Short microvilli, microridges, ruffles, or blebs were noted on almost all cell surfaces, especially over the nuclear region (Fig. 4B). Many cells were elongated. In contrast to the symmetrical rounding-up and detachment of untreated cells in preparation for cell division, many EGF-treated cells showed a tendency toward asymmetric rounding and detachment from the culture vessel surface.

IV. Discussion

Established culture methods, newly developed techniques, and attention to cellular nutrition have made it possible to isolate reproducibly normal prostatic epithelial cell cultures capable of sustained replication. These cultures are normal epithelial cells on the basis of ultrastructure, karyotype, and inability to grow in soft agar. They were not clonally isolated (Kaighn, 1974) and are therefore representative of the total neonatal prostatic gland epithelial cell population.

The "spill" method used for primary cultures was developed 20 years ago by Lasfargues and Ozzelo (1958) to cultivate human breast carcinoma. Buehring and Williams (1976) also applied this method to initiate fibroblast-free cultures of normal human mammary epithelium. The major disadvantages of the "spill" method are the poor cell yield and the difficulty in quantitation due to the mixture of free cells, clumps of cells, and tissue fragments that result. These are not insurmountable problems though, since the cultures are virtually free of fibroblastic contamination and are capable of sustained multiplication.

We have also developed replicative epithelial cultures from explanted tissue fragments. This technique has been used to initiate prostate cell cultures since 1917 (Burrows *et al.*, 1917). Commonly, epithelial cells migrate from the tissue fragments before the fibroblasts appear (Waters and Walford, 1970). We took advantage of this phenomenon and subcultured the primary culture early, before fibroblastic cells became apparent.

Previous attempts to subculture prostatic epithelial monolayers by enzymatic procedures have been unsuccessful (Stone *et al.*, 1976; Webber, 1974). Our experience using a variety of such procedures was the same. The cells attached to the culture surface but failed to multiply. Trypsin, in particular, is known to reduce both clonal plating efficiency and growth rate of human fibroblasts (McKeehan, 1977). Epithelial cells are even more sensitive to trypsin than are fibroblasts. The new nonenzymatic subculture techniques we have developed have made it possible to subculture and expand normal prostatic epithelial cells.

The mechanism by which the "K-Pass" solution brings about cell rounding is unknown. Recent studies with the SEM and TEM (data not shown) suggest that attachments both between cells and between the cell and the culture surface are not broken directly by the high osmolar "K-Pass" solution. The TEM studies suggest that the "K-Pass" solution disrupts the cytoskeleton which in turn induces cell rounding. Tension produced during the rounding process then appears to decrease the strength of the cell-to-cell and cell-to-surface contacts. The "K-Pass" procedure was used up to the fourth passage. After that, it was possible to carry out subsequent passages with high plating efficiency using a very dilute trypsin mixture (PET). When examined by phase microscopy at this time, the cells retained their epithelial morphology, but had increased in size and appeared to adhere less tightly to their neighbors. This change in cellular morphology may reflect alterations in the cytoskeleton as a result of adaptation to the culture environment (Webber, 1975). The relationship between these changes and the manner by which sensitivity to trypsin treatment is reduced is still unclear. The problem is being studied in our laboratory.

The rationale for using urea to nonenzymatically subculture cells stems from studies on cell surface proteins. Yamada *et al.* (1975) extracted cell surface proteins from chick embryo fibroblasts with 1 *M* urea. Subsequently, it was reported that surface proteins could be extracted from endothelial monolayers by urea. This treatment induced the cells to temporarily round-up. After removal of the urea, normal cell shape returned and the cells remained viable (Loren Johnson, UCSF, San Francisco, CA, personal communication, 1978). Subculture of normal prostatic epithelial cells was accomplished by gentle pipetting of monolayers after 15 minutes exposure to 1 *M* urea in HBS. The technique was improved by using 0.5 *M* urea dissolved in "K-Pass" Solution B. Cells (fifth passage) dissociated by 5 minutes exposure to urea/B cloned with a 25% plating efficiency. This technique has not been tested on primary epithelial cultures.

Until recently, cell nutrition had been a neglected problem (Ham, 1974; Waymouth, 1974). Most standard media were designed for the mouse L or the human HeLa cell. Both of these established lines differ from normal cells in many ways. Media based on the nutritional requirements of these or other transformed lines only marginally support clonal growth of normal human fibroblasts (Ham and McKeehan, 1978). Even this marginal growth requries supplementation with at least 10% carefully selected serum.

Ham and co-workers have recently developed a medium for human fibroblasts (MCDB 105) which, when supplemented with only 0.02% serum, supports the growth of large colonies (McKeehan *et al.*, 1978). This reduction in serum requirement was made possible by first optimizing the levels of small-molecular-weight nutrients in medium F12 (the parent formulation of MCDB 105). Trace elements were added and hypoxanthine and folic acid were replaced by adenine and folic acid. In addition, the culture surface was coated with polylysine.

This approach has been used in our laboratory to develop more adequate media for normal prostatic epithelial cells. With the exception of including HEPES buffer, we have been unable to improve the basal medium, F12K (Kaighn, 1973) for prostatic epithelial cells by variation in low-molecular-weight compounds. However, differential response of fibroblasts and prostatic adenoma cells to dialyzed serum was dependent on the basal nutrient. Whereas dialysis permitted growth of WI-38 cells at lower serum levels, unknown factors necessary for growth of prostatic epithelial cells were removed (Kaighn, 1977). Differences in nutrient requirements have been noted between normal human cell types as well. Peehl and Ham (1978) described major small-molecular-weight nutrient concentration differences between media developed for human lung fibroblasts and for skin keratinocytes.

It has become clear that, in order to formulate an effective medium for normal human prostatic epithelial cells, attention must be paid not only to the small-molecular-weight nutrients but to critical growth factors present in FBS. Our experience has demonstrated that careful selection of each lot of FBS is mandatory if satisfactory clonal growth of normal human prostatic epithelial cells is to be achieved. Mather and Sato (1977) have shown that different lines of transformed cells all required insulin and transferrin for growth in serum-free medium. In addition to these two proteins, each line was found to require a specific profile of three to five other hormones or growth factors. Gospodarowicz *et al.* (1976) have shown that, in addition to FBS, bovine vascular endothelial cells will not produce colonies unless fibroblast growth factor is added. We determined that clonal growth of normal human prostatic epithelial cells is markedly enhanced by incorporating 3–10 ng/ml EGF or 30 ng/ml of FGF into media containing FBS.

The normal epithelial and fibroblastic cell lines, although genetically identical, responded differently to serum. The epithelial cells required significantly more serum than did the fibroblasts. The principles of enzyme kinetics were used to quantitate these differences.

Michaelis–Menten kinetics have rearely been used to measure a cell's response to nutritional factors (Ellem and Gierthy, 1977; McKeehan and Ham, 1978; Lechner and Kaighn, 1979a). However, this approach facilitates study of interactions between two or more agents which affect the rate of cell proliferation. Two descriptive and comparable cell characteristics are readily derived from kinetic analysis of dose–response experiments. These are the $K_m^{mitogen}$ and R_{MAX}^T parameters.

Ellem and Gierthy (1977) and McKeehan and Ham (1978) defined the K_m parameter in terms of mitogen half-maximal binding. We defined $K_m^{mitogen}$ as that mitogen concentration which supported growth at the half-maximal rate. Since the molecular mechanisms which bring about sustained growth are unknown, we suggest that the $K_m^{mitogen}$ value may reflect more than the simple binding of a growth factor to a surface receptor. It seems likely that the cumulative half-maximal growth rate is the result of an interdependent series of reactions.

The second cell growth characteristic derived by kinetic analysis is the R_{MAX}^T parameter. We consider R_{MAX}^T to be identical to the V_m and r_{max} terms as defined by Ellem and Gierthy (1977) and McKeehan and Ham (1978), respectively. Thus, R_{MAX}^T defines the theoretical maximal growth rate.

Growth of both epithelial and fibroblastic cells was increased by EGF and FGF. On a molar basis, EGF was more potent than FGF for the epithelial cells, whereas both growth factors were equally effective for the fibroblasts. Kinetic analysis of the serum growth response data showed that serum-sparing by growth factors was correlated with a significant reduction of the K_m^{FBSP}. On the other hand, variation of the FBSP level did not significantly affect the K_m^{EGF}.

Hydrocortisone potentiated the mitogenic action of both growth factors. Overall more than a 25-fold reduction in the K_m^{FBSP} of both normal cell lines was accomplished by incorporating either EGF or FGF along with hydrocortisone into the growth medium. Glucocorticoids increase EGF binding by human diploid fibroblasts (Baker *et al.*, 1978). These workers proposed that the synergistic action of glucocorticoids was due to modulation of the number of EGF surface receptors. We found that hydrocortisone did not significantly change the K_m^{EGF}. Instead, the combination of EGF and hydrocortisone synergistically reduced the K_m^{FBSP} value. Although hydrocortisone may increase EGF binding, our results suggest that the amount of EGF required for optimal cell growth may not be a direct consequence of cell binding. Further experimentation is necessary, therefore, to fully elucidate the mechanism of synergism between EGF and hydrocortisone.

In a separate study (Lechner and Kaighn, 1978, 1979b), we showed that EGF influenced the potency of calcium as a growth-promoting agent. EGF reduced the $K_m^{Ca^{2+}}$ 100-fold. On the other hand, the Ca^{2+} concentration did not affect the K_m^{EGF}. McKeehan and Ham (1978) have shown a similar relationship between Ca^{2+} and Mg^{2+} using the same approach.

EGF has been observed to stimulate epidermal cell migration in chick embryo aggregates (Cohen 1965). Rheinwald and Green (1977) noted that EGF induced spreading and flattening of human epidermal keratinocytes. EGF stimulated human prostatic epithelial cell motility as well. On the other hand, cells grown in

FGF-supplemented media were arranged in tightly associated packets. As viewed in the SEM, the asymmetric detachment of the cells together with their elongated morphology is suggestive of rapidly moving cells (Vasilev and Gelfand, 1977). These surface characteristics were not generally noted on cells grown in the absence of EGF.

EGF increased the division potential of the epithelial cells but had no effect on the longevity of the genetically identical fibroblastic cells. The mechanism whereby EGF extends culture lifespan is not understood. Rheinwald and Green (1977) suggested that EGF extended the division potential of human skin keratinocytes by increasing cell multiplication and, thus, circumventing terminal differentiation. When division ceased, the keratinocytes differentiated into non-dividing cells which resembled stratified corneum. We have been unable to distinguish a definitive, terminally differentiated stage of the prostatic epithelial cultures. Thus, EGF may have extended the division potential in this system by other means. EGF is known to alter cell permeability and increase the rate of nutrient transport (Hollenberg and Cuatrecasas, 1975; DiPasquale *et al.,* 1978). Inadequate nutrition has often been suggested as an explanation for growth limitation in culture (Hay *et al.,* 1968; Waymouth, 1974). In our epithelial cell system, EGF may be indirectly correcting medium imbalances by altering the levels of various components necessary for growth. This EGF "tuning" of the medium may make it more efficient and thus indirectly extend the division potential of the cells. Further experimentation is necessary to test this hypothesis.

V. Perspectives

A. Current Applications and Findings

The availability of normal human prostatic epithelial cells and a fibroblast line isolated from the same subject has made possible several lines of investigation. These are:

1. Response to growth factors with respect to rate of cellular multiplication, migratory behavior, ultrastructure, and morphology. These studies are being carried out utilizing kinetic analyses which have been applied during the course of the work detailed in this chapter.
2. Interaction of growth factors with tumor promoters and cocarcinogens. Again, these interactions are being studied using kinetic analysis.
3. *In vitro* carcinogenesis by viruses and chemicals. The purpose here is to study sequential changes in growth control, tumorigenicity, and ultrastructural and biochemical characteristics.

The first project, response to growth factors, is underway and has already been partially described here. We are measuring the comparative responses of epithelial and fibroblastic cells to peptide and steroid growth factors. The effect of these different factors on the K_m values of the small-molecular-weight nutrients as well as their effect on one another are under study.

The interaction of growth factors with phorbol esters is also under investigation. The differential influence of phorbol esters on the growth rates of human cells was unexpected (Fig. 10). The growth rate of the epithelial cells was stimulated by TPA (12-*O*-tetradecanoyl-phorbol-13-acetate), whereas, the growth rate of the fibroblast line was not significantly changed (Lechner and Kaighn, 1979c). Phorbol esters have been reported to decrease EGF binding (Lee and Weinstein, 1978). For this reason, we examined the interaction of TPA and EGF on growth. The growth rates of both cell lines were decreased by TPA to levels comparable to those observed in the absence of EGF (Fig. 10). In contrast to the observations of Dicker and Rozengurt (1978), inclusion of insulin in the medium did not affect these results. These workers noted a synergistic growth

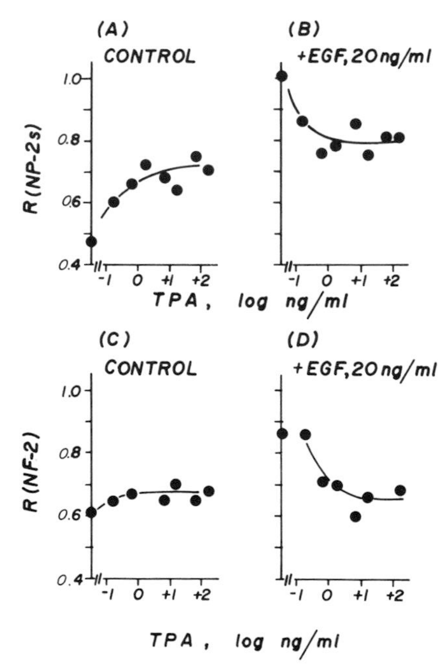

FIG. 10. Interaction of TPA with EGF. Growth rates of NP-2s (A) and NF2 (C) were determined in PFMR-4 medium supplemented with 1500 μg/ml FBSP, both without and with TPA (0.15 to 150 ng/ml). The experiment was then repeated (B, NP-2s; D, NF2) using identical media, each further supplemented with 20 ng/ml EGF. TPA stimulated growth of the epithelial (A; NP-2s) but not the fibroblastic (C, NF2) cells. On the other hand, TPA inhibited the EGF-stimulated growth of both cell types.

stimulation by insulin in the presenceof EGF and TPA. Probably this difference was due to the fact that different cell types were used. Dicker and Rozengurt worked primarily with mouse 3T3 cells. These results underline the importance of understanding that different cell types and cells from different species respond in dissimilar ways to tumor promoters.

The third project was begun at the end of October 1978. SV40 virus was used to transform both the epithelial and fibroblastic human cell lines. Cells at the twentieth population doubling level were exposed to 100-fold multiplicity of SV40 virus. The infected cells were washed and replated after 3 hours in EGF-supplemented medium at densities varying from 3,000 to 30,000 cells per plate. Large, multilayered colonies were observed 6 weeks after incubation. At this time, the uninfected control colonies were relatively small and had ceased to divide. The differential growth permitted easy identification of the transformed colonies. We scored these colonies as presumptive transformants. The approximate transformation rate in this first experiment with the epithelial line was about 10^{-3} compared to approximately 10^{-5} in the fibroblastic line. It should be emphasized that this experiment was not designed to quantitate transformation but surprisingly the rate of transformation appeared to be clearly higher in the epithelial line than in the fibroblasts.

We have partially characterized three epithelial and one fibroblastic transformed clones. All four are able to grow in soft agar. Growth rates of all four lines are stimulated by both EGF and FGF. EGF was also found to increase the clonal plating efficiency in soft agar of one of the epithelial lines 5-fold. A well-known characteristic of transformed cells is reduced serum-dependence. The K_m^{FBSP} of all four transformed clones was found to be 25-fold lower than that of the parent line, NP-2s. One transformed line, NP-2s/T2, can now be grown in our basic medium, PFMR-4, supplemented with only EGF, 5 ng/ml, hydrocortisone, 10^{-7} *M*, and crystalline bovine serum albumin, 150 μg/ml. Further characterization of these transformed lines is in progress.

B. Future Directions

We intend to continue with the three lines of investigation listed in Section V, A. In addition to work in progress, several other problems related to human prostatic cancer can be investigated in our system. First, prostatic cancer is a disease of older men and increases both in incidence and mortality rate with age. The culture lifespan of the normal neonatal prostatic epithelial cells is approximately 30–35 population doublings. This can be extended to 45 population doublings by treatment with EGF. One question that could be asked in this system is: Do the cells change in susceptibility to viral and chemical transformation and promotion as a function of *in vitro* lifespan? Information derived from these experiments could be correlated with *in vivo* changes by isolating cells from cadavers of different ages to compare donor age with susceptibility and tumor promotion. Further, cells from patients with benign prostatic adenoma could be tested to compare their susceptibility to transformation and promotion to that of normal cells. The efficacy and mode of action of tumor-inhibiting substances such as retinoids (Sporn *et al.*, 1976) could also be studied.

We have demonstrated changed growth control of SV40-transformed cells. Studies on the progression of such changes could be investigated. Further, we have observed that 48 hours after exposure to SV40, surface morphology as revealed by scanning electron microscopy is radically altered. This surface morphology is similar to that of prostatic carcinoma cells (Kaighn *et al.*, 1979), and of the SV40-transformed lines. These observations, together with similar observations reported for other cancer cells (Gonda *et al.*, 1976; Porter and Fonte, 1973), suggest that surface changes may be an early event. Perhaps these surface alterations are involved in altered growth control. Future investigations will be centered on identifying sequential changes that occur as a normal cell progresses toward the transformed state.

Another project that needs serious consideration is the study of the interaction of the stromal and epithelial components of the prostate (Franks, 1977). To do this, cells from the stromal and alveolar portions of the gland can be separately isolated. The interactions of hormones, growth factors, vitamins, and small-molecular-weight nutrients with these two types of cells would be compared. Both Moscona type (Moscona, 1962) cell mixing and feeder layer experiments can be done. The question can be asked: Do androgens and other factors act on the mixed prostatic cells differently than they do on the individual cell types? These types of experiments could also be extended to evaluate putative chemotherapeutic agents.

References

Baker, J. B., Barsh, G. W., Carney, D. H., and Cunningham, D. D. (1978). *Proc. Natl. Acad. Sci. U.S.A.* **75,** 1882–1886.

Buehring, C. G., and Williams, R. R. (1976). *Cancer Res.* **36,** 3742–3747.

Burrows, M. T., Burns, J. E., and Suzuki, Y. (1917). *J. Urol.* **1,** 3–15.

Cohen, S. (1965). *Dev. Biol.* **12,** 394–407.

Dicker, P., and Rosengurt, E. (1978). *Nature (London)* **276,** 723–726.

DiPasquale, A., White, D., and McGuire, J. (1978). *Exp. Cell Res.* **116,** 317–323.

Ellem, K. A. O., and Gierthy, J. F. (1977). *J. Cell. Physiol.* **92,** 381–400.

Franks, L. M. (1977). *Urol. Res.* **5,** 159–162.

Gonda, M. A., Aaronson, S. A., Ellmore, N., Zeve, V. Y., and Nagashima, K. (1976). *J. Natl. Cancer Inst.* **56,** 245–263.

Gospodarowicz, D., Moran, J. S., Braun, D., and Birdwell, C. (1976). *Proc. Natl. Acad. Sci. U.S.A.* **73,** 4120–4124.

Gospodarowicz, D., Mescher, A. L., and Birdwell, C. R. (1978). *Natl. Cancer Inst. Monogr.* **48,** 109–130.

Ham, R. G. (1972). *In* "Methods in Cell Physiology" (D. M. Prescott, ed.), pp. 37–74. Academic Press, New York.

Ham, R. G. (1974). *In Vitro* **10,** 119-129.
Ham, R. G., and McKeehan, W. L. (1978). *In Vitro* **14,** 11-22.
Hauschka, D. S., and Konigsberg, I. R. (1966). *Proc. Natl. Acad. Sci. U.S.A.* **55,** 119-126.
Hay, R. J., Menzies, R. A., Morgan, H. P., and Strehler, B. L. (1968). *Exp. Gerontol.* **3,** 35-44.
Hollenberg, M. D., and Cuatrecasas, P. (1975). *J. Biol. Chem.* **250,** 3845-3853.
Kaighn, M. E. (1973). *In* "Tissue Culture. Methods and Applications" (P. F. Kruse, Jr. and M. D. Patterson, Jr., eds.), pp. 54-58. Academic Press, New York.
Kaighn, M. E. (1974). *J. Natl. Cancer Inst.* **53,** 1437-1442.
Kaighn, M. E. (1977). *Cancer Treat. Rep.* **61,** 147-151.
Kaighn, M. E., and Babcock, M. S. (1975). *Cancer Chemother. Rep.* **59,** 59-63.
Kaighn, M. E., Lechner, J. F., Narayan, K. S., and Jones, L. W. (1978). *Natl. Cancer Inst. Monogr.* **49,** 17-21.
Kaighn, M. E., Narayan, K. S., Ohnuki, Y., Lechner, J. F., and Jones, L. W. (1979). *Invest. Urol.* **17,** 16-23.
Lasfargues, E. Y., and Ozzelo, L. (1958). *J. Natl. Cancer Inst.* **21,** 1131-1147.
Lechner, J. F., and Kaighn, M. E. (1978). *J. Cell Biol.* **79,** 66a.
Lechner, J. F., and Kaighn, M. E. (1979a). *J. Cell Physiol.* **100,** 519-530.
Lechner, J. F., and Kaighn, M. E. (1979b). *Exp. Cell Res.* **12,** 432-435.
Lechner, J. F., and Kaighn, M. E. (1979c). *In Vitro* **15,** 227.
Lechner, J. F., Narayan, K. S., Ohnuki, Y., Babcock, M. S., Jones, L. W., and Kaighn, M. E. (1978). *J. Natl. Cancer Inst.* **60,** 797-801.
Lee, L., and Weinstein, I. B. (1978). *Science* **202,** 313-315.
Lowry, O., Rosebrough, N. J., Farr, A. L., and Randall, R. J. (1951). *J. Biol. Chem.* **193,** 265-275.
McKeehan, W. L. (1977). *Cell Biol. Int. Rep.* **1,** 335-343.
McKeehan, W. L., and Ham, R. G. (1976). *J. Cell Biol.* **71,** 721-734.
McKeehan, W. L., and Ham, R. G. (1978). *Nature (London)* **275,** 756-758.
McKeehan, W. L., Genereux, D. P., and Ham, R. G. (1978). *Biochem. Biophys. Res. Commun.* **80,** 1013-1021.
McLimans, W. F. (1969). *In* "Axenic Mammalian Cell Reactions" (G. L. Tritsch, ed.), pp. 307-367. Dekker, New York.
Macpherson, I., and Montagnier, L. (1964). *Virology* **23,** 291-297.
Mather, J., and Sato, G. (1977). *In* "Cell, Tissue and Organ Cultures in Neurobiology" (S. Fedoroff and L. Hertz, eds.), pp. 619-630. Academic Press, New York.
Moscona, A. A. (1962). *Int. Rev. Pathol.* **1,** 371-428.
Orr, C. W., Yoshikawa-Fukada, M., and Ebert, J. D. (1972). *Proc. Natl. Acad. Sci. U.S.A.* **69,** 243-247.
Peehl, D. M., and Ham, R. G. (1978). *J. Cell Biol.* **79,** 78a.
Porter, K. R., and Fonte, V. G. (1973). *In* "Scanning Electron Microscopy" (O. Johari and I. Corvin, eds.), Vol. I, pp. 684-689. IIT Research Institute, Chicago, Illinois.
Rheinwald, J. G., and Green, H. (1977). *Nature (London)* **265,** 421-424.
Sporn, M. B., Dunlop, N. M., Newton, D. L., and Smith, J. M. (1976). *Fed. Proc.* **35,** 1332-1338.
Stone, K. R., Stone, M. P., and Paulson, D. F. (1976). *Invest. Urol.* **14,** 79-82.
Vasilev, J. M., and Gelfand, I. M. (1977). *Int. Rev. Cytol.* **50,** 159-274.
Waters, H., and Walford, R. L. (1970). *J. Gerontol.* **25,** 381-383.
Waymouth, C. (1974). *J. Natl. Cancer Inst.* **53,** 1443-1448.
Webber, M. M. (1974). *J. Urol.* **112,** 798-801.
Webber, M. M. (1975). *J. Ultrastruct. Res.* **50,** 89-102.
Yamada, K. M., Yamada, S. S., and Pastan, I. (1975). *Proc. Natl. Acad. Sci. U.S.A.* **72,** 3158-3162.

Chapter 10

The Human Placenta in Cell and Organ Culture

KURT STROMBERG

Viral Pathology Section,
Laboratory of Viral Carcinogenesis,
Division of Cancer Cause and Prevention,
National Cancer Institute,
Bethesda, Maryland

I. Introduction

The human placenta represents a gift from nature for biological investigation. As perhaps the most available human organ source, the placenta participates in nutritional, metabolic, excretory, endocrine, and immunologic functions which are pivotal to fetal growth and development. The life cycle of the placenta is brief and self-limited; explosive growth with endometrial invasion occurs early in gestation only to end with rejection at parturition. This complex array of regulated gene activity resides in a structurally simple organ. The hemomonochorial

ISBN 0-12-564140-0

placenta of man consists merely of a layer of epithelial cells (the inner cytotrophoblast and the outer syncytiotrophoblast) overlaying a basement membrane which lines mesenchymal support tissue that is arranged in a highly villus pattern. The ready availability, biochemical complexity, and structural simplicity of the placenta consequently lures many investigators of diverse backgrounds to it as a natural subject for study.

This chapter will summarize briefly the major biochemical markers of placental function, review historically the principal approaches to maintenance of human placental tissue in an *in vitro* state, evaluate the various culture methods and their important variables in respect to biochemical markers, and describe in detail those culture methods which appear to hold promise for continued use. The aim is to provide a current reference source for *in vitro* approaches to the study of placental functions, and to suggest what research opportunities are provided by placenta tissue in trophoblast monolayer and placental organ culture.

Polypeptide hormones [principally human chorionic gonadotropin (HCG) and human placental lactogen (HPL)] and steroid hormones (estrogens and progesterone) comprise the two main categories of hormonal markers of placental function. The structure, biosynthesis, and physiological role of the nonsteroid hormones have been recently reviewed (Villee, 1977; Chatterjee and Munro, 1977; Gusseck, 1977; Hobbins and Berkowitz, 1978), as have the steroid-related activities of human placenta (Saure, 1973; Levitz and Young, 1977). More than 60 enzymes of placental origin have been described (Hagerman, 1964) indicating that the placenta is capable of metabolic activities resembling in part those of the liver, kidney, intestinal mucosa, and certain endocrine glands, in addition to possessing some special enzymatic activities such as heat-stable alkaline phosphatase (Moriyama *et al.,* 1974; Sussman, 1978). A pregnancy-specific β_1 glycoprotein has also been described in human trophoblast (Bohn, 1978). Reviews on the biology of trophoblast (Billington, 1971; Padykula, 1977) and the gross, microscopic, and fine structure of human placenta (Boyd and Hamilton, 1970; Bjorkman, 1970) are also available.

II. Methods

On the occasion of her seventy-fifth birthday Festschrift, a Grande Dame of organ culture studies, Honor B. Fell, defined organ culture as "quite simply the maintenance of tissues in a differentiated functional state in a nutrient medium *in vitro.*" When asked if differentiated tissues really grow in organ culture, she replied: "Although cell division takes place, . . . the study of growth, in the sense of cell multiplication, is seldom the primary object of the organ culture technique; rather, the method is designed to provide an environment that will permit differentiated tissues to exercise their normal functions under the closely controlled conditions obtainable in an *in vitro* system'' (Balls and Monnickendam, 1976, p. 1). In an effort to maintain ''the differentiated functional state'' spoken of by Dr. Fell, the wide scope of approaches employed with human placenta have included: (*a*) perfusion of either the entire placenta or isolated cotyledons; (*b*) mechanical and enzymatic disruption of placental structure with seeding of trophoblast cells in monolayer culture; (*c*) mincing of placental villi into approximately 1 to 2 mm^3 pieces as explant culture; and finally, (*d*) the use of larger segments of about 1 cm in size as organ culture. In the transition from the *in vivo* situation to the *in vitro,* one, of course, disorganizes the tissue under study. The greater the degree of tissue processing the greater the disorder, and hence one might expect greater loss of normal functional capacity.

A. Perfusion

The simple anatomy and circulation of the placenta makes it an ideal organ for perfusion techniques. In addition, perfusion would appear to provide the least disruption of placental integrity during *in vitro* manipulation. Lerner *et al.* (1972) have reviewed the development of perfusion techniques for the placenta and the feto-placental unit for the previous 15 years. An important variable in their own system was oxygen tension. The use of an air and CO_2 mixture (approximately 23% O_2 and 4% CO_2) yielded a normal pH and nearly normal pO_2 and pCO values, whereas oxygenation with 94% O_2 and 6% CO_2 resulted in a low pH, highly elevated pO_2, and almost normal pCO_2 values. A dual closed-circulation system for extracorporeal perfusion of human placenta was described by Nesbitt *et al.* (1970). More recently, Cedard and Alsat (1975), in a review article on perfusion of either the entire placenta or isolated cotyledons, also favored a less toxic gas mixture of 60% N_2, 5% CO_2, and 35% O_2 over 5% CO_2 and 95% O_2. Utilizing a complicated dual circulation system, Guiet-Bara (1977) presented evidence that a gas mixture of 5% O_2, 5% CO_2, and 90% N_2 most closely imitates the oxygen concentration in the fetal blood circulation in the umbilical arteries. Consequently, a perfusion variable to avoid is exposure to excessive oxygen concentration which is detrimental to placental function.

Perfusion systems require fresh placental specimens in excellent condition obtained preferably by hysterotomy, the mechanics of *in vitro* maintenance are elaborate, and the duration of perfusion is rarely over 6 hours. For example, Nesbitt *et al.* (1971) have noted that only 10% of the placental material available is usually suitable for investigative purposes because of tissue damage. However, for physiologic studies of feto-placental relationships and pharmacologic studies of drug metabolism the perfusion approach most closely simulates the *in vivo* condition. That the synthetic steroids dexamethasone and betamethasone are not metabolized to an inactive state by placental minces in explant culture (Blanford and Murphy, 1977) conflicts, for example, with similar experiments performed under conditions of placental perfusion. After dual circulation was established, the addition of labeled cortisol, prednisolone, betamethasone, or dexamethasone to the fetal circuit led to equal degradation to 11-ketosteroids (Levitz *et al.*, 1978). Hence, the metabolic conversions of dexamethasone and betamethasone were far greater in the perfused than in minced placenta, as one might expect from a less disorganized tissue source.

The more simplified slice technique of perfusion used to study active transport and diffusional processes across placental epithelium was reviewed by Miller and Berndt (1975). Sodium, amino acids, acetylcholine, creatine, vitamin B_{12}, and iron were actively accumulated by placental membrances, whereas certain organic anions and cations, creatinine, urea, and antipyrine were passively transferred.

B. Explant Culture

Historically, the most frequently used approach to placental culture has involved explant fragments. A thorough review from the earliest attempts in the mid-1920s to the middle 1960s has been compiled by Gaillard and Schaberg (1965). The initial widespread use of the hanging drop technique with human blood plasma gave way in the late 1930s to the roller-tube method with human cord serum, chick and fetal bovine serum, and balanced salt solutions. This was followed by the use of a metal grid (Trowell, 1959) or a layer of glass beads (Tao and Hertig, 1965) which was covered by lens or filter paper as a platform for the 1 to 2 mm^3 sized explants. A general pattern of morphologic agreement emerged that first trimester placental outgrowths contained three principal different types of cells including fibroblastic elements from the villous mesenchyme, polygonal cells analogous to cytotrophoblast, and multinucleated cells suggestive of syncytiotrophoblast. Second, there was agreement that with the use of near or full-term placentas only the fibroblastic outgrowth was observed. Finally, the hormonal expression of first trimester placentas in explant cultures was brief, and that among the outgrowth cell population, only the cells of fibroblastic morphology proliferated in long-term culture.

The benefit of periodic exposure to alternate phases of gaseous and liquid medium was noted by Chung *et al.* (1969), in which explant cultures of first trimester placentas placed on rocker platform incubators underwent rapid proliferation and produced abundant chorionic gonadotropin compared to explants placed in static incubators. Explant culture techniques were used extensively for *in vitro* assessment of hormonal activity in human placenta in the 1970s. The conventional method was to mince first trimester placenta into 1 mm^3 pieces, place 50 to 100 mg of these pieces into either T-25 flasks or 60-mm dishes with or without a filter paper base, and agitate the container to provide alternate exposure to gaseous and to liquid phases. In this manner, for example, Benagiano *et al.* (1972) documented active placental synthesis of HCG, Saure (1973) studied the metabolism of steroids, Handwerger *et al.* (1974) demonstrated that exposure to dibutyryl cyclic AMP and theophylline enhanced HCG secretion, and Maruo (1976) presented evidence for the existence of larger precursor species to the subunits of HCG. Recently, explant culture techniques were used to follow HCG expression after exposure to prostaglandins, adrenaline and nonadrenaline, and steroids (Belleville *et al.*, 1978).

C. Monolayer Culture

Attempts to establish trophoblast cells in monolayer culture from human placenta have met with varying degrees of success (Gey *et al.*, 1938; Taylor and Hancock, 1973; Hall *et al.*, 1977; Stromberg *et al.*, 1978; Cotte *et al.*, 1980). As in explant culture, considerable effort has been devoted to correlation of the type of cell isolated with its biochemical capacities, principally in regard to gonadotropin and steroid hormone production. It has not been possible to subculture normal trophoblast and maintain cell lines derived from it. However, some success has been achieved in maintaining functional trophoblast monolayers in short-term primary culture.

The methodologic approach of two recent efforts toward trophoblast monolayer culture will be presented in detail. Using placentas acquired following Caeserian section or hysterotomy between 10 and 18 weeks gestation, Hall *et al.* (1977) observed the formation of cortisone from cortisol, the accumulation of estradiol-17β, estrone, and progesterone, as well as secretions of chorionic gonadotropin and placental lactogen, in confluent monolayer cultures obtained by the following procedure. After collection in 0.9% NaCl within 1–2 hours of surgery, chunks of placental cotyledons were cut off and rinsed in buffer to remove excess blood. A uniform suspension of villi was obtained by mincing the placental segments and periodically removing the large blood vessels. Centrifugation at 50 *g* and decantation or filtration through layers of gauze removed excess buffer. Approximately 10–20 gm of placental villi were transferred to 50–100 ml of 0.25% trypsin in calcium and magnesium-free bicarbonate buffer containing 500 units of deoxyribonuclease at pH 7.8. During incubation at 37°C for a period of 10–15 minutes, the glass serum bottles were agitated by hand several times. The supernatant was placed in tubes containing 1 ml of fetal calf serum and the cells were collected by centrifugation at 50 *g* for 10 minutes. The progress of this procedure, usually repeated four to five times, was monitored by phase contrast microscopy. Following each hydrolysis and centrifugation, the cells were resuspended in Ham's F-10 culture medium, supplemented with 10% fetal calf serum, penicillin (200 U/ml), gentamicin (10 μg/ml), and amphotericin B (5 μg/ml) and stored at 10°C. The pooled cell harvests were again centrifuged after large cellular aggregates had been discarded using a long pasteur pipet.

The objective of having confluent monolayer cultures within 24 to 36 hours was obtained by resuspending the final cell pellet in 8 to 10 ml of culture medium, plating 0.5 to 1.0 ml in a Falcon T-25 flask containing 4 ml of medium, and after 5 minutes to permit the cells to settle, estimating by phase contrast microscopy whether additional amounts were required. The set of from 8 to 16 T-75 flasks that were prepared from each placenta were placed in a CO_2 incubator, and the placental cells, in contrast to the accompanying erythrocytes, were allowed to attach for 2 to 3 hours. The flasks were then carefully rinsed with 3 ml of media, and 5 ml of fresh medium was added. The dense layer of erythrocytes were removed by this procedure, revealing the monolayer of placental cells. The culture medium was changed after 15 to 20 hours and subsequently every 24 to 48 hours. The cultures appeared erythrocyte-free within 24 to 36 hours after plating, and remained confluent for the usual 8-day duration of the experiments which concerned specific endocrine functions (Hall *et al.*, 1977).

A second and more recent technique for trophoblast isolation and cell culture examined variables that led to enhanced HCG secretion (Stromberg *et al.*, 1978). This approach used brief exposures of unminced first-trimester placental specimens to a solution of trypsin–EDTA–DNAse, and separated the dispersed cells from erythocytes by Ficoll–Hypaque centrifugation. The procedure is described in detail in the following paragraphs.

All samples consisted of first-trimester (9 to 13 weeks gestation) human placental specimens obtained after elective abortion by vacuum aspiration. Samples were selected for tissue integrity, and absence of clinical monilial vaginitis in the patient. The placental tissue was immediately immersed in a solution of sterile ice-cold Gey's balanced salt solution (BSS) in a screw-topped 8-oz specimen jar. The jar was shaken vigorously, and the tissue was drained and then transferred to a second jar also containing Gey's BSS and supplemented with penicillin (100 U/ml), streptomycin (100 μg/ml), and gentamicin (50 μg/ml). This transfer and rinsing procedure was repeated three times. For transport in ice to the laboratory the placental samples were each placed in a jar containing Leibovitz L-15 medium with 20% fetal bovine serum (North American Biological Co., Miami, FL) and penicillin, streptomycin, and gentamicin (as previously listed), as well as amphotericin B (5 μg/ml) to reduce the incidence of subsequent contamination with *Candida albicans,* an organism commonly present in the vaginal tract.

The excess fibrous membrane tissue was excised under aseptic conditions in a finger bowl containing 0.2% trypsin (200 U/mg; Worthington Biochemical Co.) and 0.02% EDTA in Dulbecco's phosphate-buffered saline (PBS) without calcium or magnesium. The remaining stalk of the placental specimen was attached to a No. 10 snap swivel (Mid-Atlantic Distributors, Columbia, MD) on the end of a string. The placental sample then was suspended in a 150-ml Erlenmeyer flask, with a side-arm spout, containing 75 ml of trypsin-EDTA solution and 1.5 ml bovine pancreatic DNAse I (2,200 Kurtz units per ml; Sigma Chemical Co., St. Louis, MO) in Dulbecco's PBS (Salomon and Sherman, 1975) with a Teflon-coated magnetic stirring bar. Generally three to five placental specimens were prepared as described and placed on a multi-magnestir (Lab-line, Melrose Park, IL). The dial of the stirrer was set on maximun so that when activated, the magnetic stirring bar moved about the flask to agitate and massage the suspended placental specimen. Successive periods (0 to 5, 5 to 10, and 10 to 20 minutes) of exposure to the trypsin-EDTA-DNAse solution at 37°C were carried out. Following each time period, the fluid from each flask was poured in equal amounts into two 50-ml Falcon plastic tubes containing 10 ml of culture medium with 20% fetal bovine serum. The fragments of tissue that settled to the bottom of the tube were withdrawn by aspiration with a 9-in. pasteur pipet. [After the larger fragments are removed by filtration through gauze, the remaining aggregates can be sedimented at unit gravity through 40% fetal bovine serum, and then cultured (see page 234 and Cotte, *et al.*, 1980).] The cells released into the fluid following exposure to the trypsin–EDTA–DNAse solution were collected by centrifugation (300 *g* for 5 minutes at room temperature). The cells from resuspended pellets from each placental sample were then counted by a hemacytometer and assessed for viability by trypan blue exclusion. Specimens of placental cells that were free of gross yeast contamination by microscopic examination (over 95%) and contained over 2×10^6 trypan blue-negative, non-RBC cells were combined. Routinely, the concentration of dispersed cells was adjusted to 1.0×10^6 per ml and 4 ml was layered over 3 ml of a Ficoll–Hypaque solution of 1.077 gm/ml density (Boyum, 1968) ("Lymphoprep," Nyegaard and Co., Oslo, Norway) in a 15- by 150-mm screw-topped tube (Falcon No. 3033). Following centrifugation at room temperature (400 *g* for 20 minutes), the cell layer at the interface was withdrawn, diluted 5-fold with culture medium, and recentrifuged (300 *g* for 5 minutes) to a pellet. After one more rinsing step, aliquots (5×10^5) of these cells were plated into 35-mm petri dishes to obtain replicate samples for experimental evaluation.

The usual duration of time between acquiring the placental specimen and plating the dispersed primary placental cells was about 4 hours. After an overnight attachment period (approximately 18 hours), the dishes were gently rinsed twice and refed with medium. Subsequently, medium was changed every 24 hours. The gradual decline in HCG level with time in monolayer culture was retarded by exposure to theophylline and cyclic adenosine monophosphate which indicated the functional nature of the trophoblast cells *in vitro* (Stromberg *et al.*, 1978).

Recently our laboratory has compared the efficacy of trypsin-EDTA, collagenase type II and type IV, and a nonspecific protease from *Streptomycin griesus* (type VI, Sigma Catalogue No. P-5130), in releasing trophoblasts from mid-term Rhesus monkey placenta obtained by hysterotomy. The use of the nonspecific protease (0.05%) and DNAse in Dulbecco's PBS according to the preceding methods gave an 8-fold higher yield of viable trophoblasts than other enzymatic treatments. Five successive 15-minute exposures to this protease–DNAse provided an accumulation of 7.5 million trophoblasts for each gram (wet weight) of starting Rhesus placental tissue. Very recently a novel method for preparation of purified monolayer cultures of human cytotrophoblast cells, essentially free of stromal and syncytial cells, has been described (Cotte *et al.*, 1980). The technique utilizes partial trypsinization and sedimentation of aggregates of cytotrophoblast cells at unit gravity for 1 hour through 40% fetal bovine serum. Our evaluation of this method, using Protease Type IV–DNAse treatment of intact first trimester human placental tissue instead of mincing and partial trypsinization, indicates this obviates the Ficoll–Hypaque centrifugation and provides an improved approach to *in vitro* growth of human cytotrophoblast. In addition, highest HCG levels have been obtained with either Dulbecco's MEM or CMRL medium supplemented with serum, vitamin, and hormones as described below for placental organ culture.

The inability to subculture normal human placental cells in monolayer which retain their functional characteristics can be overcome in part by viral transfection. Chou has obtained cell lines from term placentas (Chou, 1978a) and clonal cell lines from first trimester (Chou, 1978b) placentas, which after transformation by simian virus 40 retained their capacity for HCG secretion in extended culture. Although the chromosome number of the clonal lines was not true diploid, the regulation of synthesis of the subunits of HCG in the transformed placental cells differed from the regulation in choriocarcinoma cells (Chou, 1978b).

Unfortunately, even recent studies have not included a rigorous comparison of methodologic approaches, and optimal conditions for monolayer culture of trophoblasts are not yet established. For example, should placental villi remain intact (Stromberg *et al.*, 1978), or be finely minced (Hall *et al.*, 1977; Cotte *et al.*, 1980) in preparation for enzymatic treatment and monolayer culture? Is trypsin (Stromberg *et al.*, 1978; Hall *et al.*, 1977), collagenase (Chou, 1978a,b), Pronase (Pretlow *et al.*, 1978) or Protease Type IV the protease of choice for release of trophoblast cells from the underlying stroma? Is preferential adherence of placental cells to a plastic culture surface (Hall *et al.*, 1977), centrifugation through a Ficoll–Hypaque solution (Stromberg *et al.*, 1978) or sedimentation through 40% fetal calf serum (Cotte *et al.*, 1980) more effective in removal of erythrocytes following enzymatic dispersion of trophoblast cells? Would use of velocity sedimentation in an isokinetic gradient of Ficoll in cell culture medium provide an advantage? Would plating at high cell density stabilize trophoblast viability sufficiently to permit successful subculture? What culture medium is most appropriate, and would addition of epidermal growth factor, or as yet uncharacterized placental extracts (Chung *et al.*, 1969) enhance growth and still maintain differentiation of trophoblast cells in monolayer culture?

D. Organ Culture

Throughout this chapter, organ culture of human chorionic villi has referred to use of larger segments of placental tissue, approximately 1 to 1.5 cm in size, in contrast to placental explant culture in which minces of tissue about 1 mm^3 in size are placed in culture. The extensive work of Trowell (1959) on numerous organs in explant culture documented that specimens even 1 to 2 mm in size should project into the gas phase. If submerged, necrosis occurred in the central areas of the tissue. This was avoided by culturing on lens paper supported by a metal grid or tantalum gauze. Tao and Hertig (1965) utilized this approach with 1 mm^3 placental explants placed over strips of filter paper laid over a layer of glass beads. Following adherence of placental explants of 1–2 mm in size, Chung *et al.* (1969) enhanced maintenance in culture by use of a rocker platform incubator which intermittently exposed specimens to the gaseous phase. Perhaps, in part, because a narrow epithelial surface was cultured, segments of human bronchi of 1.5 cm^2 in size were similarly maintained by Trump *et al.* (1974) on a commercially available atmosphere chamber and rocker assembly. The highly villus structure and thin epithelial surface of the human placenta make it ideal for organ culture methods which use a platform of gelatin sponge and a rocking chamber (Leighton, 1951; Stoner *et al.*, 1978). Recently, this approach has been adapted to culture of first-trimester human placental specimens by Huot *et al.* (1979), using HCG as a differentiated marker with which to evaluate the effects of varying culture conditions.

Organ culture of human placentas was done in our laboratory in the following manner. First trimester human placentas were obtained and transported to the laboratory as previously described in this chapter under monolayer culture (Stromberg *et al.*, 1978). Within a laminar flow hood, villous placental segments approximately 1 cm in diameter were freed of excess membranes and draped over similar-sized pieces of 3-mm-thick gelatin sponge (Gelfoam, 12 cm^2 by 3 mm, The Upjohn Co., Kalamazoo, MI) premoistened with 3 ml of culture medium in 60-mm plastic culture dishes. Previously, the squares of gelatin sponge had been secured to the edge of the petri dish with 5 μl of acetone and sterilized by overnight exposure to ultraviolet light. Standard culture medium consisted of CMRL 1066 (Grand Island Biological Co., Grand Island, NY) supplemented with 2 m*M* glutamine, penicillin (100 units/ml), streptomycin (100 μg/ml), amphotericin B (5 μg/ml) (all from Grand Island Biological Co.), β-retinyl acetate (0.1 μg/ml) (Hoffmann-La Roche, Nutley, NJ), insulin (1 μg/ml), hydrocortisone hemisuccinate (0.1 μg/ml), (both from Schwarz/Mann, Rockville, MD), and 5% fetal bovine serum (North American Biological Co., Miami, FL).

The dishes were then placed into a controlled atmosphere chamber (Bellco

Glass Co., Vineland, NJ) on a rocker platform (Bellco Glass Co.) set at 10 cycles/minute to permit periodic submergence of the placental specimens. The chamber was routinely gassed with a mixture of 95% air and 5% CO_2, and maintained at a temperature of 37°C. The 3 ml of medium per 60-mm dish was changed every 48 hours during the course of culture. At the end of each experiment (12 days), specimens could be carefully removed from the gelatin sponge platform, blotted dry on absorbent paper, and weighed on an analytical balance to obtain comparisons among specimens. Routinely 1 cm^2 placental samples of 9 to 13 weeks gestation weighed about 30 mg. Depending upon specimen size about 18 dishes each containing 1 cm^2 of chorionic villi layered over a gelatin sponge platform can be prepared from each late first trimester placenta. Thus, approximately six variables in triplicate or nine variables with duplicate samples can be examined from each placental specimen.

III. Results

Despite the passage of over 40 years since the publication of Gey *et al.* (1938), success in monolayer culture of normal trophoblast cells from human placenta has unfortunately been measured in days. However, methods for short-term primary culture without cell passage are available which provide viable trophoblasts that maintain specific endocrine functions.

A. Monolayer Culture

Using trypsin dispersion of placental minces, Hall *et al.* (1977) evaluated the ability of trophoblast monolayer cultures to maintain specific endocrine functions. During the 24-hour period between day 3 and day 4 in culture, the cultures from both first trimester and term placental specimens demonstrated four placental steroidogenetic functions: (*a*) formation of cortisone from [^{14}C]cortisol, and accumulation of (*b*) estradiol-17β, (*c*) estrone, and (*d*) progesterone from labeled precursors. During the period from day 3 to day 8 in culture, the cells from first trimester placental specimens produced a peak in HCG secretion, and from term placenta a plateau of HCG secretion. A progressive decline in HPL secretion was observed during the week in culture by specimens of both early and late gestation. The formation of the steroids from appropriate radioactive precursors and the progressively increasing secretion of HCG provided evidence of actual biosynthesis of these respective hormones.

Using unminced, trypsin-dispersed first trimester placental specimens with removal of erythocytes by means of Ficoll–Hypaque centrifugation, Stromberg *et al.* (1978) examined variables in the culture procedure to identify steps that enhanced the secretion of HCG. Longer periods of trypsinization reduced the level of HCG production and morphologically correlated with a progressive effacement of trophoblast epithelium from the chorionic villi (Fig. 1). The mIU of HCG secreted from the trophoblast cells obtained from the first 5 minutes of exposure to the trypsin–EDTA–DNAse solution was twice that obtained from the 5- to 10-minute period of exposure, which was again about twice that of the 10- to 20-minute period of exposure. The functional viability of these trophoblast cells is illustrated in Fig. 2. Exposure to an optimal concentration (1 m*M*) of cAMP and theophylline retarded the decline in HCG production with time in primary culture (Fig. 2B), and, as a reagent control, resulted in a 3- to 4-fold elevation in HCG secretion of the JAR line of choriocarcinoma cells (Fg. 2A). The discrepancy in HCG production between the normal trophoblasts and their malignant counterpart is obvious with continued days in culture. However, during the initial 24 hours in cell culture (day 0 to day 1), the amount of HCG produced by the normal trophoblasts on a per cell basis (14×10^{-4} mIU HCG per cell) is equivalent to the JAR line of choriocarcinoma cells (12.5×10^{-4} mIU HCG per cell). The JAR cells have a generation time of approximately 28 hours and thus when plated at 1×10^5 per dish after 24 hours contain 2×10^5 cells that secrete 250 mIU of HCG. In comparison, the normal trophoblasts, plated at 5×10^5 cells per dish, apparently fail to proliferate and secrete about 700 mIU HCG over the first 24-hour period. This study did not attempt to extend HCG secretion by use of conditioned media, use of various lots of human placental cord serum, use of first-trimester placentas obtained by hysterotomy instead of vacuum aspiration, or plating isolated trophoblast cells at increased and perhaps more optimal concentrations. In comparison, after mincing and trypsin–DNAse treatment of first-trimester hysterotomy specimens, and plating the resulting cells at confluency, Hall *et al.* (1977) obtained an HCG peak of nearly 800 mIU/ml of supernatant fluid on days 5 and 6 of monolayer culture.

B. Organ Culture

The kinetics of HCG expression obtained by Hall *et al.* (1977) in confluent monolayer culture of first-trimester placenta are similar to that observed by Huot *et al.* (1979) in organ culture. In studies of organ cultured placenta in the author's laboratory, under optimal culture conditions, the secretion of HCG rose to reach peak levels around the sixth day in culture and then declined in the manner of a bell-shaped curve (Fig. 3). Moreover, additional pieces of placenta could be maintained in the same 60-mm petri dish with a proportional increase in the amounts of HCG produced. Culturing from one to four 1 cm^2 pieces of placenta produced a concomitant enhancement in HCG synthesis (Fig. 3B), and when HCG was adjusted on the basis of per gram of wet weight, HCG secretion was nearly comparable in all dishes (Fig. 3A). Consequently, a 4-fold increase in yield of a placental gene product was easily obtained.

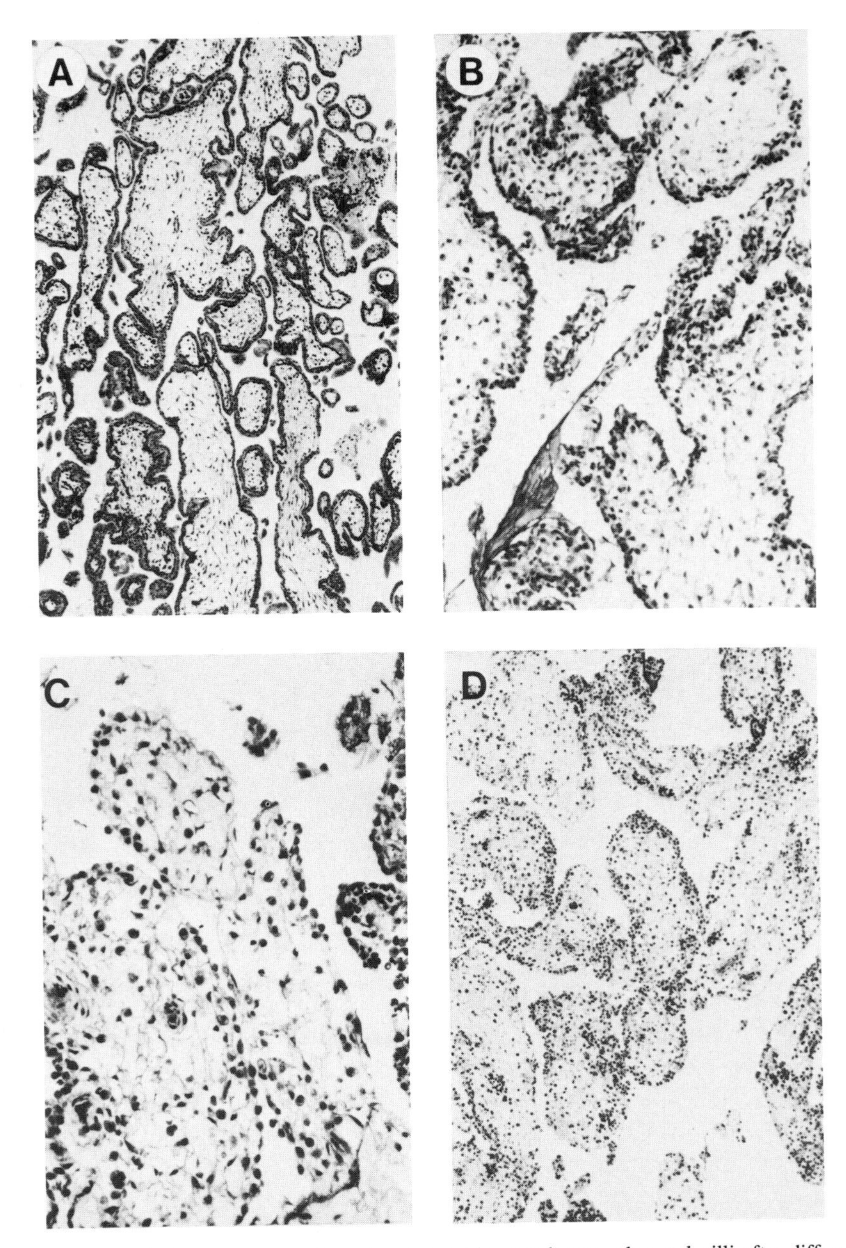

FIG. 1. Histologic appearance of unminced first-trimester human placental villi after different lengths of exposure to trypsin–EDTA–DNAse solution (see Section II). Fragments of villi before treatment (A, ×55) are compared with villi after 0 to 5 minutes (B, ×140), 5 to 10 minutes (C, ×140), and at the end of 20 minutes of exposure (D, ×55). Note the progressive loss of trophoblast epithelium.

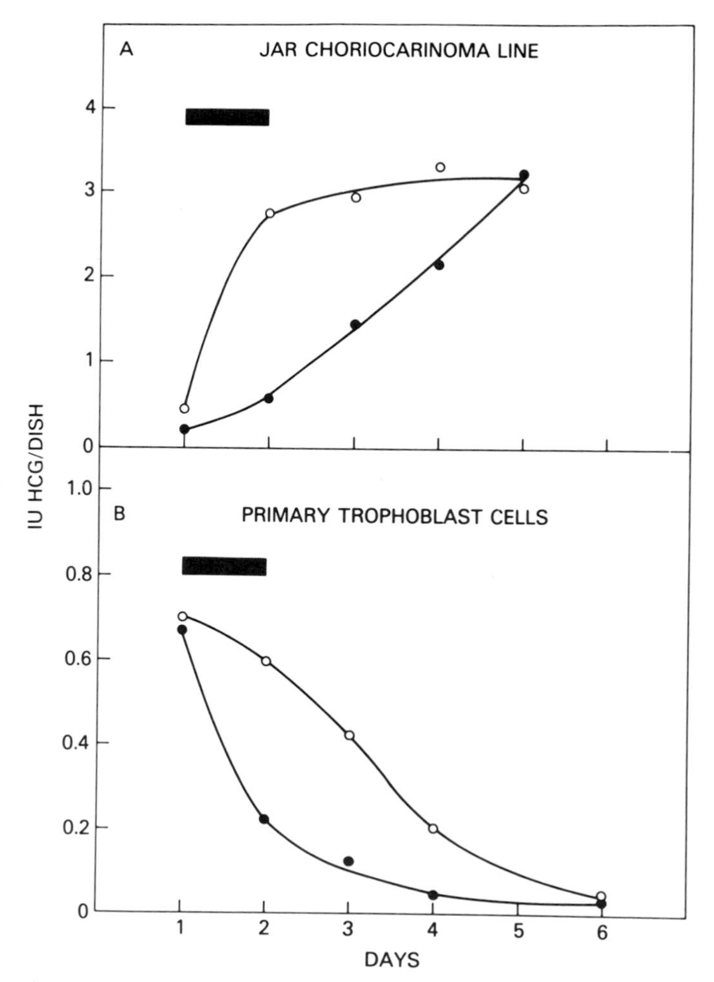

FIG. 2. Stimulation of HCG secretion by exposure to cyclic adenosine monophosphate (cAMP) and theophylline. Duplicate sets of 35-mm petri dishes, containing either 1×10^5 cells of the JAR line of choriocarcinoma (A), or 5×10^5 primary placental cells isolated as described in Stromberg *et al.* (1978) (B), were fed untreated media as a control (●) or fed media containing 1 m*M* each cAMP and theophylline (○) during the first 24 hours (black bar). After rinsing each set once, the dishes were daily given 2 ml fresh medium and the HCG content in the conditioned medium was determined by radioimmunoassay.

The role of O_2 tension as a critical variable in HCG expression in placental organ culture is documented in Fig. 4. The use of 95% O_2 was clearly toxic, with a rapid progressive decline in HCG synthesis, and without the characteristic rise in HCG secretion seen with the other O_2 concentrations. A higher HCG level could be produced with 5% CO_2 and air (approximately 20% O_2) as compared

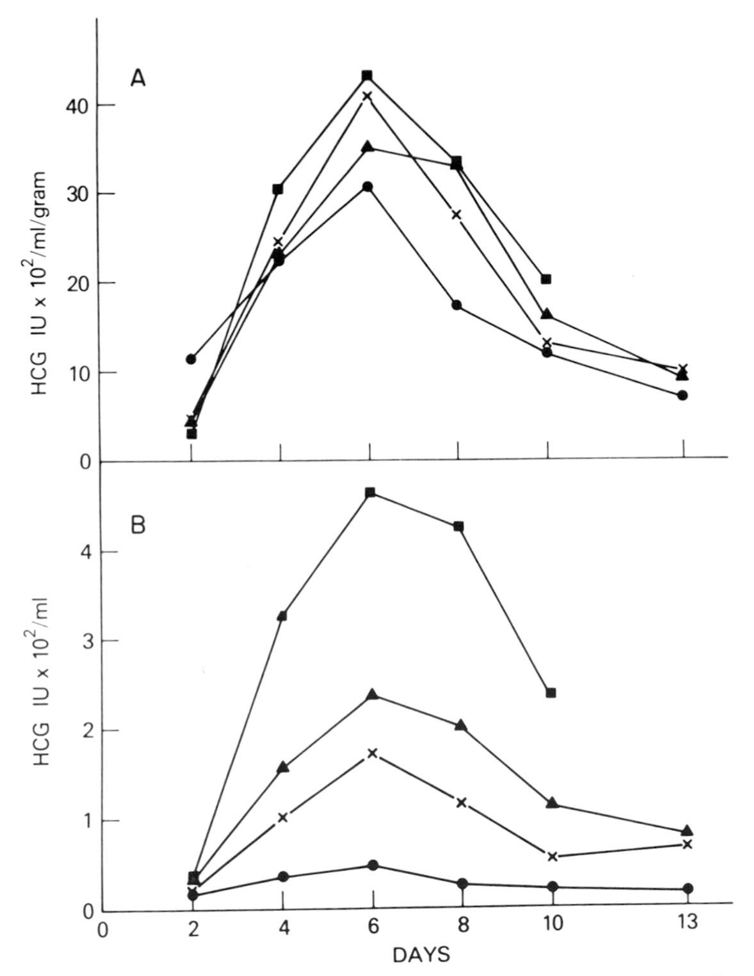

FIG. 3. Effects in organ culture of additional pieces of placenta in one dish on HCG synthesis. Organ cultures were prepared as described in Section II except that one piece (●), two pieces (X), three pieces (▲), or four pieces (■) were placed together in one dish. As seen in (B), additive amounts of HCG were produced with each additional piece in culture. As (A) illustrates, nearly similar amounts of HCG are produced by the cultures when HCG secretion is compared on a per gram of tissue weight basis.

with 5% CO_2 and either 5 or 50% oxygen. The effect on HCG concentration contributed by addition to the organ culture medium of either hydrocortisone, β-retinyl acetate, or insulin was also evaluated by Huot *et al.* (1979). Although use of serum-free CMRL 1066 medium supplemented with vitamin and hormones gave the lowest concentrations of HCG, the organ cultures could be grown in this medium and still retain kinetics of HCG secretion similar to serum-enriched CMRL 1066. Serum, hormones, and β-retinyl acetate apparently contributed independent stimuli to increased levels of HCG since they produced an additive effect when combined. Addition of theophylline and dibutyryl cyclic AMP at a concentration of 1 and 10 m*M*, respectively, for a period of 2 days elevated HCG levels by 3-fold over the untreated controls (Huot *et al.*, 1979). The levels of HCG in control and treated cultures in our organ culture study were 1000 times higher on a per gram basis than those obtained with term placenta in conventional explant culture (Handwerger *et al.*, 1973).

Huot *et al.* (1979) estimated the viability of placental tissue by measuring its

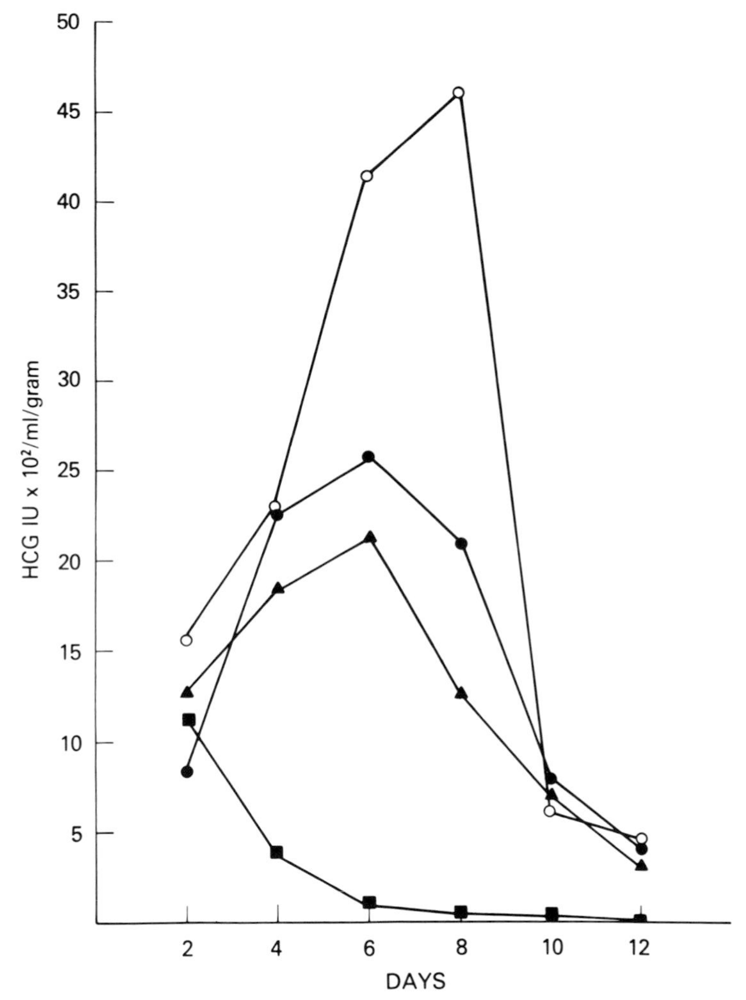

FIG. 4. Oxygen levels and HCG secretion in placental organ cultures. Cultures prepared as described in Section II were exposed to the following gas mixtures: 95% O_2 and 5% CO_2 (■); 50% O_2, 45% N_2, and 5% CO_2 (▲); 95% air and 5% CO_2 (○); and 5% O_2, 90% N_2, and 5% CO_2 (●). The supernatants were harvested every 48 hours and assayed for HCG content by radioimmunoassay.

ability to incorporate [^{125}I]iododeoxyuridine ([^{125}I]IUdR) into DNA and by measuring the lack of [^{125}I]IUdR release from the organ cultures. As an analog of thymidine, [^{125}I]UdR is incorporated exclusively into DNA of proliferating cells and, in contrast to thymidine, is not reutilized after release from dead cells and DNA degradation. Since [^{125}I]IUdR remains bound to DNA throughout the lifespan of cells, the disappearance of tissue radioactivity and its reappearance in the culture medium provides a quantitative measure of cell death. As shown in Table I, the human placental organ cultures initially incorproated [^{125}I]IUdR into their DNA, indicating an active proliferation of cells during the first several days of culture. Between 82 and 88% of the total radioactivity present in the placental samples at day 5 remained incorporated in the cell DNA after 12 days in culture and, concurrently, about 12 to 18% of the counts were released into the medium between day 5 and day 12 (Huot *et al.*, 1979).

The ability of the cells to incorporate [^{3}H]thymidine into DNA was also examined over the usual 10- to 12-day duration of culture. After successive 2-day periods in culture, placental organ cultures were incubated in the presence of standard medium supplemented with 1 μCi/ml [*methyl*-^{3}H]thymidine. The cultures were labeled for 48 hours after which the medium was discarded and the tissue was weighed and then frozen. After homogenization, an aliquot from each sample was saved for protein determination by the Lowry method, and the amount of label incorporated into 5% trichloroacetic acid (TCA)-precipitable material was determined. As shown in Fig. 5, DNA synthesis declined rapidly from day 2 to day 4 and remained constant thereafter for the duration of culture. Thus, studies with both [^{3}H]thymidine and[^{125}I]IUdR indicated that DNA synthesis occurred during the initial days in culture and that the cell population remained viable thereafter.

TABLE I

INCORPORATION AND RELEASE OF [^{125}I]IUdR FROM PLACENTAL VILLI AS AN INDEX OF CELL VIABILITY IN ORGAN CULTURE[a]

	Radioactivity released into medium (cpm) at days				
Placenta	5-6	7-8	9-10	11-12	Radioactivity in placenta at day 12 (cpm)
1	4230	2070	4110	5500	98100 (86%)
2	9710	5390	6550	4490	115910 (82%)
3	3670	1920	1290	—	49380 (88%)

[a] Three placentas were established in organ culture by the standard methods and exposed to [^{125}I]IUdR as described in Huot *et al.* (1979). The cpm of ^{125}I released into the medium at 48-hour intervals is an indicator of cell death. The radioactivity remaining in the tissue at completion of the culture period represents the viable cell population. This fraction represents 82-88% of the total radioactivity present in the placentas at day 5. About 12-18% of the counts were released into the medium between days 5 and 12.

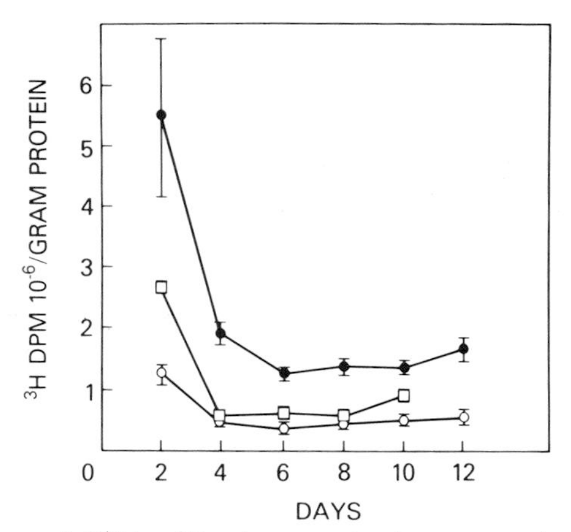

FIG. 5. Time course of [^{3}H]thymidine incorporation into organ cultures of three placentas. Specimens 1 × 1 cm in size from three placentas were cultured as described in Huot *et al.* (1979), and over the 48-hour intervals indicated above were incubated in the presence of [^{3}H]thymidine. The incorporated radioactivity into TCA-precipitable material was measured in two cultures of each placenta and the determinations were done in duplicate. Results are presented as the mean SD of the experimental values.

The morphology of the placental villi in organ culture as prepared by Huot *et al.* (1979) was evaluated by sequential light microscopy, and both transmission and scanning electron microscopy. No evidence of villus growth could be documented from photographs of selected villi taken through the dissecting microscope every other day over the usual 12-day duration of culture. After 12 days in culture, no extension of villus size or outgrowth of new syncytial sprouts (Boyd and Hamilton, 1970) was observed in particular visual fields when compared to the photomicrographs taken at day 0. However, after extended periods of organ culture (up to 3 weeks), growth of epithelial-like cells extending out from the villus onto the gelatin sponge has been observed by light microscopy of paraffin-embedded specimens (Fig. 6). In Fig. 6A the close association of the gelatin sponge and viable-appearing trophoblastic epithelium with frequent syncytial cell clusters is illustrated. The outgrowth of the placental epithelium into and over the gelatin matrix is shown in Fig. 6B. In cultures such as these, conventional trypsinization of the gelatin sponge after removal of the placental speciman occasionally yielded populations of only epithelial-like cells in monolayer culture. Scanning electron microscopy suggested a slight blunting of villus size and delicacy with increasing duration in organ culture. However, the surface of the individual trophoblasts remained very rich in microvilli. At the level of transmission electron microscopy an interesting finding was the presence of budding type C particles in human placenta immediately prior to organ culture (Fig. 7A), and with Rhesus monkey placental specimens, the persistence of budding type C particles in individual trophoblast cells (Fig. 7B) after isolation

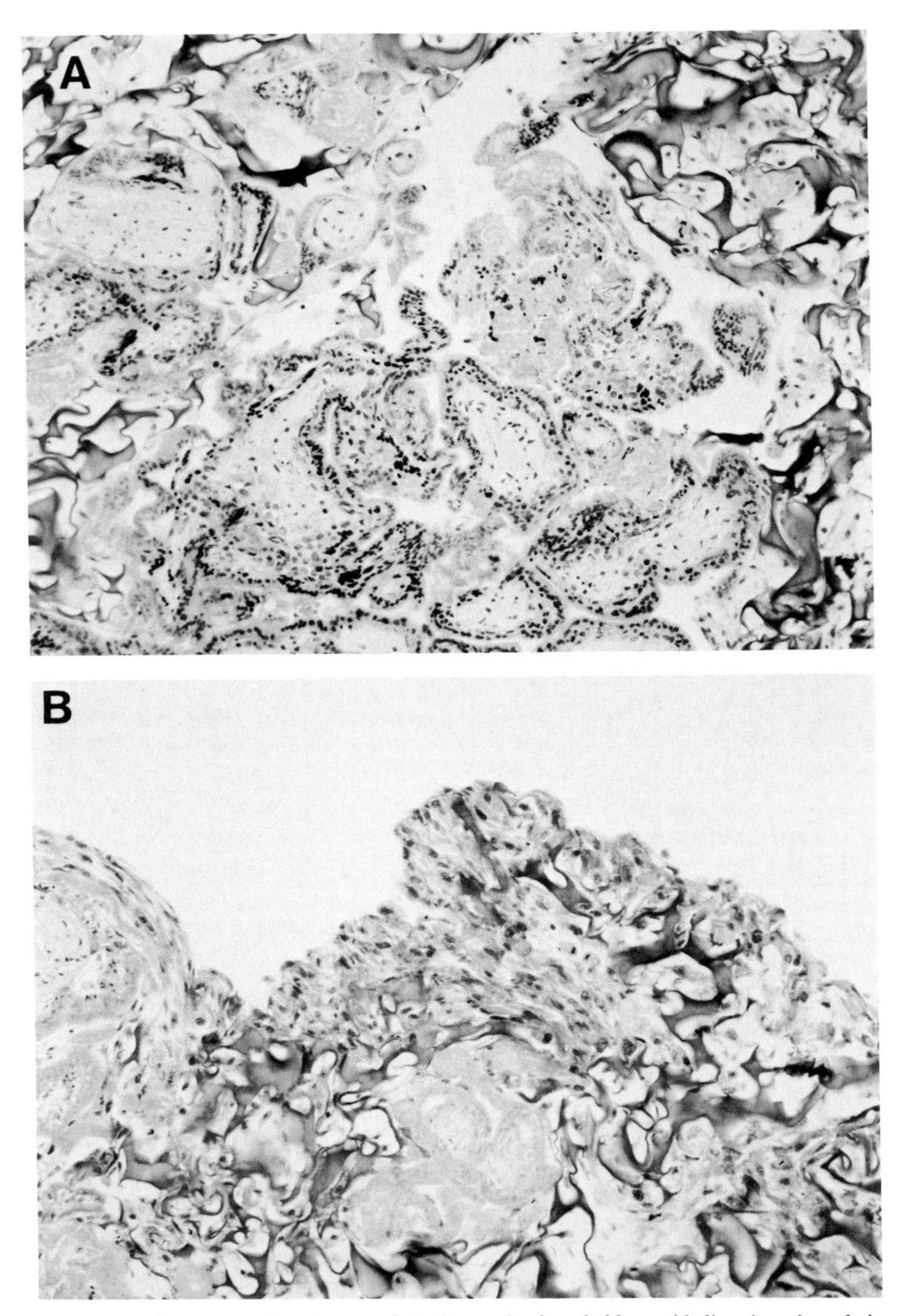

FIG. 6. Light micrographic evidence of (A) ingrowth of trophoblast epithelium into the gelatin sponge after 3 weeks in organ culture, and (B) overgrowth of epithelial-like cells along the gelatin surface. ×130.

as in Stromberg *et al.* (1978), and in organ culture after 24 hours (Fig. 7C) using the methods of Huot *et al.* (1979).

In an effort to enrich the usual CMRL 1066 medium to further enhance HCG secretion, recently putrescine dihydorchloride at 1 m*M* and 10-fold higher concentrations of insulin (10 μg/ml) and hycrocortisone (1 μg/ml) were added. The consequence, shown in Fig. 8, is both a 4-fold elevation in peak HCG secretion and a prolongation of the high plateau until day 8 in organ culture.

IV. Discussion

In vitro investigations of human placental functions can be carried out over a span of many days, but not many weeks, because either in monolayer or organ culture, techniques are not at hand which duplicate the *in vivo* feto-placental unit and therefore permit extended periods of placental growth and differentiation. Solace for this lamentable state of affairs can be found in the remarks of Honor B. Fell (see Section II) that the aim of culture studies is seldom growth in the sense of cell multiplication but is rather an attempt to provide an environment that permits already differentiated tissues to exercise their normal functions. Within this context then, an evaluation of the trophoblast monolayer and placental organ culture techniques presented in this chapter is appropriate.

Several critical steps in the monolayer culture of human trophoblast should be emphasized. General agreement exists that early gestation placentas obtained by hysterotomy with prompt isolation of trophoblast cells is the most suitable approach to monolayer culture. However, the placental function under investigation influences the choice of gestation of the placental specimen; for example, secretion of human placental lactogen rises to peak at term (Sato, 1973) while the expression of human chorionic gonadotropin is highest at the end of the first trimester. Use of placentas of any gestational age obtained following hysterotomy avoids the specimen damage and microbial contamination that frequently accompanies placentas in early gestation obtained by vacuum aspiration. Second, a troublesome consequence at times during trypsinization of intact, unminced placental tissue is the accumulation of a clear viscus gel (probably DNA) which reduces the yield of trophoblast cells and accounts for the addition of DNAse to the treatment solution (Salomon and Sherman, 1976; Stromberg *et al.*, 1978). The presence of DNAse, along with trypsin, in the treatment method of Hall *et al.* (1977) suggests a similar concern in minced specimens as well. Resolution of this technical problem would be useful. Interestingly, this has not occurred in first trimester Rhesus monkey placentas obtained by hysterotomy and processed as in Stromberg *et al.* (1978), in which the trophoblast yield on a per gram basis of beginning tissue is routinely eight to ten times greater than that obtained from first trimester human placentas obtained after vacuum aspiration.

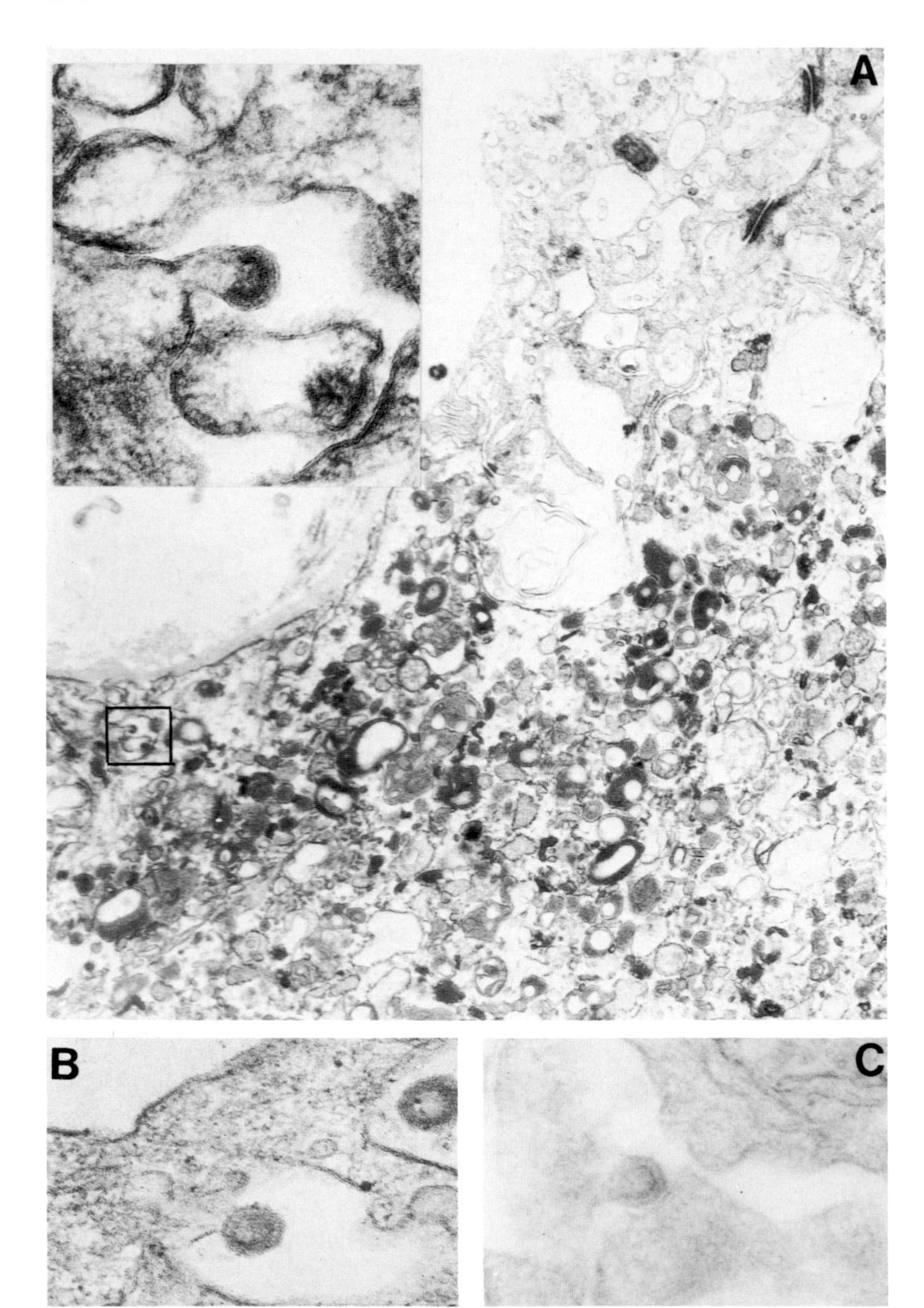
A
B
C

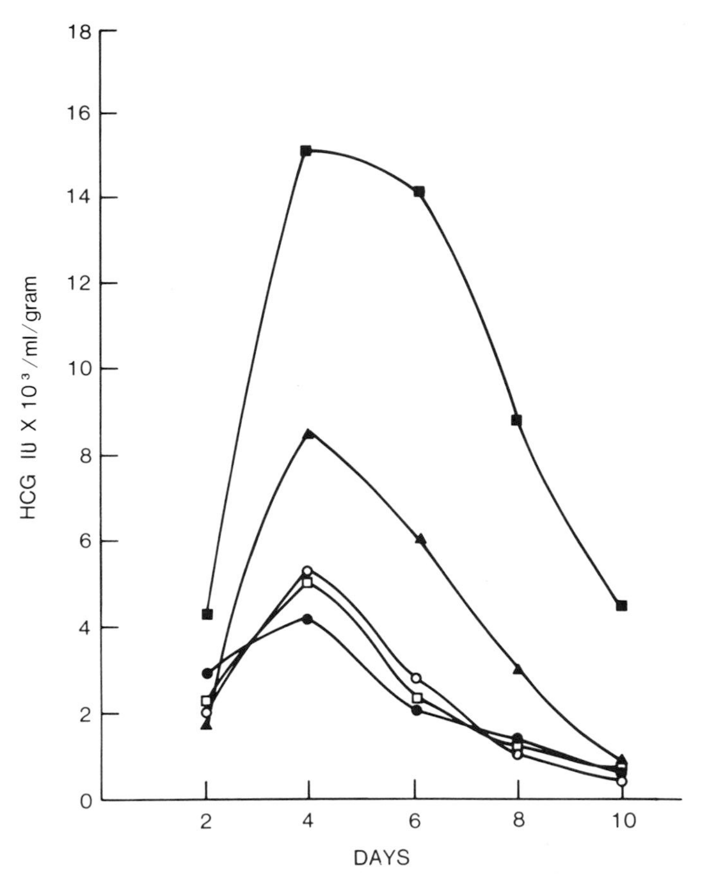

FIG. 8. Influence of enriched media on HCG secretion. Organ cultures were prepared as described in Section II except that CMRL 1066 medium supplemented with 5% fetal bovine serum, 1 μg/ml insulin, 0.1 μg/ml hydrocortisone, antibiotics, and β-retinyl acetate (●) was altered to contain 0.1% bovine serum albumin instead of 5% fetal bovine serum (○), 1 μg/ml hydrocortisone (▲), 10 μg/ml insulin, 1 μg/ml hydrocortisone, and 1 m*M* putrescine (■), and 10 μg/ml insulin, 1 μg/ml hydrocortisone, 1 m*M* putrescine, and 0.1% bovine serum albumin in place of 5% fetal bovine serum (□). The supernatants were harvested from two placental specimens every 48 hours. The results of assay for HCG content by radioimmunoassay are presented as the average of the two specimens. The enhanced HCG secretion in (■) is likely due to addition of 1 m*M* putrescine because use of insulin alone at a concentration of 10 μg/ml did not elevate HCG levels above (●) (data not shown). Note that 0.1% bovine serum albumin cannot replace 5% fetal bovine serum.

FIG. 7. Electron micrographic evidence of budding type C retroviral particles in human and Rhesus monkey placental tissue. First trimester human villi prior to organ culture with type C particle budding (A) near basement membrane (×8000; inset, ×120,000). Rhesus type C particles budding from monkey trophoblasts after isolation as in Stromberg *et al.* (1978)(B), and after 1 day in organ culture (C) as described in Huot *et al.* (1979; both ×106,000). Micrograph (C) provided by M. J. Walling.

Third, removal of principally erythrocytes, and to a lesser extent granulocytes, and contaminating bacteria and fungi from dispersed trophoblasts is promptly achieved by brief centrifugation through ''Lymphoprep'' (Stromberg *et al.*, 1978), a commercial preparation of Ficoll–Hypaque. This would appear more appropriate than use of the preferential adherence of trophoblasts to the plastic substrate of a culture vessel for several hours with removal of erythrocytes by rinsing (Hall *et al.*, 1977). On the other hand, because of the intimate association of type IV (basement membrane) collagen with trophoblast, perhaps coating the plastic culture surface with this type of collagen would enhance attachment, growth, and differentation of trophoblast as has been described for epidermal cells of skin (Murray *et al,* 1979). Fourth, in addition to the use of specimens from hysterotomy, the plating of trophoblasts at high cell density so that confluency is obtained at 2 to 3 days in culture may have contributed to the higher levels, and longer duration, of HCG expression obtained by Hall *et al.*, (1977). Lastly, and most importantly, the use of Protease Type IV–DNAse treatment of intact early trimester human placental tissue (see Section II, C) with sedimentation of cell aggregates through 40% fetal bovine serum (Cotte *et al.*, 1980) appears to provide the most satisfactory approach to growth of cytotrophoblast in cell culture.

The similarity of the kinetics of HCG expression (a bell-shaped curve which peaks during day 4 to 6 in culture) obtained by Hall *et al.* (1977) in monolayer culture of first trimester placenta, and by Huot *et al.* (1979) in first trimester organ culture, may suggest an interpretation related to DNA synthesis. If in fact, under the benefit of high-density cell plating, trophoblastic cells initially proliferate during the first days in monolayer culture, the rate of DNA synthesis would decline when confluency is reached at day 3 and would remain stable (Hall *et al.*, 1977). Concomitantly, the HCG secretion increases to peak on day 5 and begins to decline on day 7. In organ culture a striking decline in DNA synthesis, as measured by [^{3}H]thymidine incorporation, occurs between day 2 and day 4 and reaches its lowest level at day 6; concomitantly, a sharp rise in HCG expression is observed between day 2 and day 4 and peaks at day 6 (Compare Fig. 5 with Figs. 3 and 4). A similar correlation of increase in HCG secretion after decrease in DNA synthesis has been documented in detail in the BeWo line (Speeg *et al.*, 1976) and the JAR line (Azizkhan *et al.*, 1979) of choriocarcinoma cells in continuous cell culture.

V. Perspectives

The application of methods for isolation of trophoblast cells and their culture in monolayer permits a more precise definition of their function in normal and abnormal states. Isolated trophoblast cells from either condition could for example be further separated into subsets of cells, such as cytotrophoblast, intermediate cells, or syncytiotrophoblast cells by established procedures (Pretlow and Pretlow, 1977) using velocity sedimentation in an isokinetic gradient of Ficoll in tissue culture medium. By this approach identification of particular cell functions, such as HCG secretion, with particular trophoblast cell types could be resolved. For example, does cytotrophoblast and syncytiotrophoblast (Loke *et al.*, 1972) or only syncytiotrophoblast (Midgley and Pierce, 1962; Yorde *et al.*, 1979) elaborate HCG? In addition, isolated trophoblast cells can be used for cell fusion studies of placental somatic cell genetics. Moreover, replicate samples can be quantitatively obtained from isolated trophoblasts whereas sample variation is more apt to occur with explants or organ cultured specimens of chorionic villi.

Enrichment of certain subsets of trophoblast cells also might enhance isolation of selected trophoblast gene products. For example, there are persistent reports by electron microscopy of budding type-C particles in subhuman primates, as well as human placental specimens (see Schidlovsky, 1978, for review, and Fig. 7). In addition, use of molecular hybridization indicates that Rhesus monkey placenta contained increased proviral sequences in their DNA to the Mason–Pfizer virus (MPV) genome as well as RNA complementery to the entire MPV 60 to 70S RNA genome (Drohan *et al.*, 1979). After enrichment of certain classes of trophoblast cells, retroviral particles might be more likely to to observed and isolated by conventional cocultivation approaches (Todaro *et al.*, 1978) toward eventual growth and characterization in cell culture (Bryant *et al.*, 1978). Using a radioimmunoassay for the major structural protein (*p*26) of this recently isolated Rhesus type C virus (MAC-1), the tissue-specific preference for MAC-1 *p*26 expression in placental tissue, and placental cell and organ culture, has become apparent (manuscript in preparation). The primate placenta thus affords an opportunity to study retroviral expression in differentiated cells which physiologically express endogenous vital gene products. Consequently, various inducing agents, such as halogenated pyrimidines (Lowy *et al.*, 1971), can be utilized during the *in vitro* culture to enhance the likelihood of retroviral expression. Isolation of subsets of trophoblast cells also provides an opportunity to stabilize production of their respective gene products in cell culture following transfection in primary culture with the early, or transforming, region of the SV40 genome. For example, trophoblast cells from Rhesus monkeys after SV40 gene A transfection have the capacity for long-term growth in monolayer culture and continue to express differentiated products, such as Type IV (basement membrane) collagen, alkaline phosphatase, and desmosomes by EM (manuscript in preparation).

In short-term organ culture, the enzymatic capacity of the human placenta to convert promutagens and procarcinogens to ultimate mutagens (Jones *et al.*, 1977) can be studied on an *in vitro* basis. For example, the influence on carcinogen metabolism of tumor promotional agents such as tetradecanoylphorbol acetate (TPA) or growth hormones such as nerve growth factor in placenta (Goldstein *et al.*, 1978) can be evaluated. Moreover, by *in vitro* experimentation, the mechanism to account for the 70-fold interindividual variation in placental hydrocarbon hydroxylase activity among women smokers (Conney and Levin, 1974) could be addressed.

An additional area of opportunity in placental organ culture is the study of cell senescence (Vincent *et al.*, 1976). The life span of the placenta is condensed and offers an excellent system to evaluate the intrinsic cellular processes involved in the *in vitro* geriatrics of a transient organ.

Another area of potential application is to use placental organ culture to study the mechanism of absorption across a microvillus surface. The rich endowment of microvilli over the trophoblast cell membrane, reminiscent of mesothelium, reflects the incredible absorptive surface of the human placenta which at term is reported to have a surface area of 11 m^2 (Padykula, 1977). Thus, for example, the role of coated and pitted vesicles in pinocytosis and endocytosis could be evaluated *in vitro* using placental organ culture.

However, the field of placental organ culture should be seen as in its adolescence, especially when measured with regard to the length of substantial production of its differentiated gene products. After 2 weeks in culture, for example, HCG expression is less than one-fifth of its peak *in vitro* expression on days 4 to 6. Moreover, the duration of viability in organ culture as measured by such elemental parameters as incorporation of labeled precursors of protein or nucleic acid synthesis has not even been assessed beyond 2 weeks in organ culture. Consequently, although the ready availability, the biochemical complexity, and the structural simplicity of the placenta provide promise as an *in vitro* model system to study many biological processes, much fundamental work remains to be done to realize its potential.

Acknowledgments

I wish to acknowledge my colleagues Rachel Huot, Jean-Michel Foidart, K. V. Speeg, Jr., and Jane Clifford Azizkhan, and the support of George J. Todaro, as well as the secretarial help of Doris Little.

References

Azizkhan, J. C., Speeg, K. V., Jr., and Stromberg, K. (1979). *Cancer Res.* **39,** 1952.

Balls, M., and Monnickendam, M. A. (1976). "Organ Culture in Biomedical Research." Cambridge Univ. Press, Cambridge, England.

Belleville, F., Lasbennes, A., Nabet, P., and Paysant, P. (1978). *Acta Endocrinol.* **88,** 169.

Benagiano, G., Pala, A., Meirinho, M., and Ermini, M. (1972). *J. Endocrinol.* **55,** 387.

Billington, W. D. (1971). *Adv. Reprod. Physiol.* **5,** 27.

Bjorkman, N. (1970). "An Atlas of Placental Fine Structure." Williams & Wilkins, Baltimore, Maryland.

Blanford, A. T., and Murphy, B. E. (1977). *Am. J. Obstet. Gynecol.* **127,** 264.
Bohn, H. (1978). *Scand. J. Immunol.* Suppl. **7** (6), 119.
Boyd, J. D., and Hamilton, W. J. (1970). "The Human Placenta." Heffer, Cambridge, England.
Boyum, A. (1968). *Scand. J. Clin. Lab. Invest.* Suppl. **21,** 97.
Bryant, M. L., Sherr, C. J., Sen, A., and Todaro, G. J. (1978). *J. Virol.* **28,** 300.
Cedard, L., and Alsat, E. (1975). *Methods Enzymol.* **39,** 244.
Chatterjee, M., and Munro, H. N. (1977). *Vitam. Horm.* **35,** 149.
Chou, J. Y. (1978). *Proc. Natl. Acad. Sci. U.S.A.* **75,** 1409.
Chou, J. Y. (1978). *Proc. Natl. Acad. Sci. U.S.A.* **75,** 1854.
Chung, H. K., McLimans, W., Horoszewicz, J., and Hreshchyshyn, M. M. (1969). *Am. J. Obst. Gynecol.* **104,** 945.
Conney, A., and Levin, W. (1974). *In* "Chemical Carcinogenesis Essays" (R. Montesano and L. Tomatis, eds.), pp. 3-24. IARC Scientific Publ. No. 10, Lyon.
Cotte, C., Easty, C. G., Neville, A. M., and Monaghan, P. (1980). *In Vitro* **16** (in press).
Drohan, W., Colcher, D., and Schlom, J. (1979). *Cancer Res.* **39,** 1696.
Gaillard, P. J., and Schaberg, A. (1965). *In* "Endocrine Glands in Cells and Tissues in Culture" (E. N. Willmer, ed.), pp. 674–694. Academic Press, New York.
Gey, G. O., Seegar, G. E., and Hellman, L. M. (1938). *Science* **88,** 306.
Goldstein, L. D., Reynolds, C. P., and Perez-Ford, J. R., (1978). *Neurochem. Res.* **3,** 175.
Guiet-Bara, A. (1977). *Adv. Exp. Med. Biol.* **94,** 479.
Gusseck, D. L. (1977). *Gynecol. Invest.* **8,** 162.
Hagerman, D. D. (1964). *Fed. Proc.* **23,** 785.
Hall, C. St. G., James, T. E., Goodyer, C., Branchaud, C., Guyda, H., Giroud, C. J. P. (1977). *Steroids* **30,** 569.
Handwerger, S., Barrett, J., Tyrey, L., and Schomberg, D. (1974) *J. Clin. Endocrinol. Metab.* **36,** 1268.
Hobbins, J. C., and Berkowitz, R. L. (1978). *Clin. Obstet. Gynecol.* **21,** 363.
Huot, R. I., Foidart, J-M., and Stromberg, K. (1979). *In Vitro* **15,** 497.
Jones, A. H., Fantel, A. G., Kocan, R. A., and Juchau, M. R. (1977). *Life Sci.* **21,** 1831.
Leighton, J. (1951). *J. Natl. Cancer Inst.* **12,** 545.
Lerner, U., Saxena, B. N., and Diczfalusy, E. (1972). *Acta Endocrinol. (Copenhagen)* Suppl. **185,** 310.
Levitz, M., and Young, B. K. (1977). *Vitam. Horm.* **35,** 109.
Levitz, M., Jansen, V., and Dancis, J. (1978). *Am. J. Obstet. Gynecol.* **132,** 363.
Loke, Y. W., Wilson, D. V., and Borland, R. (1972). *Am. J. Obstet. Gynecol.* **113,** 875.
Lowy, D. R., Rowe, W. P., Teich, N., Hartley, J. W. (1971). *Science* **174,** 155.
Maruo, T. (1976). *Endocrinology (Jpn.)* **23,** 119.
Midgley, A. R., and Pierce, G. B. (1962). *J. Exp. Med.* **115,** 289.
Miller, R. K., and Berndt, W. O. (1975). *Life Sci.* **16,** 7.
Moriyama, I., Ushioda, E., Ishibashi, S., Hiroako, K., and Shimozato, N. (1974). *Acta. Obstet. Gynaecol. (Jpn.)* **21,** 86.
Murray, J. C., Stingal, G., Kleinman, H. K., Martin, G. R., and Katz, S. I. (1979). *J. Cell Biol.* **80,** 197.
Nesbitt, R. E., Jr., Rice, P. A., Rourke, J. E., Torresi, V. F., and Souchay, A. M. (1970). *Gynecol. Invest.* **1,** 185.
Padykula, H. A. (1977). *In* "Histology" (L. Weiss and R. O. Greep, eds.), 4th Ed., pp. 951–978. McGraw-Hill, New York.
Pretlow, T. G., II, and Pretlow, T. P. (1977). *In* "Biologicial Separations. Methods of Cell Separation" (N. Catsimpoolas, ed.), Vol. 1. Plenum, New York.
Pretlow, T. P., Stinson, A. J., Pretlow, T. G., and Glover, G. L. (1978). *J. Natl. Cancer Inst.* **61,** 1431.

Salomon, D., and Sherman, M. I. (1975). *Dev. Biol.* **47,** 394.
Sato, Y. (1973). *Endocrinology* (*Jpn.*) **20,** 291.
Saure, A. (1973). *Ann. Acad. Sci. Fenn.* (*Med.*) **159,** 1.
Schidlovsky, G. (1978). *In* "Recent Advances in Cancer Research: Cell Biology, Molecular Biology, and Tumor Virology" (R. Gallo, ed.), Vol. 1. CRC Press, Cleveland, Ohio.
Speeg, K. V., Jr., Azizkhan, J. C., and Stromberg, K. (1976). *Cancer Res.* **36,** 4570.
Stoner, G. D., Harris, C. C., Autrup, H., Trump, B. F., Kingsbury, E. W., and Myers, G. A. (1978). *Lab. Invest.* **38,** 685.
Stromberg, K., Azizkhan, J. C., and Speeg, K. V., Jr. (1978). *In Vitro* **14,** 631.
Sussman, H. H. (1978). *Scand. J. Immunol.* Suppl. **7,** (6) 127.
Tao, T. W., and Hertig, A. T. (1965). *Am. J. Anat.* **116,** 315.
Taylor, P. V., and Hancock, K. W. (1973). *J. Obstet. Gynecol.* (Br. Commonwealth) **80,** 834.
Todaro, G. J., Benveniste, R. E., Sherwin, S. A., and Sherr, C. J. (1978). *Cell* **13,** 775.
Trowell, O. A., (1959). *Exp. Cell Res.* **16,** 118.
Trump, B. F., McDowell, E. M., Barrett, L. A., Frank, A. L., and Harris, C. (1974). *In* "Experimental Lung Cancer" (E. Karbe and J. F. Park, eds.), pp. 458–563. Springer-Verlag, Berlin and New York.
Villee, C. A. (1977). *Gynecol. Invest.* **8,** 145.
Vincent, R. A., Jr., Huang, P. C., and Parmley, T. H. (1976). *In Vitro* **12,** 649.
Yorde, D. E., Hussa, R. O., Garancis, J. C., and Pattillo, R. A. (1979). *Lab. Invest.* **40,** 391.

Chapter 11

Overview: Endocrine Systems

M. EDWARD KAIGHN

Pasadena Foundation for Medical Research, Pasadena, California

I. Overview of Endocrine Systems

Although considerable work has been done on cultured animal endocrine tissues, very little information is available with human material. Many human tumors are hormonally related. Because of the difficulty in extrapolating from animal to human endocrinology, it is important to develop human culture models.

The chapters in this section show that normal human endocrine cells can be profitably studied. Studies of polypeptide hormones have been most productive. Consistent growth (Chapter 9) and morphological (Chapter 4) and metabolic effects (Chapter 6) have been observed both in cell and in organ culture. Interactions between polypeptide hormones and glucocorticoids on growth (Chapter 9) and ultrastructure (Chapter 5) have also been detected. On the other hand, sex steroids which induced growth and morphological effects in organ culture (Chapter 1) had little effect on these parameters in cell cultures (Chapters 2, 3, 5, 8, and 9). However, specific metabolism of sex steroids was seen in cell cultures (Chapter 9).

The contrasting patterns of steroid and polypeptide activities in organ and cell cultures may reflect different requirements of these two groups of hormones for cell interactions. Thus, the two kinds of systems, cell and organ culture, yield complementary information about the function of individual and coordinated cellular components of endocrine tissues.

The problem of obtaining suitable samples of human tissue is common to both

ISBN 0-12-564140-0

cell and organ culture. Because of delays imposed by the necessity to obtain informed consent, many samples are lost or nonviable when received. For this reason, probably the most reliable is tissue obtained at "immediate autopsy" as employed by Dr. Trump's group (Chapter 8).

Methods used to establish primary cultures are discussed in most chapters. Dr. Franks' (Chapter 7) excellent discussion of primary culture of the prostate is applicable to most cell culture systems. In general, mechanical subdivision of the tissue is more successful than enzymatic digestion. Various means were used to separate stromal and epithelial components including density gradient centrifugation (Chapters 5 and 9). The standard organ culture system in which tissue fragments are supported on a membrane filter still appears optimal.

Cell culture has a different objective than organ culture. To be useful, cell cultures must be able to undergo at least 15–20 population doublings. This is rarely achievable when subculture of normal epithelial cells is attempted by conventional enzymatic dissociation. Dr. Lechner in my laboratory, using nonenzymatic methods, has obtained sustained growth of neonatal prostate epithelial cells (Chapter 9).

A major concern is the nature of the nutrient medium and associated parameters. Obviously, the requirements are quite distinct for cell and organ cultures and for different cell types regardless of the system. The design of media tailored for specific cell types was described by Dr. Ham. A persistent problem is how "optimal" media are chosen. The lack of quantitation, particularly in measuring specialized properties, is a major problem. The role of hormones in growth and differentiation and their inclusion in "defined medium" formulations is a basic rather than a technical problem. The culture systems included in this section will be useful in illuminating this area.

II. Differentiation vs. Growth

The question of whether a culture is authentic usually involves whether it has tissue-specific function. In the case of cell cultures, this requirement would be superimposed on the need to undergo a useful number of population doublings. In embryonic development both processes are essential and "terminal differentiation" precludes further cell division. Cell fusion in myogenesis and nuclear pycnosis in erythropoiesis are classic examples of this process. In both cases, further differentiation depends on the existence of a "stem cell" population. Stem cells may be involved in other populations in a less obvious way. They may, for example, give rise to new layers of epidermis or new secretory cells. On the other hand, they may cease to exist after embryogenesis as is probably the case with neural tissues. The question of the existence and continued function of such cells is a basic unsolved problem of development which is central to the understanding of carcinogenesis. To what extent are such cells the target of carcinogenic attack? Are "fully differentiated" cells capable of transformation? Experimental evidence from chemical, viral, and radiation-induced carcinogenesis appears to support the idea that continued capability for cell division is a prerequisite for carcinogenesis. On the other hand, in many cases the origin of a given cancer can be deduced by residual differentiated properties, for example, the "minimum-deviation hepatoma" or the melanoma. When and how carcinogenic stimuli interact with "stem cells" or differentiated cells is still unknown. The basic process is change in growth control. The relation of specialized functions to this basic change is completely unknown.

The chapters included here are encouraging. Many technical problems have been at least partially overcome. The life expectancy of organ cultures has been dramatically extended and replicative epithelial cultures appear attainable. Information is being acquired on cell-specific factors controlling both growth and differentiation. It is also clear that exogenous factors do not necessarily act individually but interact. These interactions must be understood and controlled if meaningful *in vitro* carcinogenesis studies are to be carried out. There is more to carcinogenesis than enzyme activation or binding as significant as each may be. An exciting approach is the hormonal cycling experiments reported for uterine and breast tissue in which cyclic hormonal changes during the menstrual cycle are mimicked *in vitro*. An understanding of the influence of such changes on susceptibility to carcinogenesis could help provide approaches to prevention as well as cure. With the optimism and enthusiasm generated by this timely conference comes the news of the untimely death of one of the principal speakers, Dr. Douglas Janss, who has made significant contributions to our understanding of the endocrinology of cultured mammary tissues. I respectfully dedicate this overview to his memory.

ACKNOWLEDGMENT

I am deeply indebted to Dr. David Kirk for his critical comments and useful suggestions.

Chapter 12

Organ Culture of Normal Human Bladder: Choice of Starting Material and Culture Characteristics

M. A. KNOWLES,[1] R. M. HICKS,[2] R. J. BERRY,[1] AND E. MILROY[3]

[1]Department of Oncology, [2]School of Pathology, and [3]Department of Urology, Middlesex Hospital Medical School, London, England

---

I. Introduction

Several bladder carcinogens, some of which are epidemiologically implicated in the induction of human bladder tumors, have now been identified and studied

ISBN 0-12-564140-0

in animal models. Our work on human bladder is designed to develop an *in vitro* model in which the process of carcinogenesis can be studied and with which predictive tests of environmental chemicals can be made on the human target tissue. Organ cultures rather than cell cultures have been selected as a basis for this model because normal urothelial cell differentiation cannot be maintained in cell culture. Differentiation of the urothelium *in vivo* is influenced by the supporting stroma and has been shown to retain its normal pattern in rat bladder organ cultures where the normal stromal/epithelial interrelationship is retained (Hodges *et al.*, 1977). Development of comparable culture techniques for human bladder explants should provide a model in which the rate of cell turnover and the pattern of urothelial differentiation resembles the *in vivo* state. The histopathological response of such organ cultures to carcinogens may be expected to reflect that of human bladder *in vivo* and will enable the roles of stromal and epithelial cell types in the activation of carcinogens to be assessed.

Rat bladder has been successfully maintained in organ culture by two groups of workers and, in both cases, a histologically normal, differentiated transitional epithelium was maintained. Hodges *et al.* (1976, 1977) cultured adult rat bladder up to 100 days using a grid technique and Waymouth's medium MB 752/1 supplemented with newborn calf serum, ferrous ions, ascorbic acid, and hydrocortisone. Reese *et al.* (1976) used chemically defined media based on Waymouth's MB 752/1 and Ham's F-12, and maintained cultures for up to 21 days in the absence of serum. It was found necessary to supplement the Ham's medium with hydrocortisone or calcium to prevent urothelial hyperplasia and endophytic growth (Reese *et al.*, 1977; Reese and Friedman, 1978). Other supplements are apparently nonessential in these and rodent bladder cell cultures (e.g., Elliot *et al.*, 1975; Summerhayes and Franks, 1979) though the polyamines spermine, spermidine, and putrescine have been used to prolong the culture lifespan and increase cell proliferation in rat cultures (Roszell *et al.*, 1977).

Human bladder cell cultures from normal newborn urine (Sutherland and Bain, 1972), adult urine (Linder, 1976), and many human bladder carcinomas (e.g., Jones, 1967; Rigby and Franks, 1970; Toyoshima *et al.*, 1976; Elliott *et al.*, 1976; Rasheed *et al.*, 1977; Malkovsky and Bubenik, 1977; O'Toole *et al.*, 1976, 1978) have been maintained in Eagle's minimal essential medium, medium 199, or RPMI-1640 medium supplemented only with calf or fetal calf serum ± L-glutamine.

An essential prerequisite for this study is a supply of normal human bladder tissue. Patients of widely varying age and clinical history who are undergoing investigative cystoscopy or prostatectomy, some of whom have a history of previous neoplastic disease of the bladder, are available as a source of material. Previously published reports (e.g., Melicow, 1952; Eisenberg *et al.*, 1960; Simon *et al.*, 1962; Cooper *et al.*, 1973; Schade and Swinney, 1973; Skinner *et al.*, 1974) have shown that macroscopically visible transitional cell carcinomas are frequently accompanied by widespread histological abnormalities throughout the urothelium. Patients with previous neoplastic disease therefore may not be appropriate donors of "normal," but rather of "preneoplastic" urothelium. In this chapter, the assessment and selection of suitable normal human bladder tissue from available patients is discussed and the characteristics of short-term (35 day) cultures described.

II. Materials and Methods

A. Specimens

Patients with a wide range of clinical histories were sampled in order to locate a satisfactory source of histologically normal bladder tissue. Samples were obtained from patients undergoing cystoscopic investigation or retropubic prostatectomy under general anesthesia. Cup biopsies (approximately 5 mm^2) from cystoscopy patients were generally taken from an area postero-lateral to the ureteric orifices and consisted of urothelium, lamina propria, and a variable depth of smooth muscle. In tumor-bearing bladders, or bladders with focal inflammation, cup biopsy samples were taken from those areas in the bladder away from the trigone, which were judged by the surgeon to be most "normal" by macroscopic appearance. Glycine solution (1.1%) was used as the medium for cystoscopy as water proved less satisfactory. From patients undergoing retropubic prostatectomy, a larger (1 × 2 cm) piece of bladder wall was removed close to the bladder neck.

Specimens were handled aseptically and immediately immersed in ice-cold Waymouth's MB 752/1 medium (Gibco-Biocult, Ltd., London, England) with HEPES buffer for transport to the laboratory. Some specimens for histology were fixed in theater, in ice-cold phosphate-buffered formalin.

B. Preparation of Explants

Tissue was stored for up to 2 hours in the transport medium. It was then submerged in ice-cold Waymouth's medium MB 752/1 with HEPES buffer in a dissection dish. The latter consisted of a glass petri dish containing a 3-mm sheet of silicone rubber (Esco Rubber, London, England) which was sealed to the glass with Silastic 738RTU sealant (Dow Corning, Michigan). Cup biopsies were secured to the rubber by a single steel pin through the muscle, and the larger specimens by several pins around the periphery. The specimens were dissected in a sterile cabinet using a dissection microscope with a cold light source, at a working magnification of ×16. The urothelium and underlying stroma to a depth of 0.8–1.0 mm were cut free from the remaining stroma and muscle with Pascheff Wolff's scissors (Moria, Paris, France), DeWecker's iris scissors, or smaller types of iris scissor. This yielded a sheet which was trimmed to remove damaged edges before cutting into explants 2–3 mm^2. With the larger specimens, the sheet was stretched between pins, urothelial face uppermost, for the final cutting of explants. Parallel rat bladder explants were prepared in a similar way. The bladder was removed from animals killed by cervical dislocation, placed in medium in a dissection dish, and cut open along one side from urethra to dome. The sheet was then pinned out with the urothelium uppermost and cut into explants as described. The human explants thus consisted of intact urothelium and some stroma, with cut edges and a cut lower surface, in contrast to the rat bladder explants which were the full thickness of the bladder with cut margins, but intact mesothelium on the lower surface.

C. Culture Supports

Several culture substrates were used. These included RA 1.2-μm cellulose acetate filters (Millipore) 0.45-μm Metricel filters (Gelman), gelatin sponge (Sterispon No 1., Allen and Hanbury's, London), slices of rat tail collagen gel, and collagen-coated Millipore filters. Millipore filter squares (approx. 1 cm^2) are now used routinely. Using a small spatula the explants were placed, urothelial surface uppermost, on filters which had been soaked in warmed medium. The cultures were supported on steel mesh grids in plastic organ culture dishes (Falcon Plastics, Oxnard, CA).

D. Media

Several commercially available media were tested for their ability to support the survival of human bladder tissue. These included Waymouth's medium MB 752/1, Ham's F-10 and F-12 nutrient mixtures, CMRL 1066, RPMI-1640, medium 199, Dulbecco's modification of Eagle's medium, and NCI medium (Gibco-Biocult, London, England). Waymouth's medium MB 752/1, Ham's F-12, RPMI-1640 and CMRL 1066 are now used routinely. The following supplements were used alone or in combination: Heat-inactivated fetal or newborn calf sera (Gibco-Biocult, London, England), human group AB serum, patients' homologous serum, L-glutamine (2 m*M*), hydrocortisone sodium succinate (1 μg/ml), insulin (1 IU/ml), ascorbic acid (300 μg/ml), $FeSO_4$ (0.45 μg/ml), pyruvate (1 m*M*), urea (0.05%), dimethyl sulfoxide (0.05–0.1%), penicillin (100 units/ml), and streptomycin (100 μg/ml). Medium was added to just cover the explant surface. Cultures were maintained in a constantly monitored humidified atmosphere of 5% CO_2 in air at 36.5–37°C. Medium was renewed every 2–3 days.

E. Histology

Tissue was fixed in phosphate-buffered formalin (pH 7.4) for at least 24 hours, dehydrated, and embedded in paraffin wax. Sections were stained with hematoxylin and eosin and alcian blue at pH 1.0.

F. Electron Microscopy and High-Resolution Histology

Tissue was fixed for 1 hour in 1% osmium tetroxide in 0.1 *M* cacodylate buffer at pH 7.4. Specimens for transmission electron microscopy were then dehydrated through graded alcohols and embedded in Spurr resin (Agar Aids, Stanstead, England). Sections cut at 1 μm were stained with toluidine blue for high-resolution histology and thin sections were stained with uranyl acetate and lead citrate and viewed in a Jeol 100B electron microscope. Specimens for scanning electron microscopy were dehydrated through graded ethanols, critical point dried, mounted on specimen stubs, and coated with gold. They were examined in a Jeol JSM-35 scanning electron microscope with an accelerating voltage of 20 or 25 kV.

G. Autoradiography

Explants cultured for various periods were exposed for 24 hours to medium containing 1 μCi/ml [*methyl*-^{3}H]thymidine (sp act 5 Ci/m*M*, Radiochemical Centre, Amersham, England) before being fixed and processed for histology. Dewaxed sections were dipped in K5 liquid nuclear emulsion (Ilford Ltd., Basildon, England) diluted 1:1 with water, exposed for 2 weeks at 4°C, developed with D19 developer (Kodak, London, England), and stained with Ehrlich's hematoxylin.

III. Results

A. Morphology of Human Bladder Biopsy Specimens

1. Damage of Tissue during Biopsy Procedure

We have observed both mechanical and osmotic damage to samples. Shearing and mechanical abrasion of cells inevitably occur at the edge of all biopsies since the cup biopsies are pulled from the mucosa rather than cut. Occasionally more extensive damage has been seen (Fig. 1) with the loss of one or more layers of epithelial cells over wide areas. This is aggravated by prolonged cystoscopic

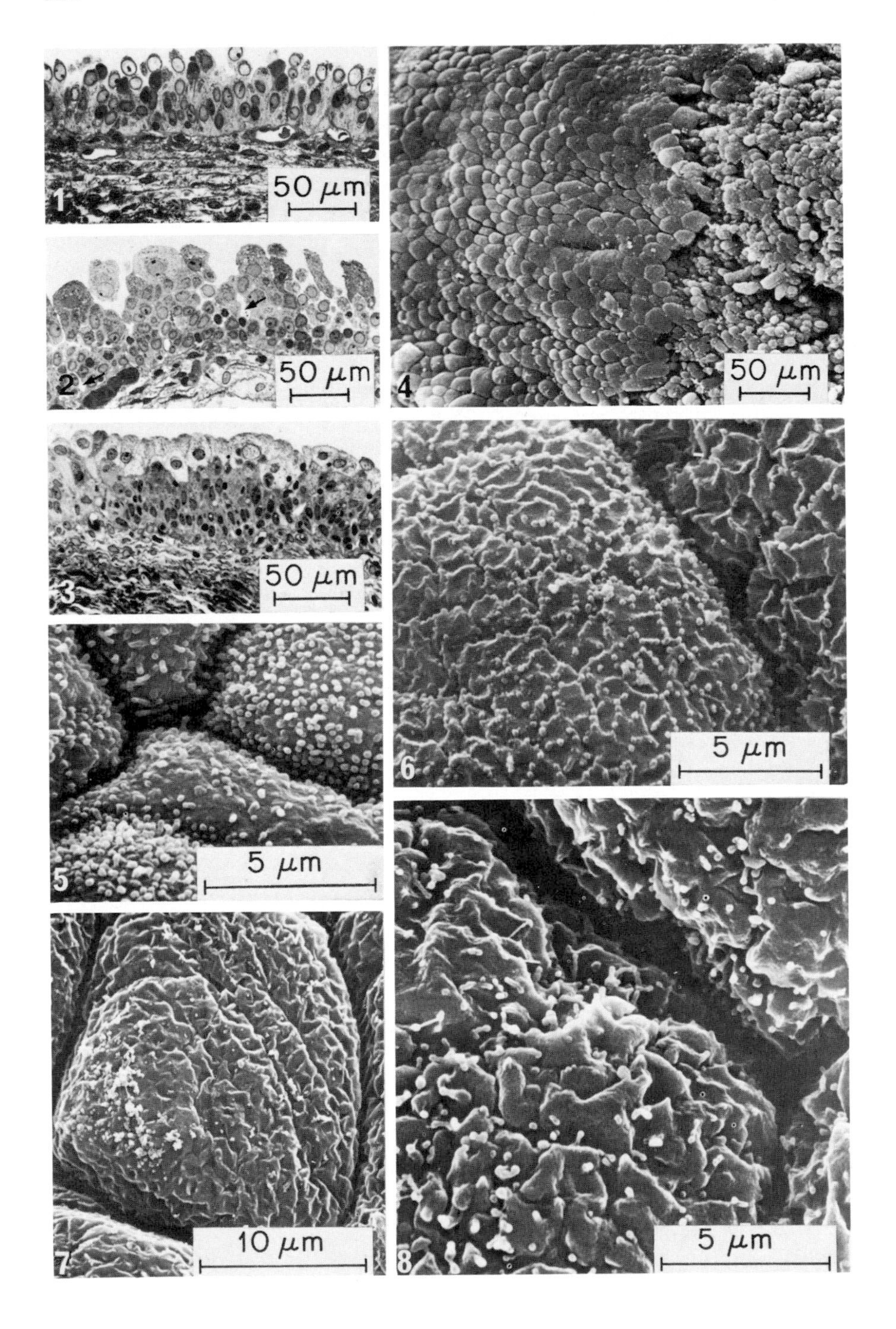
50 μm
1
50 μm
2
50 μm
3
50 μm
4
5 μm
5
5 μm
6
10 μm
7
5 μm
8

observation before the urothelium is biopsied and damage increases in proportion to the amount of irrigation fluid which has passed through the bladder. To minimize damage, specimens are now taken before any other surgical procedure is carried out.

Osmotic damage, indicated by dilation of intercellular spaces (Fig. 2), was found to be most severe when water was used as the cystoscopy medium. Glycine solution (1.1%) though hypotonic produces less damage and is now used routinely.

2. Assessment of Biopsies

Biopsies have been taken from patients of widely varying age and clinical history (Table I), and in such a diverse population some degree of histological variability is to be expected. We have therefore subdivided the patients into three main groups based on their previous history of neoplastic disease and/or conditions where urine retention may have lead to an increased exposure of the urothelium to any urine-borne carcinogens. Our primary aim is to exclude from organ culture, tissue samples with a labile urothelium showing dysplastic or possible neoplastic change. Samples from 93 patients distributed among these groups have been assessed histologically for suitability as "normal" tissue for culture (Table II).

Signs of inflammation occurred in the stroma of some specimens in all groups.

FIGS. 1-8. Figures 1-3 are light micrographs of toluidine blue-stained, resin-embedded sections. Figures 4-8 are scanning electron micrographs.

FIG. 1. Urothelial surface of human bladder biopsy. The biopsy procedure has mechanically abraded the superficial and some intermediate cells.

FIG. 2. Urothelial surface of biopsy with osmotic damage. The superficial cells are still present but dilatation of the intercellular spaces has occurred (arrows).

FIG. 3. Normal undamaged human bladder biopsy showing large pale superficial cells and smaller more darkly staining intermediate and basal cells. Note the large nucleus, indicative of polyploidy in the superficial cell layer.

FIG. 4. The urinary surface of a typical human bladder biopsy. The edge of the biopsy (at the right of the field) has sustained some mechanical damage revealing the surface of cells at various levels in the epithelium.

FIG. 5. Immature basal cells in a human bladder biopsy. Small globular processes are present in varying numbers on the cell surface.

FIG. 6. Maturing intermediate cells in a human bladder biopsy. Chains of globules and microridges cover the cell surface.

FIG. 7. A mature superficial cell on the surface of a human bladder biopsy. The membrane is thrown into angular folds. Deep clefts between adjacent cells indicate that the urothelium was in a contracted state when fixed.

FIG. 8. Superficial cells with small globular processes between mature membrane ridges.

TABLE I

SOURCE OF MACROSCOPICALLY "NORMAL" HUMAN BLADDER MUCOSAL BIOPSIES

Patient group	Number of cases	Total
1. Present or previous neoplastic disease of the bladder		
a. Check cystoscopy—no recurrence		27
b. Check cystoscopy—recurrence at distance site in bladder		9
2. Benign prostatic hypertrophy		
a. Retropubic resection		33[a]
b. Transurethral resection		27
3. Miscellaneous conditions		
a. History of urine retention		16
Urethral dilatation/bladder neck incision (male)	10	
Urethral dilatation/bladder neck incision (female)	5	
Hysterical retention	1	
b. No history of urine retention		39
Tumor of prostate or kidney	3	
Check following chronic cystitis	4	
Investigation of increased frequency	5	
Lithiasis	2	
Ureterolithiasis	5	
Interstitial cystitis	2	
Prostatic biopsy	4	
Nephrogenic adenoma elsewhere in bladder	2	
Reimplantation of ureters	1	
Vesical pain	1	
Investigation of hematuria—no bladder lesion found	8	
Trigonitis	2	
Ureteric obstruction	1	

[a] Specimens taken from bladder neck ± body of bladder. All other biopsies were taken from body of bladder.

These included edema with infiltration by polymorphonuclear leukocytes, lymphocytes, eosinophils, plasma cells or macrophages, and sometimes fibrosis. Such changes were not associated with any particular epithelial changes though they were common in tumor-bearing bladders regardless of urothelial histology. Their significance is not known at present. Our assessment is based on the histology of the epithelial component only, and samples are assessed as histologically normal, hyperplastic, or dysplastic.

a. Normal Urothelium The definition "normal" is based on what is known of the embryological development of the bladder in man, a limited amount of information on the adult human urothelium, and studies of other mammalian urothelia (a comparative review is given by Hicks, 1975). In man, the urothelium

TABLE II

HISTOLOGICAL FINDINGS IN HUMAN BLADDER BIOPSY SPECIMENS

			Hyperplasia			von Brunn's nests or cystitis glandularis		Dysplasia			
Patient groups	Number of patients	Absence of normal superficial cells	Mild 6–9 cell layers	Moderate to severe ≥10 cell layers	Irregularity of basal lamina	<6 Nests	≥6 Nests	Mild	Moderate	Severe	Inflammation ± fibrosis
1. Present or previous neoplastic disease of the bladder											
a. Check cystoscopy. No recurrence of tumor	17	5	2	1	2(0)[a]	1	2	3	2	0	7
b. Check cystoscopy. Recurrence at distant site in bladder	8	2	3	2	3(1)	2	1	1	1	0	5
2. Benign prostatic hypertrophy											
a. Retropubic prostatectomy	10	5	3	5	6(4)	1	5	5	0	0	3
b. Transurethral resection of prostrate	22	4	7	5	9(5)	1	5	3	2	1	8
3. Miscellaneous conditions											
a. (History of urine retention)	13	1	5	2	2(2)	3	0	2	1	1	9
b. (No history of urine retention)	23	3	7	1	5(2)	2	2	3	0	0	9
Total	93	20	27	16	27(14)	10	15	17	6	2	41

[a] Figures in parentheses indicate number of specimens where down-growth or irregularity was associated with the formation of epithelial cell nests.

consists of four to five epithelial cell layers which rest on a flat basal lamina (Koss, 1975). There is a regular pattern of differentiation from base to luminal surface and the component cells are termed "basal," "intermediate," or "superficial" according to their position (Fig. 3). It is not clear whether human urothelium represents a true stratified or a pseudostratified epithelium since elongate cytoplasmic processes extend from some of the superficial cells to the basal lamina. Such processes are regularly observed in man (Petry and Amon, 1966) and guinea-pig (Martin, 1972) but are seldom seen in other species. Basal and intermediate cells are more basophilic than the superficial cells which may appear pale and vacuolated. There is a concomitant increase in ploidy from diploid in the basal layer to tetraploid or more in the superficial cell layer. This has been shown to involve cell fusion in the guinea pig (Martin, 1972) and this probably also occurs in man since the nuclear to cytoplasmic ratio is similar in cells at all stages of maturation.

Mechanical damage to the urothelium around the edges of the biopsies (Fig. 4) allowed the surface ultrastructure of cells at different levels in the epithelium to be studied by scanning electron microscopy. Cells of the basal layer were small and had numerous globular processes on their surface which increased in number with maturity (Fig. 5). These globular processes were present in larger numbers on immature intermediate cells and they gradually fused to form globular chains and ridges on the maturing intermediate cells (Fig. 6). The membrane on the surfaces of the large mature superficial cells formed a series of angular ridges, varying in depth according to the degree of distention of the mucosa at the time of fixation (Fig. 7). Occasionally, these mature surface cells retained some globular processes between the mature ridges (Fig. 8). Transmission electron microscopy showed that the ridges correspond to the thin interplaque regions of the mature cell; the rigid plaques consist of specialized asymmetric plasma membrane as described previously (Warren and Hicks, 1973; Newman and Hicks, 1977).

Mature superficial cells were present on the majority of specimens (73 of 93). The best yield of mature urothelium was from the miscellaneous groups of patients including those with retention (Groups 3a and b, Table I). Normal urothelial maturation was least evident in retropubic prostatectomy patients.

The basal lamina showed no irregularities in 68 specimens. Of the remaining 27 specimens, over half showed von Brunn's nests or cystitis glandularis and the remainder (13) had some degree of urothelial dysplasia. Normal, flat basal laminae were most frequently found both in the two miscellaneous groups (Groups 3a and b, Table II) and in patients who had had a previous bladder tumor successfully removed (Group 1a, Table II).

Forty specimens showed all the features of normal urothelium, i.e., not more than five cell layers thick, with mature superficial cells, flat basal lamina, and no epithelial cell nests or dysplasia. Of these, 14 were found in the 23 miscellaneous patients with no urine retention (Group 3b, Table I), but only one was from the 10 retropubic prostatectomy patients (Group 2a, Table I).

b. Hyperplastic Urothelium. There are certain proliferative lesions of the urothelium which are considered to indicate an unstable urothelium. These include benign hyperplasia, von Brunn's nests, cystitis glandularis, and cystitis cystica (Koss, 1975). Although tumors can arise in such areas, these patterns of epithelial growth are generally regarded as benign.

We have defined all specimens with six or more cell layers as hyperplastic. Most of these were mildly hyperplastic (six to nine cells thick), had all the features of normal urothelial cell differentiation, and showed no evidence of histological atypia. Severe hyperplasia (i.e., greater than 10 cells thick) was common in retropubic prostatectomy patients (5 of 10) and more common in tumor-bearing bladders and transurethral prostatectomy patients than in other groups. Six of 16 severely hyperplastic epithelia showed no signs of normal superficial cell maturation. Occasionally, nonmaturing hyperplastic urothelia with no other histological signs of cytological abnormality were found by transmission electron microscopy to have surface microvilli with a well-developed glycocalyx. These subcellular features have been found in human and rat bladder tumors. They are also characteristic of the preneoplastic rat bladder (Hicks, 1976; Hicks and Wakefield, 1976; Newman and Hicks, 1977) and have been suggested as markers of preneoplasia in man (Newman and Hicks, 1977). The incidence of von Brunn's nests (Fig. 9) and cystitis glandularis was highest in retropubic prostatectomy patients and lowest in patients with no urine retention or previously resected tumors.

c. Dysplastic Urothelium. Dysplasia is defined as the presence of some or all of the following cytological characteristics in the epithelium: deviation of nuclear size and shape from the normal pattern, prominent nucleoli, nuclear hyperchromasia, loss of nuclear or cell polarity, and increased nuclear/cytoplasmic ratio resulting in the appearance of cell crowding. According to the severity of disturbance, dysplasia was arbitrarily designated mild, moderate, or severe (Figs. 10 and 11). The lowest overall incidence was in the miscellaneous group (Group 3b, Table II) of patients with no history of retention (3 of 23). Hyperplasia was not a prerequisite for dysplastic change, but severe hyperplasia was associated with 5 out of 17 cases of mild dysplasia, 3 out of 6 cases of moderate dysplasia, and 1 of 2 cases of severe dysplasia.

B. Characteristics of Human Bladder in Organ Culture

1. Culture Morphology

Before the effects of various media and supplements are discussed, the morphology of explants cultured for up to 35 days under optimal conditions will be described as a baseline for comparison. Cultures were derived from "normal" biopsy specimens unless otherwise indicated, and were maintained in Ham's

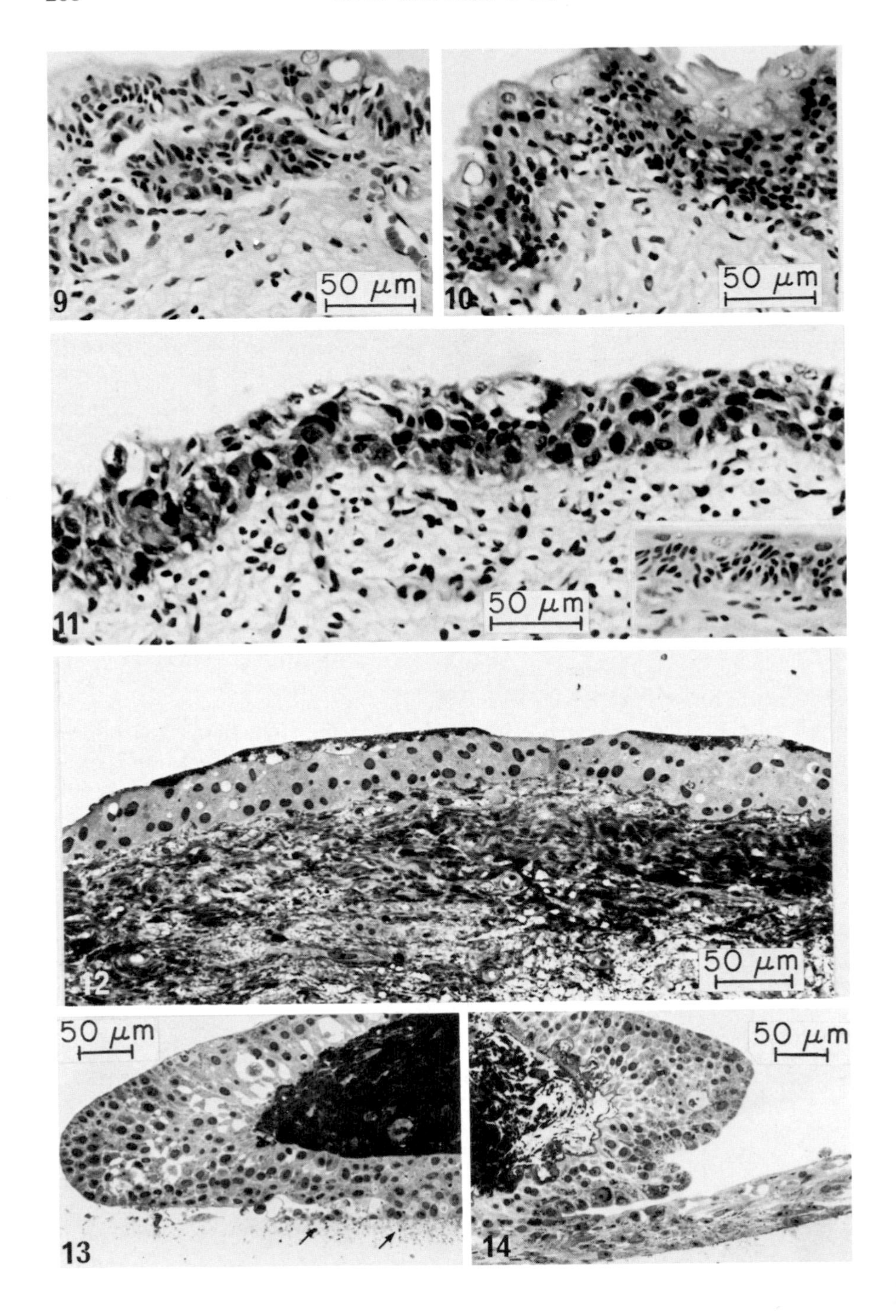
50 μm
9
50 μm
10
50 μm
11
50 μm
12
50 μm
13
50 μm
14

F-12 or Waymouth's MB 752/1 media supplemented with fetal calf serum, hydrocortisone, L-glutamine, $FeSO_4$, penicillin, and streptomycin.

a. Histology. During the first 48 hours in culture, the superficial epithelial cells died and could be clearly seen in toluidine blue-stained sections as dark squame-like cells attached to the epithelial surface (Fig. 12). In contrast to rat cultures, these dead cells sometimes remained attached for up to 35 days although they usually sloughed during the first 21 days *in vitro*. Osmotic damage sustained at the time of biopsy is reversible and disappeared during the first 48 hours in culture, so that the margins of epithelial cells again appeared contiguous.

Within 24 hours, cells of the basal epithelial layer began to migrate over the cut edges of the explant and by 48 hours these and the explant base were usually covered by a single layer of cells. This is equivalent to the process of epiboly as has been described in epidermal cultures (Sarkany *et al.*, 1965). Bare areas sometimes persisted at the base of the culture. The extent of these non-epithelialized areas appears to depend on the degree of adherence of the cut stroma to the filter and not on the size of the explant. During the first 7 days in culture, several more cell layers were added to the base of the culture and any small crevices in the stromal base which had arisen during the trimming of the biopsy were filled by cells. In some cultures a larger space existed between partially lifted explants and the filter and this became lined by epithelium which surrounded a central lumen (Fig. 16). Small cytoplasmic processes from the epithelial cell layer in contact with the filter support projected into the pores in the filter (Fig. 13).

In the rat cultures at this stage, epithelial outgrowths one to two cells thick began to extend over the filter. In human cultures this rarely happened and any outgrowth was always small (Fig. 14). More commonly, the epithelium at the edge of the culture where it met the filter support became thickened to form a

---

FIGS. 9–14. Figures 9–11 are sections of paraffin wax-embedded material, stained with hematoxyin and eosin. Figures 12–14 are toluidine blue-stained resin sections.

FIG. 9. Bladder biopsy with maturing surface epithlium and underlying von Brunn's nests.

FIG. 10. Biopsy showing mild urothelial dysplasia. The urothelium has mature surface cells, but is slightly hyperplastic and shows "crowding" of nuclei in the basal cell layers.

FIG. 11. Severe urothelial dysplasia in a biopsy. The urothelium is only three to five cells thick but there is gross variation in nuclear size and shape. No mature surface cells are present. Inset shows the appearance of normal mucosa at the same magnification, for comparison.

FIG. 12. The urothelium on the surface of a human bladder organ culture showing dead, darkly staining superficial cells attached to the living cells beneath.

FIG. 13. Edge of a human bladder organ culture after 14 days *in vitro* showing cytoplasmic processes extending from the cells at the base of the explant into the filter support (arrows).

FIG. 14. Edge of a 28-day human bladder organ culture with a small multilayered epithelial outgrowth.

bulbous rim around the explant (Fig. 15). After 7 to 14 days there was no further enlargement of this rim of cells.

By 14–21 days most of the dead superficial cells sloughed off and between 14 and 28 days histologically mature superficial cells could be identified on the surface of the culture and also on the stromal side of any lumina which formed beneath partially lifted cultures (Fig. 16). The time of appearance of mature cells was not uniform and probably reflected the relative amount of initial damage sustained by the tissue. Cultures have been maintained in this histologically mature state for up to 35 days. Longer term cultures have not yet been assessed.

As discussed in Section III, A, some starting material showed epithelial abnormalities. When such material was cultured, pre-existing conditions of benign hyperplasia, von Brunn's nests, cystitis glandularis, severe dysplasia, and irregularities of the basal lamina persisted and have been maintained for 35 days.

b. Ultrastructure. Transmission electron microscopy showed the cultured epithelium to bear a close resemblance to normal urothelium *in vivo*. The epithelium underwent a process of maturation *in vitro* to replace cells lost or damaged at the time of biopsy or early in culture.

At 7 days, the cells beneath the explant and those in the basal layer of the surface epithelium were undifferentiated and similar to basal cells *in vivo*. The cells of the intermediate layer beneath the dead superficial cells showed signs of maturation and often contained a well-developed Golgi complex and numerous smooth vesicles (Fig. 17). The surface epithelium was separated from its stroma by an intact basal lamina (Fig. 18) and a continuous layer of newly synthesized basal lamina separated the cells on the newly epithelialized surfaces from the stroma at the sides and base of the explant. No basal lamina was produced at the interface between the micropodia of the lowest epithelial cells and the Millipore filter (Fig. 19). By 7 days, the peripheral rim of epithelium around the culture consisted of cell aggregates which often contained enlarged intercellular spaces. The cells bounding these spaces had short, sparse microvilli on their free surfaces.

Between 7 and 21 days, the epithelial cells on the surface of the explant showed all the features of *in vivo* maturation (Fig. 20). The large, lysosomal bodies characterisitic of normal urothelial cells, and thought to be concerned *in vivo* with membrane turnover (Hicks, 1966, 1975), increased in number from the basal cells upward. Cells at all levels in the epithelium also accumulated glycogen in the form of single particles distributed throughout the cytoplasm. This is a characteristic feature of human but not of rat bladder (Monis and Dorfman, 1967). The features of specialized luminal membrane synthesis became increasingly prominent in the upper cell layers, and by 14–21 days the luminal membrane of the superficial cells showed the characteristic rigid plaques of asymmetric membrane and nonrigid "hinge" regions of the mature cell membrane. Fusiform vesicles containing this specialized membrane were also present in the

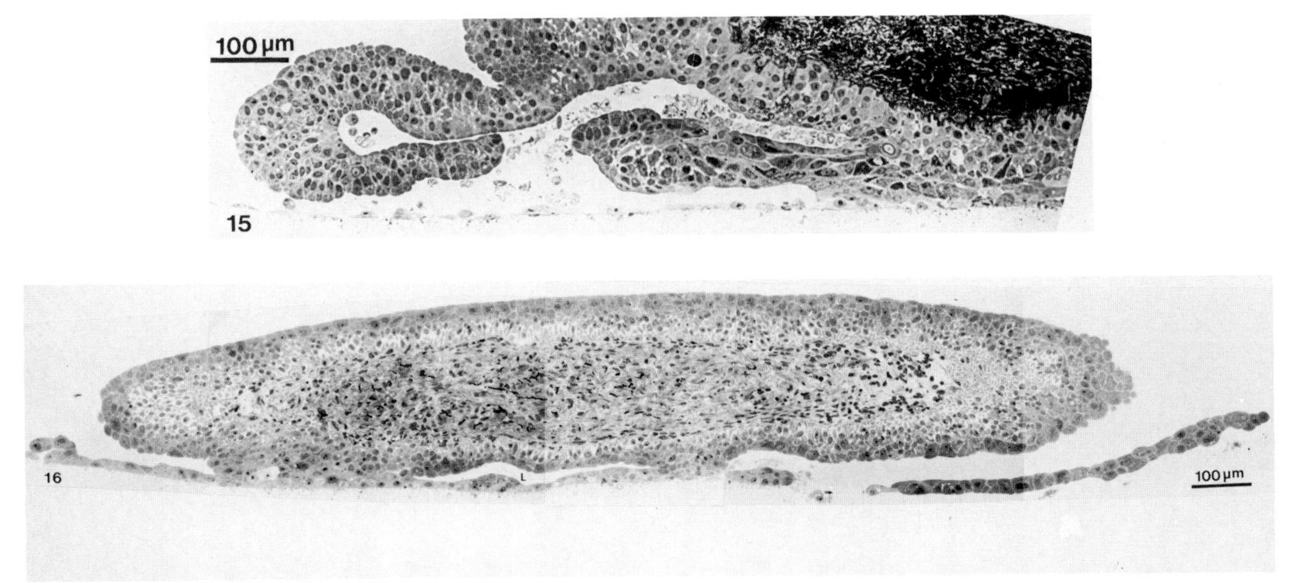

FIGS. 15 and 16. Toluidine blue-stained resin sections.

FIG. 15. Edge of a bladder organ culture with a solid rim of epithelial cells. Scattered single cells have migrated over the filter support.

FIG. 16. Section across an entire human bladder organ culture after 14 days *in vitro*. The epithelium has formed a lumen beneath the base of the culture (L) and a small epithelial cell outgrowth is present. On the surface of the explant the mature epithelium is four to five cells thick, but it is much thicker at the edges. Normal, mature superficial cells are present on the outer epithelial surface on the top and sides of the explant and on the stromal side of the lumen at the base of the explant.

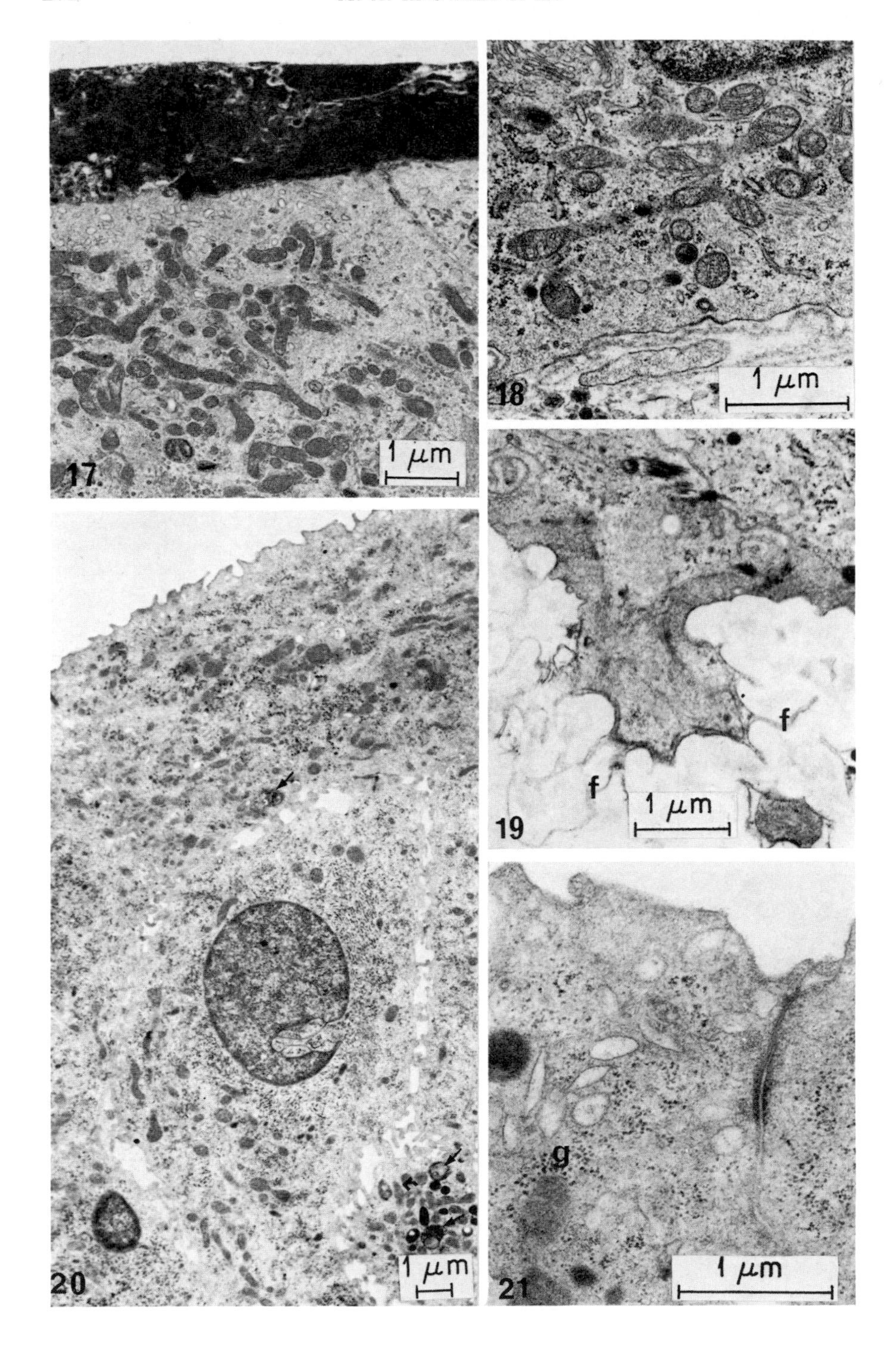
1 μm
17
18
f
f
19
20
g
21

cytoplasm though in smaller numbers than would be expected in a contracted bladder *in vivo* (Fig. 21). The time scale of epithelial cell maturation showed considerable variation between explants from different individuals. In most cases, maturation was more advanced on the sides of the explant, possibly since there were no adherent dead cells there. Ultimately, the top and sides of the explants, the bulbous rim around the culture, and the free surface of stroma-supported epithelium over lumina which had formed under the explant all showed ultrastructural features of maturing transitional epithelium. Free surfaces which developed around spaces within the large bulbous rims did not show maturation.

Scanning electron microscopy allowed a more detailed analysis of urothelial maturation and confirmed that the sequence of surface differentiation on stroma-supported urothelium was similar *in vivo* and *in vitro*. In the first few days, observations of the upper culture surface were hampered by adhering dead cells but where these had sloughed off and also on the sides of the culture, small immature cells with scattered globular microvilli characteristic of normal basal cells could be seen (Fig. 22). By 7 days, some maturation of the surface was usually present with fusion of these globular structures to form early microridges as seen in normal intermediate cells *in vivo*. All stages of maturation could, in time, be found on the same culture (Fig. 23) with more mature forms resembling late intermediate or superficial cells of the *in vivo* urothelium predominating by 21–28 days (Figs. 24–26). The bulbous rims showed the same pattern of differentiation on their outer surfaces as the upper surface of the culture. As will be discussed later, there was sometimes a tendency in Waymouth's medium for the surface epithelium to become thin, and then these rims tended to be more mature than the surface epithelium. The variation between explants from different individual donors was striking when the cultures were examined in the scanning

---

FIGS. 17–21. Electron micrographs of resin-embedded material stained with uranyl acetate and lead citrate.

FIG. 17. Culture surface after 14 days *in vitro* showing a dead superficial cell overlying a viable cell which has many smooth membrane-bounded vesicles in the apical cytoplasm. A junctional complex is present between the adjacent viable cells.

FIG. 18. The basal lamina below part of a basal cell in the epithelium covering the surface of an organ culture after 14 days *in vitro*. Part of the Golgi apparatus is shown at the top left of the field.

FIG. 19. Cytoplasmic process of a cell on the base of a bladder organ culture. The process extends into gaps in the Millipore filter. The cellulose fibers (f) and the cell membrane can be clearly seen but no basal lamina is present.

FIG. 20. Section of the surface epithlium of the culture shown in Fig. 16. The cells contain some lysosomes (arrows) and the surface membrane shows characteristic rigid plaques.

FIG. 21. Detail of surface cells of the culture shown in Figs. 16 and 20. The apical cytoplasm contains fusiform vesicles, a junctional complex is present between adjacent superficial cells, and glycogen granules (g) can be seen in the cytoplasm.

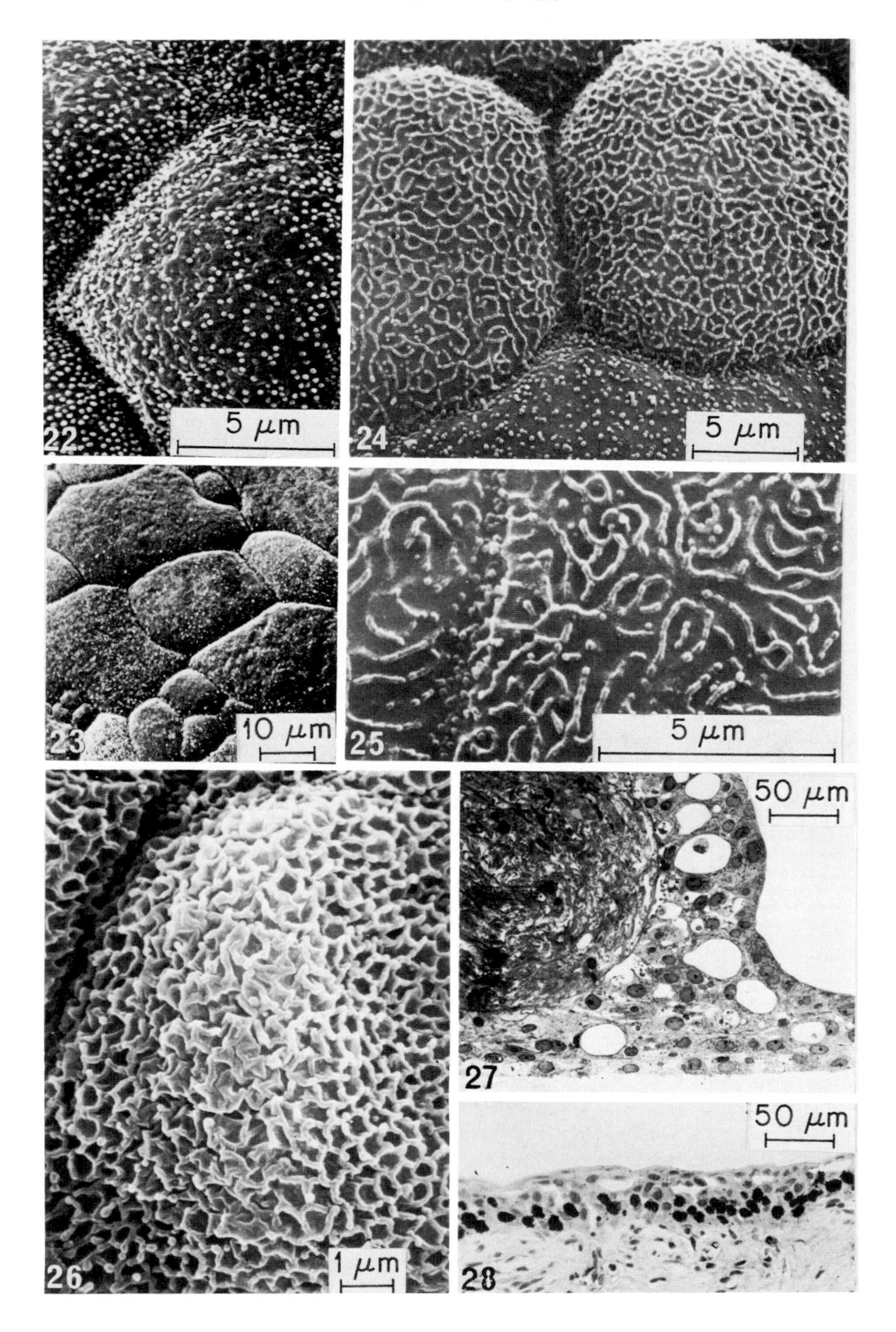
5 μm
22
5 μm
24
10 μm
23
5 μm
25
50 μm
1 μm
26
27
50 μm
28

electron microscope. Cultures from all normal patients developed characteristics of membrane maturation, but quantitatively the membrane ridges were far more profuse in some individuals than in others. Figure 29 summarizes the pattern of development in human bladder organ cultures and parallel rat cultures, based on histological and ultrastructural studies.

2. THE EFFECTS OF DIFFERENT MEDIA

We have examined several basal media, sera, and other supplements to determine the optimum conditions for survival and differentiation of bladder mucosa. It is clear that interpatient variation is of overriding importance in determining tissue survival. Thus, divided biopsies from some patients survived well in all media while biopsies from others failed in all but the most complex media. The survival of most samples lies between these two poles and shows a variable response to certain supplements. Variation in viability of the starting material has to some extent made assessment difficult, but certain general trends have been established. Material from retropubic prostatectomy patients from whom multiple cultures could be established was used for media comparisons, though the urothelium was not always absolutely normal.

Our assessment of the medium is based on the extent to which it will maintain the culture with the same epithelial thickness, and size, polarity, and staining properties of the cells as is found in the starting material. Ham's F-10 medium, medium 199, Dulbecco's modification of Eagle's medium, and NCI medium proved unsatisfactory in early experiments. Ham's F-12, Waymouth's MB 752/1, CMRL 1066, and RPMI-1640 were all satisfactory media and are now

---

FIGS. 22–26. Scanning electron micrographs.

FIG. 22. Surface cells on a bladder organ culture after 14 days *in vitro*. The surface is covered by small globular processes, as are basal cells *in vivo* (cf. Fig. 5).

FIG. 23. Maturing urothelial surface after 21 days *in vitro* culture. The larger more mature cells are interspersed in places with smaller less mature cells.

FIG. 24. Cells on the side of a culture after 21 days *in vitro* showing maturing intermediate type cells and an immature cell at the lower edge of the field.

FIG. 25. Detail of the surface membrances of maturing intermediate-type cells after 21 days *in vitro* culture. A few single globular processes, some chains of fusing globules, and many microridges can be seen. The boundary between the cells is characterized by the presence of rows of globular processes.

FIG. 26. Mature superficial cell at the edge of a 28-day culture. The cells on the edges of the culture are less flattened than those on the surface and show more membrane ridges.

FIG. 27. Section through the edge of a culture grown in medium without $FeSO_4$. Many intraepithelial acini are present. Light micrograph of a toluidine blue-stained resin section.

FIG. 28. [^{3}H]Thymidine autoradiograph of the surface epithelium of a 15-day bladder organ culture. The epithelial labeling index is high but labeled areas are confined to the basal and lowest intermediate cell layers. Wax section, stained with hematoxylin.

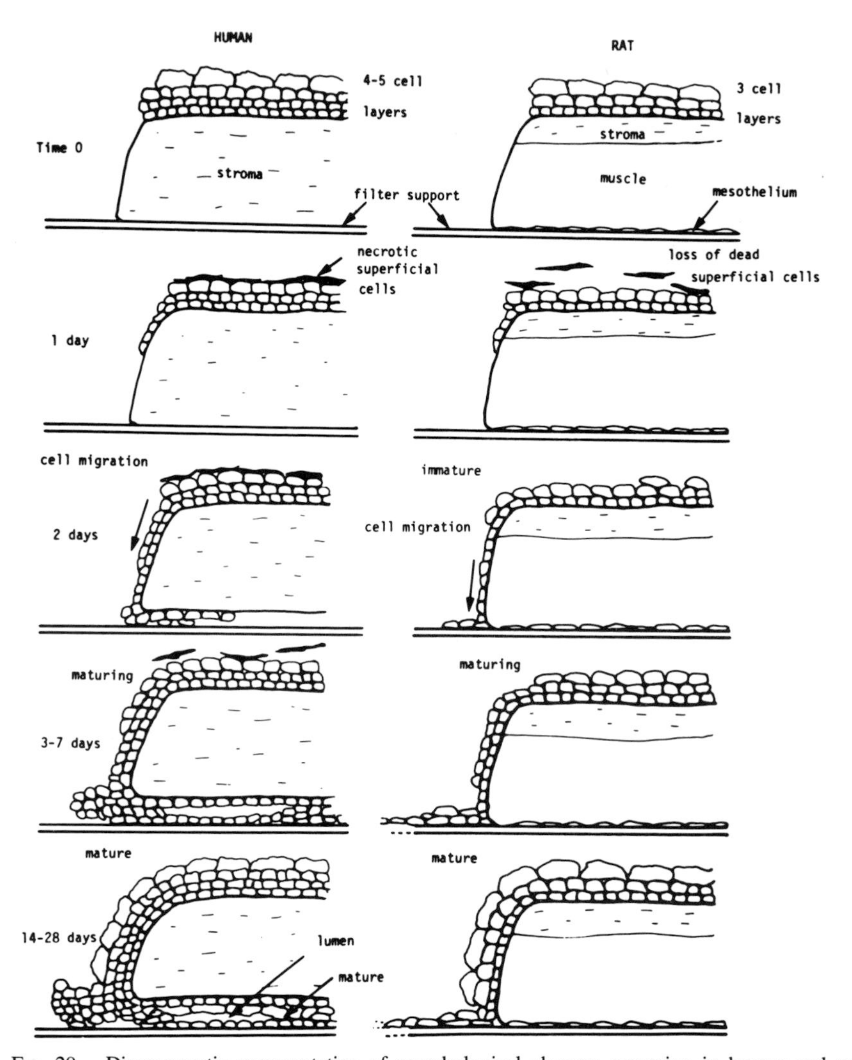

FIG. 29. Diagrammatic representation of morphological changes occurring in human and rat bladder organ cultures during the first 28 days *in vitro*.

used routinely. The effect of varying the supplements was assessed in Ham's F-12 and Waymouth's MB 752/1. As a basal medium, Ham's F-12 preserved the initial epithelial thickness better than Waymouth's medium, in which, particularly if few supplements were added, the surface epithelium tended to thin to one or two cells thick and to lose its normal nuclear polarity. Human material cultured in Ham's medium did not become hyperplastic as has been reported in rat cultures (Reese *et al.*, 1977).

Serum-free cultures were maintained for 28 days, and again Ham's F-12 proved to be the better medium. In general, cultures survived better with the addition of 10% newborn or fetal calf serum. In several instances we used pooled human group AB serum or the patient's own serum but the cultures fared no better than those with fetal calf serum.

The specific effect of hydrocortisone in this system is not well defined though its inclusion was associated with a general increase in success rate and viability. Glutamine and insulin both helped to maintain the thickness of the epithelium but the latter did not appear to act synergistically with hydrocortisone as previously reported in other systems (Reed and Grisham, 1975; Topper and Oka, 1971). Of the other supplements tested, urea was toxic and pyruvate and DMSO did not affect the histology of the explants. Ascorbic acid, like hydrocortisone, apparently increased survival but no specific histological effects could be characterized.

In cultures grown in poor or unsupplemented media intraepithelial lumina often developed in all areas of the epithelium (Fig. 27). Transmission electron microscopy showed them to be small acini surrounded by cells with glycocalyx-covered microvilli on their free surfaces. These cells were joined at the acinar edges of their lateral membranes by junctional complexes. The acini often contained a substance which stains metachromatically with toluidine blue, is alcian blue positive, and is probably an epithelial acid mucin. Large mucous droplets were not seen in the cells but large numbers of apical vesicles were usually present. This probably represents a metaplastic change in the culture and the ultrastructural changes are comparable to those seen during mucous metaplasia of the urothelium *in vivo* (Hicks and Chowaniec, 1978). In many cases this change was apparently prevented by the addition of $FeSO_4$ to the medium and in some cases ascorbic acid had the same effect. In some experiments insulin increased the numbers of acini though it is not clear whether this is a direct effect or the indirect result of increasing epithelial thickness.

3. Cultures from Stored Material

Tissue can be stored for up to 48 hours in Waymouth's medium with HEPES buffer at 4°C before initiation of cultures. The stored tissue showed no histological evidence of autolysis but by 48 hours some dilatation of the intercellular spaces was evident. Viable cultures obtained from tissue stored for 48 hours were indistinguishable from cultures of unstored material.

4. Cell Turnover

There appears to be a marked interindividual variation in the incorporation of [^{3}H]thymidine into human urothelial cultures. However, in our preliminary experiments the distribution of DNA-synthesizing cells was similar in all cases. At first, the labeling index (LI) measured over a 24-hour period was high in the developing bulbous rims (up to 10%) but fell to a low level by 14 days. The labeling index of the surface epithelium was very variable (1–50%) between 7 and 15 days and, in general, lower indices were obtained in Waymouth's than in Ham's medium, which was often reflected by the gradual thinning of the epithelium in Waymouth's medium. Synthesizing cells were confined to the basal or lowest intermediate cell layer (Fig. 28) except where the epithelium had thinned to three cells or less. The labeling index of epithelium at the base of the explants bore no obvious relationship to that at the surface but varied within the same range. This may reflect variation in the volume of the space beneath partially lifted cultures which was available for the cells to fill.

IV. Discussion

A. Selection of Normal Bladder Tissue for Culture

In most studies of human tissue *in vitro,* little attention has been paid to the objective assessment of the "normality" of the starting material. Tissue is often defined as normal because the patient appears symptom-free with regard to the organ in question. In the bladder at least, it is now apparent that a wide range of symptom-free histopathological conditions can exist in patients in whom the mucosa appears macroscopically normal at cystoscopy. In man, bladder cancer is known to have a long, "silent" latent period and since carcinogenesis in the bladder is probably a multistage process (Hicks *et al.,* 1978), it is to be expected that tissue from patients of different ages may be at different stages of neoplastic development. Whether any of the abnormal histology we have observed is preneoplastic remains open to conjecture, but for our prospective work on carcinogenesis, in which one of our objectives is to identify experimentally induced neoplastic change, it is essential to use normal tissue as our starting material. At present we are using morphological criteria to define "normality" but other criteria will be added to give more information.

Normal human urothelium is very similar to that of the rat, and is characterized by large superficial cells limited by a luminal membrance with specialized substructured plaque regions; these cells also contain cytoplasmic vesicles formed from the same membrane (Fulker *et al.,* 1971; Newman and Hicks, 1977; Koss, 1977). Our present studies confirm this to be the typical ultrastructural appearance of nonhyperplastic, mature human transitional epithelium. A report that the specialized luminal membrane, though present in infants, is not found in normal adult bladders (Merk *et al.,* 1977) is not supported by our observations. These authors described microvilli on the luminal face of "normal" superficial cells in specimens derived from patients with benign prostatic hypertrophy or bladder neck obstruction. As we demonstrate here (Table II) the urothelium often fails to mature normally in such patients. Kjoersgaard *et al.* (1977) described both normal superficial cells and also surface cells with numerous microvilli in nine patients. Seven of these, however, had cystitis or a tumor elsewhere in the bladder. In our experience the presence of small globular microvilli alone indicates an immature cell, though small microvilli or blebs sometimes persist between the ridges on otherwise normal superficial cells. Judged by these criteria, in none of our arbitrary groups was every patient a source of entirely normal urothelium (Table II). The retropubic prostatectomy patients clearly had more abnormal urothelia than any other group. This is unfortunate, since this type of patient provides the most readily accessible source of bladder tissue and has been used to provide human control material for a number of published studies. The site of biopsy, near to the bladder neck in these patients, may influence the histological findings since this is the area msot likely to be traumatized during catheterization and/or forced micturition. However, the site of biopsy was not the only factor involved since biopsies from the body of the bladder of transurethral prostatectomy patients also showed a higher incidence of epithelial abnormalities than specimens from patients in other groups. This confirms the findings of Barlebo and Sorensen (1972) who found some epithelial atypia in biopsies from trigone, lateral wall and dome of the bladder in 27 out of 28 prostatectomy patients. Nine of these were graded as malignancy grade 2 (according to the scheme of Bergkvist *et al.*, 1965) which corresponds with our mild or moderate dysplasia, and 15 of these patients also showed positive urine cytology. Urine retention may play a role in the development of these various changes, and our finding of more abnormalities in patients with some history of retention (group 3a) than in the group with no retention (group 3b) seems to support this. The lower incidence of abnormalities in patients with retention but no prostatic enlargement may be related to a shorter history of retention.

The mucosa of tumor-bearing bladders has been widely reported to show multifocal abnormalities (Melicow, 1952; Cooper *et al.*, 1973; Schade and Swinney, 1973; Skinner *et al.*, 1974). We found five examples of urothelial hyperplasia and two of mild to moderate dysplasia in macroscopically normal looking biopsies from eight tumor-bearing bladders. This is rather less than the 86% incidence of dysplasia reported by Schade and Swinney (1973) but probably reflects our choice of the most normal looking areas to biopsy in contrast to the multiple biopsies taken and described by other workers, some of which were close to the tumor. Biopsies adjacent to a tumor are usually the most abnormal (Cooper *et al.*, 1973). Although relatively few abnormalities were present in macroscopically normal looking biopsies from patients in whom tumors had previously been resected (group 1b), these may be manifestations of multifocal neoplastic disease. Since bladder carcinogens are urine-borne, the fact that neoplastic disease has developed at one site in the bladder indicates that the entire urothelium has at some time been exposed to carcinogens and we must therefore regard the urothelium from such patients with a history of bladder tumors as potentially neoplastic, irrespective of its histopathological appearance.

The significance of the pathological changes we have seen in the other group in terms of preneoplasia is not clear. Severe dysplasias can give rise to metastatic disease (Utz *et al.*, 1970) but do not always do so. Hyperplasia does not appear to be a prerequisite for dysplastic change and may or may not be preneoplastic, though it is frequently associated with tumor-bearing epithelia as previously discussed. Microvilli with a glycocalyx have been found in some specimens from prostatectomy patients, none of whom showed gross hyperplasia or dysplasia. Since the presence of such microvilli is regarded as a preneoplastic marker in the rat bladder (Hicks and Wakefield, 1976) and they have been repeatedly observed in human tumors (Fulker *et al.*, 1971; Newman and Hicks, 1977; Koss, 1977) they may well be indicative of early neoplastic change in these bladders. Clearly this aspect of the work merits further study and it is hoped that future culture experiments and xenotransplantation studies may help to elucidate the status of epithelium bearing this marker. Our observations suggest that patients with no history of neoplastic disease or of prostatic hypertrophy are the best source of ''normal'' material for culture.

B. Culture of Human Bladder Explants

Both normal and abnormal urothelia can be maintained *in vitro* for 35 days. The behavior of human tissue is very similar to that of rat bladder (Hodges *et al.*, 1976, 1977; Reese *et al.*, 1977). We find that the early death of the rat superficial cells can be prevented by decreasing the time between killing the animal and incubating the cultures and by using HEPES-buffered medium for dissection. Similar speed and maintenance of environmental pH is in theory possible for human specimens taken during open surgery on the bladder, but so far has not been achieved.

The process of epithelialization of the sides and base of the human explants resembles the process of epiboly in epidermal organ cultures (Sarkany *et al.*, 1965) and demonstrates the propensity of bladder epithelium to cover all areas of exposed stroma. In the rat cultures, the base of the explant is covered by mesothelium and only the sides of the explant become epithelialized; the urothelium then continues to grow out over the filter support. This may represent an attempt by the epithelium, which usually lines a cavity, to form a contiguous sheet rather than a blind-ended contact with the mesothelium. The human cultures, which have no mesothelium on the base of the cut stroma, are completely surrounded by urothelium which is polarized toward the medium. They essentially mimic an inverted bladder with a continuous sheet of urothelium and may functionally resemble the *in vivo* tissue more closely than do rat cultures.

There are no qualitative differences between maturation of the urothelium *in vivo* and *in vitro,* though a comparison between the time course of these processes is difficult. Lund (1969) reported that regeneration of stripped urothelium in man takes about 6 weeks, but surface maturation of the cells was not examined. High [^{3}H]thymidine-labeling indices occur *in vivo* only during repair of damaged urothelium, and it is hoped that the high labeling indices recorded in our cultures will fall in mature explants once all surfaces are epithelialized and a stable state has been attained.

The effect of hydrocortisone on the human tissue *in vitro* is less clear cut than it is on rat tissue. Hydrocortisone inhibits DNA synthesis and cell division in rat bladder cultured in Ham's F-12 medium (Reese *et al.,* 1977). In the human cultures, hydrocortisone had no effect on urothelial thickness, but the rate of cell turnover has not yet been assessed. Both the rates of cell division and cell loss will affect the maintenance of normal thickness urothelium over long periods in culture.

Our results suggest that $FeSO_4$ inhibits the formation of intraepithelial lumina and the associated mucous metaplasia. Fe^{2+} ions and ascorbic acid are both necessary cofactors in collagen synthesis (Hutton *et al.,* 1966) and it is possible that their effect on the epithelium is the indirect result of stromal modification, but further studies are required to elucidate the role of the stroma in modulating epithelial function.

Though most of the human bladder tissue survived in culture, our greatest problem in this study has been the great difference in tissue viability from different patients. Differences in tissue handling might be expected to affect its viability, but our methods are standardized and we have shown that storage at 4°C for up to 48 hours does not affect the survival of viable specimens. It is possible that the type of serum in the medium might adversely affect some specimens more than others, but no conclusive improvement was obtained when pooled human group AB serum or the patient's own serum was used in some cultures. Bovine serum albumin is now being investigated. Many cystoscopy specimens were bacterially contaminated, probably from infected urine, especially in women, but most organisms have been sensitive to penicillin and streptomycin and overt culture contamination has not occurred. It is possible however, that low levels of bacterial growth may still occur in the presence of these antibiotics and cause an apparent lack of viability in all specimens from some patients. The possibility of viral infections must also be considered and such factors have not yet been excluded. A more likely explanation is an inherent individual variation in viability, perhaps hereditary, in the tissue samples. Man, unlike the laboratory rat, is a random-bred species and human tissue is known to vary in many ways from individual to individual, for example, in its ability to bind certain chemical carcinogens (Harris *et al.*, 1976).

Since viable cultures in the optimum media have shown no signs of degeneration at 35 days, there is no reason to doubt that much longer periods of culture will be possible. This is now being attempted.

V. Perspectives

The human bladder organ culture system presents many possibilities for future study. Our assessment so far has been morphological, but functional and cell kinetic studies are now clearly needed. The system has been developed primarily for use in carcinogenesis studies and our aim is to use it to identify potential human bladder carcinogens and to characterize the changes they induce, so that a more complete picture of the biology of bladder cancer can be obtained. It is hoped that in the future this will allow individuals "at risk" of developing tumors to be identified more readily.

Cultures of human bronchus (Harris *et al.*, 1977a), colon (Autrup *et al.*, 1978), and pancreatic duct (Harris *et al.*, 1977b) have already been used successfully to study the metabolism and binding of several carcinogens, and it is now clear that several carcinogens can be metabolized *in situ* by the epithelia in these tissues. We shall use our human bladder cultures to test the effect of several carcinogens which have been shown to induce tumors in laboratory rodents *in vivo*. These will include methylnitrosourea (MNU) and metabolities of dibutylnitrosamine, butylbutanol nitrosamine, and 2-napthylamine. *In vivo,* saccharin and cyclamate have been shown to promote tumor development in the bladder after initiation of neoplastic change with MNU (Hicks, *et al.*, 1978) or with the carcinogenic nitrofuran FANFT (Cohen *et al.*, 1978). In man, these artificial sweeteners are excreted unchanged in the urine, and in the presence of low levels of urine-borne carcinogens may present a considerable health hazard (Hicks and Chowaniec, 1977). These sweeteners and other potential cocarcinogens will be tested on our *in vitro* human bladder model. Similarly, retinoids which have been reported to inhibit or delay rat bladder tumor formation *in vivo* (Sporn *et al.*, 1977) can now be tested on human material.

The main advantage of using organ cultures is the maintenance of a normal three-dimensional relationship both within the urothelium and between the epithelium and its supporting stroma. Such interrelationships may crucially affect the metabolism of carcinogens, since stromal cells may be directly involved in carcinogen metabolism once the carcinogenic molecule has entered the bladder wall. Furthermore, the large mature superficial cells which devleop only in the presence of the stroma may have different enzymes concerned with the metabolism of carcinogens by comparison with the smaller undifferentiated cells which predominate in cell cultures. Certainly the differentiated superficial cells contain more lysosomes, which include in their complement of hydrolases the enzyme β-glucuronidase which can release within the cell the free carcinogen from a glucuronide such as *N*-hydroxy-2-napthylamine *N*-glucuronide.

In vivo studies on carcinogenesis in the rat bladder have identified the presence of microvilli, often with pleomorphic morphology and a well-developed glycocalyx, as markers of neoplasia (Hicks and Wakefield, 1976). These markers also develop *in vitro* after previous treatment *in vivo* of the animals with bladder carcinogens (Hodges *et al.*, 1976). Cell cultures of rodent bladder epithelium can be neoplastically transformed in culture (Hashimoto and Kitagawa, 1974; Summerhayes and Franks, 1979) and it might therefore be expected that direct *in vitro* carcinogen treatment of organ cultures will also produce identifiable markers of neoplasia. This can now be attempted in parallel rat and human organ cultures. We have maintained organ cultures of several human bladder tumors in the media used for normal tissue and these will provide positive controls for identifying markers of neoplastic growth. In addition to morphological markers of neoplastic change, altered growth potential and biochemical markers such as loss of specific cell surface proteins (Pearlstein *et al.*, 1976) and alterations in fibrinolytic activity can be studied *in situ* in organ cultures. By using cell cultures derived from carcinogen-treated organ cultures alterations in growth potential can be studied as described for rat tracheal cultures by Marchok *et al.* (1977). Cell cultures derived from organ cultures will also allow chromosomal studies to be made and markers of neoplasia associated with cell growth, such as the ability to grow in semisolid agar or in suspension, to be assessed. Our final test of neoplastic change will involve xenotransplantation into a suitable animal host, at present under development.

ACKNOWLEDGMENTS

The authors would like to thank A. Finesilver, A. Harvey, and H. Ogbulu for skilled technical assistance, Mr. P. Copland, Mr. A. Pengelly, and Mr. K. Vaughton for the provision of surgical biopsies, and Dr. J. Newman for discussion of the pathology.

The work was supported by NCI Contract NO1 CP 75955.

REFERENCES

Autrup, H., Barrett, L. A., Jackson, F. E., Jesudason, M. L., Stoner, G., Phelps, A. B., Trump, B. F., and Harris, C. C. (1978). *Gastroenterology* **74,** 1248.

Barlebo, H., and Sorensen, B. L. (1972). *Scand. J. Urol. Nephrol.* Suppl. **6** (15), 121.

Bergkvist, A., Ljungquist, A., and Moberger, G. (1965). *Acta Chir. Scand.* **130,** 371.

Cohen, S. M., Arai, M., and Friedell, G. H. (1978). *Proc. Am. Assoc. Cancer Res.* **19,** 15.
Cooper, P. H., Waisman, J., Johnston, W. H., and Skinner, D. G. (1973). *Cancer* **31,** 1055.
Eisenberg, R. B., Roth, R. B., and Schweinsberg, M. H. (1960). *J. Urol.* **84,** 544.
Elliott, A. Y., Stein, N., and Fraley, E. E. (1975). *In Vitro* **11,** 251.
Elliott, A. Y., Bronson, D. L., Stein, N., and Fraley, E. E. (1976). *Cancer Res.* **36,** 365.
Fulker, M. J., Cooper, E. H., and Tanaka, T. (1971). *Cancer* **27,** 71.
Harris, C. C., Autrup, H., Connor, R., Barrett, L. A., McDowell, E. M., and Trump, B. F. (1976). *Science* **194,** 1067.
Harris, C. C., Autrup, H., Stoner, G. D., McDowell, E. M., Trump, B. F., and Schafer, P. (1977a). *Cancer Res.* **37,** 2309–2311.
Harris, C. C., Autrup, H., Stoner, G., Yang, S. K., Lentz, J. C., Gelboin, H. V., Selkirk, J. K., Connor, R. J., Barrett, L. A., Jones, R. T., McDowell, E., and Trump, B. F. (1977b). *Cancer Res.* **37,** 3349.
Hashimoto, Y., and Kitagawa, H. S. (1974). *Nature* (*London*) **252,** 497.
Hicks, R. M. (1966). *J. Cell Biol.* **30,** 623.
Hicks, R. M. (1975). *Biol. Rev.* **50,** 215.
Hicks, R. M., (1976). *In* "Progress in Differentiation Research" (N. Müller-Bérat, C. Rosenfeld, D. Tarin and D. Viza, eds.), pp. 339–353. North-Holland, Amsterdam.
Hicks, R. M., and Chowaniec, J. (1977). *Cancer Res.* **37,** 2394.
Hicks, R. M., and Chowaniec, J. (1978). *Int. Rev. Exp. Pathol.* **18,** 199.
Hicks, R. M., and Wakefield, J. St. J. (1976). *Cancer Res.* **36,** 2502.
Hicks, R. M., Chowaniec, J., and Wakefield, J. St. J. (1978). *In* "Carcinogenesis, Vol. 2. Mechanisms of Tumour Promotion and Cocarcinogenesis" (T. J. Slaga, A. Sivak, and R. K. Boutwell, eds.), pp. 475–489. Raven, New York.
Hodges, G. M., Hicks, R. M., and Spacey, G. D. (1976). *Differentiation* **6,** 143.
Hodges, G. M., Hicks, R. M., and Spacey, G. D. (1977). *Cancer Res.* **37,** 3720–3730.
Hutton, J. J., Jr., Tappel, A. J., and Undenfriend, S. (1966). *Biochem Biophys. Res. Commun.* **24,** 179.
Jones, G. W. (1967). *Cancer* **20,** 1893.
Kjoersgaard, J., Starklint, H., Bierring, F., and Thybo, E. (1977). *Urologia Int.* **32,** 34.
Koss, L. G. (1975). *In* "Atlas of Tumour Pathology," Fascicle II, Series 2. U.S. Armed Forces Institute of Pathology, Washington, D.C.
Koss, L. G. (1977). *Cancer Res.* **37,** 2824.
Linder, D. (1976). *Somatic Cell Genet.* **2,** 281.
Lund, F. (1969). *Urol. Nephrol.* **3,** 204.
Malkovsky, M., and Bubenik, J. (1977). *Neoplasma* **24,** 319.
Marchok, A. C., Rhoton, J. C., Griesemer, R. A., and Nettesheim, P. (1977). *Cancer Res.* **37,** 1811.
Martin, B. F. (1972). *J. Anat.* (*London*) **112,** 433.
Melicow, M. M. (1952). *J. Urol.* **68,** 261.
Merk, F. B., Pauli, B. U., Jacobs, J. B., Alroy, J., Friedell, G. H., and Weinstein, R. S. (1977). *Cancer Res.* **37,** 2843.
Monis, B., and Dorfman, H. D. (1967). *J. Histochem. Cytochem.* **15,** 475.
Newman, J., and Hicks, R. M. (1977). *Histopathology* **1,** 125.
O'Toole, C., Nayak, S., Price, Z., Gilbert, W. H., and Waisman, J. (1976). *Int. J. Cancer* **17,** 707.
O'Toole, C., Price, Z. H., Ohnuki, Y., and Unsgaard, B. (1978). *Br. J. Cancer* **38,** 64.
Pearlstein, E., Hynes, R. O., Franks, L. M., and Hemmings, V. J. (1976). *Cancer Res.* **36,** 1475.
Petry, G., and Amon, H. (1966). *Z. Zellforsch.* **69,** 587.
Rasheed, S., Gardner, M. B., Rongey, R. W., Nelson-Rees, W. A., and Arnstein, P. (1977). *J. Natl. Cancer Inst.* **58,** 881.

Reed, G. B., and Grisham, J. W. (1975). *Lab. Invest.* **33,** 298.
Reese, D. H., and Friedman, R. D. (1978). *Cancer Res.* **38,** 586.
Reese, D. H., Friedman, R. D., Smith, J. M., and Sporn, M. B. (1976). *Cancer Res.* **36,** 2525.
Reese, D. H., Friedman, R. D., and Sporn, M. B. (1977). *Cancer Res.* **37,** 1421.
Rigby, C. C., and Franks, L. M. (1970). *Br. J. Cancer* **24,** 746.
Roszell, J. A., Douglas, C. J., and Irving, C. C. (1977). *Cancer Res.* **37,** 239–243.
Sarkany, I., Grice, K., and Caron, G. A. (1965). *Br. J. Dermatol.* **77,** 65.
Schade, R. O. K., and Swinney, J. (1973). *J. Urol.* **109,** 619.
Simon, W., Cordonnier, J. J., and Snodgrass, W. T. (1962). *J. Urol.* **88,** 797.
Skinner, D. G., Richie, J. P., Cooper, P. H., Waisman, J., and Kaufman, J. J. (1974). *J. Urol.* **112,** 68.
Sporn, M. B., Squire, R. A., Brown, C. C., Smith, J. M., Wenk, M. L., and Springer, S. (1977). *Science* **195,** 487.
Summerhayes, I. C., and Franks, L. M. (1979). *J. Natl. Cancer Inst.* **62,** 1017–1023.
Sutherland, G. R., and Bain, A. D. (1972). *Nature* (*London*) **239,** 231.
Topper, Y. J., and Oka, T. (1971). *In* "Effects of Drugs on Cellular Control Mechanisms" (B. R. Rabin and R. D. Friedman, eds.), pp. 131–150. Macmillan, London.
Toyoshima, K., Valentich, J. D., Tchao, R., and Leighton, J. (1976). *Cancer Res.* **36,** 2800.
Utz, D. C., Hanash, K. A., and Farrow, G. M. (1970). *J. Urol.* **103,** 160.
Warren, R. C., and Hicks, R. M. (1973). *Micron* **4,** 257.

Note Added in Proof

Since submission of this article, cultures maintained for up to 70 days have been examined. There are no significant morphological differences between these and the 35 mature cultures described here.

Chapter 13

Histophysiologic Gradient Culture of Stratified Epithelium

JOSEPH LEIGHTON, RUY TCHAO, ROBERT STEIN, AND NABIL ABAZA

Cancer Bioassay Laboratory,
Department of Pathology,
Medical College of Pennsylvania,
Philadelphia, Pennsylvania

I. Introduction

The surgical pathologist strives to apply new knowledge in various fields of biology and new techniques to help him fulfill his responsibility. He must interpret as precisely as possible, by examination of resected tissue, the subsequent course of each patient's illness. Recent additions to the pathologist's resources have come from histochemistry and electron microscopy. The idea that tissue culture techniques might be of use has a long history. In 1910 Carrel and Bur-

ISBN 0-12-564140-0

rows reported for the first time the cultivation of a human malignant tumor in the laboratory. As a consequence of their new technique, in which growth took place in a fibrin clot, they expected "to study in vitro the growth of various human malignant as well as benign tumors and to follow all the morphologic characteristics and changes of the cancerous and other cells during life" (Carrel and Burrows, 1910). Although a great amount of information has been derived from tissue culture studies over the years, their expectations have not been realized. We suspect that the reason has been the failure to reproduce in culture the histophysiologic organization of epithelial membranes so important in the intact individual, and in our recent experience equally essential in cultures of epithelial tissues (Leighton, 1979; Leighton *et al.*, 1979). We have developed a novel procedure for the investigation of clinical and experimental cancer in tissue culture, which is based on a key principle in the organization of certain normal tissues in the body. Stratified epithelium in mammals, tissue such as that lining the skin, vagina, cervix, bladder, oral cavity, and esophagus, is always constituted so that the plane of attachment to the underlying connective tissue is also the path of nutritional exchange including oxygen. In conventional tissue culture, however, the explant is attached to the glass or other support on its underside, while the culture medium and atmosphere provide nutrition from above.

Studies in our laboratory leading to histophysiologic gradient culture have always been directed toward the goal of growth in culture taking place in a manner intelligible to the surgical pathologist. Architecture of growth in culture should be related to the structure of cancer tissue taken directly from the body. This required at least two conditions, three-dimensional growth and the presence of defined spatially recognizable metabolic gradients in the cultured tissue. Since the substance of a sponge, its matrix and interstices, provide a three-dimensional context for growth, we examined several types including natural sea sponge, Gelfoam (Upjohn, Kalamazoo, MI), and fine pore cellulose sponge (DuPont, Wilmington, DE). We chose a combined matrix of cellulose sponge and fibrin clot for most of our subsequent investigation (Leighton, 1951). At a later time we used a natural matrix, frozen thawed umbilical cord (Leighton *et al.*, 1960). A semisolid gel composed of a mixture of agar and fibrin in several ratios of combination was also useful for studies on aggregation and outgrowth as budding to form new aggregates (Leighton *et al.*, 1962). After Ehrmann and Gey (1956) introduced collagen as a substrate for cell culture, we settled on a matrix of collagen-coated cellulose sponge, with or without a supplement of fibrin clot in the interstices. Our laboratory has acquired extensive experience with this matrix, most recently on a large series of clinical bladder cancers (Leighton *et al.*, 1977; Abaza *et al.*, 1978).

Our experience with gradients appears in a review we published in 1975 (Leighton and Tchao, 1975), in which we discussed aspects of metabolic gradients in three-dimensional matrix culture and in monolayer culture. Our real concern is the coordinated interrelation of cells in solid tumors in the context of their indigenous metabolic gradients. Toward this goal, a variety of studies have been conducted with continuous gradients of oxygen tension and of temperature using monolayer cultures (Leighton and Katsuta, 1973; Takeuchi *et al.*, 1974; Geisinger *et al.*, 1978). We found that a squamous cell carcinoma derived from rat urinary bladder responded to vitamin A with inhibition of keratinization, increased motility, and decreased attachment of cells to one another (Toyoshima and Leighton, 1975; Tchao and Leighton, 1979).

An approach to histophysiologic gradient culture was illuminated for us recently through examination of histologic preparations of tissues from the intact body, and of matrix cultures. Subcutaneous cysts lined by epithelium of epidermal origin are a common clinical occurrence. Histologic examination of such cysts shows a cavity filled with dead cellular products and a lining of normal appearing or hyperplastic stratified epithelium. Invariably, the proliferating layer of the epithelium is adjacent to the supporting connective tissue (Fig. 1). Cysts lined with epithelia are readily produced experimentally in the subcutaneous tissue of adult rats by combined injection of a large volume of air and a fine mince of epithelial tissue of the same donor strain, whether the inoculum is adult bladder, fetal stomach, or fetal colon. With bladder tissue as the inoculum, the

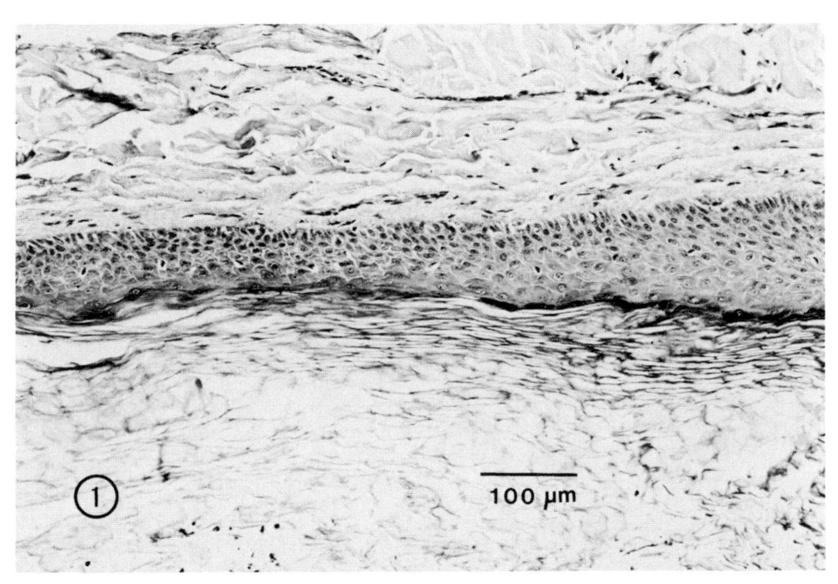

FIG. 1. Histologic section of the wall of an epidermal cyst from the lower back of a white adult male. The subcutaneous fibrous tissue is in the upper part of the field, and the viable stratified epithelium in the center. Occasional mitoses, though not illustrated here, are present in the layer of epithelium in contact with the dense fibrous tissue. The nonviable inspissated residue of desquamation of cells, filling the cavity of the cyst, is seen below. Hematoxylin and eosin.

resulting heterotopic urinary bladder was of particular interest (Roberts *et al.*, 1974). In all these cysts, the proliferating stratum of the epithelium was most peripheral and in contact with the stroma. Maturation of epithelium proceeded in an orderly sequence from this basal zone to the lining of the cavity (Fig. 2).

Cysts have also been noted repeatedly in matrix cultures of human bladder cancers combined with chick embryonic tissue. They have occurred in cultures where the matrix supporting the growth was a combination of collagen-coated cellulose sponge and plasma clot. The cysts were the consequence of imperfections in technique in the preparation of the cultures, where one or more large bubbles of air were trapped in the clotting mixture within the sponge matrix and could not be removed before clotting. Where such optically refractile air ''cysts'' occurred, they disappeared in a day or two as their gaseous content was replaced by liquid. A spherical space remained, and tissue in proximity to the fluid-filled sphere migrated to line the entire globular structure. In histologic sections of some of these cysts a signet ring pattern of growth was seen. Growth was eccentric and was most dense at the segment of the ring adjacent to a free surface of the sponge (Fig. 3).

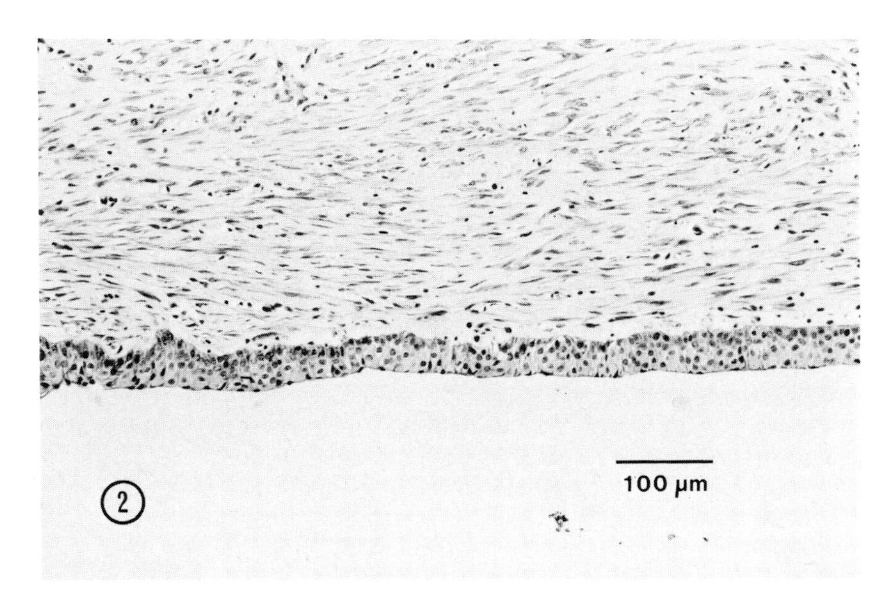

FIG. 2. Histologic section of a 1-week-old ''heterotopic urinary bladder'' prepared by injecting a young adult rat subcutaneously with a combination of two or three minced normal rat urinary bladders and 10 to 15 ml of air. The air is spontaneously replaced by fluid in a day or two. One week later the cyst is lined partially or completely with an orderly layer of stratified uroepithelium, which is slightly hyperplastic but otherwise not remarkable. The changing nuclear/cytoplasmic ratio from the base to the free surface reflects the fact that the cell layer in contact with the connective tissue is the zone of proliferation, and that the superficial cells are maturing. Hematoxylin and eosin.

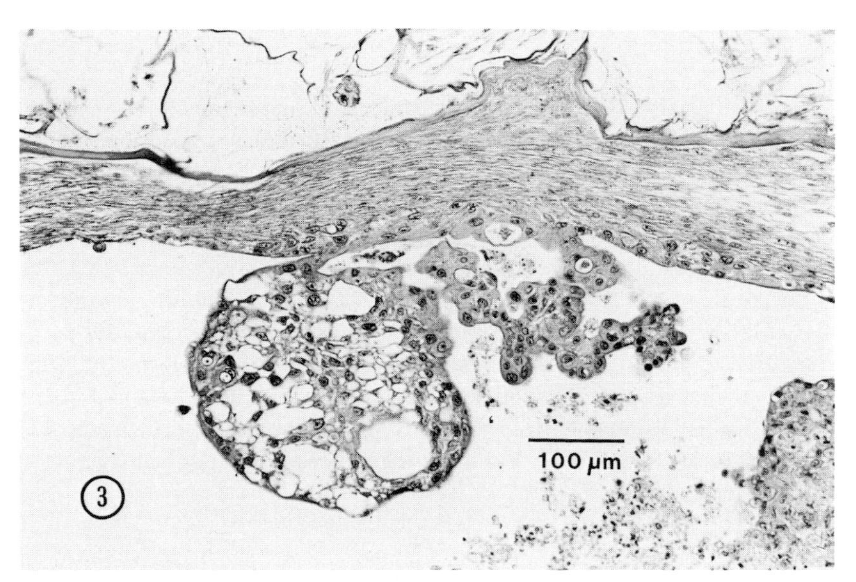

FIG. 3. Histologic section of a 2-week-old sponge matrix culture in which the inoculum consisted of a fine mince of clinical bladder cancer and chick embryo heart. The inoculum was held in place initially with a chick plasma clot. In the preparation of some cultures, imperfections in technique resulted in the formation of air bubbles in the clotting mixture that could not be eliminated prior to clotting. The air in the bubbles was replaced by fluid-filled cavities in a day or two. Where fragments of the inoculum were near these cysts, cells of the inoculum migrated to and proliferated on the fibrin–fluid interface, forming eventually cysts completely lined by tissue. The density and complexity of growth around the cavity was not uniform, but was greatest, as seen here, in proximity to a medium-bathed surface of the sponge. Sponge is seen in the upper part of the figure. Hematoxylin and eosin.

From these considerations of epithelial-lined cysts in the body and in culture, it was evident that a procedure for histophysiologic gradient culture could be achieved. Inocula of normal stratified epithelium usually do not survive long. We theorized that histophysiologic diffusion gradients, as occurring in the body, are also essential for prolonged life of stratified epithelium *in vitro* in a form that will be intelligible equally to the surgical pathologist and the cell biologist. To provide this obligate arrangement in culture, we have placed inocula in macrocysts, using as the cavity a shallow cylindrical chamber relatively impermeable to the culture medium except for one of the two planar surfaces of the cylinder. Then both attachment and nutrition could take place only at the permeable surface. We have tested the model with inocula of normal chick embryo skin, normal adult rat bladder, a rat bladder carcinoma, and a few clinical tumors. The data presented here, in keeping with the plan of the conference, are limited to growth characteristics of inocula of normal origin, adult rat bladder, and chick embryo skin.

II. Methods and Materials

To provide conditions in which both attachment and metabolic exchange including oxygen occur on the same surface of the tissue, cylindrical holes in cellulose sponge are closed at one end with a thin collagen membrane, and explants are attached to this permeable floor. The sponge is inverted and placed against the glass side of the culture tube to produce a chamber with a collagen roof through which nutrition reaches the tissue. Living cultures are studied effectively with an inverted microscope. Following incubation, cultures are prepared with routine procedures for histologic and electron microscopic study. The method used to prepare the chamber is a development from an earlier procedure for collagen-coated cellulose sponges (Leighton *et al.*, 1967; Leighton, 1973). Since this is a new approach to the culture of tissue, some technical details of our model system are undergoing a continuing process of modification and improvement (Fig. 4).

A. Preparation of Chambers with Collagen Diaphragms

1. Preparation of Sponges

Cut dry, fine-pore cellulose sponge (DuPont), such as is used in photography, into slices 4 mm thick with an electric meat slicer. Cut the slices of sponge into rectangles measuring 25 × 20 mm. With a 6-mm-diameter paper punch, make

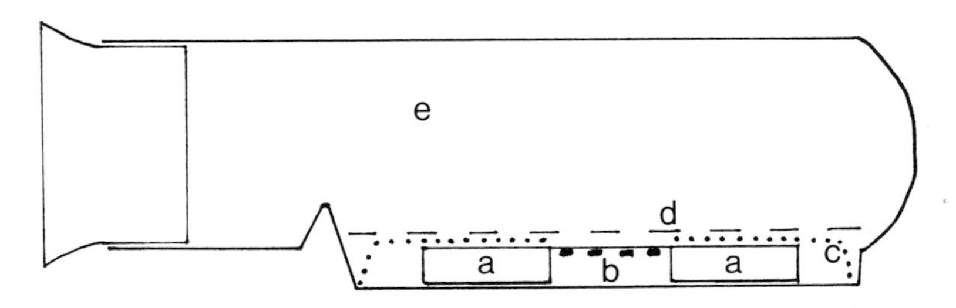

FIG. 4. Diagram of a culture tube ready for incubation. The sponge and culture chamber are not in proportion to the tube in this illustration; specific dimensions are provided in the text. The components of the diagram are (a) the sponge with a hole, (b) the cylindrical hole with tissue attached by fibrin to the thin collagen membrane that forms the roof, (c) the stainless-steel wire mesh, (d) the level of the nutrient medium, and (e) the atmosphere in the culture tube.

two holes in each rectangle, 1 mm apart. Wash the sponges thoroughly by boiling twice in glass-distilled water for a total of 1 hour. Immerse at room temperature in acetone, ethyl ether, and absolute alcohol for 30 minutes each. Return the sponge slices to water and repeat boiling in distilled water twice as before. Spread sponges in petri dishes to dry at room temperature.

2. Primary Impregnation of the Sponge with Collagen

Using a polyethylene box, the kind used at home for storage in the refrigerator, pour full-strength collagen dispersion (Ethicon, Inc., Sommerville, NJ) into the box to a height of approximately 6 mm. Place dry sponges on the dispersion. The sponges sink readily and completely into the dispersion. Allow several days for complete evaporation of the dispersing medium. During the drying period, cover the container loosely to prevent dust from falling on the sponges, and yet to permit evaporation of the dispersing medium. Remove sponges from plastic box, trim, and discard loose membranous collagen. Cure (cross-link the collagen) by placing sponges in 50% methanol containing 0.5% NH_4OH for 1 day. Remove residual NH_4OH with two consecutive rinses of 1 hour each in 50% methanol. Return to a fresh box for complete drying.

3. Lining the Walls of the Cylinders with Several Layers of Collagen

To minimize the diffusion of medium into the chamber through the sides of the sponge, the walls of the cylindrical chambers are heavily coated with collagen. Each sponge will be found to have a more prominent coating of collagen on one surface, the one originally touching the bottom of the box. Place sponges in plastic box with coated surface up. Dispense full-strength collagen dispersion into each hole to fill the hole. Allow dispersion to evaporate completely, and repeat the passage through 50% methanol with NH_4OH, the 50% methanol washes, and drying. Place sponges in plastic box for second application, in inverted position from first coating. Repeat the cycle of putting collagen dispersion into the cylinder, evaporation, curing, washing, and drying. Repeated lining of the cylinder with alternation of the upper surface of the sponge results in a cylinder with relatively impermeable walls. Our present procedure is to process sponges through four applications of dispersion. Whether a lesser number is sufficient, or a greater is desirable, has not yet been determined.

4. Attaching Diaphragms to Cylinders

This procedure involves two steps, preparation of a thin collagen diaphragm and cementing a sponge with collagen lined holes to it. The membrane is prepared by pouring a dilute collagen dispersion into a plastic box and allowing the dispersing medium to evaporate. The dispersion may be diluted with two parts of 50% methanol or 100% methanol. Evaporation is more rapid with 100% methanol. The thickness of the final membrane is determined by the initial volume of undiluted collagen dispersion. When the thin membrane is dry, immerse prepared sponges in full-strength collagen dispersion, remove excess dispersion especially from the cylindrical holes, and place sponges on the preformed collagen membrane in the plastic box. After evaporation peel membranes and sponges from plastic box as a unit, cut out the sponges containing diaphragms, and cure as in previous steps. Sterilization is accomplished by placing sponges with diaphragms for 1 hour in 100% methyl alcohol. We obtain diaphragms 15μm thick ($\pm$ 5μm) using 5 ml of full-strength dispersion (diluted with 50 or 100% methyl alcohol) to cover the base of a plastic box measuring approximately 100 $\times$ 110 mm. The thickness is increased by using larger volumes of dispersion.

5. Preparation of Wire Mesh Grids

1. Cut stainless-steel wire cloth, wire diameter 0.015 in. with 49% open area (CX-20, Small Parts, Inc., Miami, FL), into rectangles 3.25 $\times$ 1.25 in. using metal cutting shears.
2. Bend all four edges of the rectangle to make an open box 1 $\times$ 3 in. at its base, with sides 1/8 in. high.
3. Cut an opening approximately 20 $\times$ 9 mm in the center of the base. Bend the bare wire ends of the opening inward to provide a number of prongs on which to impale the sponge and keep it in place.
4. Check each box and opening for uniformity. The large surface of the grid must be flat to prevent the sponge from moving in the tube when it is on the rocker.
5. Boil the completed grids in a cleaning agent suitable for tissue culture such as CONTRAD 70 (Harleco, Gibbstown, NJ) for 30 minutes. Rinse thoroughly in boiling distilled water. Dry the grids and sterilize in dry heat.

B. Setting up Cultures

Orient the sponges so that the diaphragms are at the base of the chambers. The inocula are attached to the thin collagen membrane with or without fibrin clot.

a. Fibrin clot: Add two drops of heparinized plasma to each chamber. Position explants on the diaphragm. Add two drops of chick embryo extract, mix with plasma, remove excess of the clotting system, and allow clot to form. After clotting, moisten sponge with medium, and position sponge on an inverted stainless-steel wire grid, impaling substance of sponge on several wire ends.

Invert, and place the grid and attached sponge in a large Leighton tube (Bellco, Vineland, NJ). Add medium just sufficient to cover the wire mesh, usually 14 ml, stopper, and incubate tube on a rocking platform.

b. Alternatively, place inocula with a few drops of medium on the thin collagen membrane. Remove excess medium from chamber. Place several sponges in culture container, with medium to a depth of 1 mm, maintaining the arrangement with the diaphragm as the floor of the chamber overnight. Tissue may be adherent to the diaphragm by this time. The sponge is mounted in the stainless-steel mesh, the mesh and sponge inverted, and placed in a large flat tube as above.

C. Observation and Maintenance of Living Cultures

Cell and tissue movement is studied best using an inverted microscope. Medium is replaced as required by the degree of proliferation and the decrease of pH in the medium. Medium used in all these studies was Waymouth's MB 752/1 and 20% fetal bovine serum (Microbiological Associates, Bethesda, MD).

D. Histologic Processing of Cultures

Any fixative used in the standard practice of surgical pathology is satisfactory. We use Stieve's Fixative, which requires subsequent treatment with Lugol's iodine and 1% $Na_2S_2O_3$. After fixation the cultures are trimmed with a razor so that the chambers are tangential to a common line. The sponge is passed through the routine sequence of dehydration, clearing, and embedding in paraffin on edge, with the nearly exposed edges of the two cylindrical chambers down, so that they are toward the knife when mounted on the microtome. Good histologic sections require a perfectly sharp knife. Our usual stain is hematoxylin and eosin.

E. Preparation for Autoradiography

For convenience in autoradiography, use slides with a frosted segment at one end for labeling. The orientation of each slide is recognized easily to the touch if the surface at one end is frosted. Modify a 100-ml graduated glass measuring cylinder by cutting off the top at the 40 ml mark and smoothing off the cut surface.

Our routine has been to add [^{3}H]TdR to the culture medium 1 day before fixation. Immediately before fixation the culture is washed in medium without [^{3}H]TdR for 0.5 hour at 36°C. Fixation with Stieve's and processing are done as above. The subsequent procedure is described here.

1. Immerse deparaffinized sections in (*a*) Lugol's iodine for 10 minutes, rinse in water; (*b*) 1% $Na_2S_2O_3$ for 3 to 4 minutes; (*c*) distilled water for 30 minutes and air dried.

2. Steps 2 through 9 are done in the darkroom with a Wratten #2 safe light. Dilute NTB II emulsion (Kodak) 1:1 with distilled water. Put about 25 ml emulsion gel in the 40-ml cylinder. Place the cylinder in a water bath at 40°C for about 15 minutes to melt the gel. Pour the melted emulsion gently down the side of a clean glass bottle. Add an equal amount of warmed distilled water to the emulsion. The mixture is ready and can be used for three to four batches of slides (36 in each batch).

3. Pour the emulsion into the special container (Ikemoto Chemical Industrial Co., Ltd., Tokyo, Japan) about one-half way up the bowl.

4. Keeping the container in a water bath during dipping, first dip a few blank slides to make sure there are no air bubbles in the emulsion. To remove air bubbles, use a clear slide to scoop up the surface of the emulsion. Then let the emulsion stand for 10 to 20 minutes.

5. Arrange all the slides so that the sections face the same way. Holding the frosted end of the slide, dip each slide with uniform speed and time. Drain off excess emulsion by touching one corner of the slide on the bowl.

6. Drain the dipped slide by pressing one end of the slide on sheets of tissue paper for 30 to 40 seconds.

7. Put the slide in the drying rack with the sections facing up, allowing the emulsion to dry for about 20 to 30 minutes.

8. Put the slides in the dark slide box containing some Drierite (W. A. Hammond Drierite Co., Xenia, OH), and tape the box with black tape. Mark the date on the box.

9. Keep the box with slides at 4°C. Position the box so that the slides are kept horizontal, and allow 1 to 2 weeks for the exposure of emulsion to ^{3}H.

10. In the darkroom, transfer the slides to small stainless-steel racks. It is important that the developer, rinse solution, and fixer be kept at 15°C. A large tray with chilled water at 15°C is usually sufficient to maintain the temperature for 15 to 20 minutes, to process two racks of slides.

11. Use D-19 diluted 1:1 to develop the autoradiographs for 4 minutes at 15°C.

12. Rinse in water and fix for 5 minutes with Kodak fixer.

13. After washing, the slides can be stained with H & E. If the slides have become dried, dip them in water for 5 minutes before staining. Hematoxylin staining should be for less than the routine time. Recognition of silver grains is easier when the nuclear background stain is light.

III. Results

In cultures of the skin of 11-day-old chick embryos and of normal young adult rat bladder, growth was dense in some chambers and irregular in others. Growth was densest where the fragments were evenly distributed on the diaphragm, and where the interfragmentary distance did not exceed 1 mm. Expressed in histologic terms, dense growth refers to thick multilayers of cells, 200 μm or more, and sparse growth to monolayers or at most trilayers of cells.

A. Chick Embryo Skin

In preparing the inoculum skin was removed from the thorax and abdomen of 11-day-old chick embryos, rinsed in medium, and minced with crossed scalpel blades to fragments of 1 mm diameter or less. Particles of skin, in random orientation, were cemented with fibrin to the diaphragm, and the subsequent procedure was as described.

We were able to observe the migration of both spindle-shaped cells and sheets of epithelium for up to 1 week, but as multilayers developed observations as to cell type were difficult. Dense growth was prominent or complete in most chambers. On histologic examination it was seen to consist of a distinct anatomic relationship, a mat of dermal connective tissue separating the collagen diaphragm from a sheet of well-differentiated stratified squamous epithelium. The multilayered epithelium had two distinct zones of equal thickness. Nucleated cells about five cells thick, were adjacent to the dermis and a superficial zone of keratinized epithelium without nuclei faced the empty chamber (Fig. 5). With [^{3}H]TdR autoradiography heavy labeling was found in the nucleated epithelium in contact with the dermis, the basal epithelium (Fig. 6).

In some cultures stratified epithelium was also seen within the dermis as multicellular aggregates. The centers of these masses consisted of mature keratin and the periphery of nucleated stratified epithelium, producing in effect an experimental counterpart to epidermal cysts seen clinically in subcutaneous tissue (Fig. 1). Labeling with [^{3}H]TdR was confined to the peripheral, i.e., basal, cells of the cyst wall, and was obviously greatest in the segment of the basal layer closest to the collagen diaphragm. Apparently, opportunity for metabolic exchange at one pole of the sphere favored selective proliferation of one segment of the basal layer.

Sections in which growth was sparse usually consisted only of atrophic contiguous epithelial cells. These sheets of abnormal epithelium were in direct contact with the diaphragm and no stroma was evident.

B. Normal Rat Bladder

Normal young adult male Fischer rats were killed; their bladders were removed and minced to a pool of small fragments. Cultures were prepared in the same way as for chick embryo skin. Unlike the chick embryo skin, minimal proliferation of connective tissue was evident in the living cultures, and none was

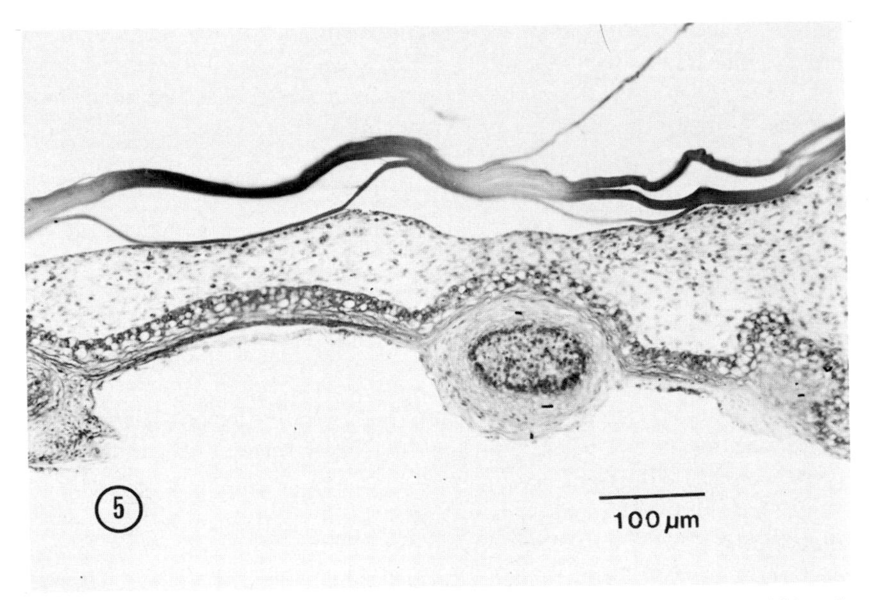

FIG. 5. Histologic section from a 2-week-old culture of a fine mince of 11-day-old chick embryo skin. Normal architecture has been restored; a cellular dermis separates the collagen diaphragm above from the stratified epithelium below. The layers of epithelial cells are nucleated adjacent to the dermis, and keratinized near the cavity of the chamber. The large oval structure in the epidermis is an incompletely formed feather bud. Hematoxylin and eosin.

recognized on histologic study. Remnants of the connective tissue of the bladder wall appeared as foci of dense collagen containing a few connective tissue cells, and served to identify the sites of inoculation.

Epithelial movement was evident after a few days of culture, and by the tenth day most diaphragms were covered with a continuous sheet of epithelium. In living cultures the epithelium had two appearances. In one, dramatic in living cultures, the diaphragm was covered by a monolayer of large thin epithelial cells with large nuclei. In the other the diaphragm was covered with closely packed small epithelial cells. These two types of growth were seen within a week, sometimes at discrete sectors of the same diaphragm, and both types were present through a 9-week period of culture.

The zones with small cells, in some cases occupying entire diaphragms, were of particular interest on histologic examination. After 2 weeks, as well as after longer periods of cultivation, they consisted of stratified urothelium. In the 2-week-old cultures this stratified epithelium varied in thickness from 8 to 10 cells near the sites of inocula, tapering gradually to strata of 2 or 3 cells at the greatest distance from the foci of the inocula. The cells adjacent to the diaphragm were small and stained prominently with hematoxylin, suggesting high prolifera-

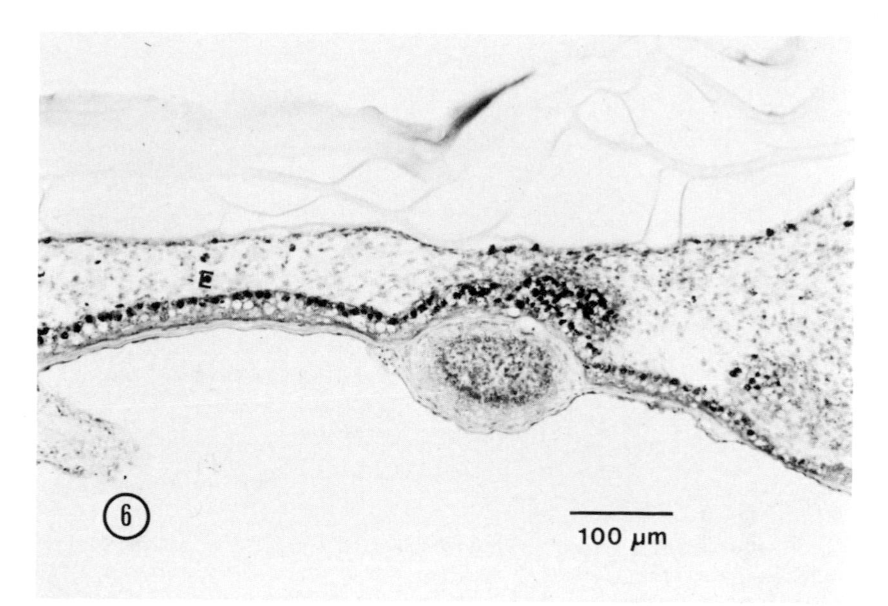

FIG. 6. [^{3}H]Thymidine autoradiograph of chick embryo skin in culture. The heaviest labeling, reflecting cells in proliferation, is in the layer of epidermal cells adjacent to the dermis.

tive activity. A few mitotic figures were evident. An orderly sequence of maturation to the free surface was suggested, with cells toward the surface being larger, the cytoplasmic/nuclear ratio increased, and the cytoplasm having an eosinophilic stain (Fig. 7). This impression was supported by the appearance of [^{3}H]TdR autoradiographs, where nuclear labeling was concentrated in the zone of cells in contact with the diaphragm (Fig. 8). The overall appearance of the thickened epithelium was one that in the intact rat would be interpreted as diffuse hyperplasia.

Cultures terminated after 9 weeks had additional features. The decision to terminate after 9 weeks was made for two reasons. After the seventh week, in many areas of compact small cells, we observed with incredulity the appearance of nodules of epithelium first in one, and then in several cultures (Fig. 9). As observed with the inverted microscope, they appeared to change their patterns of arrangement slightly from day to day, and to increase in number. We wanted very much to see them on histology. A second reason for the decision to fix at 9 weeks was the appearance of small perforations in the diaphragms of a few cultures. Our concern was that cells would migrate through the holes to the outer surface of the diaphragms and alter unfavorably the evolution of the nodules. We found on histologic examination that the nodules were globular eruptions from the basal layer of the stratified epithelium, extending up into an interstitium between the collagen diaphragm and the stratified epithelium (Fig. 10). [^{3}H]TdR

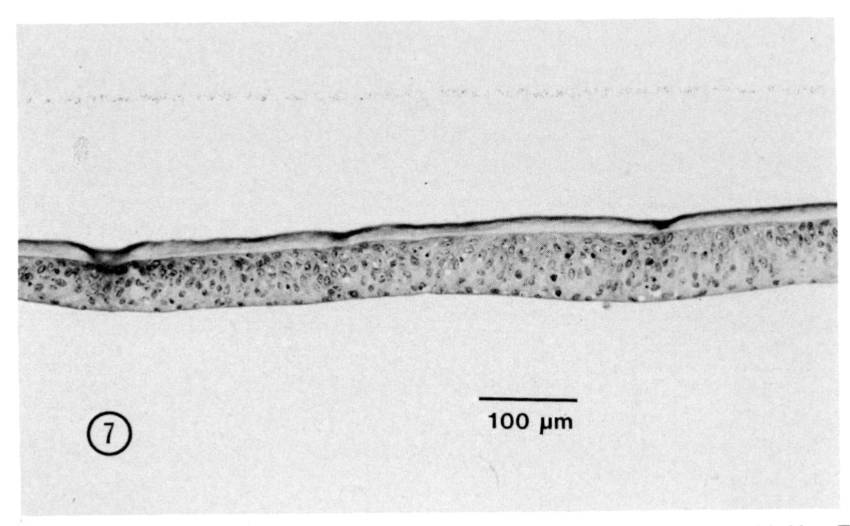

FIG. 7. Histologic section from a 17-day-old culture of a mince of normal rat bladder. The epithelial component of the bladder wall migrates widely on the collagen diaphragm as a sheet of stratified epithelium, but there is no stromal outgrowth. The epithelium seen here, near the explant, is diffusely hyperplastic, six to eight cells thick. As in Fig. 2, where rat bladder has been "cultivated" *in vivo,* the cells adjacent to the source of nutrient have relatively prominent nuclei and scanty cytoplasm. In contrast cells adjacent to the free surface (below) are more mature, with abundant cytoplasm. Hematoxylin and eosin.

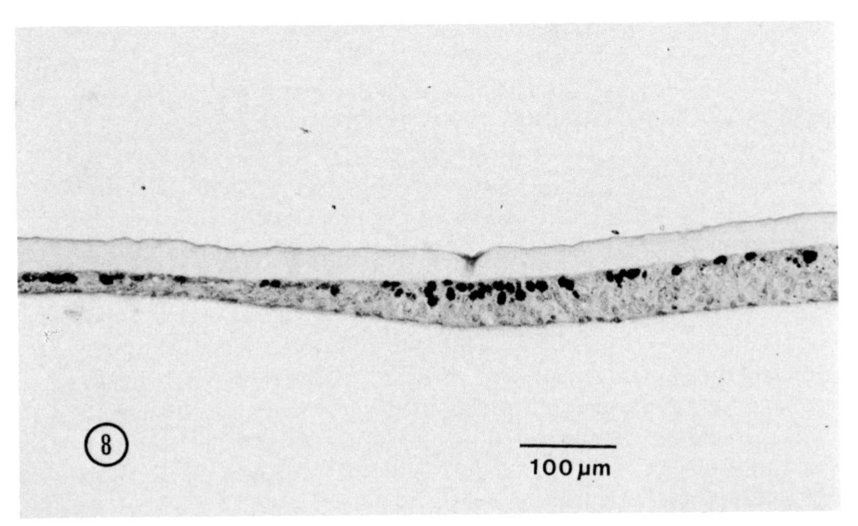

FIG. 8. A [^{3}H]TdR autoradiograph from a 17-day-old culture of a mince of normal bladder. Labeling is confined to those cells adjacent to the collagen diaphragm.

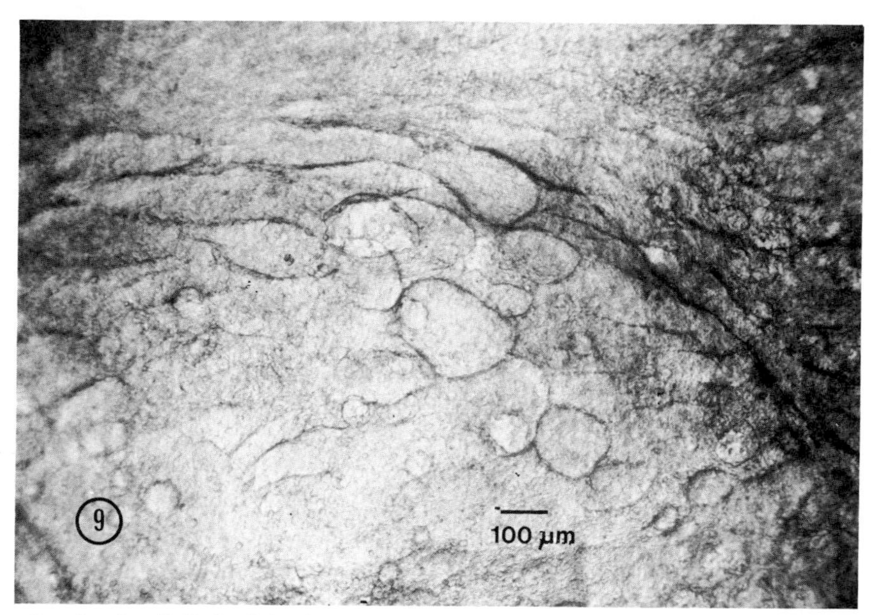

FIG. 9. One field of a living 8.5-week-old culture, photographed with an inverted microscope from the open lower end of the chamber resting on the glass wall. Many rounded or elongated nodules of various size are seen, as well as other rounded translucent structures. There is great variation in the size and shape of the nodular structures from one field to another.

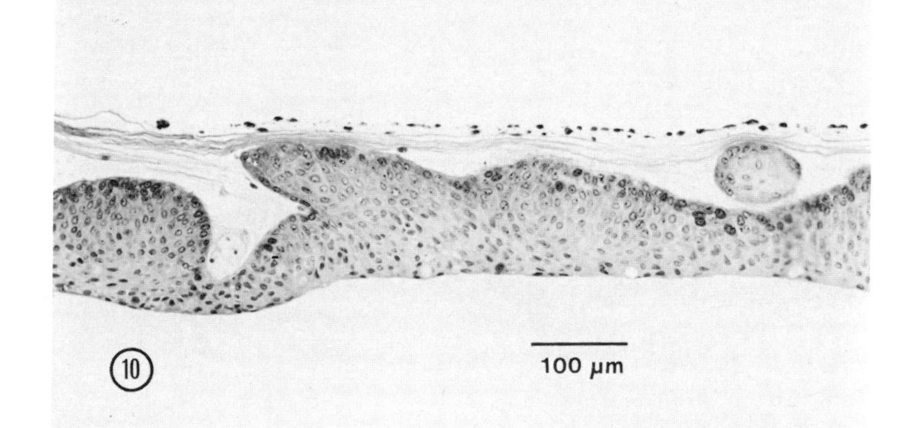

FIG. 10. Histologic section of a 64-day-old culture of normal rat bladder. Several nodules have erupted from the basal layer of epithelium into the "interstitial space" between the stratified epithelium and the collagen diaphragm. One protrusion of epithelium appears to be dissecting a path through the diaphragm. Elsewhere, not in this field, there was a small perforation of the collagen membrane. We presume that the few cells on the nutrient side of the membrane gained access to the outside through the perforation. Hematoxylin and eosin.

autoradiographs demonstrated DNA synthesis and proliferation in these nodules, confirming our observations in the living cultures (Fig. 11). The parent stratified epithelium overlying the buds of nodular growth was thick, and the normal sequence of maturation from the basal to the superficial surfaces of the epithelium was in disarray but without the presence of any significant cellular atypia. The appearance of this epithelium with its erupting nodules is one that, when found in the bladder of the intact rat, is interpreted by many pathologists as nodular or focal hyperplasia.

On autoradiography, proliferative responses of very different magnitude were observed where stratified epithelium was sandwiched between the collagen diaphragm and the collagenous residues of the explants. Extensive labeling occurred in relation to the diaphragm, and occasional labeling in relation to the explant (Fig. 12). Each plane provided evidence of contact-dependent growth, but the magnitude was altered markedly by available metabolities. This recalls the apparent role of nutrition in selective proliferation of epidermal inclusions in cultures of chick embryo skin previously described.

IV. Discussion

Throughout our work with histophysiologic gradient cultures of normal inocula and of neoplasms, we have observed almost complete absence of necrosis

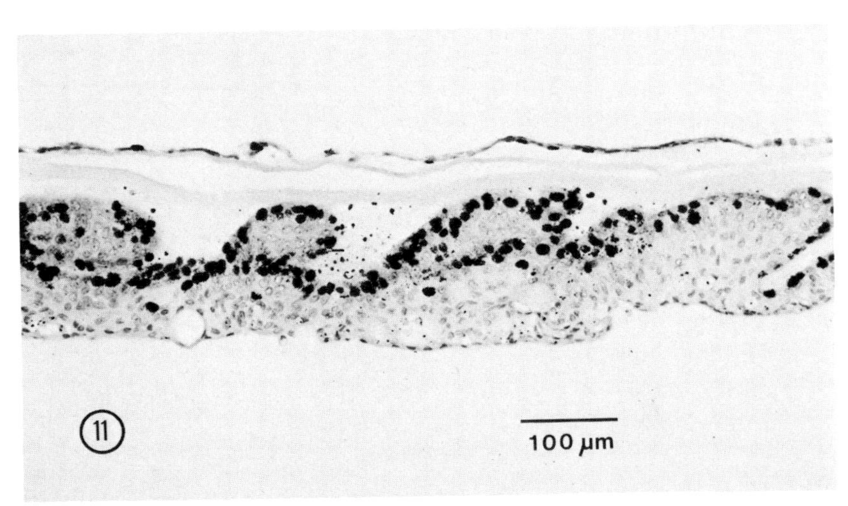

FIG. 11. Autoradiograph of a 64-day-old culture. Proliferation, as evidenced by heavy labeling, is most prominent at the periphery of the nodules that are emerging into the interstitium between the stratified epithelium and the collagen membrane.

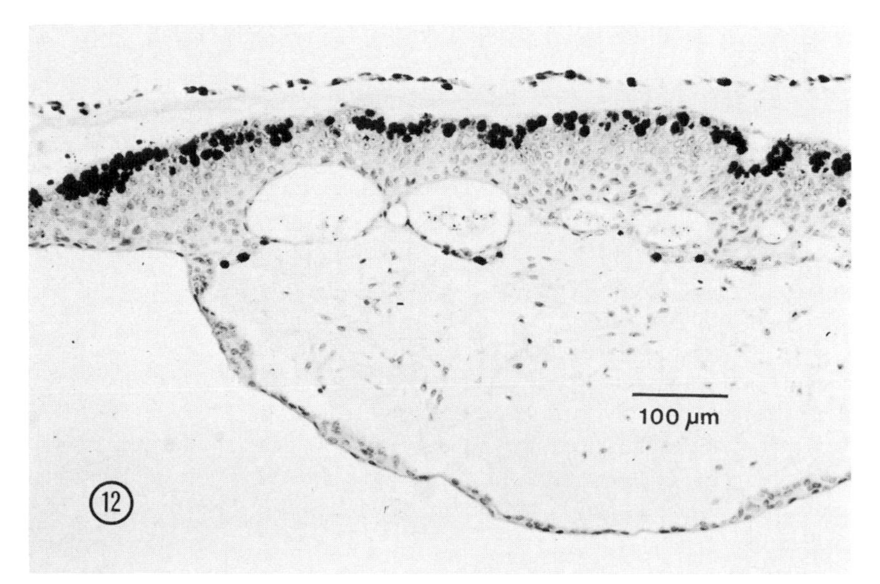

FIG. 12. Autoradiograph of a 64-day-old culture of normal rat bladder, in which a stromal residue of the inoculum is seen. A stratified epithelium separates this collagenous residue from the diaphragm. A few cells are labeled at the interface of the residue and the epithelium. Labeling is intense, however, at the plane of contact of the collagen diaphragm and the epithelium, the interface of nutritional exchange.

in all histologic sections. This is in contrast to conventional organ culture where one expects and accepts central necrosis of organ fragments. We think the absence of necrosis is easily explained. When necrosis occurs in this system, the tissue involved is farthest away from the diaphragm. Necrotic debris falls to the bottom of the chamber, leaving behind a viable surface where histokinetic activities occur freely, unhampered by masses of necrosis. Rather than describing this process of cellular rearrangement as "morphogenesis," we prefer the term "histokinetic activity" since it applies precisely to the demonstrable movements of cells in the dynamic processes of changing structure and organization, without a subliminal implication of differentiation where none has actually been observed.

The hyperplastic reactions of bladder epithelium, from an early diffuse response to a later nodular one, may be explained directly by the growth-enhancing effect of the single medium we have used thus far, or indirectly as a consequence of the failure of a stromal mat to form and to intervene between the collagen membrane and the basal layer of the epithelium. The preliminary data now available have another significance. Since nodular hyperplasia has appeared in the absence of either stroma or blood vessels, it is clear that such proliferative responses of rat bladder epithelium do not have an obligate dependence on stroma or vessels; a histophysiologic gradient is a setting sufficient for the response. We have not observed anything of this sort in control sister cultures of normal bladder maintained in the conventional gestalt, with the diaphragm as the floor of the well resting against glass. Instead, after 2 months, cells were extremely sparse, at best incomplete fragmented monolayers.

The laboratory model with the stainless-steel grid isolates largely, but not absolutely, the cylindrical chamber from exchange with the medium by any route other than transmembrane passage. Sponges sometimes become loosened from the stainless wire anchor pins. There is slight seepage of medium under the sponges as the wells oscillate on the rocker, probably more in some cultures than in others. To improve the precision of the model, we are now actively collaborating with manufacturers of biological plastic ware to encase each sponge in a transparent container whose only opening will be a window through which the collagen diaphragm alone will be exposed to the medium.

Collagen can exhibit selectivity in the diffusion of macromolecules across it, thus changing the relative ratios of serum components as they reach the tissues. Some of these variables may be manipulated by modifying the composition or the thickness of the diaphragm.

V. Perspectives

Transmembrane histophysiologic gradient culture provides a model with which to investigate the role of diffusion gradients in tissue responses. These gradients may be of preeminent importance as regulators of normal architecture in metazoans since the direction of diffusion gradients from blood vessel to tissue is a constant, but the absolute concentration of any metabolite may vary at any moment. The concentration of metabolites that enter the tissues from the blood vessels is always highest at the vascular–interstitial interface and lower at a distance from that interface, whereas the concentration entering the tissues is not always the same.

Stratified epithelium is regulated very effectively in nature. The homeostasis of the normal membrane with its balance of proliferative gain and desquamative loss has been studied in animals, but inevitably with the ever-present constellation of unknown interventions by nervous system, blood vascular system, and endocrines. Stratified epithelium in culture is now available for new kinds of analysis.

Our current interest is a comparison of stratified epithelia from donors of different ages, for example bladders of newborn, young adult, and senescent rats. We are interested as well in confrontation cultures, where epithelial inocula of different origin meet and interact. In the simplest, an inoculum of normal epithelium is placed in one quadrant, and one of carcinoma in the opposite quadrant of the same diaphragm. We will determine whether the events at the time the two meet and subsequently are the same for all combinations of normal and neoplastic cells. We expect to define the biologic "rules" of such encounters, and the changes in the rules that follow alteration of the medium. Our interest in these studies is in improving our understanding of the destructive replacement of normal tissues by cancer. The mechanism may sometimes be accounted for by "neoplastic blockade" (Leighton, 1968), a phenomenon that has been suggested in which tumor cells destroy normal cells by growth and movement so that the tumor component becomes a barrier between normal parenchymal cells and the nutritional source.

The normal development of many glandular epithelial organs involves the appearance of specialized progenitor cells as distinct structural aggregates, and the development of the organ by a sequence whereby the primary aggregate gives rise to secondary ones, and these in turn to others, etc. The final structure is one where all the aggregates or acini are connected by a sequence of branching ducts. The factors regulating the replication of aggregates in normal development, and the determinants of their eventual size and internal structure are not understood. Many carcinomas grow and invade as aggregates. Connections of neoplastic cells between aggregates may sometimes occur, but, unlike the ducts of normal acinar tissues, connections are frequently not present. As cancers grow there are more and more cancer cells, and more and more nests of cells. As in normal development, the factors regulating the replication of aggregates of carcinoma cells are completely unknown, although there have been pilot studies on this subject in tissue culture (Leighton, 1959; Leighton *et al.*, 1960, 1962). Early models were unsatisfactory. The future of studies on the replication of aggregates may now be brighter using histophysiologic gradient procedures. This is suggested by our observations of cultures more than 2 months old, prepared with inocula of normal adult rat bladder, where multiple nodules of epithelium erupted from the basal layer of epithelium into the interstitial space, and proceeded to make secondary nodules by a kind of budding. Such phenomena of growth can be modified by chemicals in the living culture, and followed with the inverted microscope. Cultures can be fixed and examined in great detail with light and electron microscopy at selected intervals.

The unexpected appearance in long-term cultures of normal rat bladder of a "lesion" reminiscent of that appearing *in vivo* after exposure to carcinogens provides a compelling basis for studying carcinogenesis in histophysiologic gradient culture. There is every reason to expect that such investigation can go beyond the rat bladder, and can even include human epithelium.

The use of tissue culture procedures as an adjunct to surgical pathology diagnosis was the hope of those pathologists and surgeons who were major contributors to the conception and birth of tissue culture in the early years of this century. Our new model system provides a context that is recognizable to the surgical pathologist, as well as being rational to the cell biologist. As we acquire experience in histophysiologic culture with some commonly occurring types of cancer, we look forward to contributing to more effective management of individual patients. When the first tissue diagnosis is made on the basis of a metastasis and the site of origin is in doubt, we may be able to recognize in culture the organ of origin, and thereby provide the clinical oncologist with the information required to choose a mode of treatment. Another diagnostic problem of interest to us is dysplasia of stratified epithelium such as may be seen in the uterine cervix. We expect that in histophysiologic gradient culture these lesions will present different patterns of growth and response to chemicals, and thus may permit us to differentiate between dysplasias that will become invasive cancer and those that will not.

ACKNOWLEDGMENTS

This investigation was supported by grant numbers CA 14137 and CA 17772, awarded by the National Cancer Institute, DHEW.

REFERENCES

Abaza, N. A., Leighton, J., and Zajac, B. A. (1978). *Cancer* **42,** 1364.

Carrel, A., and Burrows, M. T. (1910). *J. Am. Med. Assoc.* **55,** 1732.

Ehrmann, R. L., and Gey, G. O. (1956). *J. Natl. Cancer Inst.* **16,** 1375.

Geisinger, K. R., Leighton, J., and Zealberg, J. (1978). *Cancer Res.* **38,** 1223.

Leighton, J. (1951). *J. Natl. Cancer Inst.* **12,** 545.

Leighton, J. (1959). *Science* **129,** 466.

Leighton, J. (1968). *In* "Cancer Cells in Culture" (H. Katsuta, ed.), pp. 143-156. Univ. of Tokyo Press, Tokyo.

Leighton, J. (1973). *In* "Tissue Culture, Methods and Applications" (P. K. Kruse, Jr. and M. K. Patterson, Jr., eds.), pp. 367-372. Academic Press, New York.

Leighton, J. (1979). *Lab. Invest.* **40,** 32.

Leighton, J., and Katsuta, H. (1973). *In* "Chemotherapy of Cancer Dissemination and Metastasis" (S. Garattini and G. Franchi, eds.), pp. 31-43. Raven, New York.

Leighton, J., and Tchao, R. (1975). *In* "Human Tumor Cells in Vitro" (J. Fogh, ed.), pp. 241-265. Plenum, New York.

Leighton, J., Kalla, R. L., Turner, J. M., and Fennell, R. H. (1960). *Cancer Res.* **20,** 575.

Leighton, J., Siar, J., and Mahoney, M. J. (1962). *In* "Henry Ford Hospital International Symposium Biological Interactions in Normal and Neoplastic Growth" (M. J. Brennan and W. L. Simpson, eds.), pp. 681-702. Little, Brown, Boston, Massachusetts.

Leighton, J., Justh, G., Esper, M., and Kronenthal, R. L. (1967). *Science* **155,** 1259.

Leighton, J., Abaza, N., Tchao, R., Geisinger, K., and Valentich, J. D. (1977). *Cancer Res.* **37,** 2854.

Leighton, J., Tchao, R., and Abaza, N. (1979). *Proc. Am. Assoc. Cancer Res.* **20,** 2.
Roberts, D. D., Leighton, J., Abaza, N. A., and Troll, W. (1974). *Cancer Res.* **34,** 2773.
Takeuchi, J., Tchao, R., and Leighton, J. (1974). *Cancer Res.* **34,** 161.
Tchao, R., and Leighton, J. (1979). *Inv. Urol.* **16,** 476.
Toyoshima, K., and Leighton, J. (1975). *Cancer Res.* **35,** 1873.

Chapter 14

Cell and Explant Culture of Kidney Tubular Epithelium[1,2]

B. F. TRUMP, T. SATO, A. TRIFILLIS, M. HALL-CRAGGS, M. W. KAHNG, AND M. W. SMITH

Department of Pathology, Maryland Cancer Program and Maryland Institute for Emergency Medicine, University of Maryland School of Medicine, Baltimore, Maryland

---

I. Introduction

Although there are many reasons to develop cell and organ culture of human kidney including studies on metabolism, carcinogenesis, and transplantation, there has been relatively little progress in terms of long-term explant or cell culture. In the 1930s Carrel and Lindbergh (1938) developed methods for perfusion of animal kidneys for 2 to 3 days with apparent preservation of histologic structure. They utilized mammalian systems—predominantly calf kidney explants. Chambers (1932) described the explant culture of chicken mesonephrons and noted that tubular explants often sealed at the ends and formed cystic structures which actively transported organic anions including phenol red into the

[1]This is contribution #777 from the Cellular Pathobiology Laboratory.
[2]This work was supported by PHS grant no. AM 15440-07 from NIH.

ISBN 0-12-564140-0

lumen. Forster (1948), using teased preparations of teleost fish kidneys, described short-term explant cultures of functioning kidney tubules with which he and his associates studied the characteristics of active transport and its metabolic relationships. Later, Trump and Bulger (1967, 1968a,b, 1969, 1971), Trump and Ginn (1968), and Sahaphong and Trump (1971) utilized similar systems and techniques to explore the fundamental cellular and subcellular reactions to injury of epithelial cells and correlated structural and functional alterations.

Although a number of cell lines presumably derived from normal or neoplastic kidney have been studied (Berman *et al.,* 1979; Perantoni and Berman, 1979), there has been little or no reported success with long-term explant or cell culture of human kidney.

II. Methods and Results

A. Fish

Utilizing studies based on the follow-up of Forster's original work (1948) with flounder kidney tubules, we have developed short- and long-term culture methods for teleost nephrons (Trump and Jones, 1977; Jones *et al.,* 1978). Currently we are most actively studying explant cultures of the euryhaline teleost, the hog choker (*Trinectes maculatus*), and the fresh water teleost channel catfish (*Ictalurus punctatus*). These studies are of some interest because the teleost nephron consists of two main proximal segments, a distal nephron and a collecting tubule segment (in fresh water and euryhaline species only) (Hickman and Trump, 1969). In fresh water and marine glomerular teleosts, a short-ciliated neck segment is also present. The major or second region of the proximal tubule is the longest segment of the nephron and appears to be capable of secreting organic anions and probably divalent cations (Trump and Jones, 1977). This segment seems to be analogous to the terminal proximal tubule or "pars recta" of mammals which also in mammalian systems possesses the organic anion and cation transport system as well as apparently being the site of a number of hydroxylases including those for steroids and for polycyclic aromatic hydrocarbons. In mammals this segment seems to be the one which gives rise to adenocarcinomas of the kidney following treatment with chemical carcinogens such as *N*-(4′-fluoro-4-biphenylyl)acetamide (Dees *et al.,* 1976; Heatfield *et al.,* 1976).

For the culture of fish kidney explants, 1–2 mm cubes of kidney removed under the sterile conditions from freshly opened fish are placed in petri

[1]Abbreviations used: MEM, minimum essential medium (Gibco); CMRL 1066, Connaught Medical Research Laboratory medium; PAS, periodic acid–Shiff; EDTA, ethylenediaminetetraacetic acid; ATP, adenosine triphosphate; ADP, adenosine diphosphate; AMP, adenosine monophosphate; PBS, phosphate-buffered saline; PTC, proximal tubule cells.

dishes containing CMRL 1066[1] supplemented with glutamine, hydrocortisone, insulin, streptomycin, and penicillin. As described elsewhere (Trump and Jones, 1977), the tissues are placed in air-tight chambers on rocking platforms and gassed with 45% oxygen. Tissues are sampled at serial time intervals.

The results of these experiments to this time show that the tubular morphology and function can be maintained in the explants for at least 3 months. After an initial period of lethal and sublethal injury, tubules within the explant maintain good morphology at both the light and electron microscopic levels. Furthermore, epithelial outgrowth of tubules occurs over the explant surface until by 1–2 months the surface is completely covered with cuboidal or columnar polarized epithelium with basal lamina toward the center explant and microvilli along the free surface. The cells are joined by junctional complexes. The entire epithelium resembles the epithelium of the second region of proximal tubule similar or identical to that occurring at zero time (Figs. 1 and 2).

B. Mammalian Kidneys

Prompted by previous work on thin free-hand or Stadie–Riggs slices of rat kidney cortex, we undertook a systematic study of the behavior of cells in such slices maintained at 37°C or 0–4°C (Trump *et al.*, 1974). When such slices are maintained at 37°C in an oxygenated salt solution, such as Robinson's buffer,

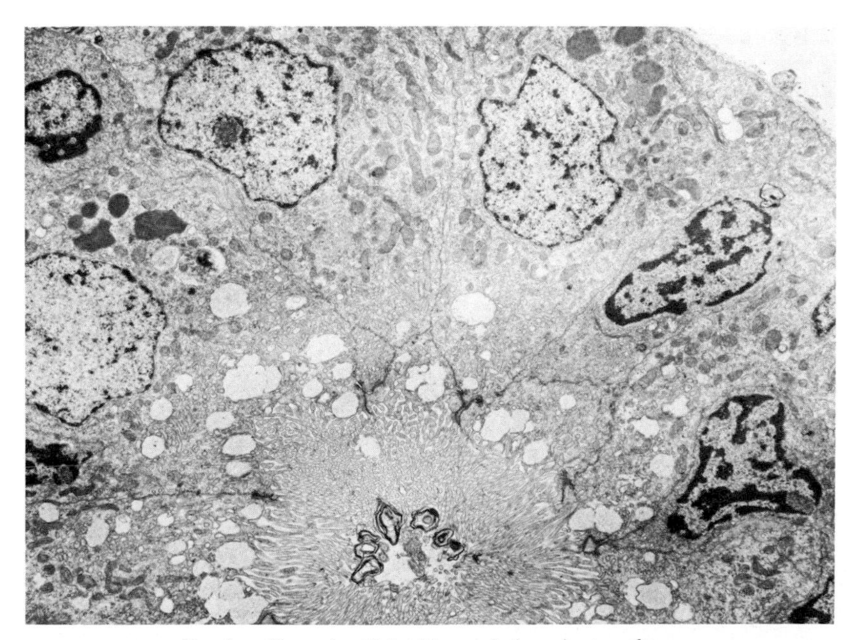

FIG. 1. Channel catfish kidney tubules prior to culture.

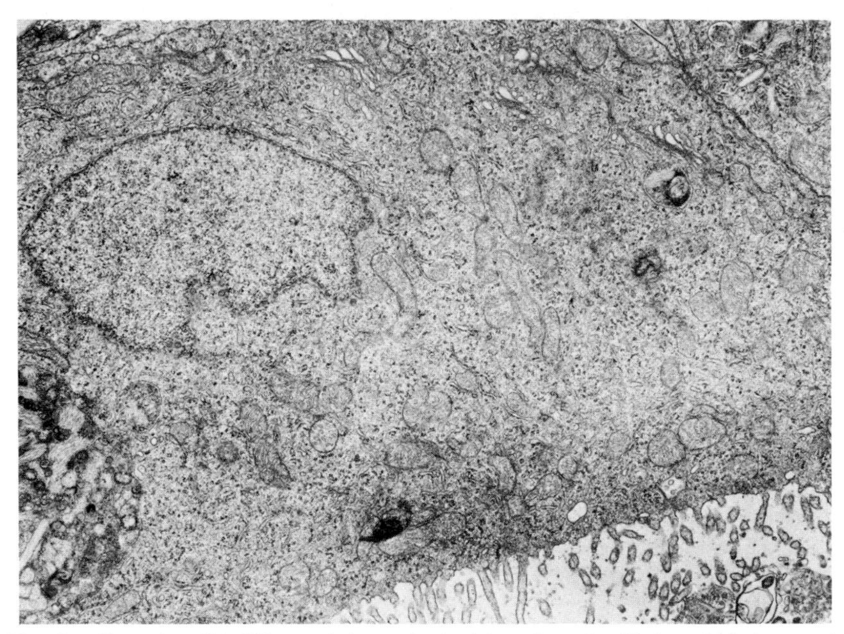

FIG. 2. Channel catfish kidney tubules maintained in culture for 27 days. This is a section through the second proximal segment. Notice the excellent preservation of the ultrastructural morphology.

viability and cell volume regulation can be maintained for only a few hours, probably due to combinations of mechanical damage from the slicing, shaking, and the lack of proper nutrients in the medium. When stored at 0–4°C, however, capability for volume regulations can be maintained for up to 48 hours (Fig. 3). Such studies are, however, useful only for short-term investigations of drug metabolism, active transport, and cell injury. Similar studies could readily be carried out on human tissue, however, such data have not been forthcoming.

1. ISOLATION AND CULTURE OF PROXIMAL TUBULAR EPITHELIUM FROM RAT AND HUMAN KIDNEY

Because of the complexity of the kidney in terms of regional nephron differentiation, it is highly desirable to obtain purified cell fractions from selected nephron regions. This has been studied over the past few years and an improved method has recently been developed in our laboratory employing density gradient centrifugation following enzymatic perfusion. Either kidney tubules or kidney cells can be easily isolated from the kidney with the help of enzyme digestion (De Oca, 1973) or mechanical sieving (Cade-Treyer and Tsuji, 1975). But difficulty lies in the purification of individual cell types in order to give the investigator the

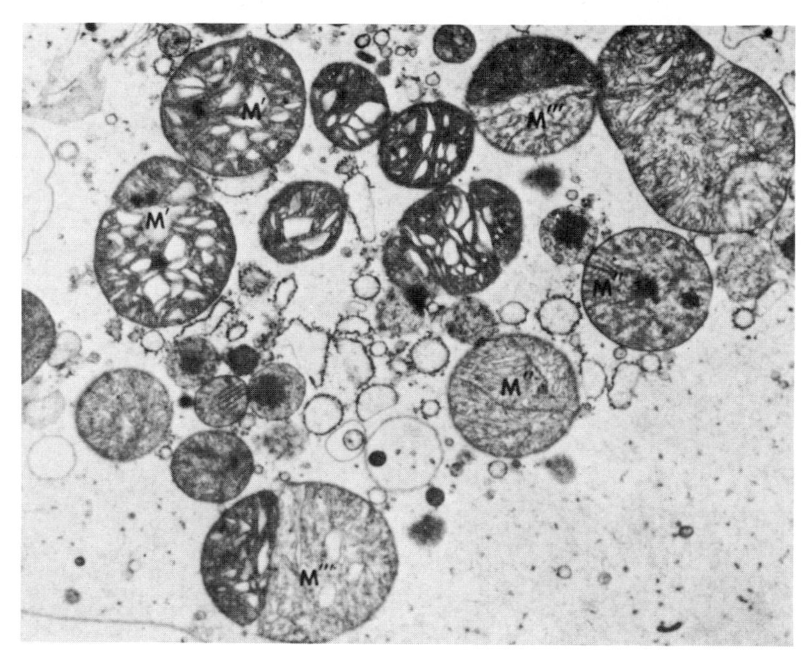

FIG. 3. Rat kidney slice incubated in Robinson's buffer for 48 hours at 0–4°C. M', condensed mitochondria; M'', swollen mitochondria; M''', M' + M''.

capacity for precise quantitative and qualitative assessment of the metabolic or chemical characteristics of particular cells.

Electrophoresis of cells is relatively ancient among methods for cell separation having been used for more than half a century (Seaman and Cook, 1965). Recently, attempts have been made to employ equipment of simple design for the sterile separation of viable mammalian cells (Sherbet *et al.*, 1972) and an apparatus was developed for the vertical density-gradient electrophoresis of viable mammalian cells (Boltz *et al.*, 1973). This allows cells to be separated on the basis of small differences in their electrophoretic mobilities and sedimentation rates, giving homogeneous populations of viable cells.

In our laboratory proximal tubule cells are prepared either by enzymatic digestion giving a high yield (1.5×10^7 cells per rat kidney) and 75% purity or by enzymatic digestion followed by electrophoresis or density gradient centrifugation to give a preparation of up to 95% proximal tubule cells.

a. High-Yield Preparation of Human and Rat Kidney Proximal Tubule Epithelial Cells. Male Sprague–Dawley rats, 200–250 gm, were fasted overnight and anesthetized by intraperitoneal injection of sodium pentobarbital (50 mg/kg). Kidneys were perfused for 5 minutes with PBS containing 0.04% EDTA at 120 mm Hg via the abdominal aorta and drained through the vena cava. This was followed by 2 minutes of perfusion with PBS to wash away the EDTA and subsequently by a few milliliters of PBS containing 0.05% collagenase and 0.125% trypsin. The perfusion solutions were pregassed with 95% O_2–5% CO_2 and kept at room temperature. The renal cortex of human kidney or perfused rat kidney was dissected and minced in the enzyme solution. Digestion was performed with three 20-minute incubations at 37°C with stirring in a trypsinizing flask. Cell suspensions from the first digestion were discarded and those from the second and third digestions were collected in MEM (Gibco) containing 5% heat-inactivated calf serum and washed free of digestive enzymes by washing three times in MEM-calf serum and centrifuging at 83 *g* for 5 minutes. The pellet was then suspended in the spinner flask at 30°C for 30 minutes and gassed with 95% O_2–5% CO_2. This suspension, allowed to settle at 0–4°C for 15 minutes, was then filtered through 40-μm nylon mesh to remove tissue fragments. The filtered cells were collected and washed three times in MEM without calf serum by centrifugation at 83 *g* for 5 minutes. Cell viability was over 90% as determined by exclusion of trypan blue.

b. High-Purity Preparations of Human and Rat Proximal Tubule Epithelial Cells. For high-purity preparations of proximal tubule epithelial cells human or rat kidneys perfused free of blood were placed in PBS at 4°C and the renal cortex was minced into 2–3 mm or smaller pieces. To prepare cell suspensions for electrophoresis the minced cortex was either dispersed by mechanical sieving or by enzymatic digestion as previously mentioned. Mechanical dispersion was accomplished by forcing the minced tissue through 100-μm stainless-steel mesh with a spatula, washing in Joklik-modified MEM with centrifugation, and straining through 40-μm nylon mesh. Viability of cells prepared by this method was 60% by trypan blue exclusion.

Stabilization of cells during electrophoresis was achieved by a continuous 2.0–4.0% Ficoll (Sigma 400) gradient which was also an inverse 6.46–6.12% sucrose gradient (Boltz *et al.*, 1973) contained in a cooled (5°C) column. The floor solution was 50 ml of 20% sucrose in phosphate buffer. The ceiling solution was 80 ml of 6.8% sucrose in phosphate buffer. The gradient was a volume of 100 ml formed on top of the floor solution by an automatic gradient maker. Cells were carefully layered on the top of the gradient, covered with the ceiling solution and electrophoresed at a constant current of 15 mA for 2 hours during which time the original cell sample separated into three or four bands.

Figure 4 shows the successful separation of viable kidney cells. Twenty fractions were collected and proximal tubule cells identified by brush border were concentrated in the fifth (up to 95% purity) and the seventh fraction (75% purity), indicating more mobility of the proximal tubule cells than of the other cell types. Cell debris and cells smaller than proximal tubule cells tended to remain in the earlier fractions (less mobility). On the other hand, small tissue fragments or clumps of cells were found in the later fractions (more mobility and/or sedimenta-

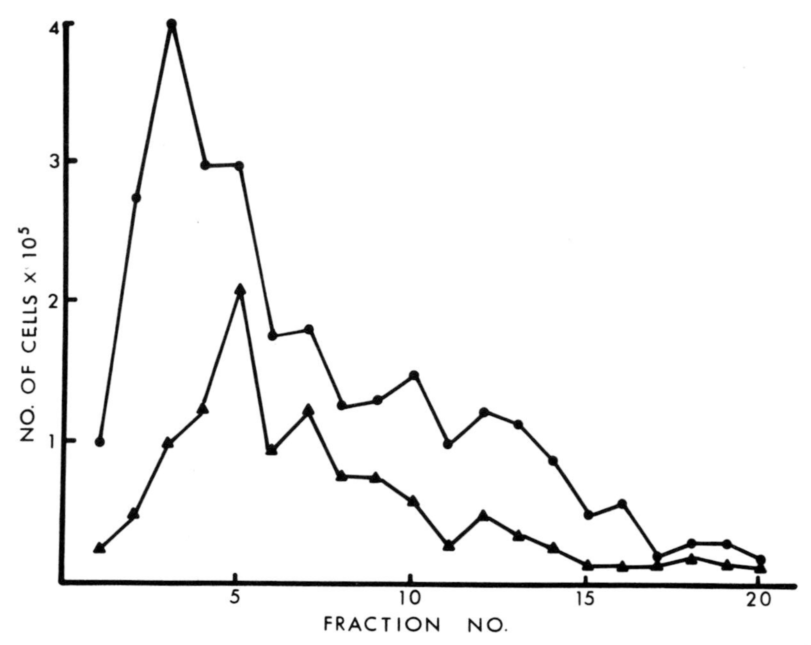

FIG. 4. Separation profile of viable rat kidney proximal tubule cells: ●, total cells; ▲, proximal tubule cells.

tion). In the same experiment but without any electrical field, only tissue fragments and clumps of cells had a significant sedimentation rate.

The cells prepared by enzymatic dispersion showed generally less mobility than those prepared by mechanical sieving. This may be due to alteration on cell surface charge possibly due to changes in surface glycoproteins during enzymatic digestion. Indeed, after being in MEM overnight, the cells prepared by digestion recovered more mobility, which was still less than those prepared by sieving. This may be due to regeneration of glycoproteins on the cell surface.

2. IDENTIFICATION OF PROXIMAL TUBULE CELLS

Glycoprotein on the brush border was demonstrated by the PAS reaction. Alkaline phosphatase (Burstone, 1958), predominantly in the proximal tubule cells, was histochemically demonstrated by the azo dye method. Electron microscopic examination of isolated cells showed microvilli surrounding the entire cell instead of being only at the apical surface (Fig. 5).

The cells from the three purest fractions were cultured in petri dishes or spinner bottles in MEM culture medium with 5% fetal calf serum at an atmosphere of 95% air and 5% CO_2. Rat proximal tubule cells have been maintained

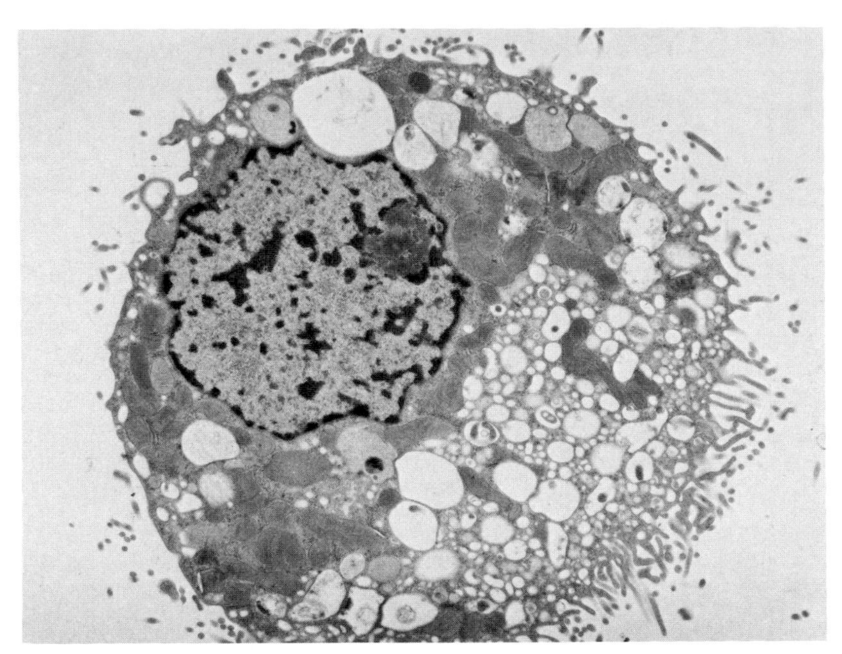

FIG. 5. Proximal tubule cell isolated by enzymatic digestion. Original magnification ×70.

in culture for 3 months and human proximal tubule cells have been maintained for 1 month. Dividing cells have been demonstrated by autoradiography prepared from cultured cells exposed to 1 μCi of [^{3}H]thymidine per ml culture medium for 6 hours (Figs. 6 and 7).

3. METABOLIC STUDIES ON ISOLATED PROXIMAL TUBULE CELLS

Energy metabolism of isolated proximal tubule cells has been studied by measuring metabolities under anoxic conditions in the presence and absence of glucose for 1, 5, 15, and 30 minute time intervals. Cells suspended in PBS containing glucose and prebubbled with oxygen were used for controls at corresponding time intervals.

As key metabolities of the energy-producing pathway, adenine nucleotides (ATP, ADP, AMP) have been measured in isolated proximal tubule cells following anoxic treatment (Table I). Adenine nucleotides were measured fluorometrically by established methods (Lowry and Passonneau, 1972).

At this time, adenine nucleotides have been measured on two human kidney cell preparations immediately after preparation by the high-yield method at zero time.

The presence of glucose in anoxic incubation enabled cells to maintain adenine

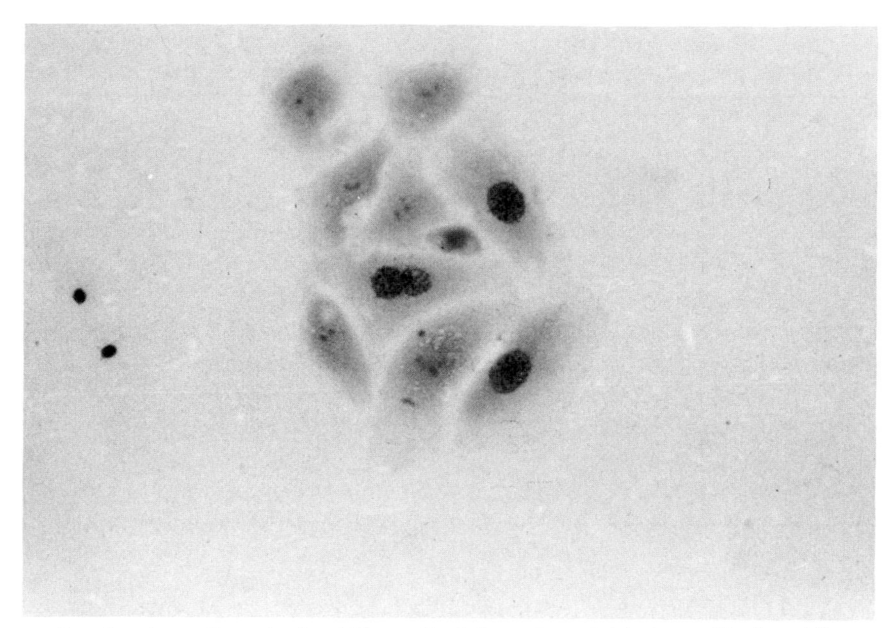

FIG. 6. Autoradiograph of rat kidney cells: 3 day culture. Original magnification ×70.

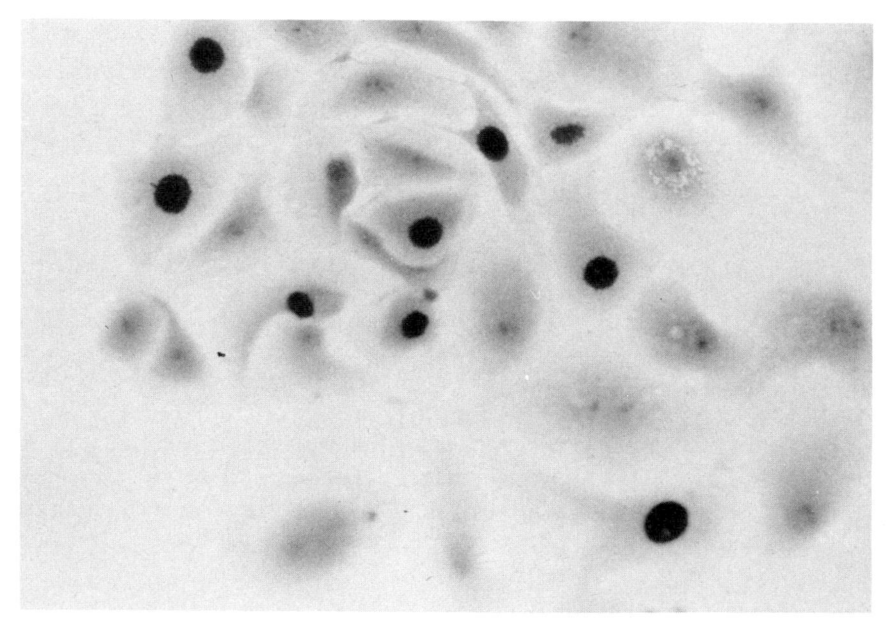

FIG. 7. Autoradiograph of human kidney cells: 4 day culture. Original magnification ×70.

TABLE I

ADENINE NUCLEOTIDE CONTENTS IN THE ISOLATED PROXIMAL TUBULE CELLS

Incubation time (minutes) 37°C	Control (μmoles/gm protein)				Anoxia plus glucose (μmoles/gm protein)				Anoxia minus glucose (μmoles/gm protein)			
	ATP	ADP	AMP	AXP	ATP	ADP	AMP	AXP	ATP	ADP	AMP	AXP
Rat Kidney PTC												
0	3.62[a]	2.88	2.36	8.86	—	—	—	—	—	—	—	—
1	2.10	3.83	1.69	7.62	2.40	3.29	1.83	7.52	1.56	3.78	2.71	8.05
5	2.69	2.56	2.06	7.31	2.13	2.79	2.64	7.56	1.50	2.01	3.04	6.55
15	1.85	2.94	1.15	6.83	1.06	3.28	1.38	5.80	0.96	1.55	2.08	4.60
30	1.53	0.92	1.95	4.40	1.77	2.54	1.61	5.91	0.35	1.32	1.24	2.91
Human PTC												
0	2.00	1.26	0.63	3.89	—	—	—	—	—	—	—	—
	3.33	1.26	0.55	5.14								

[a] All values for rat PTC are averages of four to five experiments.

nucleotide levels at all time intervals studied. However, the absence of glucose caused variable decreases in nucleotide levels depending upon the endogenous glucose present in the cells. Changes in energy metabolites in isolated proximal tubule cells under anoxic conditions minus glucose (simulation of ischemia *in vivo*) are reminiscent of the metabolite changes observed in whole kidney slices (Kahng *et al.*, 1978).

C. Human Renal Preservation

1. BACKGROUND

Current methods of preservation of human cadaver kidneys for subsequent transplantation employ either simple hypothermic storage, hypothermic pulsatile perfusion, or a combination of the two.

Since the first report of successful 17-hour preservation of a human kidney and transplantation by Belzer and his co-workers in 1968 (Belzer *et al.*, 1968a), a vast and often contradictory literature has developed. Indeed every step in the preservation procedure has become a matter of controversy. Yet, 10 years after reporting a series of transplantations of dog kidneys, which had been preserved for 72 hours using cryoprecipitated plasma and pulsatile perfusion at 10°C and had produced 100% graft survival (Belzer *et al.*, 1967), Belzer (1977) had to admit that "these results have not been improved upon." Those who favor pulsatile perfusion preservation state that because the kidney is provided with both substrate and oxygen, longer periods of preservation are possible, thus changing an emergency operation into an elective one and allowing greater time to find suitable matching recipients (Scott *et al.*, 1970). There is also the potential to monitor renal function and, therefore, predict post-transplantation performance, while the kidney is on the perfusion machine (Baxby *et al.*, 1974; Sampson *et al.*, 1977).

Proponents of simple hypothermia point to its simplicity and economy (Collins, 1977) and say that the satisfactory preservation time has not been significantly increased by perfusion preservation. They also maintain that post-transplantation function cannot be accurately assessed by the perfusion characteristics of the kidney nor by biochemical measurements of the perfusate (Sampson *et al.*, 1977; Magnusson and Stowe, 1976). Recently, there have also been reports of perfusion injury to the kidney (Spector *et al.*, 1976; Limas *et al.*, 1977), while others report that immune injury may be associated with pulsatile renal preservation (Filo *et al.*, 1974; Light *et al.*, 1975).

2. Procedure for Preservation for Renal Transplant

The procedure followed at the University of Maryland Transplantation Center, following certification of death and donor organ removal, involves flushing the kidney with a solution composed of 6% Dextran 70 in 0.9% sodium chloride to which is added 4,000 U heparin, 40 mg lidocaine, and 20 meq sodium bicarbonate per liter.

The kidney is immediately flushed at room temperature until the effluent is free from blood. Thereafter the chilled Dextran solution is used and flushing continued until the core temperature is considered between 5 and 10°C. At the same time the kidney is immersed in flushing fluid at 4–7°C. If the kidney is to be preserved by simple hypothermia it is then sealed in a double polyethylene bag along with the flush solution, and packed in ice slush until required.

Those organs receiving pulsatile hypothermic perfusion are transferred to a Waters perfusion machine. The perfusate consists of 750 ml of 5% plasma protein fraction (Hyland or Cutter) to which is added 50 ml of 25% albumin solution, 10 ml of 50% glucose, 50 U insulin, 6 mg phenolsulfonphthalein, 4 meq potassium chloride, 500,000 U penicillin G, 3 meq calcium chloride, 250 mg methyl prednisolone, and 4000 U heparin. After filtration, sodium bicarbonate and mannitol are added to bring the pH to 7.3–7.4 (at 8°C) and the osmolality to about 340 mOsm/kg.

Perfusion temperature is kept within the range of 4–10°C, perfusate flow is approximately 100 ml/minute, and pressure is adjusted to keep the flow stable, although a pressure of less than 45 mm Hg is considered optimal.

The practice at the University of Maryland is to use simple hypothermia if a preservation time of less than 12 hours is anticipated. If a longer period is necessary, the kidney is transferred to the perfusion machine as soon as possible. Pulsatile preservation times up to 48 hours are acceptable.

3. Discussion

In experimental work, kidneys are removed from beating-heart donors. In the clinical setting this is not always possible and, as a result, there is often a period of ischemia at body temperature—the warm ischemia—which may be prolonged. The human kidney may be considered viable, that is, to have suffered only reversible ischemic damage, after up to 60 minutes of warm ischemia (Sterling *et al.*, 1972), and Sterling has reported (Sterling *et al.*, 1977) nonfunction in a series of kidneys after only 45 minutes of warm ischemia. Less than 30 minutes is probably desirable (Friedman *et al.*, 1954). In our hospital, warm ischemia of more than 15 minutes is not considered acceptable. Belzer and his co-workers (Belzer *et al.*, 1968b) working with dogs have shown a fall in renal survival from 40 to 100% if 24 hours of perfusion is preceded by 1 hour of warm ischemia. Similarly several clinical series (Calne *et al.*, 1968; Pletka *et al.*, 1969; White *et al.*, 1968) have reported good immediate renal function, a low incidence of post-transplantation ‘‘acute tubular necrosis,’’ and decreased need for post-transplantation dialysis in those cases with short warm ischemia times.

Fluids used for flushing the kidney prior to hypothermic storage may either have an electrolyte composition resembling intra- or extracellular fluid. Variations of osmolarity within these two groups have also been reported.

In his original work, Belzer reported flushing with Ringer's lactate solution (Belzer *et al.*, 1968a) but, on the other hand (Acquatella *et al.*, 1972), this was found to result in ‘‘immediate and dramatic cell swelling with marked Na^+ and Cl^- gain and K^+ loss.’’ Belzer's team subsequently (Downes *et al.*, 1973) added sucrose to Ringer's lactate, thus raising the osmolarity and decreasing the water retention.

An alternative solution (C_3 solution) which more nearly resembles extracellular fluid was recommended by Collins *et al.* (1969) and with this he reported good renal preservation for 30 hours. The C_3 solution is probably the most widely used in transplantation practice.

A modification of Collins solution was suggested by Sacks *et al.* (1974). This solution raises the osmolarity by addition of mannitol and in Sacks' hands produced good results. A subsequent report (Chatterjee and Berne, 1975) has failed to confirm the superiority of Sacks' solution. Finally a hypertonic citrate solution has been suggested and reported to give good results (Bishop and Ross, 1978). But according to one report, no improvement over C_3 or Sacks' solution (Gunter *et al.*, 1978) was achieved.

Despite the rationale for placing an intracellular type solution around cells of minimal metabolic activity, thus nearly eliminating electrolyte gradients, successful preservation seems more closely related to osmolality of the flush solution. Dextran is commercially available, easy to use, and has produced good results in our hands up to 24 hours. Motivation to change to a more difficult and expensive solution is slight.

The perfusate solution in pulsatile preservation might ideally be considered to be blood, in order that maximal tissue oxygenation might occur. However, at low temperatures oxygen requirements are reduced and at the temperature of perfusion (4–10°C) blood becomes viscous. Cryoprecipitated plasma as advocated by Belzer *et al.* (1968b) is widely used. This provides adequate oxygenation and the lipoprotein fraction which may be denatured by the oxygenator has been removed by prior freezing, thawing, and filtration.

The more recent use of plasma protein fraction, such as is used at the University of Maryland, theoretically reduces the risk of immunologic injury by excluding the gamma globulin fraction of the plasma. It is also cheaper, has a long shelf life, is simple to prepare, and reduces the risk of hepatitis (Mendez-Picon *et al.*, 1976).

Measurements of future viability and function are divided into those which assess the hemodynamic qualities and those measuring various metabolic parameters. The aim during perfusion is to keep the flow on the machine at about 100 ml/minute while maintaining pressure below 60 mm Hg (Belzer *et al.*, 1968a). Increasing the perfusion pressure to maintain a constant flow has been considered to indicate poor future prognosis. However, Sampson and his associates (Sampson *et al.*, 1977) believe that the perfusion characteristics bear "no relation to the final outcome of transplantation" and that many potentially functioning kidneys are lost to transplantation if the hemodynamic qualities of the kidney are considered.

Metabolic criteria used to predict the function of the kidney have included the perfusate pH during the first hours of perfusion (Magnusson and Kiser, 1971). Theorizing that tissue hypoxia may lead to acidosis and cell death, they report that a drop in perfusate pH is a reliable indicator of cell death, and a rising pH indicative of potential good function. Johnson *et al.* (1973) reported that they could predict precisely those kidneys having immediate or delayed function on the basis of lactic acidosis and lactate dehydrogenase measurement in the perfusate. Magnusson and Stowe (1976) reject this work and claim to have measured a large number of metabolic parameters without "compelling evidence to show that any of these measurements is a reliable criterion of subsequent function."

The consensus at this time appears to be that in the absence of warm ischemia time, and for a period not exceeding 15 hours, simple hypothermic preservation is satisfactory and has not been shown to cause positive damage to the kidney. Pulsatile perfusion can undoubtedly be used for longer periods and has been shown to be positively advantageous in those kidneys with prolonged warm ischemia time (Halasz and Collins, 1976).

Combinations of both types of preservation are probably used most commonly in clinical practice, but Sterling has reported (Sterling *et al.*, 1972) that prolonged cold ischemia prior to pulsatile preservation is associated with increased incidence of acute tubular necrosis. Similar results were reported by Light *et al.* (1975). A recent study in our laboratory (Hall-Craggs *et al.*, 1980) on the ultrastructural changes in three human kidneys preserved for transplantation also showed that the structural damage was greatest in the kidney with the longest simple hypothermic preservation prior to pulsatile perfusion. In that kidney irreversible cell injury was seen in the pars recta of the proximal convoluted tubule and in the ascending thick limb of the loop of Henle (Fig. 8). Other segments of the nephron in this one kidney and all segments in the other two kidneys studied showed evidence of reversible injury only. On the other hand, a short-term hypothermia followed by pulsatile perfusion maintained the ultrastructure compatible with reversible injury (Fig. 9). The control human kidney obtained from the immediate autopsy is shown for comparison (Fig. 10).

The problem in assessing these methods of preservation is that the ultimate measure of function can probably only be measured in a nephrectomized animal with an autotransplanted kidney. In these circumstances the animal is "fit" both

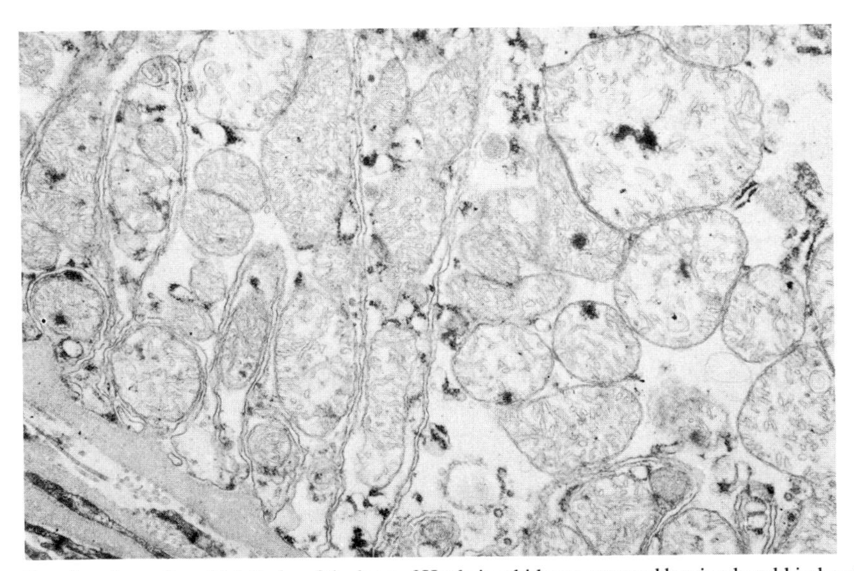

FIG. 8. Ascending thick limbs of the loop of Henle in a kidney preserved by simple cold ischemia for 27 hours followed by 14 hours of pulsatile perfusion. Mitochondria show high amplitude swelling and flocculent densities, indicating irreversible cell injury. Cell cytoplasm is lucent, rough endoplasmic reticulum is dilated, and ribosomes dispersed.

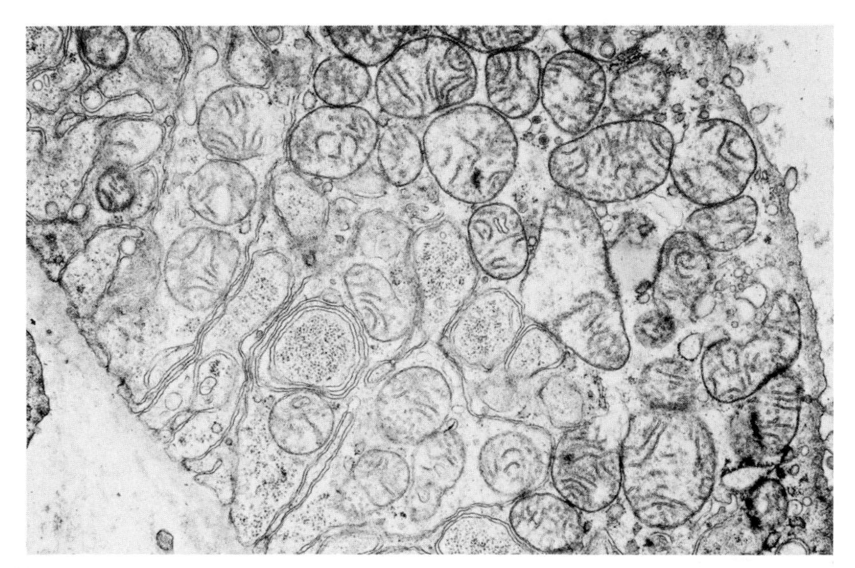

FIG. 9. A segment of the distal tubule in a kidney preserved by simple hypothermia for 32 minutes followed by 47 hours of pulsatile perfusion. Mitochondria are swollen but flocculent densities are not present.

at the time of original unilateral nephrectomy, so that agonal damage experienced in the donor in the clinical situation is avoided, and warm ischemia may also be reduced to a minimum in the heart-beating donor. With autotransplantation no immunologic injury is experienced by the transplanted organ.

In the clinical situation no such control of circumstances is possible, even though medication may be given to maintain terminal blood pressure and urine flow. Probably the best definition of immediate function is that used by Marshall (Marshall *et al.*, 1977) which requires early diuresis with a spontaneous and continuing fall of creatinine and urea in the first days after transplantation, with no dialysis being necessary in the first week after transplantation.

It seems that at this time immediate flushing with cold solution of an intercellular type followed by simple cold ischemia is considered to be the method of choice for up to 12 hours in those kidneys which have not been subjected to prolonged warm ischemia.

For preservation longer than 12 hours and up to 36 hours and in those kidneys with prolonged warm ischemia, pulsatile perfusion using either cryoprecipitated plasma or partial protein fraction is currently the method of choice.

In a review of transplantation "past, present and future," Belzer (1977) suggested that the present lack of progress in renal preservation may be due to the practice of using hypothermic preservation in nonhibernating tissues.

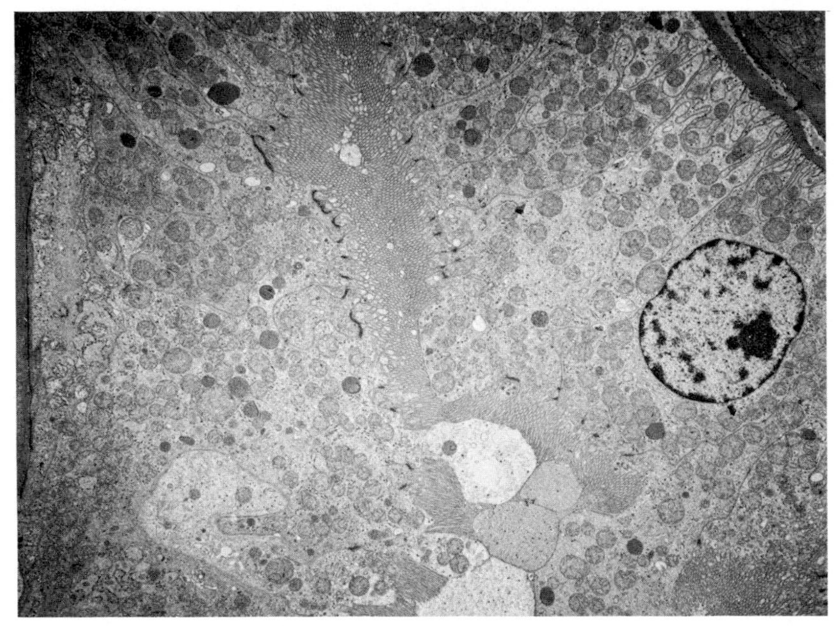

FIG. 10. Human kidney tubule from an immediate autopsy on a patient who died following a motor vehicle accident. Notice the normal ultrastructure of the cells, indicating that the patient had not suffered shock.

He feels that until more knowledge is available in basic biochemical principles at low temperatures further progress in kidney preservation is unlikely.

III. Perspectives

The kidney has proven to be a difficult subject for both organ and cell culture but on the other hand has been extensively studied in terms of preservation for transplantation. Little work so far has been done on cell and explant culture of human kidney. However, work in our laboratory suggests that methods developed for isolation and culture of rat kidney cells can also be applied to humans. Some of the principles developed for renal preservation *in vitro* may also work with other organs. Preservation of human kidneys for up to 48 hours with cold solutions correlates well with the maintenance of rat kidney slices for 48 hours at 0–4°C. It is, however, difficult to understand why the kidney can be maintained only for 2 days at 0–4°C while bronchial pieces can be maintained at the same temperature for over 1 week.

REFERENCES

Acquatella, H., Perez-Gonzalez, M., Morales, J. M., and Whittembury, G. (1972). *Transplantation* **14,** 480.

Baxby, K., Taylor, R. M. R., Anderson, M., Johnson, R. W. G., and Swinney, J. (1974). *Lancet* **2,** 977.

Belzer, F. O. (1977). *Transplant. Proc.* **IX,** 1543.

Belzer, F. O., Ashby, B. S., and Dunphy, J. E. (1967). *Lancet* **2,** 536.

Belzer, F. O., Ashby, B. S., Gulyassy, P. F., and Powell, M. (1968a). *N. Engl. J. Med.* **278,** 608.

Belzer, F. O., Ashby, B. S., and Downes, G. (1968b). *Surg. Forum* **19,** 205.

Berman, J. J., Perantoni, A., and Jackson, A. (1979). *Exp. Cell Res.* **121,** 47.

Bishop, M. C., and Ross, B. D. (1978). *Transplantation* **25,** 235.

Boltz, R. C., Todd, P., Streibel, M. J., and Louie, M. K. (1973). *Prep. Biochem.* **3,** 383.

Burstone, M. S. (1958). *J. Natl. Cancer Inst.* **20,** 601.

Cade-Treyer, D., and Tsuji, S. (1975). *Cell Tissue Res.* **163,** 15.

Calne, R. Y., Evans, D. B., Herbertson, B. M., Joysey, V., McMillan, R., Maginn, R. R., Millard, P. R., Pena, J. R., Salaman, J. R., White, H. J. O., Whitcombe, J. F. R., and Yoffa, D. E. (1968). *Br. Med. J.* **2,** 404.

Carrel, A., and Lindbergh, C. A. (1938). "The Culture of Organs." Hoeber, New York.

Chambers, R. (1932). *Anat. Rec.* Suppl. **54,** 36.

Chatterjee, S. N., and Berne, T. V. (1975). *Transplantation* **19,** 441.

Collins, G. M. (1977). *Transplant. Proc.* **IX,** 1529.

Collins, G. M., Bravo-Shugarman, M., and Terasaki, P. I. (1969). *Lancet* **2,** 1219.

De Oca, H. M. (1973). *In* "Tissue Culture: Methods and Applications" (P. F. Kruse and M. K. Patterson, eds.), pp. 8–12. Academic Press, New York.

Dees, J., Reuber, M., and Trump, B. F. (1976). *J. Natl. Cancer Inst.* **57,** 779.

Downes, G., Hoffman, R., Huang, J., and Belzer, F. O. (1973). *Transplantation* **16,** 46.

Filo, R. S., Dickson, L. G., Suba, E. A., and Sell, K. W. (1974). *Surgery* **76,** 88.

Forster, R. P. (1948). *Science* **108,** 65.

Friedman, S. M., Johnson, R. L., and Friedman, C. L. (1954). *Circ. Res.* **2,** 231.

Gunter, W. P., Linke, C. A., Pranikoff, K., and Frido, C. W. (1978). *Transplantation* **25,** 279.

Halasz, N. A., and Collins, G. M. (1976). *Arch. Surg.* **111,** 175.

Hall-Craggs, M., Little, J. R., Sadler, J. H., and Trump, B. F. (1980). *Human Pathol.* **11,** 23.

Heatfield, B. M., Hinton, D. E., and Trump, B. F. (1976). *J. Natl. Cancer Inst.* **57,** 795.

Hickman, C. P., Jr., and Trump, B. F. (1969). *In* "Fish Physiology" (W. S. Hoar and D. J. Randall, eds.), Vol. 1, pp. 91–239. Academic Press, New York.

Johnson, R. W. G., Anderson, M., Taylor, R. M. R., and Swinney, J. (1973). *Br. Med. J.* **1,** 391.

Jones, R. T., Sanefuji, H., and Hinton, D. E. (1978). *Am. Zool.* **18,** 671.

Kahng, M. W., Berezesky, I. K., and Trump, B. F. (1978). *Exp. Molec. Pathol.* **29,** 183.

Light, J. A., Annable, C., Perloff, L. J., Sulkin, M. D., Hill, G. S., Etheredge, E. E., and Spees, E. K. (1975). *Transplantation* **19,** 511.

Limas, C., Spector, D., and Wright, J. R. (1977). *Am. J. Pathol.* **88,** 403.

Lowry, O. H., and Passonneau, J. V. (1972). "A Flexible System of Enzymic Analysis." Academic Press, New York.

Magnusson, M. D., and Kiser, W. S. (1971). *Surg. Clin. N. Am.* **52,** 1235.

Magnusson, M. D., and Stowe, N. T. (1976). *Urol. Clin. N. Am.* **3,** 491.

Marshall, V. C., Ross, H., Scott, D. F., McInnes, S., Thomson, N., Atkins, R. C., Mathew, T. H., and Kincaid-Smith, P. S. (1977). *Med. J. Aust.* **2,** 353.

Mendez-Picon, G., Belle, C., Pierce, J. C., Thomas, F., Murai, M., Wolf, J., and Lee, H. M. (1976). *Surgery* **79,** 364.

Perantoni, A., and Berman, J. J. (1979). *In Vitro.* **15,** 446.
Pletka, P., Cohen, S. L., Hulme, B., and Kenyon, J. R. (1969). *Lancet* **1.**
Robinson, J. R. (1949). *Biochem. J.* **45,** 68.
Sacks, S. A., Petritsch, P. H., and Kaufman, J. J. (1974). *Transplant Proc.* **VI,** 283.
Sahaphong, S., and Trump, B. F. (1971). *Am. J. Pathol.* **63,** 277.
Sampson, D., Kauffman, H. M., Peters, T., and Walczak, P. (1977). *Transplant. Proc.* **IX,** 1551.
Scott, D. F., Martin, D. R., Kountz, S. L., and Belzer, F. O. (1970). *Med. J. Aust.* **2,** 1157.
Seaman, G. V. F., and Cook, G. M. W. (1965). *In* "Cell Electrophoresis" (C. J. Ambrose, ed.), pp. 48-65. Little, Brown, Boston, Massachusetts.
Sherbet, G. V., Lakshmi, M. S., and Rao, K. V. (1972). *Exp. Cell Res.* **70,** 113.
Spector, D., Limas, C., Frost, J. L., Zachary, J. B., Sterioff, S., Williams, G. M., Rolley, R. T., and Sadler, J. H. (1976). *N. Engl. J. Med.* **295,** 1217.
Sterling, W. A., Pierce, J. C., Lee, H. M., Hume, D. M., Hutcher, N. E., and Mendez-Picon, G. (1972). *Surg. Gynecol. Obstet.* **135,** 98.
Sterling, W. A., Turner, M. E., Aldrete, J. S., Morgan, J. M., Shaw, J. F., and Diethelm, A. G. (1977). *Transplantation* **23,** 98.
Trump, B. F., and Bulger, R. E. (1967). *Lab. Invest.* **16,** 453.
Trump, B. F., and Bulger, R. E. (1968a). *Lab. Invest.* **18,** 721.
Trump, B. F., and Bulger, R. E. (1968b). *Lab. Invest.* **18,** 731.
Trump, B. F., and Bulger, R. E. (1969). *J. Ultrastruct. Res.* **28,** 301.
Trump, B. F., and Bulger, R. E. (1971). *Fed. Proc.* **30,** 22.
Trump, B. F., and Ginn, F. L. (1968). *Lab. Invest.* **18,** 341.
Trump, B. F., and Jones, R. T. (1977). *J. Exp. Zool.* **199,** 365.
Trump, B. F., Strum, J. M., and Bulger, R. E. (1974). *Virchows Arch. B. Zellpathol.* **16,** 1.
White, H. J. O., Evans, D. B., and Calne, R. Y. (1968). *Br. Med. J.* **4,** 739.

Chapter 15

Summary—Urinary Tract

BENJAMIN F. TRUMP

Department of Pathology and Maryland Institute for Emergency Medicine,
University of Maryland School of Medicine,
Baltimore, Maryland

The urinary tract is, of course, extremely important in a variety of neoplastic and non-neoplastic diseases and also from the standpoint of normal cell structure and function. Currently we know relatively little about long-term maintenance of any part of this system in organ or cell culture. Additionally, there are some special problems of this system, particularly in the kidney. The kidney is a solid organ with epithelial cells lining the tubules which is, therefore, quite different from other epithelial organs such as the bladder, the bronchus, or the espohagus, where free-flattened epithelium bordering the lumens is present. The kidney, along with the liver, therefore, presents special problems in establishing both explant and cell cultures. There are many possible applications to research in this area including: carcinogenesis, acute renal failure, congenital disorders, mechanisms of environmental toxicity, studies of antigen–antibody and/or complement attack on glomeruli, and the function of the juxtaglomerular apparatus in relation to the renin angiotensin system and prostaglandin metabolism. Furthermore, there are some unique problems of carcinogenesis in the kidney where environmental agents such as nitrotriacetic acid and saccharin are associated with carcinogenesis in animals and could be explored in the human. Some of these agents may be acting other than as mutagenic initiators by unknown mechanisms which may include promotion and/or chronic irritation.

Dr. Knowles and co-workers (Chapter 12) present their detailed studies of explant culture of human bladder. They have successfully maintained cultures for up to 35 days. In our laboratory, we have recently successfully cultured human bladder explants with maintenance of viability for over 2.5 months. The bladder thus seems to be imminently suitable at the present time for explant culture and this system is now ready for extensive studies on carcinogenesis, toxicity, tumor

ISBN 0-12-564140-0

promotion, or other problems. One of the problems pointed out by Knowles and co-workers is extremely important in the bladder as well as in all other epithelia. In bladder specimens obtained from patients with neoplastic disease, there is the likelihood of the presence of initiated cells in wide-spread scattered regions. This makes the interpretation of carcinogenic effects very difficult. We, therefore, in our laboratory are utilizing only immediate autopsy specimens from young individuals who do not have neoplastic disease and who predominantly have died from irreversible head injury following acute automobile accidents. Another controversial area in the bladder, commented on by Knowles *et al.*, is the possible significance of microvillous change in the cell surface epithelium as a preneoplastic marker. Such villi do develop *in vitro* after previous treatment *in vivo* with bladder carcinogens and it has been suggested by other workers studying human material that they represent a type of preneoplastic change. It should be pointed out, however, that recent studies in our own laboratory indicate that such villi readily occur in totally normal bladders in culture, probably the result of cell division and, therefore, such markers evidently do not represent markers of cells committed to neoplasia. Many other studies, including possible ion and water transport, cell turnover, bladder calculus formation, and effects of infectious agents, can now begin utilizing these epithelia.

Leighton and co-workers (Chapter 13) present new concepts of histophysiologic gradient culture of stratified epithelium, including chick embryo skin and normal rat bladder. These techniques can now be easily utilized in human tissues. This novel method, which is a significant departure from usual explant culture techniques, does provide a system in which the role of diffusion gradients in tissue responses can be examined. Such gradients, Leighton asserts, may have marked significance in differentiation, especially in multilayer epithelia, where a gradient of phenotypic expression occurs from the side nearest the blood vessel to the side nearest the environment or lumen. The regulation of cell turnover, as well as the patterns of differentiation, may be largely affected by the gradient. Obviously, such considerations could also be extremely important in the modeling and differentiation that occur during organogenesis in fetal development. Little is known about confrontation phenomena where normal and tumor cells meet. Biologic rules of such encounters should be examined. Recent work, for example, in our laboratory by Iseri (*Human Pathology,* 11:66–69, 1980), indicate that various types of differentiated specialized cell junctions can develop, at least *in vivo,* between epithelial tumor cells in metastasis and cells of the organ during the metastasis. The use of this gradient system in studying invasion metastasis and other elements of tumor progression might result in significant contributions. Little is presently known about progression from so-called carcinoma *in situ* to invasive metastasizing neoplasms. However, this does imply that the neoplastic cells change the normal response to the histophysiologic gradient.

The work from our laboratory on cell and explant culture of tubular epithelium summarizes the progress to date. The kidney lags far behind all the organ systems in terms of *in vitro* explant or cell culture. We have, however, recently described relatively simple methods for isolation and purification of proximal tubule epithelial cells from either rat or human kidneys. These have been maintained for more than a week in cell culture and show differentiation properties and polarities resembling those seen in the tubule. We are now experimenting with explant cultures of the kidney and are greatly encouraged by our extensive experience with long-term maintenance of fish kidney explant cultures, which is possible for up to several months. The kidney epithelium, especially that of the proximal tubule, has pronounced capabilities for division, regeneration, and differentiation and also is an important target organ for carcinogens in both animals and humans. In the kidney, with its different nephron segments, it is difficult to study metabolic, biochemical, and other properties in homogenates of whole kidneys because of the great diversity of function between different specialized nephron regions. Explant and/or cell culture, therefore, may well provide excellent systems with which we can further substantially improve our knowledge of normal kidney structure and function. Dr. Striker and his group have preliminary data on maintenance of isolated glomeruli *in vitro*. This is a technique which has hardly been explored at all but which is theoretically possible with both human and experimental animals. Such glomerular cultures could provide the opportunity to learn much more about maintenance of the filtration barrier, role of podocyte shape in its function, and the effects of antigen–antibody or complement interactions on cells of the glomerular capillary wall. Related studies could develop where cultures of the juxtaglomerular region, including the afferent, arterial, and macula densa, would yield considerable new information on the synthesis, turnover, and function of the renin angiotensin system. The renal physiologists have made great progress by dissecting isolated segments of nephrons which are maintained and profused for relatively short periods of time permitting studies of ion flux, transport, and other properties. Extension of these studies, using methods now available, could result in long-term maintenance or establishment of cell lines from particular regions of the nephron.

METHODS IN CELL BIOLOGY, VOLUME 21B

Chapter 16

Human Esophageal Organ Culture Studies[1]

E. A. HILLMAN AND M. J. VOCCI

Department of Pathology,
University of Maryland School of Medicine,
Baltimore, Maryland

W. SCHÜRCH

Department of Pathology,
Hotel-Dieu Hospital,
Montreal, Canada

C. C. HARRIS

Human Tissues Studies Section,
National Cancer Institute,
Bethesda, Maryland

B. F. TRUMP

Department of Pathology and Maryland Institute for Emergency Medical Services,
University of Maryland,
Baltimore, Maryland

---

[1]This is contribution #796 from the Cellular Pathobiology Laboratory, University of Maryland School of Medicine, Baltimore, Maryland. This work was supported in part by NCI Contract No. NO1-CP-75909.

ISBN 0-12-564140-0

I. Introduction

At the University of Maryland we are interested in studying the organ culture conditions necessary for maintaining several different types of human epithelia. With the development of the Immediate Autopsy Program at the University of Maryland, we have been in a unique position to obtain a variety of human tissues since autopsy was performed within minutes of death. The excellent quality of ultrastructural studies that can be obtained with this technique has been previously documented (Trump *et al.*, 1973). In addition, other sources of human tissues have been obtained at the time of surgery from either neoplastic or non-neoplastic disease. Therefore, we embarked on a detailed morphological, culture, and transplant study of the conditions necessary for maintaining normal human esophageal epithelium in long-term organ culture. The overall goal of this project was to develop a model to study carcinogenesis *in vitro*.

A review of the literature on conditions for maintaining normal esophageal epithelium using the organ culture technique reveals very few previous investigations. Rosztoczy *et al.* (1975) maintained normal human fetal and adult esophageal epithelium for up to 7 days in organ culture. Organ culture of human fetal esophageal epithelium supported influenza virus replication for 4 days. The viability and the morphology of these cultures were not discussed. Similarly there have been very few reports concerning organ cultures of esophageal mucosa derived from animals. Aydelotte (1963) maintained embryonic chicken esophageal epithelium for up to 13 days. The epithelium was dissected from 13-day-old chick embryos, cut into slices, and supported on a rayon net and cultured in sealed watch glasses. Histological examination of the cultures was performed. Lasnitzski (1963) demonstrated that embryonic rat esophageal epithelium could be maintained for 14 days in explant culture. The esophageal epithelium was grown as tubular explants of tissue on strips of fabric placed on clots. During the incubation period, the explants were kept in a chamber which was perfused daily for 30 minutes with a mixture of 95% O_2 and 5% O_2. More recently, there was one report of a short-term organ culture of the adult mouse esophagus in chemically defined medium (Stenn and Stenn, 1976). Using light microscopy and autoradiographic criteria, it was shown that these cultures remained viable for 3 days. Thus, the complete range of conditions for maintaining the esophagus in long-term explant culture has yet to be fully explored. Therefore, we set about defining the culture conditions for maintaining normal human esophageal epithelium in organ culture. In brief, our results represent the first successful attempt to maintain normal human esophagus epithelium in a defined organ culture system for an extended period of time.

II. Materials and Methods

A. Source of Normal Tissue

Normal human esophageal mucosa was obtained primarily from immediate autopsy specimens or from surgical resections for carcinoma. In the case of malignancy, particular attention was paid to obtain the samples sufficiently removed from the tumor in order to obtain mucosal explants which appeared normal on gross examination. Immediately after removal, the esophageal mucosa was dissected from the underlying muscularis using aseptic conditions. In this way, 6 to 18 explants of 1 cm^2 consisting of the epithelium with 1 to 2 mm stroma were obtained from each case. The explants were taken from the proximal, middle, and distal third of the esophagus.

B. Culture Conditions

Explants were placed in cold (4°C) L-15 medium prior to processing in the culture laboratory. Explants with the epithelium facing the gas phase were submerged in 60-mm plastic petri dishes (Falcon Plastics, Oxnard, CA) containing with 3 to 5 ml culture medium (Hillman *et al.*, 1979). The culture medium, CMRL 1066 was supplemented with insulin, hydrocortisone, and 5% fetal calf serum. Culture dishes were placed in an environmentally controlled chamber on a rocker platform and rocked at 5 to 10 cycles/minute causing medium to flow intermittently over the epithelial surface. The chambers were kept at 36.5°C, in an environment of 45% O_2, 50% N_2, and 5% CO_2. The culture medium was replaced three times a week.

C. Morphology

Morphologic studies were done on each case at zerotime, before culture, and from the cultured explants at regular intervals. In the early experiments esophageal cultures were sampled twice per week. As experience was gained with this system, samples were taken weekly and finally every 2 weeks.

Zerotime samples and cultured explants were fixed in 4F-IG (McDowell and Trump, 1976) and were processed for light microscopy, transmission (TEM), and scanning electron microscopy (SEM). For TEM, pieces of tissue were embedded with particular attention to their orientation in an attempt to obtain sections of the epithelium perpendicular to the basal lamina. In every instance all three techniques were performed on sequential sections taken from the same specimen.

III. Results

A. Morphology of Esophageal Mucosa at Zerotime

The light and electron microscopic (TEM and SEM) characteristics of the human esophagus *in vivo* have been described previously by others (Weinstein, *et al.*, 1975; Hopwood *et al.*, 1978; Ackerman *et al.*, 1976). Thus, the morphological characteristics of zerotime samples which were routinely examined will be presented briefly to serve as controls for the organ culture studies.

In the human esophagus, the stratified squamous epithelium was firmly anchored to the underlying lamina propria. The muscularis externa of the esophagus at the upper segment was composed primarily of striated muscle, the midsection of striated and smooth muscle, and the distal portion primarily of smooth muscle. Examination of tissue obtained from an immediate autopsy specimen taken from the distal segment of the esophagus prior to culturing (Fig. 1) demonstrates the excellent viability of tissues obtained in this manner. Likewise, tissue obtained from surgical specimens was still quite viable even after the extended period of ischemia caused by prolonged surgical procedures (6–8 hours). On the basal surface of the distal segment (Fig. 2), the cuboidal epithelial cells contained prominent, large nucleoli. There was a gradual gradation in the cell shape to a more elongate and slender morphology. Ultrastructural examination of the epithelial cells taken from the midportion of the esophageal mucosa (Fig. 3) revealed that cell cytoplasm contained unattached ribosomes and a few mitochondria and was quite electron lucent. In the more superficial layers

FIG. 1. Survey light micrograph of zerotime human esophagus showing typical stratified squamous epithelium. H+E. Bar = 30 μm.

FIG. 2. On the basal surface, the cuboidal epithelial cells rest on dense connective tissue. The nuclei contain large, active nucleoli which are multiple in a few cells. Zerotime sample. T+P. Bar = 25 μm.

FIG. 3. Esophageal epithelial cells appear quite electron lucent and are joined by prominent desmosomes. Zerotime sample, TEM. Bar = 2 μm.

FIG. 4. Epithelial cells are flat and have a uniform polygonal shape. Prominent intercellular microridge separate the cells. Zerotime, SEM. Bar = 2.5 μm.

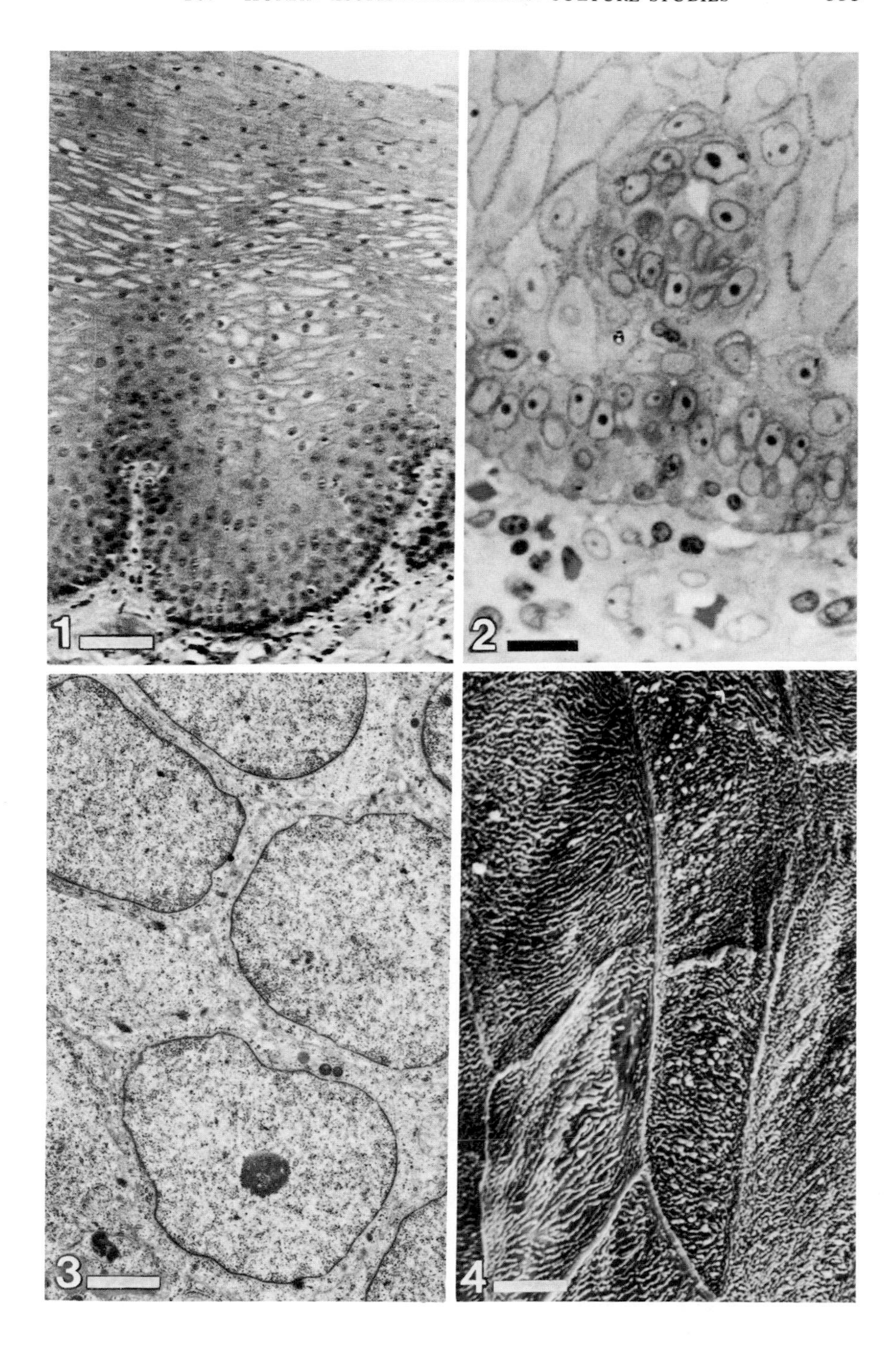
1
2
3
4

of the mucosa, the cell nuclei were elongated and oriented parallel to the surface. The SEM appearance of the surface of the normal esophageal mucosa (Fig. 4) was characterized by flat, rather uniform polygonal epithelial cells. Each cell was bordered by prominent intercellular ridges or terminal bars, most likely corresponding to juxtapositions of the surface microridges at the margins of adjacent cells. The surface of individual mucosa cells was convoluted by numerous densely packed microridges which frequently formed whorls. Occasionally, very short microvilli seemed to emerge from the cellular microridges and, infrequently, there were also short microvilli interspersed between the microridges. The cellular shedding seen in Fig. 5 was present in all the samples of the esophageal mucosa examined before culture.

B. Organ Culture Studies

To date, we have cultured 18 cases, 11 from immediate autopsy and 7 from surgical specimens. The time in culture has ranged from 5 weeks to 6 months; however, most specimens were cultured 6–16 weeks. Seven of 18 esophagus cultures were lost after 6–9 weeks due to microbial contamination.

During the first week in culture the multilayered, stratified, and noncornified epithelium of the esophagus showed a rapid desquamation of the superficial layers. This increased shedding of epithelial cells could easily be observed by SEM (Fig. 5), and in the process of shedding, cells were only partially attached to the underlying epithelial cells. The overall surface ultrastructure was somewhat irregular with clefts between the desquamating cells, and the detaching cells showed a flat polygonal surface which was convoluted by numerous microridges. In areas where the epithelial cells were completely attached to the underlying cells, prominent intercellular ridges were still present. The surface ultrastructure of these individual epithelial cells showed occasionally a mixture of microridges and microvilli, the latter being less numerous than the microridges.

After 7 days in culture the overall surface ultrastructure of the esophagus was relatively regular, and further shedding of epithelial cells was rarely observed. At this time in culture, the epithelium was three to five cells thick as seen with the light microscope. Mitotic figures were observed in the basal layer of epithelial cells, and, after 12 days, the epithelium became multilayered. The cells of the superficial and intermediate layers, which were oriented with their long axes parallel to the surface, were joined by desmosomes and rested on a prominent

---

FIG. 5. Scanning electron micrograph demonstrating desquamation of the surface epithelial cells. Seven-day culture, SEM. Bar = 4 μm.

FIG. 6. Multilayered squamous epithelium resting on a uniform basal lamina. The epithelial cell cytoplasm contains numerous free ribosomes, slender mitrochondria, and a few tonofilaments. Fifteen-day culture, TEM. Bar = 2.4 μm.

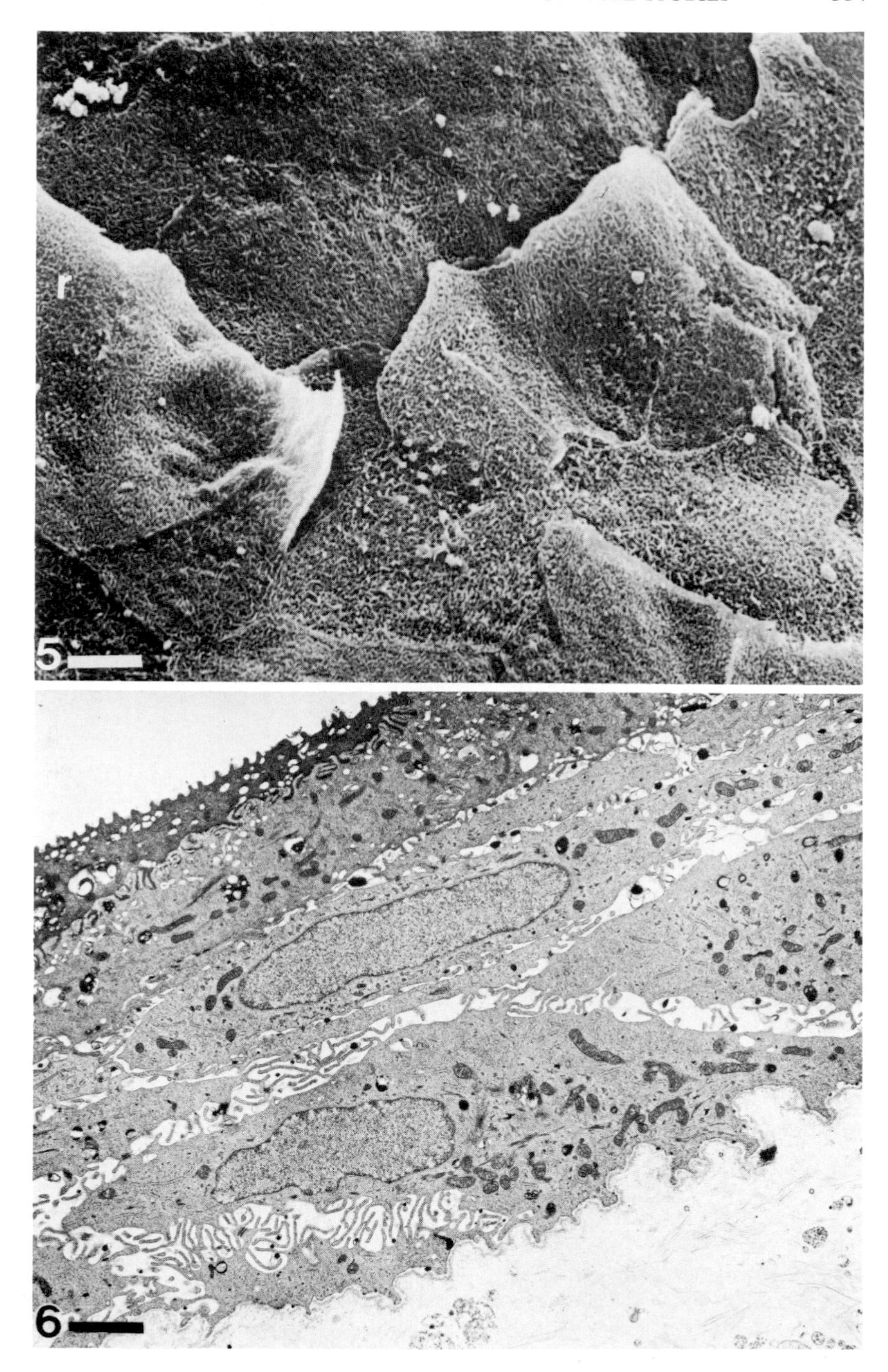
r
5
6

basal lamina (Fig. 6). At the ultrastructural level, the excellent viability of the squamous epithelium was observed. Only minor degenerative changes were visible with a few individual cells.

In most esophagus specimens cultured 2–3 weeks, a typical stratified squamous epithelium was reformed (Fig. 7) and rested on a viable stroma. The cell nuclei throughout the layers contained large active nucleoli. Occasionally, mitotic figures were observed in the basal cells. The ultrastructural characteristics of these basal cells after 71 days in culture are demonstrated in Fig. 8. Long, slender cell processes joined by junctional complexes were observed. The cells contained both abundant free and bound ribosomes and tonofilaments, and they rested on a prominent basal lamina. With continued time in culture (98 days), the basal epithelial cells migrated out and surrounded the entire explant (Fig. 9) where they often formed cyst-like cavities. At higher magnification, one observed that the basal cells had become more irregular in shape (Fig. 10). After 160 days in culture (Fig. 11), these long slender epithelial cells were joined by multiple desmosomes. On the lateral surface, long slender microvilli are present in the intercellular space. The epithelial cellular cytoplasm was rich in free ribosomes, tonofilaments, mitochondria, and rough endoplasmic reticulum. With continued time in culture, there was an increase in the numbers of tonofilaments present. In addition, on the superficial surface of the explant the cells had primarily short stubby microvilli. The basal lamina was no longer continuous; however, it was present.

In addition to the ultrastructural changes in esophageal epithelial cells noted, morphological changes detected by SEM were seen in explants maintained from 1 week to 6 months. From 1 week to 6 months, individual cells showed a progressively less uniform polygonal surface shape, and were interrupted by the presence of cells with an elongate form. The latter cells were more numerous at the edges of the explant where the cells appeared to be growing around the edge of the explant as in the 101-day culture (Fig. 12). Frequently there is a mixture of short microvilli and microridges in the same cell. A few cells showing only microridges were present up to 16 weeks. Infrequently some cells present bouquets of long slender microvilli in addition to the short microvilli (Fig. 13). In a few instances, cells with a flat smooth surface were observed. The intercellular ridges also were changing. They became progressively less prominent. In areas of cellular outgrowth, the cell borders were sometimes hard to identify, consisting of more densely packed microridges at the margins of individual cells.

FIG. 7. Reformation of typical stratified squamous epithelium. Twenty-seven-day culture, semithin section, T+P. Bar = 80 μm.

FIG. 8. Electron micrograph demonstrating the basal surface of the explant. Epithelial cells contain numerous tonofilaments and rest on a continuous basal lamina. Seventy-one-day culture, semithin section, T+P. Bar = 2 μm.

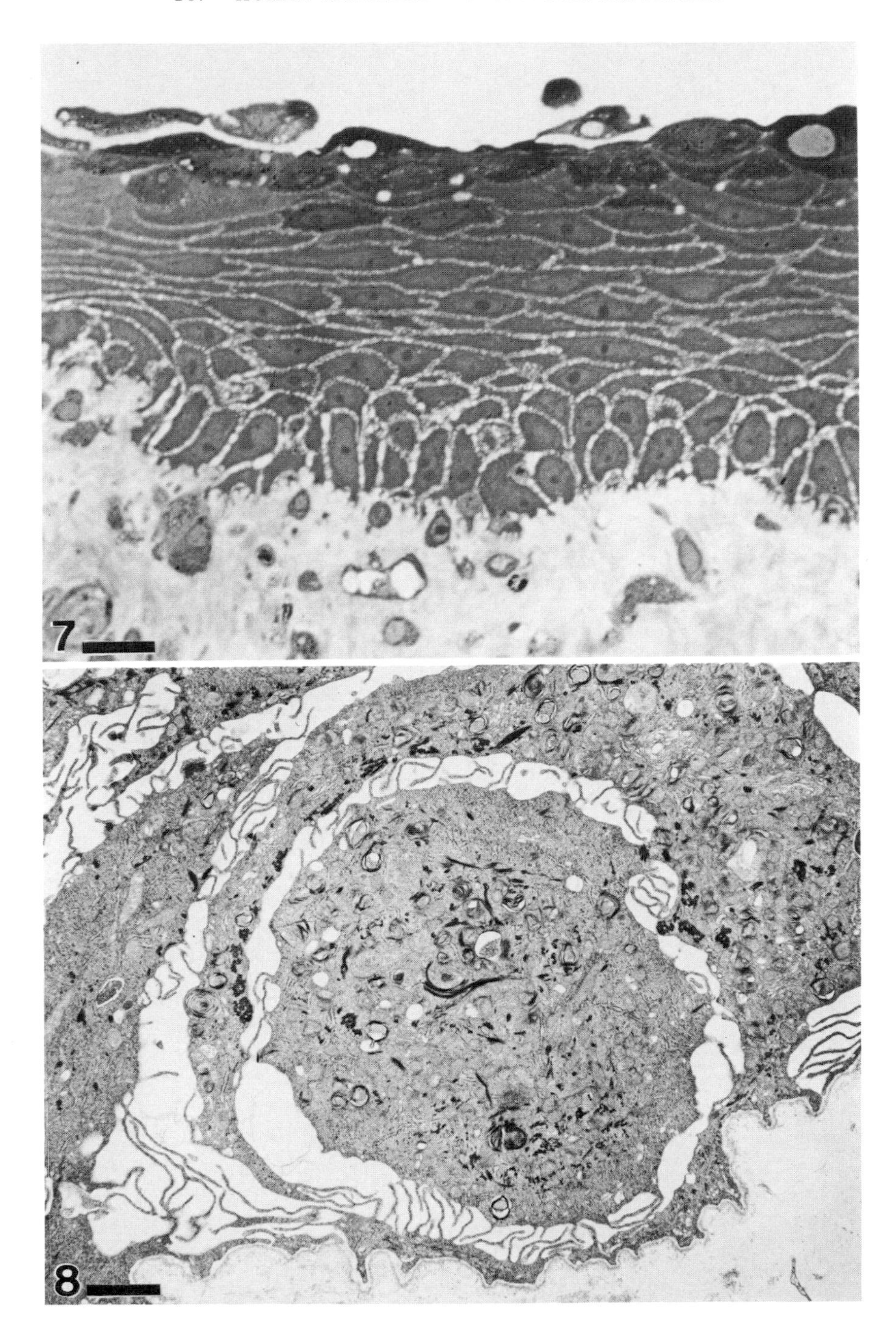
7
8

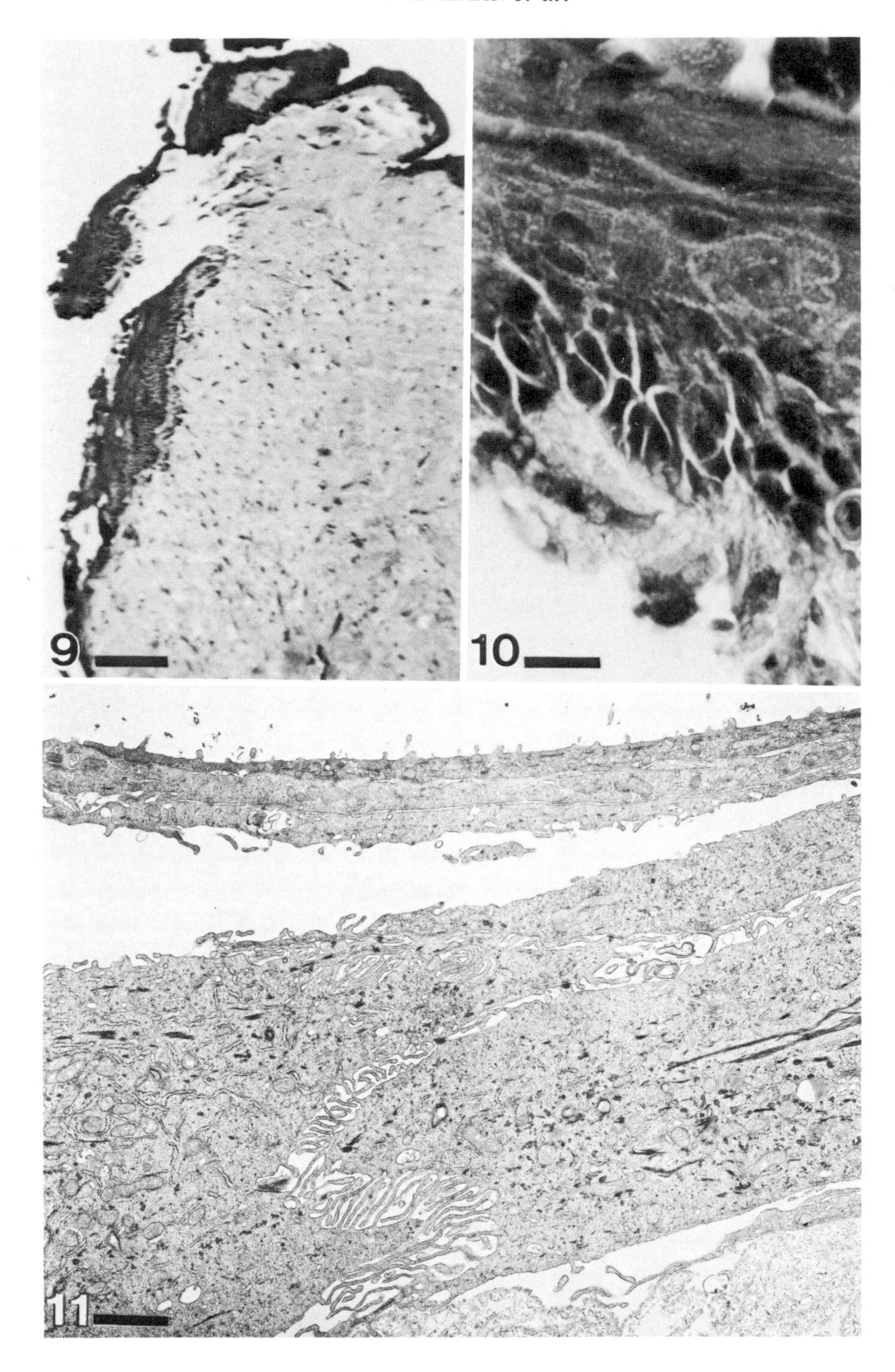
9
10
11

In conclusion, we have demonstrated that normal human esophageal epithelium can be grown in organ culture for several months, in some cases for 6 months. In this culture system, the squamous epithelial characteristics were retained as demonstrated by light and electron microscopy. These cells contained numerous tonofilaments and were joined by prominent desmosomes. Throughout the culture period, the basal lamina was retained. Also, the normal maturation of squamous epithelium continued in this organ culture system.

IV. Discussion

At the initiation of the study, the proximal, middle, and distal portions of the esophageal epithelium were cultured separately to determine if there were any regional differences in the viability or differentiation in our culture system. At this juncture, we have found no differences in the viability of any anatomical portion of the human esophagus in culture. In this regard, Harris *et al.* (1979) reported that the different anatomical segments of human esophagus show only a 2-fold variation in production of metabolites of benzo[*a*]pyrene and that the binding level of carinogens (such as dimethylnitrosamine and benzo[*a*]pyrene) to DNA and cellular macromolecules in the human esophagus *in vitro* were similar in magnitude among the three major anatomical segments. In summary, the three major anatomical segments of the human esophagus *in vitro* are similar in viability, differentiation, and metabolism.

Successful initiation of esophagus explant cultures was found to be dependent upon placing the epithelial mucosa uppermost when the tissue was placed in the petri dish. This assured that the epithelial tissue was bathed in culture media and permitted adequate diffusion of the nutrients to the cells. The proper orientation in the culture dish might also be important in maintaining the normal epithelial polarity. Another important factor in the successful maintenance of esophageal epithelium culture is the inclusion of insulin and hydrocortisone in the chemically defined medium. These two hormones used in our culture system have been successfully utilized singly and in combination in a variety of explant culture systems to maintain mammalian epithelia. Ceriani *et al.* (1972) demonstrated

---

FIG. 9. Survey light micrograph demonstrating epithelial outgrowth which surrounds the explant. Ninety-eight-day culture, H+E. Bar = 83.3 μm.

FIG. 10. With continued time in culture, the basal cells of the epithelium become more squamous and less cuboidal. Ninety-eight-day culture, H+E. Bar = 19 μm.

FIG. 11. On the uppermost surface the epithelial cell processes are extremely long and slender. The basal cells are much larger and contain numerous tonofilaments. One hundred and sixty-day culture, TEM. Bar = 16.6 μm.

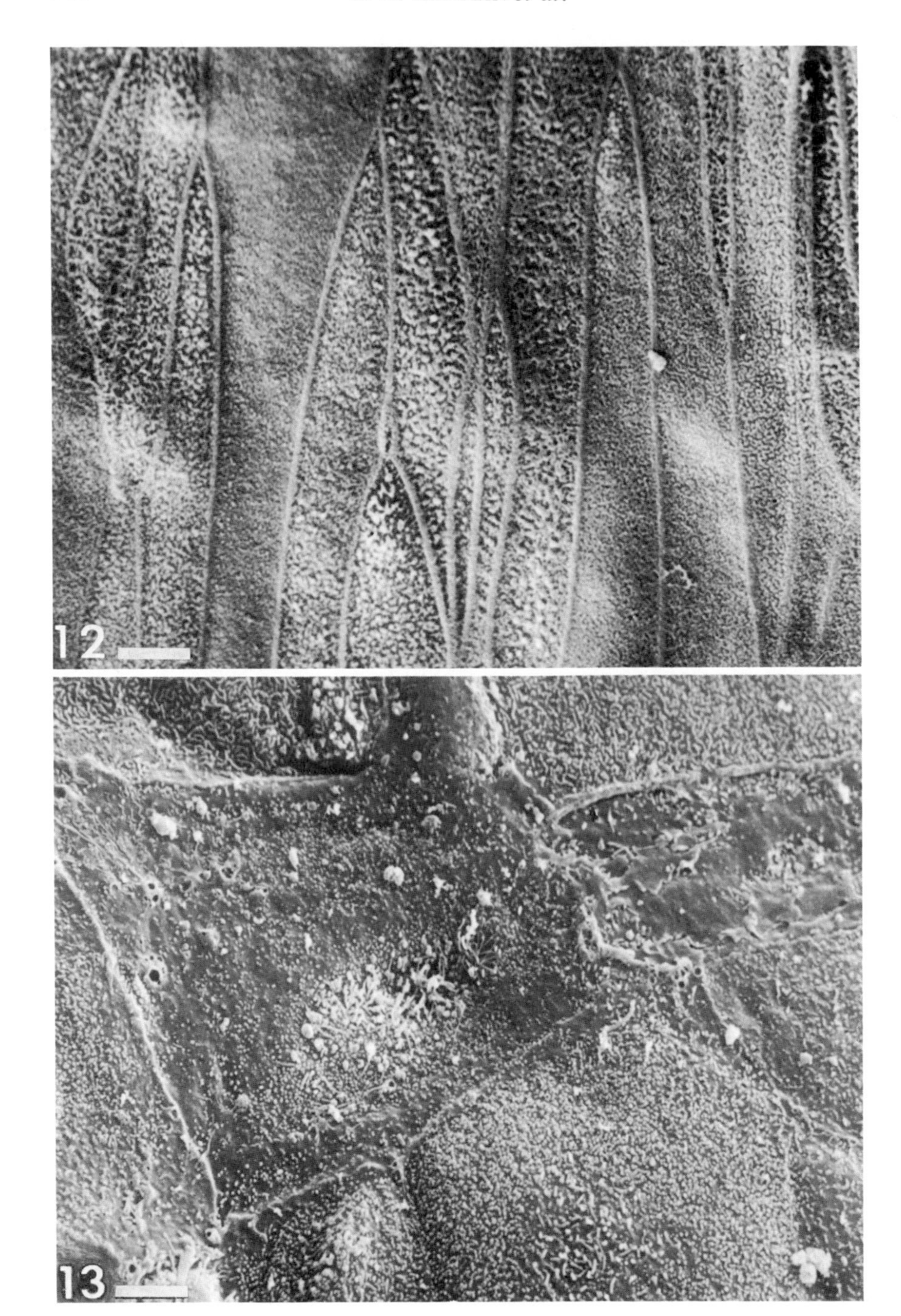
12
13

that insulin added to Waymouth's medium at a final concentration of 5.0 μg/ml was more effective than the control medium in maintaining viability of the human mammary gland in organ culture. Welsch *et al.* (1976), using the same final concentration of insulin in medium 199, documented a consistently increased incorporation of [^{3}H]thymidine into the DNA of mouse, rat, and human mammary slices. Reed and Grisham (1975) provided evidence that both insulin (5.0 μg/ml) and hydrocortisone (1.0 μg/ml) in combination were more effective than insulin or hydrocortisone singly in maintaining viability and glycogen stores in explant cultures of postnatal rat liver. These investigators measured glycogen synthesis and storage as deposition of [^{14}C]glucose and determined viability by both light and electron microscopic criteria. Insulin caused glycogen storage and synthesis, whereas hydrocortisone was ineffective in this regard. The results are noteworthy in that these investigators showed that these hormones in combination acted synergistically to enhance viability and glycogen synthesis as determined by histologic, ultrastructural, and biochemical observations. These results concur with those of Lucas *et al.* (1970), who demonstrated increased viability of rat salivary gland in explant culture using medium supplemented with insulin and hydrocortisone as compared to control media alone. Investigations of the organ culture of human bronchus (Barrett *et al.*, 1976), human pancreatic duct (Jones *et al.*, 1977), and human uterine endocervix (Schürch *et al.*, 1978) demonstrated the sufficiency but not the necessity of insulin and hydrocortisone to maintain viability of these human tissues in organ culture as determined by light and electron microscopy. These investigators used the same chemically defined medium (CMRL 1066) and concentrations of insulin (1.0 μg/ml) and hydrocortisone (0.1 μg/ml) that were employed in our culture system for the human esophagus explants.

To date we have maintained normal human esophagus in organ culture as a stratified nonkeratinizing squamous epithelium routinely for 4 and in some cases, for 7 months. During the first week in culture, the epithelial mucosa was much thinner than the starting zero time sample; however, later cultures became multilayered and had morphological characteristics of fresh explants. Stenn and Stenn (1976) likewise noted that mouse esophageal explant surface by 6 days was one to two cells in thickness; however, their studies lacked long-term culture observations. It would be interesting indeed to know if the mouse epithelial surface would have been regenerated to a typical multilayered squamous epithelium by 3 weeks as in the human esophageal cultures. After several weeks

---

FIG. 12. Long slender epithelial cells are distinctly separated by prominent intercellular ridges. One hundred and one-day culture, SEM. Bar = 8 μm.

FIG. 13. The types of microvilli present change with prolonged culture. Within the same epithelial cell, both long slender and short stubby microvilli are often observed. One hundred and thirty-day culture, SEM. Bar = 1.8 μm.

in culture the human epithelial cells migrated out and surrounded the explant surface. These extremely long and slender epithelial cells were primarily one to two layers thick. The epithelial nature of the outgrowth was verified by transmission electron microscopy. The maturation and differentiation of the cultured human esophageal epithelium observed in our laboratory were similar to the *in vivo* human epithelium as studied by Hopwood *et al.* (1978). In both our culture system and in the biopsies studied by Hopwood *et al.* (1978) the basal cells contained the most organelles, few tonofilaments, and were joined by numerous desmosomes while the superficial cells contained few organelles, numerous tonofilaments, and were joined by few desmosomes.

V. Perspectives

A. Current Applications and Findings

1. Carcinogenesis Metabolism Studies

It is important to develop *in vitro* models to help understand the detailed pathogenesis of the human lesion *in vivo*. In this respect, epidemiological studies have shown that both the smoking of tobacco and the drinking of alcoholic beverages are important risk factors in the etiology of esophageal cancer (Wynder and Stellman, 1977). In the former case, at least two major classes of carcinogens found in tobacco smoke, polycyclic aromatic hydrocarbons and nitrosamines, are suspected carcinogens in this target organ. Since polycyclic aromatic hydrocarbons and nitrosamines are the precursors of carcinogens (procarcinogens) and must be metabolically activated to carcinogenically active compounds (electrophiles) in order to bind to cellular macromolecules, it is important to study the metabolism of these chemical carcinogens by human esophagus cultures. Recently Harris *et al.* (1979) demonstrated that human esophageal explants, which had been cultured in a chemically defined medium for 7 days, incorporated [^{3}H]benzo[*a*]pyrene and [^{14}C]dimethylnitrosamine in both the mucosal protein and DNA of these explants. In addition, the major carcinogen–DNA adducts were identified.

2. Xenotransplantation Studies

A number of human tumors have been found to be transplantable into the nude mouse; however, the preservation of normal human tissues as a xenotransplant is not as well documented (Rygaard and Povlsen, 1977). Before studying the morphology and replication of normal human esophagus exposed to chemical carcinogens in an immunologically anergic recipient animal, it is important to document the behavior of normal esophageal explants under similar conditions. Zerotime human esophageal epithelium and control human esophageal organ cultures have been surgically implanted subcutaneously in nude mice. To date, baseline samples of normal human esophageal epithelium have remained viable for 8 weeks. Histologically, the implants have retained their typical stratified squamous characteristics. The cultured control explants after 2 days and 2 weeks culture period are currently being evaluated (Dr. M. Valerio, personal communication).

B. Future Studies

1. Carcinogenesis Studies

Some 60–90% of human cancers are believed to have an environmentally related etiology (Higginson and Muir, 1976) In addition carcinogens may act in conjunction with other exogenous stimuli such as promotors, hormones, nutrition, and life style which may influence neoplasia in the esophagus. Using morphologic and ultrastructural information on normal human esophagus, we hope to be able to follow the neoplastic process *in vitro* using suspected organotrophic chemical carcinogens (Druckery, 1972).

In addition, the long-term esophageal organ culture model would seem to be a good system for studying the phenomena of promotion (Berenblum, 1941; Chouroulinkov and Lasne, 1978). For example, recent studies by Weber and Hecker (1978) suggest that a tea made from croton flavens may be associated with the high incidence of esophageal cancer on the island of Curacao, Dutch Antilles. These investigators identified the three major active components of a methanol extract of the roots to be 12,13, disubstituted phorbol esters structurally similar to tetradecanoyl phorbol acetate. The phorbol esters were shown to be capable of tumor promotion in the standard mouse skin assay after "initiation" with a small dose 7,12-dimethylbenz[*a*]anthracene. Mirvish *et al.* (1979) more recently demonstrated that *Bidews pilosa* might be a promoting factor in the etiology of esophageal cancer in Transkei, South Africa. Rats when fed a mixture of the dried leaves of *Bidews pilosa* in a semisynthetic diet enhanced the [^{3}H]thymidine incorporation into esophageal epithelial DNA. Thus, the human esophageal organ culture may provide a worthwhile *in vitro* model for the study of tumor promotors in a "2-stage carcinogenesis" system utilizing human rather than animal tissues.

Our studies will concentrate on evaluating the results of chemical carcinogens and procarcinogens with both intermittent and continuous exposures. The treated tissues will be maintained in culture and will be examined periodically by light and electron microscopy and compared to untreated controls. To document the effect of organotropic carcinogens further, human esophagus, control, and treated tissue explants will be xenografted in nude mice. It is hoped that altered morphologic properties of human esophagus *in vitro* can be duplicated *in vivo*. To this end, nude mice will be implanted with control and treated tissues and these will be examined periodically. If palpable tumors are formed, they will be removed and analyzed histologically.

2. Cellular Kinetics Studies

Mitotic figures were observed in the basal layers of the human esophageal explants at that time, 2–3 weeks, in which the epithelium was being regenerated; however, autoradiographic analyses were not performed. Stenn and Stenn (1976) in their studies on mouse esophageal epithelium in organ culture demonstrated that the synthesis of DNA was limited to the basal layer. Leblond *et al.* (1964) and Greulich (1964) concluded that this was the true from *in vivo* labeling studies in mouse esophageal epithelium. In addition, Leblond *et al.* (1964) determined that all the basal cells in rodent esophagi are proliferative since they found 100% labeling of the basal cell layer nuclei after the chronic administration of [^{3}H]thymidine. Messier and Leblond (1960) defined the cellular migratory pattern in the rat and mouse esophagus as follows: the two labeled nuclei of the radioactive pairs seen after 8 hours in mice and 12 hours in rats were sometimes side by side in the basal layer, or sometimes one laying in the basal layer and the other above the basal, and at other times both labeled cells were seen in the prickle cell layer. By 3 days, heavily labeled cells in both rodent esophagi were seen in the prickle cell layer, while only a few grains persisted in the basal layer. These investigators interpreted these data to show that the parent and daughter cells in the basal layer may remain there, or that either the parent and/or daughter can migrate through the prickle cell layer.

Bell *et al.* (1967) studied the *in vivo* replication and cell cycle of the human esophagus of a patient with a treated pernicious anemia and a normal control. Four hours after an intravenous bolus of [^{3}H]thymidine, biopsies were taken from the esophagus, gastric mucosa, and jejunum. The [^{3}H]thymidine labeling index was found in the esophagus to be 8.7% for the basal layer and 0.95% for the entire mucosa; the mitotic index was 0.69% for the basal layer and 0.08% for the entire mucosa. The ratio of the labeling index to mitotic index was 12.6:1, indicating an approximate ratio for the duration of DNA synthesis and mitotic phases. Using the labeling index to approximate the DNA synthesis rate and mean cell cycle of time, Bell *et al.* estimated the S phase to be 10–15 hours in duration. With the assumption that all basal cells are proliferating, if the duration of the S phase is 10 hours, then the rate of proliferation is 8.7/10 or 0.87

cells/100 cells/hour. Using this figure as the proliferative rate, the duration of the mitotic phase was estimated as 0.69%/0.87 cell/hour = 0.79 hours.

What is unknown at this time is the relationship of the *in vivo* kinetic parameters and the yet to be determined *in vitro* kinetic parameter of the human esophageal explant in our organ culture system. Knowing that hormones such as insulin and hydrocortisone can increase thymidine incorporation into both rodent and human mammary gland slices (Welsch *et al.*, 1976), it is necessary to determine the effect of these hormones on the kinetic parameters of the human esophageal organ cultures. Specifically, experiments will be performed to determine the necessity and optimum concentrations of insulin and hydrocortisone for the human esophageal organ culture. The cultures will be evaluated serially for differentiation, maturation, and viability using morphologic criteria and for cellular proliferation using the [^{3}H]thymidine labeling index, the mitotic index, and fraction of label mitoses as pertinent indexes of the turnover rate *in vitro*.

3. Refinement of the Organ Culture System

We are currently testing several different media to determine whether we can further improve viability, maturation, and differentiation of the esophageal epithelium. One addition under consideration is vitamin A based upon its reported effects upon animal esophageal explant culture. Lasnitzsky (1963) reported that vitamin A modifies the differentiation of the embryonic rat esophageal epithelium in organ culture. At the end of 7 days culture with vitamin A, the rat esophagus remained multilayered but the cells were undifferentiated and become neither stratified nor keratinizing, while the control cultures show increased keratinization and proliferation of the cells of the intermediate layers with an increase in the precornified elements. By 14 days, the vitamin A-treated rat culture is still a multilayered, undifferentiated, nonkeratinizing epithelium covered by goblet cells while the control culture has thinned to two cell layers. Similarly, Aydelotte (1963) demonstrated that vitamin A and/or citral had antagonistic effects on the normal differentiation of chicken esophagus in organ culture. Treatment of human esophageal organ cultures with vitamin A might help maintain a multilayered, undifferential squamous epithelium which could be more susceptible to transformation by a carcinogen of suspected organotrophicity for the esophagus.

References

Ackerman, L., Piros, J., de Carle, D., and Christensen, J. (1976). *In* "Scanning Electron Microscopy" (O. Jahari and R. P. Becicer, eds.), Vol. II, pp. 247–252. Wiley, New York.

Aydelotte, M. B. (1963). *J. Embryol. Exp. Morphol.* **11,** 621.

Barrett, L. A., McDowell, E. M., Frank, A. L., Harris, C. C., and Trump, B. F. (1976). *Cancer* **36,** 1003.
Bell, B., Almy, R., and Lipkin, M. (1967). *J. Natl. Cancer Inst.* **38,** 615.
Berenblum, I. (1941). *Cancer Res.* **1,** 44.
Ceriani, R. L., Contesso, G. P., and Nataf, B. M. (1972). *Cancer Res.* **32,** 2190.
Chouroulinkov, I., and Lasne, C. (1978). *Bull. Cancer* **65,** 255.
Druckery, H. (1972). *In* "Topics in Chemical Carcinogenesis" (W. Nakahare, S. Takoyama, T. Sugimene and S. Odashima, eds.), pp. 73–77. Univ. Park Press, Baltimore, Maryland.
Greulich, R. C. (1964). *In* "The Epidermis" (W. Montagne and W. C. Lobitz. eds.), pp. 117–133. Academic Press, New York.
Harris, C. C., Autrup, H., Stoner, G. D., Trump, B. F., Hillman, E. A., Schafer, P. W., and Jeffrey, A. M. (1979). *Cancer Res.* **39,** 4401.
Higginson, J., and Muir, C. S. (1976). *Cancer Detect. Prevent.* **1,** 79.
Hillman, E. A., Schürch, W., Vocci, M. J., and Trump, B. F. (1979). *Cancer Res. Proc.* AACR Abstr. **20,** 848.
Hopwood, D., Logan, K. R., and Bouchier, I. A. D. (1978). *Virchows Arch. B. Cell Pathol.* **26,** 345.
Jones, R. T., Barrett, L. A., van Haaften, C., Harris, C. C., and Trump, B. F. (1977). *J. Natl. Cancer Inst.* **58,** 557.
Lasnitzsky, I. (1963). *J. Exp. Med.* **118,** 1.
Leblond, C. P., Greulich, R. C., and Pereira, J. P. M. (1964). *In* "Advances in Biology of Skin" (W. Montagna and R. E. Billingham, eds.), Vol. V, pp. 39–67. Macmillan, New York.
Lucas, D. R., Peakman, E. M., and Smith, C. (1970). *Exp. Cell Res.* **60,** 262.
McDowell, E. M., and Trump, B. F. (1976). *Arch. Pathol. Lab. Med.* **100,** 405.
Messier, B., and Leblond, C. P. (1960). *Am. J. Anat.* **106,** 247.
Mirvish, S. S., Rose, E. F., and Sutherland, D. M. (1979). *Cancer Lett.* **6,** 159.
Reed, G. B., and Grisham, J. W. (1975). *Lab. Invest.* **33,** 298.
Rosztoczy, I., Sweet, C., Toms, G. L., and Smith, H. (1975). *Br. J. Exp. Pathol.* **56,** 322.
Rygaard, J., and Povlsen, C. O., eds. (1977). "Bibliography of the Nude Mouse, 1966–1976." Fisher Verlag, Jena and New York.
Schürch, W., McDowell, E. M., and Trump, B. F. (1978). *Cancer Res.* **38,** 3723.
Stenn, K. S., and Stenn, J. O. (1976). *J. Invest. Dermatol.* **66,** 302.
Trump, B. F., Valigorsky, J. M., Dees, J. H., Mergner, W. J., Kim, K. M., Jones, R. T., Pendergrass, R. E., Garbus, J., and Cowley, R. A. (1973). *Human Pathol.* **4,** 89.
Weber, J., and Hecker, E. (1978). *Experimentia* **34,** 679.
Weinstein, W. M., Bogoch, E. R., and Bowes, K. L. (1975). *Gastroenterology* **68,** 40.
Welsch, C. W., de Ituri, G. C., and Brennan, M. J. (1976). *Cancer* **38,** 1271.
Wynder, E. L., and Stellman, S. D. (1977). *Cancer Res.* **37,** 4608.

Chapter 17

Organ Culture of Gastric Mucosa: Advantages and Limitations

ROBERT M. DONALDSON, JR. AND CYRUS R. KAPADIA

Department of Internal Medicine,
Yale University School of Medicine,
New Haven, Connecticut

I. Introduction

In addition to serving as a reservoir for ingested nutrients and as a means of modulating delivery of these nutrients to the small intestine, the stomach secretes several biologically important substances including hydrochloric acid, pepsinogen, gastrin, intrinsic factor, immunoglobulins, and mucus. For more than a century, investigators have employed a remarkably diverse array of experimental approaches designed to elucidate the mechanisms involved in gastric secretion, particularly gastric secretion of acid (Emas *et al.*, 1967).

In recent years *in vitro* techniques have been increasingly applied in efforts to

ISBN 0-12-564140-0

examine gastric mucosal function under more rigidly controlled conditions. The earliest *in vitro* preparations consisted of everted sacs prepared from frog or mouse stomach in a way that allowed the mucosa to come into direct contact with the nutrients and oxygen in the incubation medium. Although never widely employed, these everted gastric sacs were actually in use for several years before the much more popular technique employing everted sacs of small bowel was described. Another *in vitro* preparation which has been used to a limited extent is the excised frog or mouse stomach immersed in an incubation medium within a hyperbaric chamber. In this preparation adequate oxygenation is achieved by the development of high oxygen tensions. To date the most widely used *in vitro* technique involves strips of amphibian gastric mucosa mounted in lucite chambers (Rehm, 1962). This technique permits the luminal and serosal surfaces of the gastric mucosa to be bathed separately so that ionic movements and potential differences can be directly measured.

Investigators have also attempted to scrutinize gastric function more precisely at the cellular level by studying suspensions of isolated gastric mucosal cells (Soll, 1978). This approach aims at the development of methods for making relatively pure preparations of parietal cells or chief cells. The responses of these isolated cell preparations to known stimulants and inhibitors of gastric secretion are currently being examined in some detail, and in many instances the observed responses have been correlated with laterations in cellular cyclic nucleotides.

Although these various *in vitro* approaches have yielded information concerning some aspects of gastric mucosal function which cannot be obtained *in vivo*, the *in vitro* techniques developed thus far have also had distinct limitations. In general these techniques provide some indication of the immediate, but not delayed, responses of isolated gastric mucosa because it has not been possible to maintain steady-state conditions for prolonged periods of time when these various *in vitro* techniques are employed. Thus, standard *in vitro* approaches can be used to examine ionic movements, macromolecular secretion, or other events which occur rapidly, but the limited tissue viability imposed by these approaches prevents study of more complex phenomena which occur over longer periods of time. Although suspensions of isolated gastric mucosal cells retain metabolic activity and reasonable morphologic integrity for periods of up to 2 or 3 hours, long-term maintenance of gastric epithelial cells by culture techniques used for other cell types has not yet been achieved. Tissue cultures have been developed from gastric mucosal cells (Miller *et al.*, 1973), but these cultured cells rapidly lose their epithelial cell characteristics.

Although organ culture techniques have been used to examine maturation of fetal stomach, it was not until the principles of organ culture were successfully applied to adult small bowel mucosa (Browning and Trier, 1969) that there seemed any likelihood that gastric mucosa from adult mammalian species could be maintained *in vitro* for more than an hour or two. Experience with organ culture of gastric mucosa is still limited but offers considerable promise for the future. The present chapter attempts to describe this early experience with organ culture of gastric tissue, to indicate ways in which the technique has been used to study gastric function, and to point out the kinds of investigations that are needed in the future.

II. Methods

A. Organ Culture of Fetal Stomach

Two general approaches have been employed to culture embryonic stomach. One approach that has been used often is described in detail by Philpott (1971). The entire stomach is removed from 8-day-old chick embryos. The serosa and some of the mesenchyme is separated, and the remainder of the muscular stomach (gizzard) is divided into two equal pieces. Each half is then placed, epithelial side up, on a gelatin sponge in a plastic petri dish. The gastric explants are then incubated in Eagle's minimal essential medium containing penicillin and streptomycin. The amount of medium added is just sufficient to barely cover the explant. The petri dishes containing the explants are maintained at 37°C in a water-saturated incubator in an atmosphere of 5% CO_2–95% O_2.

A somewhat different approach to organ culture of fetal tissue is exemplified in a report by Yeomans *et al.* (1976) who examined differentiation of fetal rat stomach *in vitro*. At 18 days of gestation, fetuses were removed from lightly anesthetized rats. Fetal stomachs were rapidly removed and divided into thirds under the dissecting microscope. The proximal third was discarded. The middle third was designated gastric "fundus," and the distal third was presumed to constitute "antrum." Subsequent histological examination indicated that this presumptive division was usually accurate. The "antral" and "fundic" sections of fetal stomach were then cut into full thickness explants less than 2 mm in size. As previously described by Hoorn (1966), these explants were oriented mucosal side up in plastic culture dishes previously scored with a scalpel blade to allow attachment of the explant. The dishes were then cultured for 3 days in room air at 37°C on a rocker platform that allowed culture medium to wash over the explants two times each minute. The culture medium used in these studies was Leibovitz L-15 medium (Leibovitz, 1963) supplemented with 9.2% bovine serum albumin, glutamine, penicillin, and streptomycin.

B. Organ Culture of Adult Fundic Mucosa

The first successful organ culture of adult fundic (oxyntic) gastric mucosa was achieved in rabbits (Sutton and Donaldson, 1975) by applying the principles that had previously been used (Browning and Trier, 1969) to culture small bowel mucosa. Stomachs were excised from lightly anesthetized 2- to 5-kg white New Zealand rabbits and were thoroughly washed with cold isotonic saline. Multiple mucosal biopsies were then rapidly obtained from the body of the stomach near the greater curvature with a multipurpose suction biopsy tube (Quinton Instrument Co., Seattle, WA). These biopsies were mounted, mucosal surface up, on stainless-steel screens and with a razor blade were transected into two or more explants each of which was less than 2 mm in all dimensions. The steel screens containing mounted biopsies were then floated on 1 ml of culture medium in the center well of plastic organ culture dishes (Falcon Plastics, Los Angeles, CA). The loosely covered culture dishes were then incubated at 37°C in sealed McIntosh jars (Scientific Products, Inc., Los Angeles, CA) in 95% O_2–5% CO_2 at atmospheric pressure.

Biopsies of oxyntic mucosa have also been obtained from human subjects and cultured exactly as described in the preceding paragraph. In the case of human subjects, however, biopsies were obtained with a multipurpose suction biopsy tube that was passed into the stomach under fluoroscopic control. To minimize the risk of hemorrhage the number of biopsies taken at any one time never exceeded five. For both human and rabbit mucosal biopsies, the time required to initiate organ culture was kept to a minimum. Under no circumstances were more than 20 minutes allowed to elapse between excision of rabbit stomach or collection of human biopsies and the beginning of culture in a sealed, oxygenated McIntosh jar. The jars were replenished with 95% O_2–5% CO_2 every 12 hours.

For most of our studies the culture medium has consisted of Trowell T8 medium (Trowell, 1959) containing 10% fetal calf serum to which has been added penicillin (100 units/ml) and streptomycin (100 μg/ml). For certain experiments we have cultured oxyntic mucosa over NCTC medium (Evans *et al.*, 1964) from which either methionine or leucine has been removed. In these experiments the fetal calf serum (10%) was dialyzed before being added to the synthetic medium. In other experiments we cultured rabbit fundic mucosal biopsies directly on a solid medium rather than use a wire screen floating on liquid medium. To accomplish this, 0.25% agarose in Trowell T8 medium was heated to 90°C. The medium was then cooled to 55°C and 10% fetal calf serum was added together with penicillin and streptomycin as described. The gel was then allowed to "set," and biopsies were cultured directly on this semi-solid medium.

C. Organ Culture of Adult Antral Mucosa

Harty *et al.* (1977) and Lichtenberger *et al.* (1978) have reported successful culture of mucosal explants obtained from the gastric antrum of rats. The techniques used by both groups of investigators were based on the same principles outlined for oxyntic mucosa but differed in some methodologic details. A small ring of antral mucosa was excised from rats sacrificed by cervical dislocation or transection. Antral mucosa was separated from the muscle layer by sharp dissection under a dissecting microscope. Explants measuring 1–2 mm were prepared and immediately transferred to an organ culture system similar to that described in Section II,B. The culture media used for maintaining rat antral mucosa have consisted of either Trowell T8 medium with 10% heat-inactivated fetal calf serum (Lichtenberger *et al.*, 1978) or 80% Trowell T8, 10% NCTC medium, and 10% dialyzed fetal calf serum (Harty *et al.*, 1977). Suction biopsies obtained from human gastric antrum have also been maintained in organ culture (Trier, 1976) using precisely the same methods described for human small bowel (Browning and Trier, 1969).

III. Results

A. Organ Culture of Fetal Stomach

When embryonic chick stomach is maintained in organ culture for periods up to 44 hours, cell proliferation did proceed *in vitro* and in general the cells in these explants looked healthy histologically (Philpott, 1971). Addition to the culture medium of saline extracts of embryonic gastric epithelium, gastric mesenchyme, intestinal epithelium, intestinal mesenchyme, or skin produced no obvious morphological differences although gastric epithelial extracts suppressed cell proliferation in cultured explants. Addition to the medium of adrenalin or hydrocortisone had no effect on proliferative activity. The gastric epithelium of 8-day-old embryos shows no formation of gastric glands and this epithelium did not undergo detectable differentiation during the period of organ culture.

In contrast, the organ culture studies carried out by Yeomans *et al.* (1976) were conducted on fetal rat stomach obtained at 18 days, a time when differentiation was just beginning. Uncultured explants obtained at 18 days were morphologically undifferentiated and showed a stratified epithelium. After 3 days of organ culture, however, distinct maturation of architecture occurred *in vitro*. This development was observed in antral as well as fundic mucosas and was characterized by (*a*) epithelial invagination to form small glands and (*b*) the appearance of secretory granules in occasional epithelial cells. When cortisol was added to the culture medium in micromolar concentrations, differentiation after 3 days of organ culture was more advanced as documented by increased formation of glands and larger numbers of secretory granules. This maturation induced by cortisol was completely blocked by cytochalasin B.

B. Organ Culture of Adult Fundic Mucosa

When biopsies of rabbit gastric mucosa are maintained in organ culture for periods of up to 24 hours, the isolated gastric mucosa remains viable as determined by light microscopy and by quantitative measurements of gastric mucosal function (Sutton and Donaldson, 1975). After culture for 24 hours, the overall glandular architecture of the biopsies remains generally intact. Even at the light microscopic level, there is detectable damage to some of the mucus cells lining the gastric mucosal surface, and occasional small foci of necrosis are seen within some of the deeper gastric glands. Biopsies which are cultured longer than 24 hours distinctly show extensive areas in which parietal and chief cells are necrotic and in which glandular architecture is largely lost.

These morphologic observations correlate well with functional measurements. When rabbit fundic mucosal biopsies are continuously cultured over medium containing [^{14}C]leucine, incorporation of radioactivity into tissue protein is linear for 24 hours (Fig. 1). Similarly, the secretion of labeled protein into the culture medium is also linear with time after an initial lag phase of 3 to 6 hours. Moreover, freshly collected rabbit biopsies incorporate labeled leucine into protein just as efficiently as do biopsies which have been maintained in organ culture for 6, 9, or 21 hours (Table I).

Further evidence for steady maintenance of gastric mucosal function during 24 hours of organ culture is derived from studies of the secretion of pepsinogen

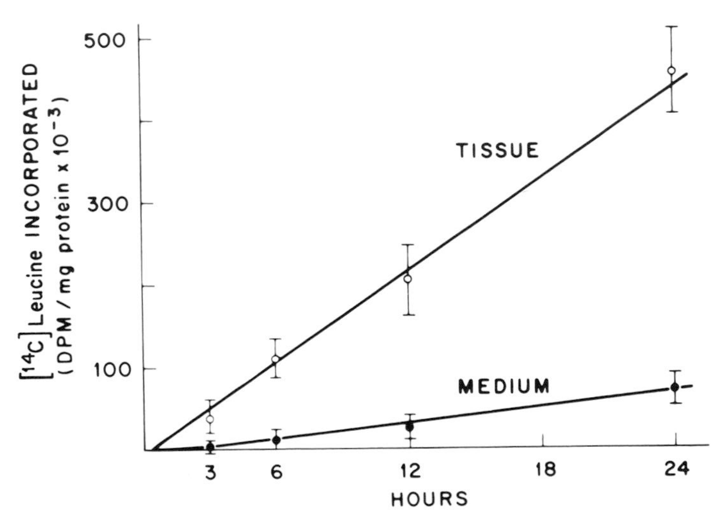

FIG. 1. Incorporation of [^{14}C]leucine into the tissue of cultured rabbit gastric mucosal biopsies and into protein recovered from culture medium. Each point represents the mean value for 11 to 30 biopsies ± SD. Reproduced from Sutton and Donaldson (1975) with the kind permission of Williams & Wilkins.

TABLE I

RATE OF INCORPORATION OF [^{14}C]LEUCINE INTO TISSUE PROTEIN DURING ORGAN CULTURE OF RABBIT GASTRIC MUCOSA[a]

Number of biopsies	Initial culture (hours)	Incorporation into tissue protein (dpm/mg protein $\times 10^{-3}$)
13	0	53.1 ± 9.2
5	6	54.8 ± 2.5
3	9	51.7 ± 2.6
4	21	54.9 ± 12.7

[a] Biopsies were initially cultured over unlabeled medium and were then cultured for 3 hours over medium containing [^{14}C]leucine. Results are expressed as mean ± SD.

(Sutton and Donaldson, 1975) and intrinsic factor (Kapadia and Donaldson, 1978). As shown in Fig. 2, total pepsinogen activity in gastric mucosal biopsy tissue remains stable during 24 hours of organ culture, while at the same time a linear increase in the amount of pepsinogen secreted into the culture medium occurs. The same phenomenon is observed with respect to intrinsic factor secretion.

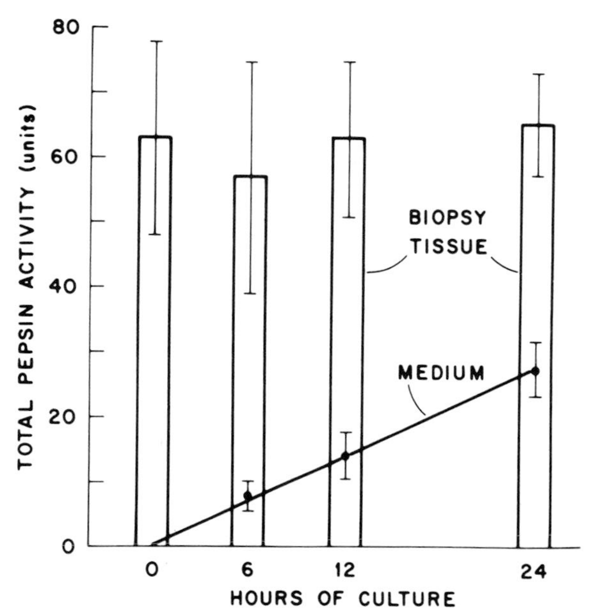

FIG. 2. Pepsin activity present in cultured rabbit gastric mucosal biopsies is shown by the bars while the solid circles indicate the amount of pepsin secreted into the culture medium. Values represent the average results obtained with 12 to 24 biopsies ± SD. Reproduced from Sutton and Donaldson (1975) with the kind permission of Williams & Wilkins.

Attempts to culture rabbit fundic biopsies for more than 24 hours have been uniformly unsuccessful, but such attempts have not as yet involved any substantive changes in the organ culture system that is used. After 24 hours of organ culture there regularly occurs, in addition to histological deterioration, a marked decrease in the incorporation of labeled amino acid into protein. Moreover, the rate of secretion of both pepsinogen and intrinsic factor declines by more than 50%. In addition, the total protein content of rabbit gastric mucosal biopsies, which remains remarkably stable during the first 24 hours of organ culture, also decreased by 25 to 45% after 24 hours.

When fetal calf serum is removed from the culture medium, viability of rabbit gastric mucosal biopsies is dramatically reduced. Steady-state incorporation of amino acid into protein or secretion of pepsinogen and intrinsic factor can usually be maintained for only 6 hours. When the fetal calf serum is dialyzed, however, mucosal viability is not impaired. Similarly, removal of a single amino acid such as leucine or methionine or removal of cobalamin from the culture medium has no effect on viability. On the other hand when the oxygen tension is increased above atmospheric pressure, we have observed an initial increase in pepsinogen secretion lasting 3 to 4 hours followed by a rapid decline in mucosal function.

The effects of different synthetic media on gastric mucosal integrity and viability have not been systematically compared. We have observed, however, that biopsies cultured for 24 hours over NCTC medium consistently produce more intrinsic factor (4.5 ± 0.3 ng units/mg biopsy protein) than do biopsies cultured over Trowell T8 medium (3.0 ± 0.4 ng units/mg biopsy protein).

As documented in Table II, rabbit fundic biopsies maintain mucosal function equally well whether cultured over standard liquid Trowell T8 medium or in direct contact with the same medium in semi-solid gel form (see Section II,B).

TABLE II

FUNCTION OF RABBIT GASTRIC MUCOSAL BIOPSIES IN ORGAN CULTURE OVER LIQUID MEDIUM OR ON AGAR GEL[a]

	Liquid medium	Agar gel
[^{14}C]Leucine incorporated (dpm/mg protein × 10^{-4})	6.4 ± 0.5	7.7 ± 0.9
Pepsinogen secretion (Mirsky units/mg protein)	23 ± 3	26 ± 3
Intrinsic factor secretion (binding units/mg protein)	217 ± 39	256 ± 44

[a] Biopsies were initially cultured for 18 hours by the standard method over liquid medium or in direct contact with a semi-solid agar gel medium. The biopsies were then cultured over liquid medium containing [^{14}C]leucine for 6 hours during which time pepsinogen and intrinsic factor secretion were measured. Values represent the mean ± SD for 12 biopsies.

TABLE III

ORGAN CULTURE OF HUMAN GASTRIC OXYNTIC MUCOSA[a]

Initial culture (hours)	$[^{14}C]$Leucine incorporated (dpm/mg protein × 10^{-4})	Pepsinogen (Mirsky units/mg protein)	Intrinsic factor (binding units/mg protein)
0	7.3 ± 1.1	29 ± 6	347 ± 62
6	7.8 ± 1.4	31 ± 7	376 ± 39
9	6.5 ± 2.1	23 ± 11	298 ± 51
12	3.1 ± 1.8	13 ± 6	154 ± 62
18	1.6 ± 1.1	4 ± 3	62 ± 31

[a] Biopsies were initially cultured over unlabeled medium and were then cultured for 6 hours over medium containing $[^{14}C]$leucine during which time pepsinogen and intrinsic factor secretion were measured. Values represent the mean ± SD for four biopsies.

After 18 hours of culture on semi-solid medium, rabbit gastric mucosal biopsies incorporated leucine into protein and secreted both pepsinogen and intrinsic factor to the same extent as did biopsies cultured for 18 hours under standard organ culture conditions.

When maintained in organ culture, biopsies of oxyntic mucosa obtained from human stomachs retain morphological integrity and functional stability for only 6 to 12 hours in contrast to the 24 hours of steady-state function observed with rabbit gastric mucosal biopsies. After either 6 or 12 hours of organ culture, human biopsies demonstrate cellular necrosis and loss of glandular architecture which is most marked in the deepest portions of the gastric glands. This histological deterioration is accompanied by a distinct decrease in the incorporation of leucine into protein as well as in the secretion of pepsinogen and intrinsic factor (Table III).

C. Organ Culture of Adult Antral Mucosa

After maintaining rat gastric antral explants in organ culture for 24 hours, Harty *et al.* (1977) observed that the explants were morphologically well preserved. By light microscopy the cellular elements of surface epithelium appeared normal, and the architecture of pyloric glands was not distorted. These workers did note scattered areas of focal degeneration and necrosis involving both surface cells and pyloric glands in "a small proportion of explants" examined at 24 hours, while these changes were "rarely present at earlier time points. A more detailed morphological assessment has been reported by Lichtenberger *et al.* (1978). These workers also noted that antral explants from rat stomach were histologically intact after 24 hours of organ culture. They noted that antral surface and glandular cells were well preserved and well differentiated and that specialized endocrine glands containing characteristic secretory granules could be seen along the basal region of antral glands.

Morphological integrity during 24 hours of organ culture was accompanied by steady state functional activity as determined by (*a*) incorporation of [^{3}H]tryptophan into antral explant protein (Harty *et al.*, 1977), (*b*) a steady rate of secretion of the antral hormone gastrin into the culture medium (Lichtenberger *et al.*, 1977), and (*c*) a linear increase in antral gastrin specific activity when explants were cultured over [^{3}H]tryptophan (Harty *et al.*, 1977).

Human antral biopsies maintained in organ culture, unlike human biopsies of oxyntic mucosa, remain histologically intact for 24 hours (Trier, 1977), but detailed morphologic and functional findings have not been reported.

IV. Discussion

Investigations of gastric mucosal function *in vitro* are severely limited by the fact that the mucosa remains viable for such short periods of time. There are no reports of long-term cultures of normal gastric epithelial cells by the usual tissue culture techniques which have been successfully employed for a wide variety of cells. Moreover, cells derived from gastric mucosal epithelium, when cultured by these techniques, rapidly dedifferentiate and lose their specialized epithelial characteristics (Miller *et al.*, 1973). When the principles of organ culture are applied to gastric explants, mucosal viability can be distinctly prolonged as determined by morphologic and functional criteria. Thus fetal stomach can be maintained in organ culture for 3 days (Lichtenberger *et al.*, 1978) and gastric mucosal explants from rabbits remain viable for periods up to 24 hours (Sutton and Donaldson, 1975).

Although organ culture substantially lengthens the time during which steady-state conditions can be maintained *in vitro*, this approach as currently applied also has distinct limitations. First, current methodology does not permit investigators to distinguish secretion from the mucosal surface of cultured explants as opposed to secretion from the serosal surface. Thus one presumes that pepsinogen and intrinsic factor are predominantly secreted from the mucosal surface while gastrin is predominantly secreted into the culture medium via the open serosal surface. The extent to which this is true, however, cannot be determined because the culture medium bathing the serosal surface freely mixes with that bathing the mucosal surface. The serosal surface of the explant is in contact with the culture medium by virtue of the fact that it lies upon an open mesh wire screen which floats on the culture medium. Capillary action permits a thin layer of this same culture medium to cover the mucosal surface. When this technique has been altered during culture of small bowel mucosa to permit complete separation of mucosal from serosal surfaces (Kagnoff *et al.*, 1972), mucosal viability was compromised. Although in the case of gastric mucosa similar attempts have not been made to modify the basic technique of organ culture, it seems likely that such changes would similarly compromise viability.

A second serious limitation of the present approach to organ culture of gastric mucosal explants is the fact that the technique is not practical for directly assessing the secretion of hydrochloric acid, a major function of oxyntic mucosa. Successful maintenance of mucosal function in organ culture presently depends at least in part on the use of buffers in the culture medium which keep the pH constant at all times. Moreover, lack of separation of mucosal and serosal compartments means that any bicarbonate secreted by the explant will also neutralize secreted HCl.

Another difficulty with the technique currently used to culture gastric mucosal explants is that, unlike standard cell culture techniques, the organ culture method does not permit addition of exogenous cells such as lymphocytes which can then come into direct contact with the cultured explants. When exogenous cells are added to culture medium on which the explant is floating, the added cells simply sediment to the bottom of the culture dish. The technique used for fetal explants which allows attachment of these explants to the scored bottom of a plastic petri dish (Hoorn, 1966) would allow for direct cell-to-cell contact, but gastric explants from adult animals have not as yet been successfully cultured by this technique. However, we have shown that oxyntic mucosal explants can be cultured on a semi-solid culture medium for 24 hours without detectable loss of function (Table II), and this approach would allow addition of cells capable of coming into direct contact with cultured explants.

A further limitation of organ culture at its present stage of development is the fact that cultured biopsies of oxyntic mucosa obtained from human stomachs do not survive well. In contrast to rabbit oxyntic mucosa which is well preserved for 24 hours, human biopsies consistently deteriorate both morphologically and functionally (Table III) after 6 hours of organ culture. The reasons for this marked interspecies difference is not at all clear, particularly since mucosal biopsies obtained from the esophagus, gastric antrum, small bowel, and colon of human subjects are all well maintained in organ culture for 24 hours (Trier, 1976).

It must be recognized that to date relatively little fundamental work has been done to determine the most favorable conditions for organ culture of gastric mucosal explants. Perhaps because of the limitations outlined above, the number of investigators currently attempting to maintain gastric mucosa in organ culture is very small. There is not even much information available concerning hormonal and other factors which might be responsible for preserving gastric mucosa for as long as 24 hours, and there is no information at all about attempts to maintain gastric mucosa *in vitro* for periods longer than 24 hours. It is clear from several investigations that some factor or factors present in serum is necessary to maintain the viability of isolated gastric mucosa and that these factors survive heat inactivation (Lichtenberger *et al.*, 1978) and dialysis (Harty *et al.*, 1977). It is also clear that removal of a single amino acid such as leucine, tryptophan, or methionine from the culture medium does not shorten tissue survival.

The variety of different chemically defined, synthetic media which have been tried for culturing gastric mucosa is extremely limited, however, and there have been no systematic investigations to determine the relative efficacy of various nutrients in maintaining isolated gastric mucosa in organ culture. Of the media employed, Trowel T8 medium is the least complex. The other media which have been used including Eagle's minimal essential medium, Liebovitz L-15 medium, and NCTC 135 medium all contain several more amino acids, vitamins, and cofactors than does Trowell T8 medium. We have observed that gastric biopsies cultured in NCTC medium have a greater capacity to secrete intrinsic factor than do biopsies cultured in Trowell T8 medium (see Section III,B), but neither we nor other workers have examined the conditions of organ culture in ways specifically designed to prolong survival of gastric mucosa beyond 24 hours. As detailed in Section V,A, organ culture of gastric mucosa has been employed to examine phenomena which can be adequately assessed in experiments which require steady-state conditions for less than 24 hours, and therefore there has been no concerted effort to improve the culture technique itself.

V. Perspectives

A. Current Applications and Findings

The rather limited information which organ culture has yielded with respect to gastric mucosal differentiation in fetal stomachs has been summarized in Section III,A and will not be repeated here. It is apparent that further progress along these lines will require successful maintenance of fetal stomachs in organ culture for longer intervals of time.

Organ culture is being used to examine in considerable detail the mechanisms involved in gastric mucosal synthesis and secretion of pepsinogen. In addition to demonstrating *de novo* production or pepsinogen by isolated gastric mucosa maintained in organ culture (Fig. 2), our early investigations (Sutton and Donaldson, 1975) demonstrated that rabbit gastric mucosal biopsies incorporated [^{14}C]leucine into tissue protein and secreted labeled protein into culture medium steadily for 24 hours. Incorporation of radioactivity was abolished by cycloheximide. When examined by sodium dodecyl sulfate gel electrophoresis, dextran gel filtration, and ion exchange chromatography, 65 to 90% of macromolecular radioactivity secreted into culture medium migrated coincidentally with enzymatically assayed pepsinogen. Pepsin activity in cultured biopsies did not decrease during 24 hours of organ culture. Nevertheless, pepsin activity increased linearly in culture medium during this period. Acetylcholine markedly stimulated secretion of labeled protein and pepsinogen by cultured biopsies. In the presence of a subthreshold concentration (10^{-10} *M*) of acetylcholine, pentagastrin, secretin, and the octapeptide of cholecystokinin, all stimulated protein secretion. Overall incorporation of [^{14}C]leucine into protein by cultured biopsies was stimulated by 10^{-9} *M* pentagastrin. These results directly demonstrate: (*a*) synthesis and secretion of protein and pepsinogen by isolated gastric mucosa, (*b*) stimulation of gastric secretion of protein by acetylcholine and polypeptide hormones, and (*c*) stimulation of gastric synthesis of protein by pentagastrin.

More recent work with cultured mucosal explants (Kapadia and Donaldson, 1978) has been concerned with comparing the secretion of pepsinogen with that of intrinsic factor (IF) when examined over prolonged periods of time in response to known stimulators and inhibitors of gastric secretion. We found that acetylcholine (10^{-8} *M*) stimulated both pepsinogen ($p<0.01$) secretion and this stimulation was blocked by atropine. Parasympatholytic agents did not alter unstimulated (basal) secretion of pepsinogen even at high concentrations (atropine, 10^{-2} *M* or propanthelene bromide, 5×10^{-3} *M*); however, at these concentrations basal IF secretion was abolished. Histamine (10^{-4} and 10^{-2} *M*) had no effect on pepsinogen secretion but stimulated IF secretion ($p<0.001$). Antagonism of H2 receptors by cimetidine reduced both basal and histamine-stimulated IF secretion, but pepsinogen secretion remained unaltered. Under the conditions of these experiments the gastric mucosal surface was not exposed to HC1 but was constantly buffered by culture medium at pH 7.4. When we applied 50 m*N* HCl to the mucosal surface of the biopsies, pepsinogen secretion doubled ($p<0.001$) but IF secretion was abolished. These studies have clearly documented that: (*a*) fundamental differences exist in the responses of pepsin and IF secreting cells; (*b*) H+ ions bathing the mucosal surface of the stomach may influence the results of experiments designed to examine the mechanisms of gastric mucosal macromolecular secretion. Current investigations are concerned with using organ culture of isolated gastric mucosa: (*a*) to prepare biosynthetically labed IF, (*b*) to determine the role of cyclic nucleotides in the cellular control of IF secretion, and (*c*) to examine the mechanisms involved in gastric synthesis of pepsinogen and IF.

Other investigators have been similarly concerned with the use of organ culture to investigate the biosynthesis and secretion of gastrin by isolated antral mucosa. Harty *et al.* (1977) have shown that both the synthesis and secretion of gastrin is stimulated by dibutyryl cyclic AMP, although the relationship between this stimulation by cyclic AMP and the physiological mechanisms known to be involved in the *in vivo* release of gastrin remains to be defined. Lichtenberger *et al.* (1978), on the other hand, have shown that both peptone and a digest of bovine serum albumin which are known to release antral gastrin *in vivo* also stimulate the release of the hormone from cultured antral mucosa. Of considerable interest was the fact that the stimulation induced by these substances could not be demonstrated as long as peptone or albumin digest was present in the culture medium. As soon as antral explants were transferred to fresh culture medium which did not contain these substances, however, a striking increase in gastrin release became apparent. These interesting findings raise the possibility that complex peptides and amino acids which stimulate gastrin release *in vivo* may actually cause in cultured antral mucosa an *in vitro* accumulation of an inhibitor of gastrin release.

B. Future Studies

It is clear that if the full potential of organ culture is to be realized, future investigations must emphasize the development of methods which will prolong the viability of gastric mucosal explants. At the present time organ culture provides a reasonable experimental approach for examining gastric biosynthesis and secretion of macromolecules and peptides for periods as long as 24 hours compared to 2 or perhaps 3 hours by other *in vitro* techniques. However, in many instances this does not prove to be a particularly great advantage. Much longer periods during which steady-state conditions can be maintained are needed for rigorously controlled *in vitro* investigations into the trophic effects of hormones and other agents, the effects of humoral and cell-mediated immune reactions on gastric mucosal structure and function, or the long-term effects of physiologic and pharmacologic stimulants and inhibitors on oxyntic and antral mucosa.

To accomplish this goal, multiple approaches will be necessary. Systematic investigations into the specific requirements for optimal organ culture are needed. The factor or factors present in serum which prolong the viability of gastric mucosa *in vitro* need to be identified. The possible usefulness of substances such as epithelial growth factor and somatostatin should also be examined. Also required are further efforts at isolating specific gastric cells such as parietal cells, chief cells, and endocrine cells combined with attempts to establish cell culture lines from these isolated cells. For example, it is possible that after organ culture of gastric mucosal explants for 24 hours isolated cell lines might then be established from cultured explants. Although results obtained thus far with organ culture of gastric mucosa are of interest and offer some promise for the future, it is also clear that the present state of the art of maintaining gastric mucosa *in vitro* is in its infancy and must still be regarded as relatively primitive.

REFERENCES

Browing, T. H., and Trier, J. S. (1969). *J. Clin. Invest.* **48,** 1423.
Emas, S., Swan, K. G., and Jacobson, E. D. (1967). *In* "Alimentary Canal" (C. F. Code and W. Heidel, eds.), Vol. II, pp. 743–758. American Physiological Society, Washington, D. C.
Evans, V. J., Bryant, J. C., Ken, H. A., and Schilling, E. L. (1964). *Exp. Cell Res.* **36,** 439.
Harty, R. F., van der Vigver, J. C., and McGuigan, J. E. (1977). *J. Clin. Invest.* **60,** 51.
Hoorn, B. (1966). *Acta Pathol. Microbiol. Scand.* **66** Suppl. **183,** 1.
Kagnoff, M. F., Donaldson, R. M., and Trier, J. S. (1972). *Gastroenterology* **63,** 541.
Kapadia, C. R., and Donaldson, R. M. (1978). *Gastroenterology* **74,** 535.
Leibovitz, A. (1963). *Am. J. Hyg.* **78,** 173.
Lichtenberger, L. M., Shorey, J. M., and Trier, J. S. (1978). *Am. J. Physiol.* **4,** E410.
Miller, L. R., Jacobson, E. D., and Johnson, L. R. (1973). *Gastroenterology* **64,** 254.
Philpott, G. W. (1971). *Gastroenterology* **61,** 25.
Rehm, W. S. (1962). *Am. J. Physiol.* **203,** 63.
Soll, A. H. (1978). *J. Clin. Invest.* **61,** 370.
Sutton, D. R., and Donaldson, R. M. (1975). *Gastroenterology* **69,** 166.
Trier, J. S. (1976). *N. Engl. J. Med.* **295,** 150.
Trowell, O. A. (1959). *Exp. Cell Res.* **16,** 118.
Yeomans, N. D., Trier, J. S., Moxey, P. C., and Markezin, E. T. (1976). *Gastroenterology* **71,** 770.

Chapter 18

Organ Culture of the Mucosa of Human Small Intestine

JERRY S. TRIER

Division of Gastroenterology,
Department of Medicine,
Peter Bent Brigham Hospital and Department of Medicine,
Harvard Medical School,
Boston, Massachusetts

I. Introduction

The maintenance *in vitro* of the epithelium of the small intestine has posed formidable problems. Suspensions of epithelial cells can be isolated following exposure of the intestinal mucosa to agents that chelate divalent cations (Stern and Jensen, 1966; Weiser, 1973), but available data indicate that the cells in these suspensions remain viable for only 2 to 3 hours at best. Efforts to grow mammalian small intestinal epithelial cells in tissue culture have been hampered by their rapid dedifferentiation and often concomitant overgrowth of mesenchymal cells such as fibroblasts (Lichtenberger *et al.*, 1973), although a recent report of culture of epitheloid cells from the intestine of germ-free rats is more promising (Quaroni *et al.*, 1979).

ISBN 0-12-564140-0

The mucosa of the small intestine is a complex and highly polarized tissue with its villi lined primarily by mature absorptive cells and mucus-secreting goblet cells and its crypts lined by proliferating undifferentiated cells and differentiated secretory cells which include Paneth, goblet, and endocrine cells. Likewise, the individual epithelial cells are characterized by highly polarized structural features. Their basal plasma membrane rests on a basal lamina which is in close contact with the mesenchymal elements of the lamina propria. Their apical surface is highly specialized with numerous microvilli enclosed by a plasma membrane which contains hydrolytic enzymes including disaccharidases and peptidases and specific receptors which are important in digestive and absorptive processes. Preservation of the unique composition of this specialized apical absorptive–digestive surface may require the presence of the intimate relationship provided by intact occluding junctions between neighboring cells (Pisam and Ripoche, 1976). Thus, maintenance of the normal structural relationship of small intestinal epithelial cells to each other, to their basal lamina, and to the underlying mesenchymal elements of the lamina propria may be essential to achieve preservation of muscosal elements that would permit study of many of the normal or abnormal intestinal synthetic, secretory, and transport functions which take place in health and disease.

Fragments of intestinal mucosa such as human peroral intestinal biopsies, and everted sacs or rings of intestine from laboratory mammals (Wilson and Wiseman, 1954) immersed in oxygenated buffers or media are useful for short-term *in vitro* studies. However, careful morphological examination shows that epithelial cell necrosis and degeneration occurs rapidly in such systems, generally within 0.5 to 3 hours (Plattner *et al.,* 1970).

The application of organ culture methodology to the maintenance *in vitro* of mucosal biopsies from the small intestine of humans and experimental animals now permits study of intestinal mucosal synthetic, secretory, and epithelial cell proliferative function for at least 24 to 48 hours (Trier, 1976). During organ culture, the relationships of epithelial cells to one another and to their underlying mesenchyme are preserved. Moreover, as with other *in vitro* systems, the environment of the tissues under study can be controlled more precisely than when *in vivo* preparations are used since exposure to the normal intestinal luminal contents, circulating humoral factors, and neurogenic stimuli are largely eliminated.

II. Methods

A. Organ Culture of Mature Intestinal Mucosa

The ideal conditions for maintaining explants of mature human small intestinal mucosa in organ culture have not been established. This is because, to our knowledge, there are no systematic studies of the influence of the many available nutrient media, various enrichment factors, incubation temperatures, or oxygen concentrations on tissue survival as judged by morphological or functional criteria. Rather, the approach to organ culture of human mucosal intestinal explants, at least in the author's laboratory, has been quite pragmatic; once it was demonstrated that the explants remained viable for a sufficient period to conduct the desired experiments (24 hours), no additional methodological studies were performed.

The method employed in our laboratory is an adaptation of the organ culture method used by Trowell (1961) for the maintenance *in vitro* of fragments of other mature organs. Explants for culture are generally obtained perorally from the distal duodenum or proximal jejunum with a multipurpose (Brandborg *et al.*, 1959) or hydraulically activated biopsy tube (Loder *et al.*, 1964). The multipurpose tube can be equipped with a four port capsule to obtain four 10 mg full thickness mucosal biopsies per intubation whereas any number of biopsies can be obtained by intubation with the hydraulic tube although no more than 10 should be taken at a given time because of the hazard of significant bleeding from multiple biopsy sites. The multipurpose tube can also be used to collect biopsies of appropriate thickness for culture from freshly excised human intestine obtained at surgery and from freshly excised small intestine from laboratory mammals such as rabbits, dogs, or monkeys.

Once obtained, the biopsies are divided immediately after their excision into 1-mm slices (usually three to four per biopsy) with a sharp razor blade or scalpel. These are mounted villi up and cut surface down on a wire mesh screen placed in the central well of sterile plastic organ culture dishes 5.5 cm in diameter (Falcon Plastics, Oxnard, CA). Sufficient medium (approximately 0.9 ml) is then added to the central well to barely float the wire mesh on which the tissue was placed. A drop of medium is also placed on the top of each explant slice (usually three to four per dish). The outer well of the dish contains a circular felt pad which is saturated with water to maintain a humid environment in the culture dish. The medium used in our studies consists of nine parts Trowell's T-8 medium [Grand Island Biological Co., Inc., (GIBCO) Grand Island, NY] and one part heat-inactivated fetal calf serum (GIBCO) to which 10,000 U of crystalline penicillin G and 0.01 g of streptomycin sulfate is added per 100 ml of medium. The addition of serum seems crucial for optimal results; in our experience small intestinal explants from mature intestine undergo degeneration within 6–12 hours when cultured in serum-free medium. Since the biopsies are obtained via the peroral route, they are already contaminated and sterile technique is not employed while orienting the explants. The medium, however, is kept sterile and added to the culture dishes using sterile technique. Once the explants have been oriented and mounted the dishes are covered, placed in an airtight container in a 37°C incubator, and gassed with a mixture of 95% 0_2 and 5% CO_2 for 20 minutes after which the container is sealed. We have found MacIntosh jars (Torsion Balance Company, Clifton, NJ) equipped with custom made four shelf lucite racks to be convenient containers for maintaining 12 culture dishes per jar but any airtight container with vents for gassing would appear suitable. Attempts to maintain the tissue at room air in our laboratory have been unsuccessful with the appearance of substantial tissue degeneration within 6 hours.

To examine tissue synthetic and secretory phenomena, appropriate radioactive precursors such as thymidine for DNA synthesis, amino acids for protein synthesis, and glucosamine, galactose, fucose, sulfur, or *n*-acetylglucosamine for glycoprotein synthesis can be conveniently added to the culture medium. The effects of continuous exposure to radioactive precursors can be studied. Alternatively, pulse-chase experiments can be carried out by exposing explants to radioactive precursors for a relatively brief period, withdrawing the labeled medium, rinsing the medium well, and then adding medium containing unlabeled precursor. Similarly, the effect of pharmacological or humoral agents can be studied by adding these to the medium at the appropriate time during the culture period. Medium should be renewed and the cultures regassed at least once every 12 hours. At the end of the culture period, explants can be prepared for histological, autoradiographic, or electron microscopic study or utilized for biochemical studies. Material released into the culture medium by the explants can also be studied by appropriate biochemical or immunochemical methods.

Certain modifications of the previously described technique have also resulted in successful short-term culture of mucosal explants from human small intestine. Other media used have included nine parts of RPMI-1640 and one part of fetal calf serum enriched with insulin (0.5 mg/100 ml medium), glutamine (30 mg/100 ml), and glucose (366 mg/100 ml) (Falchuk *et al.*, 1974), 90% NCTC-135 medium and 10% fetal calf serum (Ginsel *et al.*, 1977) and Trowell's T-8 medium, 20% NCTC 135 medium, and 15% fetal calf serum to which a peptic–tryptic–chymotryptic digest of human serum albumin (0.5 mg/ml) was added (Jos *et al.*, 1975). The latter medium has also been used successfully with roller-tube cultures in which explants were placed on 22-gauge wire mesh covered with agar which was wedged into tubes containing 1 ml of medium. The tubes were rotated at 12 rev/hour at 37°C in a 95% 0_2–5% CO_2 environment (L'Hirondel *et al.*, 1976). A roller-tube method has also been used with Eagle's medium enriched with 10% fetal calf serum (Townley *et al.*, 1973). With the methods previously described small intestinal explants from monkeys (Trier, 1976), rabbits (Kagnoff *et al.*, 1972), guinea pigs (Kedinger *et al.*, 1975), and dogs (Gebhard and Cooper, 1978) also have been successfully cultured for up to 48 hours. Organ culture of mucosal explants of mouse small intestine was successful for up to 48 hours in medium composed of nine parts of DMEM–HEPES, one part NCTC-135 enriched with 10% fetal calf serum (Chabot *et al.*, 1978).

B. Organ Culture of Fetal Intestine

Explants 1 to 4 mm^2 of small intestine from 9- to 16-week human fetuses have also been cultured for at least 21 days in plastic 50 × 15-mm petri dishes to which were added 1.25 ml of Liebovitz L-15 medium supplemented with a final concentration of 0.2% bovine albumin, 2.0 m*M* glutamine, 250 μg/ml streptomycin, 250 μg/ml chlortetracycline, and 5 μg/ml amphotericin B (Dolin *et al.*, 1970). For this method, the culture dishes are placed on a rocker platform which completes 1 cycle per minute in an environment of room air at 34°C. The medium is replaced every second day. The same method has also been used to culture successfully explants of fetal rat intestine for up to 72 hours with daily renewal of the medium (DeRitis *et al.*, 1975). Cultures of human fetal intestinal tissue appear to have been used exclusively as a substrate for the growth of viruses (Dolin *et al.*, 1970; Wyatt *et al.*, 1974), not for studies of intestinal function and development and will not be discussed further.

III. Results

A. Viability of Cultured Human Small Intestinal Mucosa

The histology of explants of normal human small intestine maintained in organ culture for 24 to 48 hours is generally well preserved (Browning and Trier, 1969; Falchuk *et al.*, 1974; Hauri *et al.*, 1975; Ginsel *et al.*, 1977). Figure 1 is a light micrograph of normal mucosa which was maintained in culture for 24 hours. After 24 to 48 hours of culture, the villi are characteristically somewhat wider and shorter than those of uncultured freshly fixed tissue. The cellularity of the lamina propria is decreased; this may represent both accumulation of fluid in the lamina propria and depletion of mononuclear cells and eosinophils which in culture are not replaced by an intact circulation. The epithelial lining in general is intact and well preserved. Along the surface of the explants there is consistently a coating of heterogeneous material composed of secreted mucus and cellular debris including exfoliated cells. Mitoses persist in the crypt epithelium although they decrease in frequency as the culture period is prolonged. Focal areas of necrosis are found in some cultured explants; they are rare before 24 hours and more frequent in explants cultured for 24 to 48 hours. Such degenerative changes are seen most often in the epithelium lining the crypts in the central region of the explants.

Electron microscopy of cultured explants has confirmed the impression gained by light microscopy that near normal epithelial cell morphology is maintained in most areas of explants cultured for up to 48 hours (Browning and Trier, 1969;

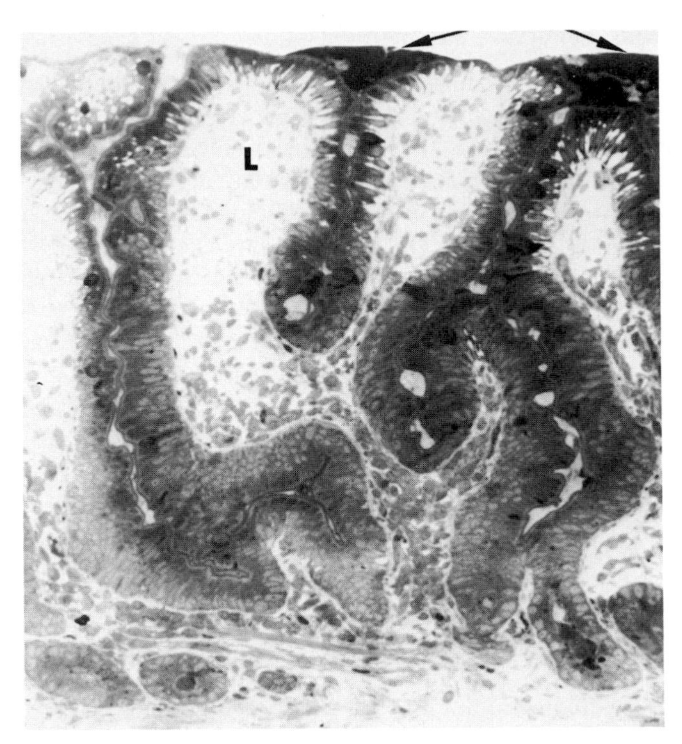

FIG. 1. Light micrograph of a duodenojejunal biopsy maintained *in vitro* in organ culture for 24 hours. The villi are somewhat shortened, and the cellularity of the lamina propria (L) is decreased. Heterogeneous material (arrows) covers some of the well-preserved epithelium. Toluidine blue stain, ×150. From Browning and Trier (1969) with permission.

Falchuk *et al.*, 1974; Hauri *et al.*, 1975; Jos *et al.*, 1975; Ginsel *et al.*, 1977). Figure 2 shows epithelium from the mid-villus region of an explant of normal mucosa maintained in culture for 24 hours. The fine structure of these absorptive cells is virtually indistinguishable from absorptive cells of freshly fixed biopsies. The cytoplasm is compact and specific cytoplasmic structures including the apical microvilli with their attached gylcocalyx, mitochondria, and Golgi complexes appear normal. In some areas of most explants, some absorptive cells appear more cubiodal than in freshly fixed normal tissue with increased numbers of lysosomes in the apical cytoplasm. The fine structure of other epithelial elements including undifferentiated crypt cells (Fig. 3), goblet cells, endocrine cells, and Paneth cells is also well preserved and these cells contain abundant secretory granules (Browning and Trier, 1969; Hauri *et al.*, 1975). In many cells of explants cultured in Trowell's T-8 medium substantial aggregates of particulate glycogen are present in the cytoplasm (Fig. 3); such aggregates of glycogen are uncommon in freshly fixed normal human intestinal mucosa. The fine struc-

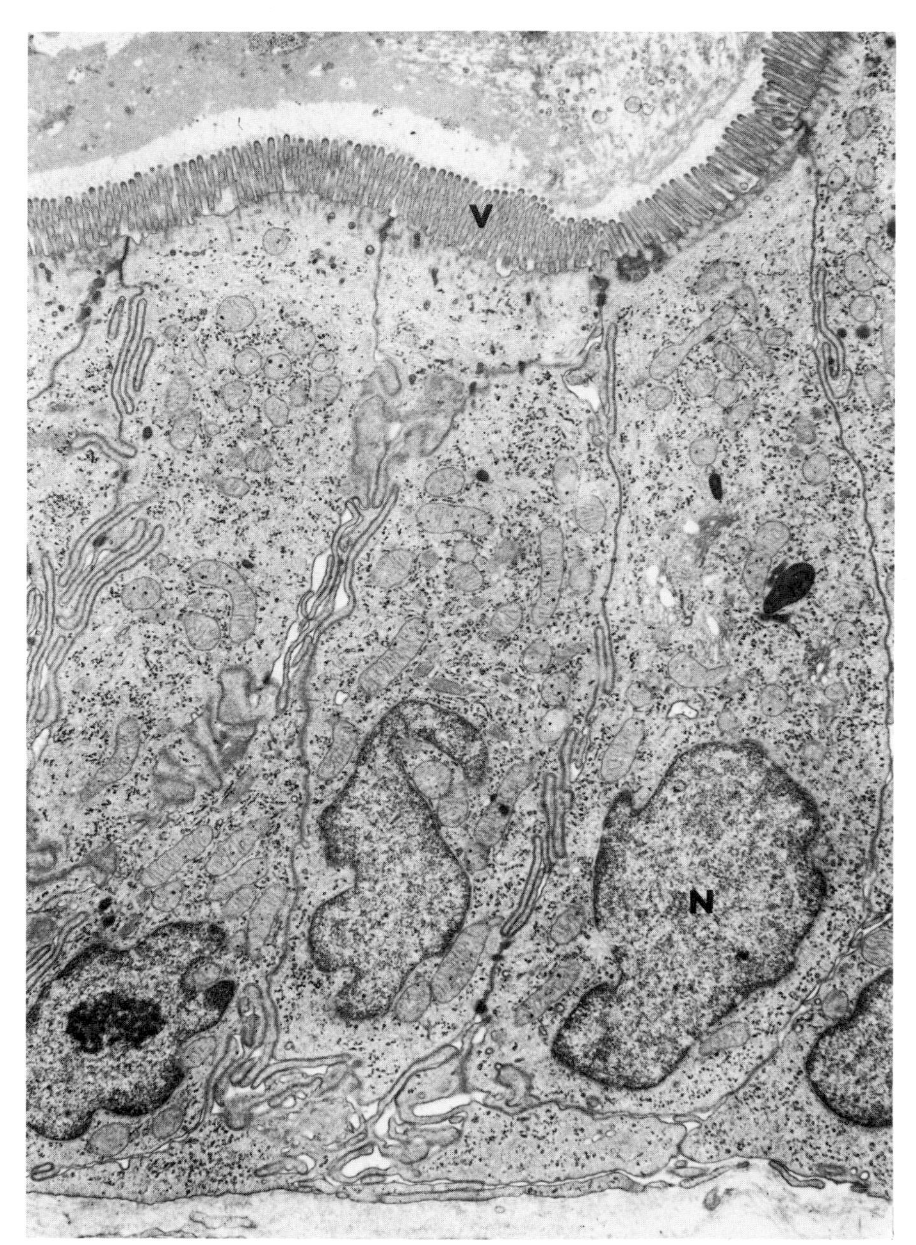

FIG. 2. Electron micrograph of absorptive cells from the mid-villous region of a normal human duodenojejunal explant maintained in organ culture for 24 hours. The cells appear essentially indistinguishable from cells of freshly fixed proximal intenstine. The microvillous border (V), cytoplasm with its various organelles, and nuclei (N) all appear normal. Mucus and cellular debris are seen extracellularly above the microvilli. ×5000.

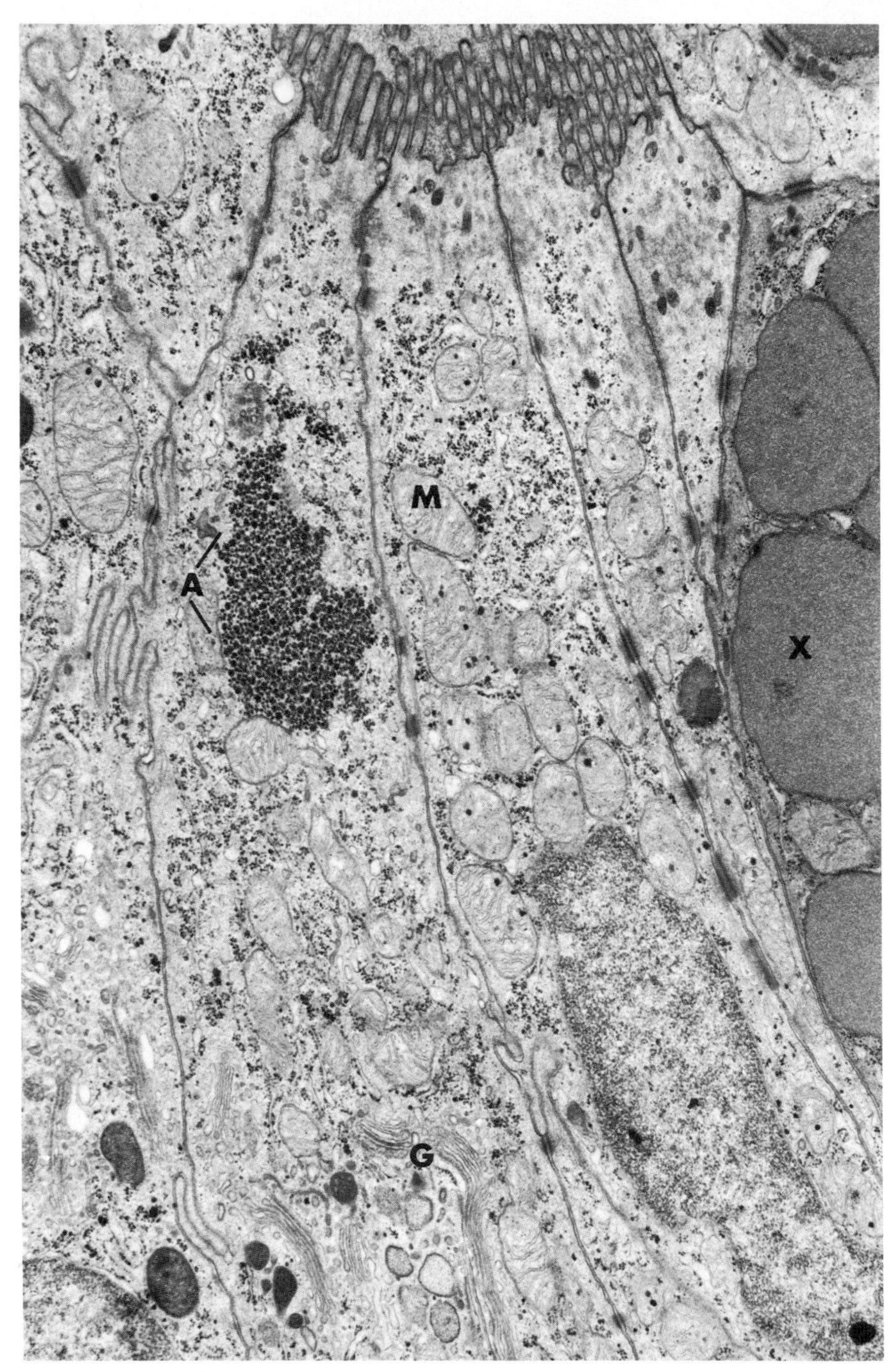

FIG. 3. Electron micrograph of crypt epithelium of a normal human duodenojejunal explant cultured for 24 hours. There are portions of several undifferentiated cells and, on the right, a goblet cell. The cytoplasmic organelles are well preserved including the mitochondria (M), Golgi material (G), and mucous granules (X). Particulate glycogen is more abundant than in uncultured crypt cells with a large aggregate (A) evident in one of the cells. ×12,000.

ture of elements of the lamina propria including fibroblasts, plasma cells, lymphocytes, eosinophils, unmyelinated nerve fibers, smooth muscle cells, and vascular channels is normal in most explants cultured for 24 to 48 hours (Fig. 4).

Studies of mucosal function support the morphological evidence for viability of explants cultured for up to 48 hours. Explants cultured over medium which contains [^{3}H]thymidine incorporated thymidine into nuclei of undifferentiated crypt cells as demonstrated by autoradiography (Browning and Trier, 1969; Trier and Browning, 1970; L'Hirondel *et al.*, 1976) and labeled epithelial cells migrated from the crypts onto the villi during culture (Trier and Browning, 1970). However, no systematic study designed to determine whether *de novo* DNA synthesis by human explants remains constant during a 24- to 48-hour culture period has been reported. Since the number of mitotic figures evident in cultured normal human intestinal explants decreases with time, it is likely that the number of cells undergoing active DNA synthesis also decreases as time in culture is prolonged.

On the other hand, exposure of cultured human explants to radiolabeled leucine revealed comparable incorporation of radioactivity into acid-precipitable tissue protein following culture for 0, 6, and 24 hours (Fig. 5) in one study (Trier, 1974) and following culture for 0, 5, 21, and 45 hours in another study (Hauri *et al.*, 1975). *De novo* protein synthesis as measured by exposing explants to radiolabeled leucine continuously during culture occurred at a constant rate for 48 hours when the labeled protein retained in the explant plus the labeled protein released into the medium were quantitated (Hauri *et al.*, 1975). Incorporation of radiolabeled glucosamine into acid-precipitable macromolecules by mucosal explants of human small intestine also occurred at a constant rate for 48 hours *in vitro* (Hauri *et al.*, 1977). Further evidence for active protein and glycoprotein synthesis during organ culture include the demonstration of newly glycosylated brush border enzymes such as lactase, sucrase–isomaltase, and alkaline phosphatase after 4 and 24 hours of culture (Hauri *et al.*, 1977). Moreover, total brush border enzyme activity in cultured explants plus the culture media was substantially higher than in uncultured fresh mucosal samples, indicating *in vitro* synthesis of these enzymes (Mitchell *et al.*, 1974; Hauri *et al.*, 1977). Total protein content of explants decreases during culture (by approximately 33% after 24 hours). This may reflect in part loss of lamina propria cells such as lymphocytes into the medium by their transmigration across the epithelium and incomplete replacement of exfoliated senescent epithelial cells during culture. In light and electron microscopic autoradiographic studies of normal human cultured biopsies pulsed with [^{3}H]fucose and [^{3}H]glucosamine, labeled macromolecules initially were most abundant in the Golgi region of absorptive cells but appeared in large amounts in the brush border region and in lysosome-like bodies by 60 to 90 minutes (Ginsel *et al.*, 1979).

Limited studies of lipid uptake, synthesis, and secretion by cultured human

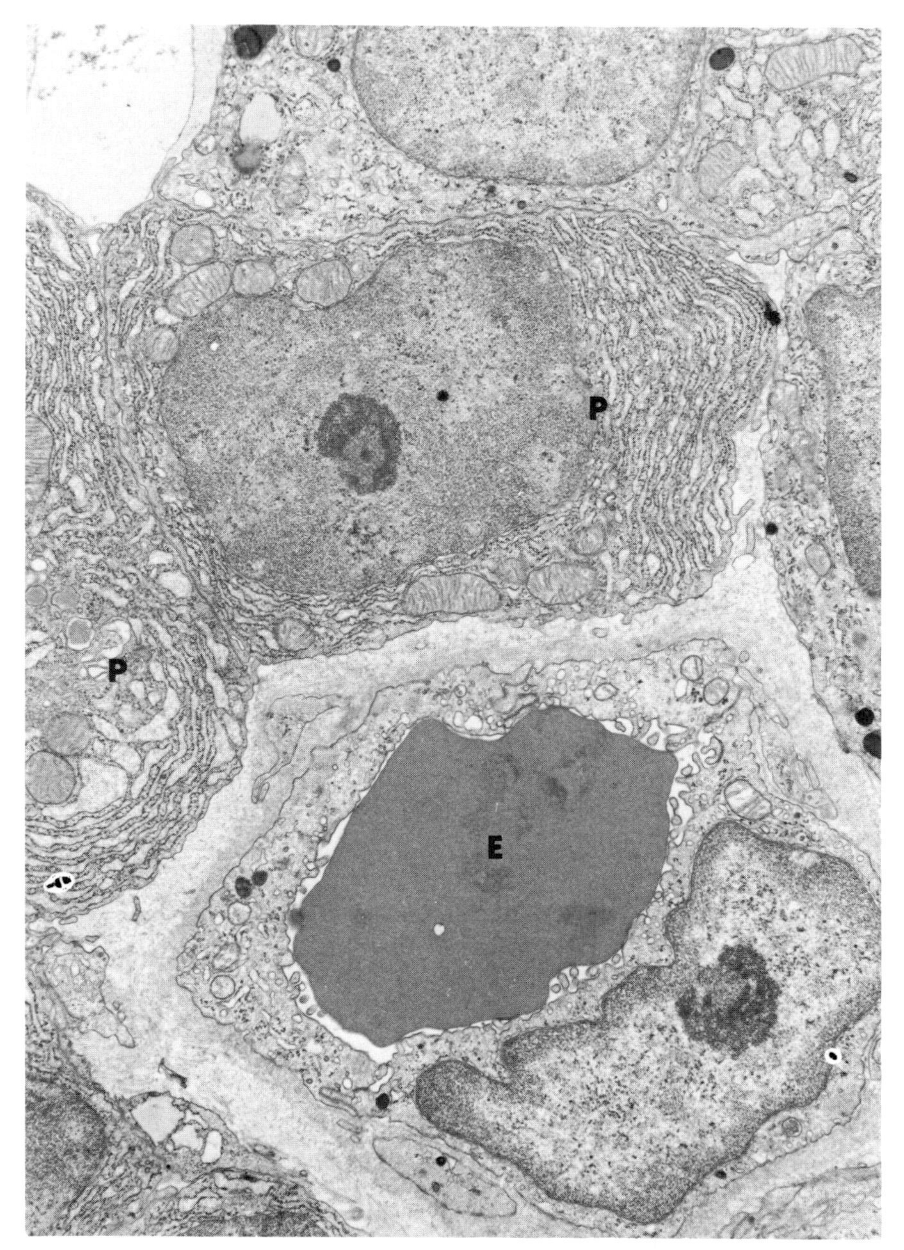

FIG. 4. Electron micrograph of lamina propria from a normal human duodenojejunal explant cultured for 24 hours. Portions of well-preserved plasma cells (P) and a capillary with an erythrocyte (E) in its lumen can be seen. ×7000.

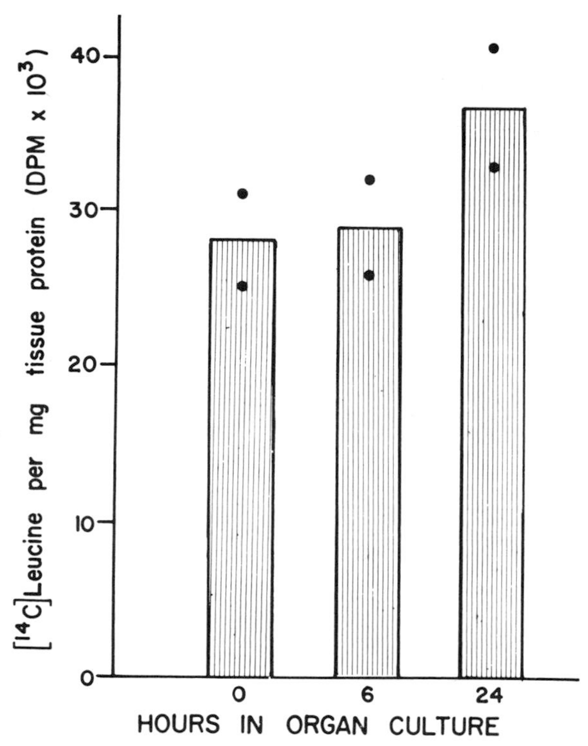

FIG. 5. Incorporation of [^{14}C]leucine into tissue protein by cultured mucosal explants from normal human jejunum. Biopsies were cultured over isotope-free medium for 0, 6, and 24 hours and then cultured over medium to which [^{14}C[leucine had been added. From Trier (1974) with permission.

mucosal explants have also been reported. Explants cultured for 3, 6, and 24 hours over isotope-free medium and then immersed for 15 minutes in an oxygenated micellar lipid solution containing radiolabeled oleic acid absorbed amounts of oleic acid comparable to uncultured mucosa (Fig. 6). Whereas uncultured biopsies incorporated 60% of the oleic acid into triglyceride as determined by thin layer chromatography, explants which had been cultured for 6 and 24 hours incorporated approximately 40 and 25% of absorbed oleic acid, respectively, into triglyceride (Fig. 6). Thus absorptive cells of explants cultured *in vitro* retain the capacity to synthesize triglyceride from absorbed fatty acid albeit at rates which decrease as time in culture increases. Absorptive cells of explants cultured for 24 hours and then exposed to micellar lipid were noted by electron microscopy to have abundant lipid droplets in the cisterns of the endoplasmic reticulum and Golgi complexes (Browning and Trier, 1969). In recent studies, mucosal explants of human small intestine were shown to incorporate radiolabeled palmi-

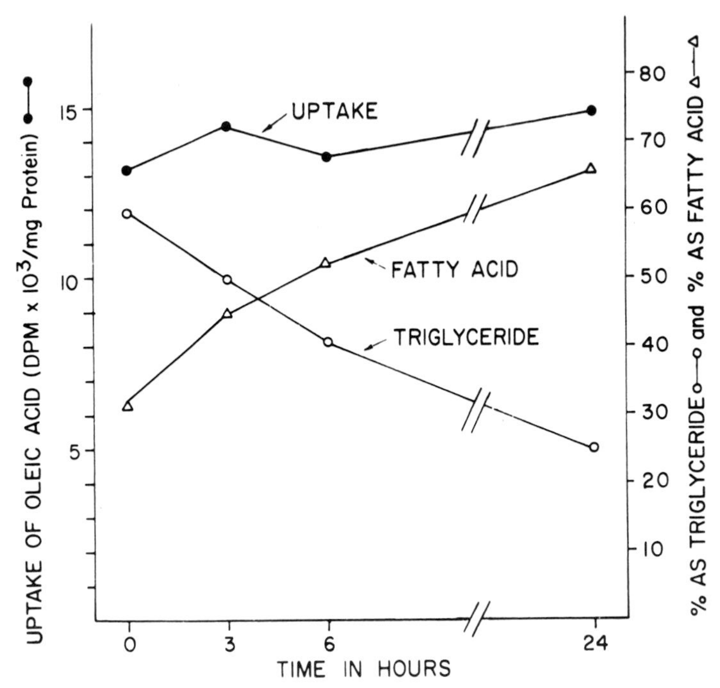

FIG. 6. Uptake of [^{14}C]oleic acid and its incorporation into triglyceride by cultured mucosal explants from normal human jejunum. After culture for 0, 3, 6, and 24 hours, explants were immersed for 15 minutes into a micellar solution containing [^{14}C]oleic acid. Uptake of fatty acid was determined by scintillation counting; lipid fractions were separated by thin layer chromatography. From Trier (1974) with permission.

tic acid into lipoprotein triglyceride and phospholipid and to secrete newly synthesized lipoprotein into the medium during a 24-hour culture period (Rachmilewitz *et al.*, 1978).

The morphological and functional studies described provide considerable evidence that mucosal explants of human small intestine remain viable in organ culture for 24 to 48 hours. However, current culture systems are far from ideal; morphological studies reveal focal areas of degeneration in many explants after 24 to 48 hours, total protein content of cultured explants decreases during culture and certain synthetic functions such as mucosal incorporation of fatty acid into triglyceride decrease as time in culture is prolonged.

As far as can be determined from published reports, comparisons of the several different media used to culture mucosal explants of human small intestine have revealed only subtle differences. For example, there is a sustained increase in specific activity of alkaline phosphatase when explants are cultured for 48 hours over RPMI-1640 medium supplemented with glucose, insulin, glutamine, and fetal calf serum whereas the specific activity of alkaline phosphatase is increased after 24 hours but returns to baseline levels after 48 hours of culture over Trowell's T-8 medium supplemented with fetal calf serum (Hauri *et al.*, 1975).

B. Organ Culture Studies of Small Intestinal Disease

In the disease, celiac-sprue, certain cereal glutens (those present in wheat, barley, rye, and probably oats) interact with the mucosa of the small intestine to produce a lesion in which the mucosa is virtually devoid of villi and the flat mucosal surface is lined by abnormal absorptive cells. As a result, patients with this disease may have serious malabsorption of normal dietary nutrients. Organ culture methods have been applied to examine the histogenesis of the intestinal lesion and to examine mechanisms by which toxic glutens may damage the intestinal mucosa in celiac-sprue.

In initial studies it was noted that the severely damaged surface absorptive cells of untreated patients with attenuated microvilli, large lysosomes, and large lipid vacuoles were replaced by tall columnar absorptive cells with near normal microvilli and more normal cytoplasmic structure (Fig. 7) in mucosal explants cultured for 24 hours in a gluten-free environment (Trier and Browning, 1970). These observations were subsequently confirmed by workers in several laboratories (Townley *et al.*, 1973; Falchuk *et al.*, 1974; Jos *et al.*, 1975; Hauri *et al.*, 1978). When mucosal explants from patients with untreated sprue were cultured over medium containing [^{3}H]thymidine, the number of labeled epithelial cell nuclei as assessed by autoradiography was almost three times greater than in identically cultured explants from normal volunteers (Fig. 8). Additionally, the migration of labeled epithelial cells from the crypts toward the exfoliation zone on surface epithelium was more rapid in explants from untreated celiac-sprue patients than in those from normal volunteers (Trier and Browning, 1970). Treatment with a gluten-free diet for 6 to 15 weeks resulted in a reduction of labeled epithelial cells (Fig. 8) and partial normalization of the rate of epithelial cell migration *in vitro* (Trier and Browning, 1970). The addition of gluten to the medium was shown in several laboratories to prevent this improvement (Townley *et al.*, 1973; Falchuk *et al.*, 1974; Jos *et al.*, 1975; Fluge and Aksnes, 1978) and addition of various gluten fractions has been used to characterize the toxic moiety in wheat gluten (Jos *et al.*, 1978).

The specific activity in mucosal tissue of the brush border enzyme, alkaline phosphatase parallels the morphology of the mucosa in celiac-sprue; biopsies with markedly abnormal morphology have very low levels of enzyme activity whereas biopsies with mild lesions or biopsies from treated patients with histological improvement have substantially higher activity. This relation has been used to assess the tissue response in culture to the presence or absence of gluten and gluten derivatives in the culture medium (Falchuk *et al.*, 1974). When mucosal explants from untreated patients are cultured in the absence of gluten,

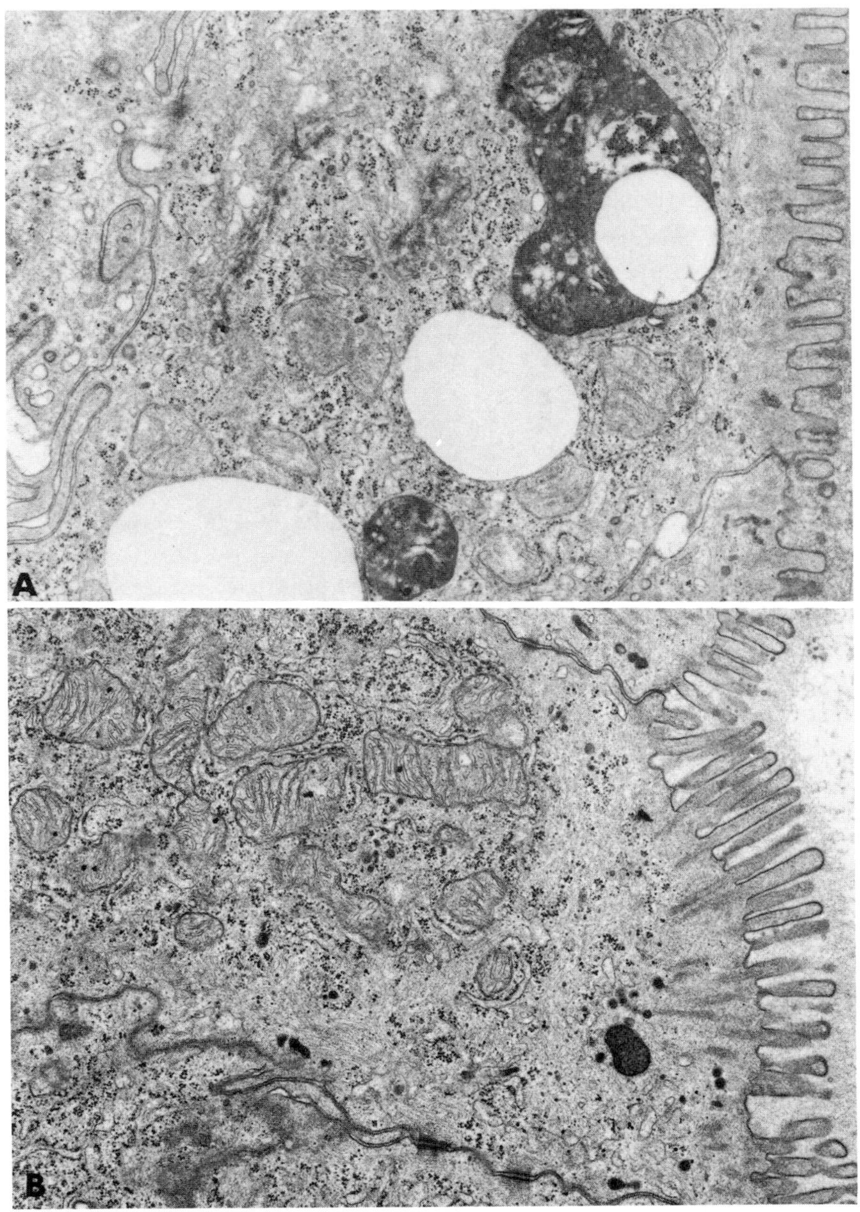

FIG. 7. Electron micrographs of the apical cytoplasm of surface absorptive cells of duodenal explants from a patient with untreated celiac-sprue cultured for 6 hours (A) and 24 hours (B). There is marked improvement in surface absorptive cell structure by 24 hours with lengthening of microvilli and disappearance of large lysosomes and lipid droplets. ×12,000. From Trier and Browning (1970) with permission from the *New England Journal of Medicine*.

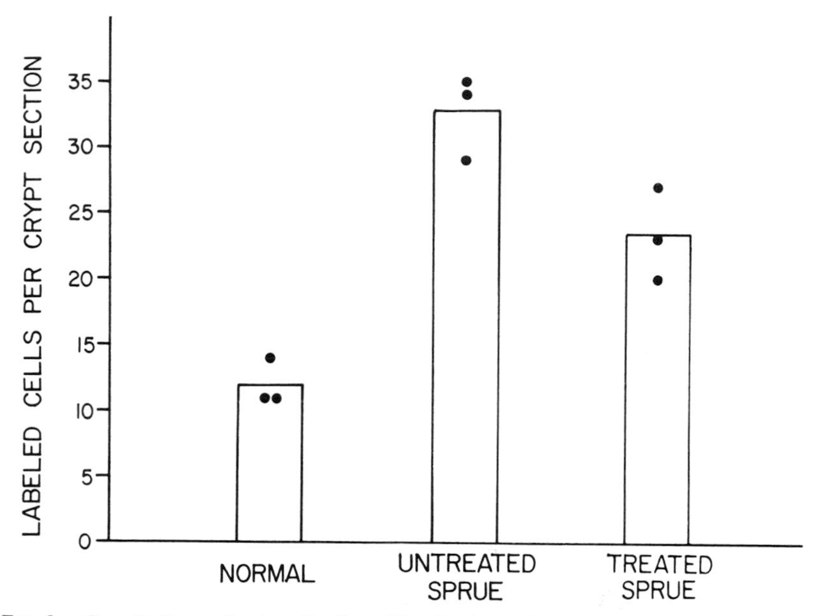

FIG. 8. Quantitative evaluation of cell proliferation in duodenal explants cultured over medium containing [^{3}H]thymidine for 6 hours. The average number of labeled cells per crypt for each subject is shown by a dot. From Trier and Browning (1970) with permission from the *New England Journal of Medicine*.

mucosal alkaline phosphatase activity increased toward normal; when cultured in the presence of gluten, this increase in alkaline phosphatase activity is prevented (Fig. 9). Moreover, the addition of cortisol to the medium inhibits the effects of added gluten with resultant normalization of alkaline phosphatase specific activity (Fig. 9). The morphological appearance of these cultured explants parallels the alkaline phosphatase activity; those cultured with cortisol plus gluten look more normal than those cultured with gluten alone (Katz *et al.*, 1976). This response to cortisol is consistent with other evidence that immunological factors may participate in the pathogenesis of the intestinal lesion in celiac-sprue (Strober *et al.*, 1975).

Thus in patients with untreated celiac-sprue, toxic glutens interact with the mucosa to damage mature absorptive cells that normally line the villi. These damaged cells are then shed prematurely into the intestinal lumen. To compensate for this rapid loss of mature cells, the proliferative compartment that contains immature crypt cells expands as cell proliferation and migration increase. This results in the development of the characteristic mucosal lesion of celiac-sprue with its shortened or absent villi and hyperplastic, elongated crypts. Whether gluten and its derivatives act directly on epithelial cells or whether

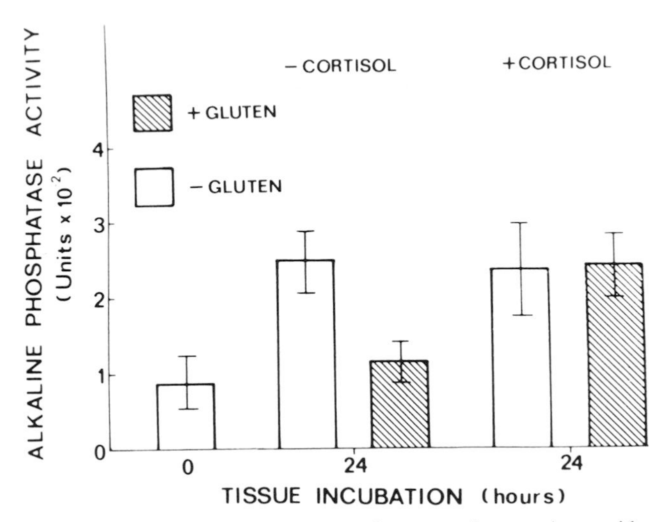

FIG. 9. Alkaline phosphatase activity of jejunal mucosa from patients with active gluten-sensitive enteropathy. Cultures were carried out in the presence or absence of gluten or cortisol. The hatched bars represent the values obtained in the presence of gluten. In the absence of cortisol, gluten prevents the increase of activity of alkaline phosphatase. Addition of cortisol ablates this effect. Each bar represents the mean ± SE of eight studies done in duplicate. From Katz *et al.* (1976) with permission from the *New England Journal of Medicine*.

epithelial cell damage is secondary to an immune response mediated at least in part by cells of the lamina propria remains to be established.

The intestinal lesion seen in patients with untreated celiac-sprue may resemble closely the lesion observed in the mucosa of the small intestine in several other diseases such as infectious gastroenteritis, tropical sprue, and intestinal lymphoma. It has recently been suggested that the response to gluten *in vitro* of the intestinal mucosa may be of value in distinguishing celiac-sprue from other diseases with similar lesions. Explants from celiac-sprue patients have higher alkaline phosphatase activity when cultured over gluten-free than gluten-containing medium whereas explants from patients with other diseases but similar lesions show no difference in alkaline phosphatase activity whether or not the medium contains gluten (Katz and Falchuk, 1978).

C. Organ Culture of Small Intestine of Laboratory Animals

The information obtained from a few studies in which small intestinal mucosal explants from laboratory mammals were cultured merit mention as examples of the types of studies which could also be performed with human explants. Mucosal explants of rabbit small intestine, like those from human small intestine, have been shown to synthesize and secrete protein and glycoprotein at a steady rate for 24 hours. In addition, it has been shown that explants of rabbit small intestine secrete intact, newly synthesized secretory IgA into the culture medium (Kagnoff *et al.*, 1972, 1973). Culture of muscosal explants of rabbit small intestine over medium to which 1 m*M* deoxycholate had been added resulted in marked morphologic damage, impaired protein synthesis, and impaired amino acid uptake whereas culture over medium containing lower concentrations of deoxycholate or 10 and 100 μg/ml of bacterial endotoxin had no significant influence on mucosal morphology, protein synthesis, or amino acid uptake (Beeken *et al.*, 1974).

Explants of canine ileal mucosa cultured for 6 hours demonstrated a 4-fold increase in the activity of 3-hydroxy-3-methyl-glutaryl coenzyme A (HMG CoA reductase), the rate-limiting enzyme of cholesterol synthesis. The increase in HMG CoA reductase activity was partially suppressed by addition of pure cholesterol, 7-ketocholesterol, 75 hydroxycholesterol, and low concentrations of taurochenodeoxycholate and could be inhibited completely by 0.2 n*M* cycloheximide (Gebhard and Cooper, 1978).

IV. Discussion and Perspectives

Although the techniques used during the past decade to maintain mucosal explants of mature human small intestine in organ culture have been useful and have contributed substantively to our understanding of the disease celiac-sprue, there is need for substantial improvement of available methodology. With current methods, reliable survival of epithelial elements of cultured explants is limited to 24 to 48 hours. Hence, only relatively short-term experiments are possible. Indeed, given present technical limitations, it is desirable to design experiments in which the proposed period of culture is as short as possible to answer the questions posed.

It is noteworthy that there are no published systematic studies designed to establish optimal growth requirements for the culture of mucosal explants of human small intestine. Only relatively few media have been employed and their advantages and disadvantages in promoting tissue survival have, by and large, not been compared with one another in a controlled manner. At present, there are no data to suggest that use of a more complete medium such as RPMI-1640 offers significant advantages over a simpler medium such as Trowell's T-8. Not only should additional media and combinations of media be tested for their influence on explant survival, but such fundamental information as the optimal concentration and type of sera to be added to the basic culture medium must be established. Optimal medium concentrations of glucose and insulin have not been established nor have the effects of other humoral additives such as corticosteroids, enteroglucagon, epidermal growth factor, or gastrin on explant survival or metabolic function been determined. Whether an environment of 95% 0_2 and 5% CO_2 is best or whether somewhat lower oxygen concentrations or hyperbaric culture conditions might offer advantages are not known. Similarly, rigidly controlled studies are needed to determine whether placing explants on wire mesh floating on medium in static dishes or placing explants directly in plastic dishes maintained on a rocking platform offer specific advantages for small intestinal mucosa. The latter method has been used successfully for relatively long-term culture of explants of human colonic mucosa (Autrup *et al.,* 1978).

Whereas it seems likely that systematic assessment of different culture conditions would result in modifications of techniques which would permit longer survival of mucosal explants of small intestinal mucosa than is possible with existing methods, a major factor which has constrained such studies is the availability of fresh normal human intestinal mucosa. In most studies of normal small intestinal mucosa, tissue has been obtained by peroral suction biopsy of the proximal intestinal of normal volunteers. Such biopsies seem ideal for explant culture but the number of biopsies which can be obtained safely and ethically from a given volunteer or patient at any one time is limited to four to ten biopsies averaging 10 mg wet weight per biopsy. Another potential source of tissue for culture is normal small intestine removed at surgery (Hauri *et al.,* 1975). Except for proximal duodenum resected in concert with gastrectomy, relatively normal small intestinal tissue is not removed frequently in most centers. Intestinal mucosa obtained at autopsy would be entirely unsuitable due to its rapid autolysis.

It is unfortunate that the requirements for successful culture of explants of small intestine vary among mammalian species (Trier, 1976; Chabot *et al.,* 1978). Because of this, results of studies of growth requirements for culture of mucosal explants from commonly used laboratory mammals cannot be applied unreservedly to the culture of human explants, but utilization of laboratory primate small intestine for such studies should be encouraged.

Despite its limitations, organ culture of intestinal mucosa should prove useful for examining synthetic and secretory functions of human intestine under controlled conditions. Many studies which would require exposure of intestinal mucosa to precursor isotopes are impossible to carry out safely *in vivo* but are feasible using cultured explants *in vitro*. The studies mentioned in the previous sections have demonstrated the application of organ culture methods to the study of synthesis of structural and secretory epithelial cell proteins and glycoproteins, lipoproteins, mucosal production, and release of immunoglobulins and mucosal epithelial cell proliferation and migration. Examination of factors which may modulate such processes in health and disease should be particularly useful in cultured mucosal explants since the diverse intraluminal stimuli, humoral factors, and autonomic influences which occur *in vivo* are eliminated whereas epithelial cell contacts, epithelial–mesenchymal cell relationships, and connective tissue support are maintained. Thus, the direct effects of individual hormones, drugs, or other blood-borne agents and neurotransmitters on mucosal metabolism and secretion can be studied to advantage with explant culture of intestinal mucosa.

On the other hand, the potential usefulness of human mucosal explant culture in the study of transepithelial transport of substances such as water, ions, monosaccharides, and amino acids is less promising. Clear cut luminal and abluminal compartments do not exist in current explant culture systems. Moreover, if test substances are added to the culture medium, adequate contact of the test substance and the absorptive surface may not always be achieved. For example, when micellar solutions of radiolabeled oleic acid were added by us to medium there was little uptake of radiolabeled oleic acid by cultured explants. In contrast, when cultured explants were subsequently immersed in a small volume of oxygenated micellar solution to facilitate its contact with the apical surface of absorptive cells, the uptake of radiolabeled oleic acid was substantial (Fig. 6).

V. Conclusions

Maintenance of explants of mucosa of human small intestine in organ culture is a promising technique and permits survival of metabolically active, differentiated, and morphologically intact mucosal elements for longer than other available *in vitro* methods. It seems likely that systematic studies designed to optimize variables in the culture method might permit survival of explants beyond the maximum 48 hours achieved by current techniques. Explant culture of the small intestine will facilitate studies of mucosal metabolism in health and disease that cannot be conducted *in vivo* in man.

Acknowledgment

The studies from our laboratory were supported by National Institutes of Health grant AM-17537 from the United States Public Health Service.

References

Autrup, H., Barrett, L. A., Jackson, F. E., Jesudason, M. L., Stoner, G., Phelps, P., Trump, B. F., and Harris, C. C. (1978). *Gastroenterology* **74**, 1248.

Beeken, W. L., Roessner, K. D., and Krawitt, E. L. (1974). *Gastroenterology* **66,** 998.

Brandborg, L. L., Rubin, C. E., and Quinton, W. E. (1959). *Gastroenterology* **37,** 1.

Browning, T. H., and Trier, J. S. (1969). *J. Clin. Invest.* **48,** 1423.

Chabot, J.-G., Menard, D., and Hugon, J. S. (1978). *Histochemistry* **57,** 33.
DeRitis, G., Falchuk, Z. M., and Trier, J. S. (1975). *Dev. Biol.* **45,** 304.
Dolin, R., Blacklow, N. R., Malmgren, R. A., and Chanock, R. M. (1970). *J. Infect. Dis.* **122,** 227.
Falchuk, Z. M., Gebhard, R. L., Sessoms, C., and Strober, W. (1974). *J. Clin. Invest.* **53**, 487.
Fluge, G., and Aksnes, L. (1978). *In* "Perspectives in Coeliac Disease" (B. McNicholl, C. F. McCarthy and P. F. Fottrell, eds.), pp. 91-99. MTP Press, Lancaster, England.
Gebhard, R. L., and Cooper, A. D. (1978). *J. Biol. Chem.* **253,** 2790.
Ginsel, L. A., van der Want, J.J.L., and Daems, W.Th. (1977). *Cell Tiss. Res.* **181,** 143.
Ginsel, L. A., Onderwater, J.J.M., and Daems, W.Th. (1979). *Virchows Arch. B Cell Path.* **30,** 245.
Hauri, H. P., Kedinger, M., Haffen, K., Grenier, J. F., and Hadorn, B. (1975). *Biol. Gastroenterol. (Paris)* **8**, 307.
Hauri, H. P., Kedinger, M., Haffen, K., Freiburghaus, A., Grenier, J. F., and Hadorn, B. (1977). *Biochim. Biophys. Acta* **467,** 327.
Hauri, H. P., Kedinger, M., Haffen, K., Gaze, H., Hadorn, B., and Hekkens, W. (1978). *Gut* **19,** 1090.
Jos, J., Lenoir. G., DeRitis, G., and Rey, J. (1975). *Scand. J. Gastroenterol.* **10,** 121.
Jos, J., Charbonnier, L., Mougenot, J. F., Mosse, J., and Rey, J. (1978). *In* "Perspectives in Coeliac Disease" (B. McNicholl, C. F. McCarthy and P. R. Fottrel, eds.), pp. 75-89. MTP Press, Lancaster, England.
Kagnoff, M. F., Donaldson, R. M., Jr., and Trier, J. S. (1972). *Gastroenterology* **63,** 541.
Kagnoff, M. F., Serfilipi, D., and Donaldson, R. M., Jr. (1973). *J. Immunol.* **110,** 297.
Katz, A. J., and Falchuk, Z. M. (1978). *Gastroenterology* **75,** 695.
Katz, A. J., Falchuk, Z. M., Strober, W., and Schwachman, H. (1976). *N. Engl. J. Med.* **295,** 131.
Kedinger, M., Haffen, K., and Hugon, J. S. (1975). *Cell Tissue Res.* **156,** 353.
L'Hirondel, C., Doe, W. F., and Peters, T. J. (1976). *Clin. Sci. Mol. Med.* **50,** 425.
Lichtenberger, L., Miller, L. R., Erwin, D. N., and Johnson, L. R. (1973). *Gastroenterology* **65,** 240.
Loder, R. M., Mueller, V. C., Trier, J. S., Dobbins, W. O., III, Barrett, B., and Rubin, C. E. (1964). *Gastroenterology* **46,** 418.
Mitchell, J. D., Mitchell, J., and Peters, T. J. (1974). *Gut* **15,** 805.
Pisam, M., and Ripoche, P. (1976). *J. Cell Biol.* **71,** 907.
Plattner, H., Klima, J., Mehnert, A., and Berger, H. (1970). *Virchows Arch. B Zellpathol.* **6,** 337.
Quaroni, A., Wands, J., Trelstad, R. L., and Isselbacher, K. J. (1979). *J. Cell Biol.* **80,** 248.
Rachmilewitz, D., Fainaru, F., and Eisenberg, S. (1978). *Gastroenterology* **74,** 1081.
Stern, B. K., and Jensen, W. E. (1966). *Nature (London)* **209,** 789.
Strober, W., Falchuk, Z. M., Rogentine, G. N., Nelson, D. L., and Klaeveman, H. L. (1975). *Ann. Intern. Med.* **83,** 242.
Townley, R.R.W., Cornell, H. J., Bhathal, P. S., and Mitchell, J. D. (1973). *Lancet* **1,** 1363.
Trier, J. S. (1974). *In* "Coeliac Disease" (W.Th.J.M. Hekkens and A. S. Pena, eds.), pp. 81-88. Stenfert Kroese, Leiden, Holland.
Trier, J. S. (1976). *N. Engl. J. Med.* **295,** 150.
Trier, J. S., and Browning, T. H. (1970). *N. Engl. J. Med.* **283,** 1245.
Trowell, O. A. (1961). *Colloq. Int. N. R. S.* **101,** 237.
Weiser, M. M. (1973). *J. Biol. Chem.* **248,** 2536.
Wilson, T. H., and Wiseman, G. (1954). *J. Physiol.* **123,** 116.
Wyatt, R. G., Kapikian, A. Z., Thornhill, T. S., Sereno, M. M., Kim, H. W., and Chanock, R. M. (1974). *J. Infect. Dis.* **130,** 523.

Chapter 19

Explant Culture of Human Colon

HERMAN AUTRUP

Human Tissue Studies Section,
Laboratory of Experimental Pathology,
National Cancer Institute,
Bethesda, Maryland

I. Introduction

Explant culture systems of mammalian large intestine have been limited by the rapid necrosis and degeneration of the mucosal epithelial cells. Short-term incubation of small fragments of human colonic tissue and everted sacs of animal colon in oxygenated buffer has permitted survival of the tissue for only a couple of hours. More recently, two different approaches to culture colonic tissue have been taken (Table I). In the stationary method, the tissue is grown on a support—either a matrix or on a steel grid. Using human fibrin foam as the matrix, Schiff (1975) was able to maintain embryonic hamster and rat colon for

ISBN 0-12-564140-0

TABLE I

EXPLANT CULTURE OF NORMAL COLON

Species	Technique	Medium	Survival	Reference
Mouse	Stationary; membrane filter on grid	Waymouth's MB 752/1 + 10% FCS	Several weeks	Defries and Frank (1977)
Hamster (fetal)	Stationary; fibrin foam	Leibovitz's L-15; Waymouth's MB 752/1 + 10% FCS	Up to 3 weeks	Schiff (1975)
Rat	Rocking	CMRL 1066 + 5% BSA	Up to 4 weeks	Autrup *et al.* (1978b)
		CMRL 1066 + 10% FCS	Up to 9 weeks	Shamsuddin *et al.* (1978)
Rabbit	Stationary; steel grid	Trowell's T-8 + 10% FCS	Up to 3 days	Mak and Chang (1978)
Human	Stationary; steel grid	Trowell's T-8 + 10% FCS	Up to 24 hours	O'Gorman and La Mont (1978)
Human	Rocking; gelatin sponge	CMRL 1066 + 5% BSA	Up to 22 days	Autrup *et al.* (1978a)

up to 3 weeks. Using a steel grid as support, adult mouse colon has been maintained in a modified form for several weeks (Defries and Frank, 1976), rabbit colon for up to 3 days (Mak and Chang, 1978), and human colon for at least 24 hours (Browning and Trier, 1969; Trier, 1976). Using a rocking culture method, we were able to maintain rat colon for at least 28 days in a chemically defined medium (Autrup *et al.*, 1978b) and for 63 days in a serum-containing medium (Shamsuddin *et al.*, 1978). Using geliatin sponge as growth support in the rocking method, we can now maintain normal human colonic mucosa for up to 28 days in explant culture. The viability of the tissue was determined by biochemical, immunological, and morphological markers.

II. Materials and Methods

A. Collection and Preparation of Tissue for Culture

Nontumorous human colonic tissues were obtained at the time of either surgery or "immediate" autopsy from patients with and without cancer of the colon (Autrup *et al.*, 1978a). The specimens were cleaned of mesentery fat to the colonic serosal layer, cut into 1 × 5-cm pieces, immersed in L-15 medium [Grand Island Biological Co. (GIBCO), Grand Island, NY] containing penicillin G (300 U/ml) and streptomycin (300 μg/ml) (GIBCO) and kept at 4°C for 3 to 20 hours until cultured.

Prior to initiation of the explant culture, the mucosal layers of nontumorous colonic specimens were dissected away from the muscularis externa with a pair of small surgical scissors and the explant was cut into pieces (5 × 5 mm) with a surgical blade (#20). All the preparation procedures were performed in L-15 medium. The explants were placed on top of gelatin sponge (Gelfoam, Upjohn, Kalamazoo, MI) squares (1 × 1 cm) in 60-mm plastic tissue culture dishes with the epithelium facing the gas-medium interface. The gelatin sponge was attached to an area of the dish scraped with a gauge 19 needle, and was located near the edge of the dish. To facilitate the diffusion of nutrients and oxygen to the total tissue, care was taken to stretch out the tissues on the top of the gelatin sponge without inflicting traumatic cell injury.

B. Growth Conditions

Three milliliters of culture medium was added to each dish. The medium was generally CMRL 1066 (GIBCO) supplemented with glucose (15 m*M*, Sigma Chemical, St. Louis, MO), methionine (7 μM, Sigma), tricine buffer (20 m*M*, pH 7.4), hydrocortisone hemisuccinate (1.5 μM, Upjohn, Kalamazoo, MI), β-retinyl acetate (1 μg/ml, Hoffman-LaRoche, Nutley, NJ), glutamine (3 m*M*), 5% bovine plasma albumin (Reheis Chemical Co., Chicago, IL), 1.5% DMSO (Pierce Chemical Co., Rockford, IL), penicillin G (100 U/ml), streptomycin (100 μg/ml), gentamicin (50 μg/ml; Schering Corporation, Kenilworth, NJ), and amphotericin B (0.25 μg/ml, GIBCO). Amphotericin B was included only in the first two medium changes. The media were changed after 24 hours in culture and then every other day.

The cultures were placed in a controlled atmosphere chamber (Bellco Glass, Vineland, NJ) and gassed with 95% O_2–5% CO_2 for 5 minutes before incubation at 36.5°C. The chamber was rocked at approximately 10 cpm so that the colonic tissue was submerged in the culture medium 50% of the time.

C. Determination of Viability of Cultured Tissues

1. Morphology

Human colonic explants were fixed either in a phosphate-buffered solution of 4% formaldehyde–1% glutaraldehyde (McDowell and Trump, 1976) before embedding in Epon-Araldite or in 10% formalin before embedding in paraffin. One-micrometer sections of the plastic-embedded material were stained with toluidine blue. The paraffin sections were stained with either hematoxylin and eosin or with periodic acid–Schiff (PAS) for detection of mucin. Ultrathin sections were cut from the Epon-embedded tissue, stained on the grid with uranyl magnesium acetate followed by lead citrate, and examined in a JEOL 100B electron microscope.

2. Immunological Markers

Paraffin sections were stained for secretory component by the immunoperoxidase technique or by immunofluorescent techniques (Brown, 1977; Katoh *et al.*, 1979).

3. Biochemical Markers

The incorporation of precursors into cellular macromolecules was determined by both autoradiographic methods (Autrup *et al.*, 1978a) and scintillation spectrophotometry. The explants were incubated with either [^{3}H]Tdr (50 μCi; 49 Ci/mmole; Amersham/Searle, Arlington Heights, IL) or [^{3}H]Leu (50 μCi; 27 Ci/mmole; Amersham/Searle) for 4 hours. At least two explants were combined for each determination.

III. Results

A. Survival of Explants

Nontumorous human colonic tissue was routinely cultured for 14 days and a few cases up to 28 days (Table II). The increase in survival time and in the number of explants having viable epithelium after 14 days compared to our previously published observations (Autrup *et al.*, 1978a) was attributed to the incorporation of 1.5% DMSO in the culture medium and the use of gelatin sponge as support for the explants. Both intraindividual and interindividual variations were observed in the survival of the explants. Attempts have been made to culture normal colonic mucosa from a total of 38 different donors using the new, modified culture conditions. Ninety percent of all the cases had viable epithelium after 7 days in culture, and 48% after 14 days. The intraindividual variation was mostly due to improper attachment of the gelatin sponge to the dishes, so the tissue was not submerged 50% of the time. The tissue could be kept in L-15 media at 4°C for up to 24 hours before initiation of explant cultures, although the best survival of the explants was achieved when the cultures were initiated immediately after removal of the tissue from the donor. There was no apparent difference in survival time of the explant culture from the different anatomical segments of the colon, a finding which is in contrast to our observations in the culture of rat colon. Tissues obtained from immediate autopsy generally survived for shorter periods of time than tissues from surgery. In addition, autopsy tissues were more prone to infection with microbial agents.

TABLE II

SURVIVAL OF HUMAN COLON IN EXPLANT CULTURE[a]

Patient number	Anatomical site	Survival time of culture	Structural morphology
156E	Sigmoid	17	Single layer (S.L.) of epithelium; few glands
156D	Ascending	21	Few glands, mucus-producing (M-P) cells entire length of crypt at day 14
156C	Descending	21	Few glands, M-P cells entire length of crypt at day 14
152A	Descending	28	S.L. of epithelium; few glands, M-P cells at day 21
152	Transverse	21	S.L. of epithelium; few glands
151	Descending	21	S.L. of epithelium; many glands; M-P cells
150C	Sigmoid	18	S.L. of epithelium
150B	Ascending	14	S.L. of epithelium; several glands; M-P cells
150A	Rectum	28	S.L. of epithelium; few glands; M-P at day 21
150	Ascending	28	S.L. of epithelium

[a] Attempts were made to culture the explants for 28 days.

B. Culture Conditions

Several different media, i.e., Eagle's minimal essential medium, Parsa, L-15, Ham's F-12, CMRL 1066, were initially tested for their ability to support the growth of the explants. CMRL 1066 was chosen because it gave more consistent results in maintaining colonic explants for longer periods of time in culture, when compared to the other media.

Several types of sera were examined for their ability to sustain explant survival, i.e., human serum, human placental cord serum, horse serum, and fetal bovine serum in concentrations of 1 to 10%. The best results were obtained by using either heat-activated fetal bovine serum (5%) or bovine albumin (5%). The only difference between the fetal bovine serum and albumin was that the bovine serum stimulated the growth of stromal fibroblasts cells in the lamina propria. Bovine albumin (fraction V) was generally used, as it allowed us to use a chemically defined system.

The vitamin and hormone concentrations in the medium were similar to those used in our bronchial explant system. No improvement in either survival or ultrastructural features of the explants was observed by varying the concentrations of hydrocortisone (0 to 3 μM), insulin (0 to 50 μg/ml), or β-retinyl acetate (0 to 10 μg/ml). Inclusion of thyroxine (0 to 1.0 μg/ml), which affects the differentiation of intestinal cells in animals, in the culture medium had no observable effect on the maintenance of human colon explants. More recently, addition of dexamethasone (25 μg/ml) and pentagastrin (15 μg/ml) to the media has been found to increase the synthesis of DNA in the mucosal cells and to improve the length of survival of the colonic explants.

DMSO (0.5–2%), which has been shown to induce maturation in cultured tumor cells (Friend *et al.*, 1971), increased the length of survival of colonic tissues in explant culture. The best results were obtained at a concentration of 1.5%. One of the initial problems in culturing human colon was a rapid lowering of the pH of the media, as indicated by the shifts of the pH indicator in the media. This was partly caused by increased production of lactic acid. Increasing the 0_2 tension from 20 to 95% resulted in a 40% decrease of the rate of production of lactic acid (Table III) and significantly reduced the ultrastructural damage in the explants. The addition of a tricine buffer (20 m*M*) also helped to stabilize the pH of the medium. Because glucose was heavily utilized during the culture (more than 50% was utilized within 48 hours), the culture medium was supplemented with additional glucose. An interesting observation was the high utilization of methionine by cultured colon, under conditions in which the other amino acids were either released into the medium or possibly formed by proteolysis of the albumin (Table IV). The concentration of methionine in the media after 24 hours in culture was about 20% of the level at the beginning of the culture. Since methionine is a major amino acid component of mucoproteins, this may reflect the requirement of methionine for the synthesis of mucus.

Two different types of growth support have been examined, gelatin sponge and Gelman GA-4 filters. Both improved the structural integrity of the explants,

TABLE III

EFFECT OF OXYGEN CONCENTRATIONS ON LACTATE PRODUCTION AND GLUCOSE UTILIZATION[a]

Day in culture	95% O_2–5% CO_2	50% O_2–45% N_2 –5% CO_2	95% Atmosphere air–5% CO_2
		Lactate (μmoles/ml media)	
0	0.01	0.01	0.01
1	2.2	3.3	5.0
2	4.3	6.3	6.7
3	6.6	6.4	9.6
4	8.8	9.1	9.3
		Glucose (μmoles/ml media)	
0	5.5	5.5	5.5
1	4.0	4.1	2.4
2	2.3	1.9	1.6
3	0.8	0.8	0.5
4	1.2	0.2	0.7

[a] Media from two explants were harvested after 1, 2, 3, and 4 days of culture without media change. The lactate concentration was determined by the LDH–NAD spectrophotometric assay and glucose by GOD–peroxidase assay. The results are the average of duplicate assays.

TABLE IV

UTILIZATION OF SOME AMINO ACIDS BY CULTURED HUMAN COLON[a]

	0 hours	24 hours	48 hours
Glutamine	1,600	1,600	800
Glycine	666	800	754
Alanine	366	774	758
Valine	286	356	300
Methionine	962	156	138
Isoleucine	170	248	201
Phenylalanine	186	278	242

[a] Media from four explants were harvested after either 24 or 48 hours incubation and the amino acid concentration was determined by ion-exchange chromatography, and expressed as nmoles/ml media.

gelatin sponge giving the best results. However, due to the release of protolytic enzymes, the sponge becomes digested after generally 21 to 28 days of culture.

C. Viability

The viability of the explants was monitored either by examining the incorporation of precursors into cellular macromolecules, DNA, and protein, or by observing the changes in ultrastructure of the colon. The explants incorporated both [^{3}H]leucine and [^{3}H]TdR at 1, 2, 4, 5, and 7 days, and the maximum levels for both were reached at the fifth or seventh day in culture (Table V). Autoradiograms revealed that the epithelial cells incorproated these precursors more than the stromal cells, and that the incorporation occurred predominately in the middle and lower part of the crypt. Cells labelled with [^{3}H]TdR for 8 hours after initiation of the culture migrated up through the glands and were observed at the luminal surface at day 7. Incorporation of radioactive precursors during culture occurred mainly in the middle and lower portion of the glands.

MORPHOLOGICAL MARKERS

The near normal appearance of the epithelium was maintained up to 14 days (Figs. 1 and 2). The luminal surface were normally lined with columnar epithelial cells. The intestinal glands were lined mostly by mucous-producing cells, some columnar epithelial cells, and undifferentiated cells at the bottom of the glands. The length of the glands was similar to normal tissue. The mucous-producing cells were also demonstrated by staining of the fixed explant with PAS. Good integrity between muscularis mucosa and the lamina propria was also observed. The lamina propria consisted of a mixed cell population,

TABLE V

INCORPORATION OF [^{3}H]THYMIDINE INTO DNA AND [^{3}H]LEUCINE INTO PROTEIN BY CULTURED HUMAN COLON[a]

Duration in culture	[^{3}H]Tdr incorporation			[^{3}H]Leucine incorporation	
	Case 169A	169B	169C	169A	169B
0	21	80	12	845	3437
1	8	17	20	789	1146
2	10	41	56	610	1063
3	95	100	28	1698	1416
4	246	553	265	760	4879
5	192	378	103	670	444
6	75	311	65	309	n.d.[b]

[a] dpm per microgram of either protein or DNA.
[b] n.d., not done.

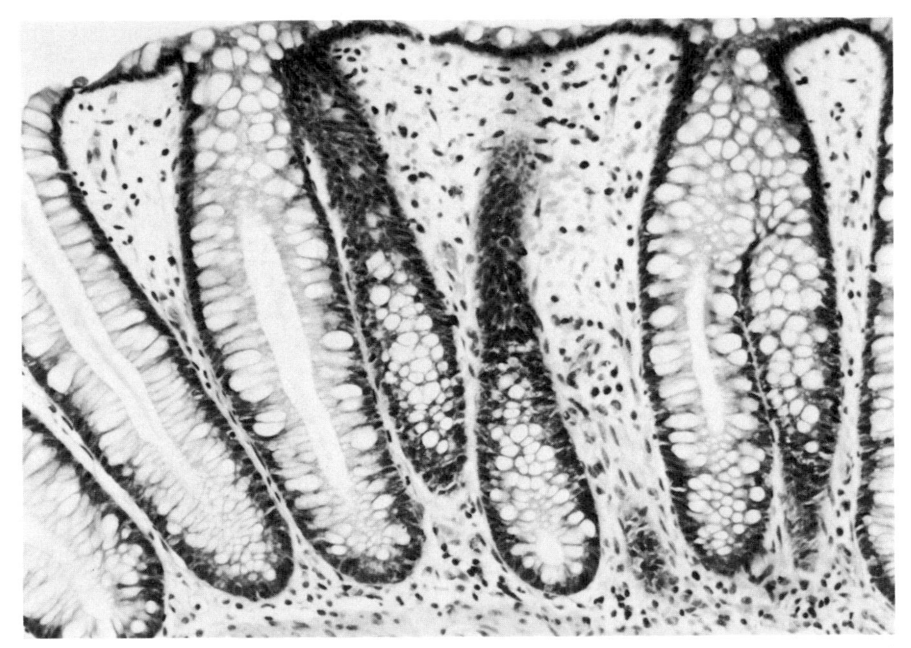

FIG. 1. Human colonic mucosa cultured for 8 days. The depth of the gland was similar to normal tissue and was lined with mucous-producing cells. H & E. ×220.

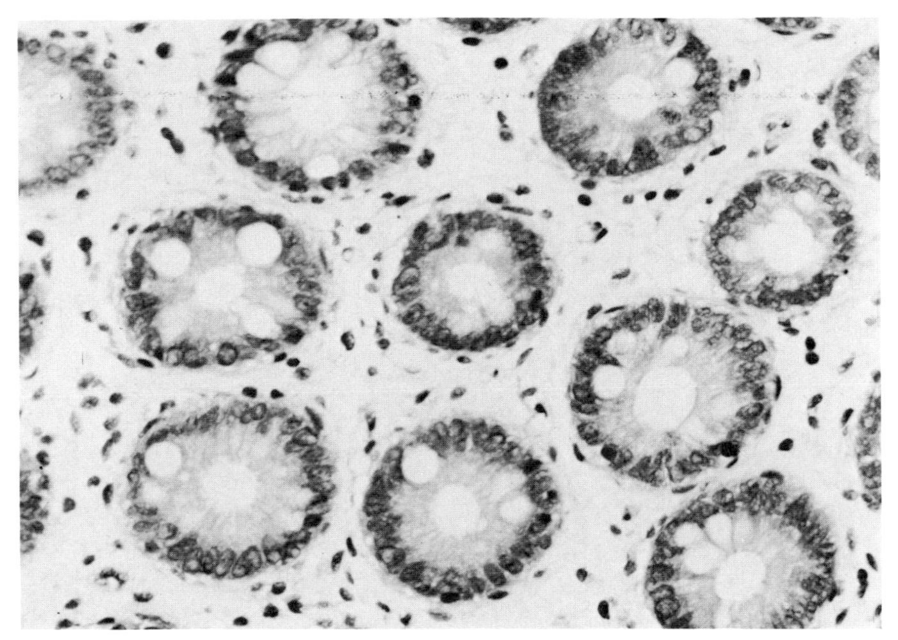

FIG. 2. Human colonic mucosa cultured for 14 days. The glands show both mucous-producing cells and normal columnar epithelium. H & E. ×330.

containing lymphocytes, macrophages, and a few fibroblastic cells. Culturing the tissue in a serum-containing medium increased the proliferation of fibroblast cells into the lamina propria. If the explants were not properly attached to the gelatin sponge, necrosis was observed in the middle of the explant, while viable tissue was still observed at the edges. With increasing time in culture a progressive decrease in the number of glands and the length of the glands was observed. A single layer of cuboidal epithelial cells and a few glands were generally seen after 21 days in culture, while the number of mucous-producing cells was dramatically reduced. Dilation of the glands and vacuolization of the surface epithelium were also noticed (Fig. 3). Migration of squamous epithelial cells around the edge of the explants and repopulation of the submucosa was observed when the explants were grown on gelatin sponge. Occasionally the surface of the submucosa was repopulated with a glandular-like structure. In a few cases, total necrosis of the explants was observed during the early period of culture, and this was followed by repopulation of muscularis mucosa with cuboidal epithelial cells after approximately 4 weeks in culture.

When the explants were cultured in a stationary system on a glass coverslip, outgrowth was observed after 3–4 weeks in culture. This outgrowth was predominantly fibroblastic, as shown by their morphology under the light microscope

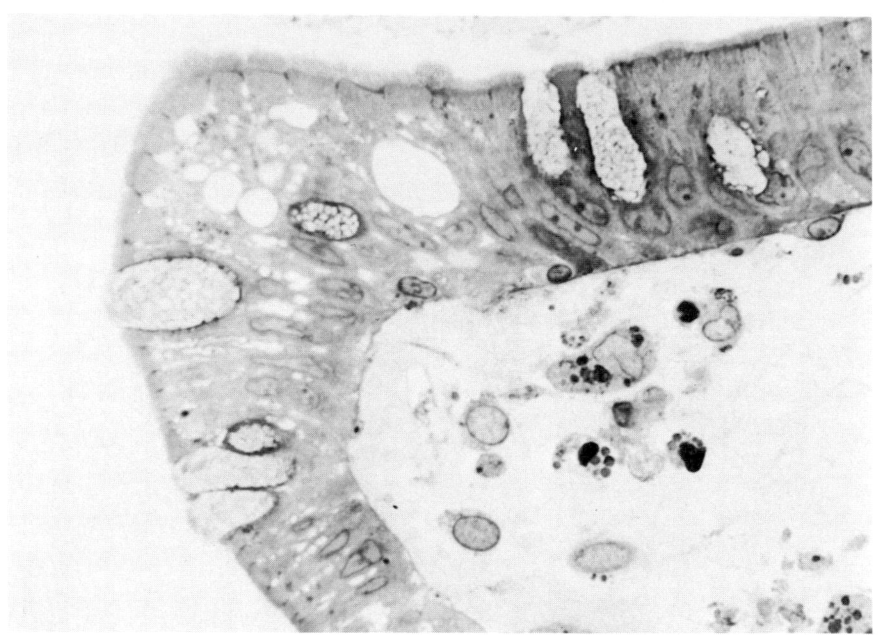

FIG. 3. Human colonic mucosa cultured for 21 days shows a decreased number of goblet cells but the cells still retain their tall columnar architecture (1-μm section). Toluidine blue. ×880.

and their lack of blood group specificity when stained by immunofluorescence (Katoh *et al.*, 1979). Blood group antigens (A, B, and H) are localized in epithelial cells of most human tissues including the colon.

Electron microscopy of the explants after 14 days in culture showed good structural and functional viability of the epithelial cells (Fig. 4). The cells had several desmosomes and a well-developed basal lamina was evidemt.

Secretory component, a protein involved in the transport of immunoglobulins across the intestinal membrane (Brown, 1977), was present in colonic epithelial cells both prior to culture and after 14 days in culture as shown by both immunofluorescent and immunoperoxidase techniques. The intensity of the stain for secretory component was reduced after 14 days in culture as compared to the staining prior to culture. The staining of the explant indicates that the colonic epithelial cells maintained some functional viability in explant culture.

IV. Discussion

Using our modified culture conditions for human colon, we have significantly increased the length of time in which colon explants can be maintained in culture,

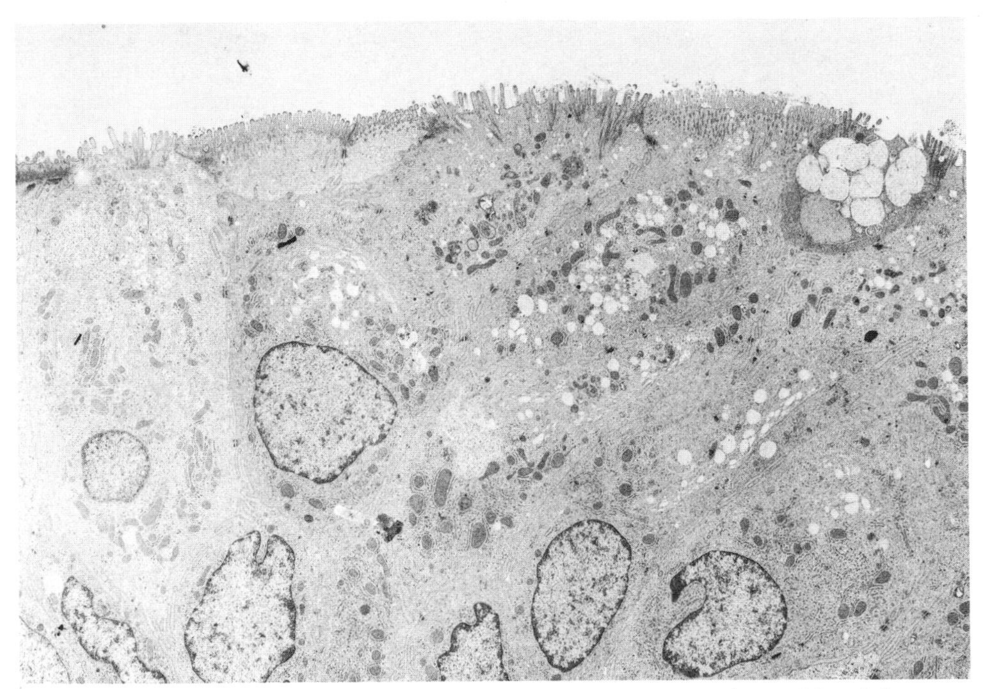

FIG. 4. Absorptive and undifferentiated cells of human colonic mucosa cultured for 14 days. ×4680.

as compared to our early reports (Autrup *et al.*, 1977, 1978a). The explants could generally be maintained up to 14 days with near normal morphology after which time we observed a decrease in both the number of intestinal glands and the number of mucus-producing cells. The improvement was especially noted after addition of DMSO (1.5%) to the media. At the moment, we do not know the mechanism of the enhanced viability of tissues in the presence of DMSO. Addition of polar solvents to tissue culture media of cultured human colon carcinoma lines has resulted in a significant change in both growth parameters and in cellular morphology (Dexter *et al.*, 1979). Introduction of gelatin sponge as a growth support also increased the viability of the explants as observed in the explant culture of peripheral lung (Stoner *et al.*, 1978). The function of the sponge is probably of a physical nature, as it allows the tissue to become stretched out and opening up of the glands to allow for diffusion of oxygen and nutrients to the bottom of the crypts. A similar effect was observed using 0.45 μm filters, but the results were not as consistent as by using gelatin sponge due to floating of the filter paper when the explants were rocked. One of the major problems in culturing human colon is the diffusion of nutrients to the glands. The coat of mucous released by the goblet cells, which lines the luminal surface, could be a major obstacle in this diffusion. Preliminary experiments on the inclusion of low concentrations of bile acids in the medium has resulted in slightly improved morphology of the explants. The bile acids have been shown to decrease the viscocity of mucous (Martin *et al.*, 1978) which could then be washed away from the luminal surface by the rocking motion. Physical removal of the mucous has been unsuccessful.

The phenomenon of encircling of the explant with epithelial cells has also been observed in cultured human bladder (Knowles *et al.*, 1980) and in human bronchus (Harris *et al.*, unpublished observation). The outgrowth of cells onto the surface of glass or plastic tissue culture dishes from explants was slow compared to the outgrowth from human bronchus (Stoner *et al.*, 1980), and the cells did not stain strongly for the blood group antigens. Presently, the effect of several different hormones and nutritional supplements on the stimulation of growth and on improvement of the structural morphology of human colonic explants in culture is being investigated.

V. Application and Perspective

Explant cultures of human colon have been used in the study of biochemical parameters in both normal and diseased colonic tissues and for studies on colon carcinogenesis (Table VI).

A. Study of Carcinogen Metabolism

One of the purposes for the development of a colon explant system was to create a system in which we could correlate carcinogenesis data obtained from

TABLE VI

APPLICATION OF EXPLANT CULTURE OF HUMAN COLON

Application	Tissue	References
Glycoprotein synthesis and secretion	Normal and cancerous	O'Gorman and La Mont (1978)
Glycoprotein synthesis and secretion	Normal and cystic fibrosis	Neutra *et al.* (1977)
Immunoglobin pattern	Hirschsprung's disease	Halpin *et al.* (1978)
Prostaglandin E_2	Ulcerative colitis (remission)	Rachmilewitz *et al.* (1978)
Unscheduled DNA synthesis	Normal	Mak *et al.* (1979)
Carcinogen metabolism	Normal	Autrup *et al.* (1978a–d, 1979a,b)
Xenotransplantation	Normal	Valario *et al.* (1980)

TABLE VII

METABOLISM OF CARCINOGENS BY CULTURED COLON[a]

Carcinogen	Concentration (μM)	Human	CD rat
Benzo[*a*]pyrene	1.5	5.2 ± 4.8 (0.2–20.0; 87 cases)	3.5 ± 0.3 (7 exp.)
7,12-Dimethyl-benzanthracene	3.2	7.8 ± 3.8 (4.0–13.8; 6 cases)	5.2 ± 0.5 (3 exp.)
Aflatoxin B_1	0.5	2.0 ± 2.1 (0–6.9; 16 cases)	125 ± 2 (3 exp.)
1,2-Dimethyl-hydrazine	100	762 ± 761 (109–3273; 55 cases)	1,353 ± 32 (7 exp.)
N,N-Dimethyl-nitrosamine	100	123 ± 156 (12–570; 14 cases)	845 ± 59 (3 exp.)

[a] After 24 hours in culture radioactively labeled carcinogen was added to the media. After incubation for an additional 24 hours with the carcinogen, the tissue was washed three times with Dulbecco's phosphate-buffered saline. The mucosa from four and five explants was scraped from the supporting stroma with a surgical blade and pooled for each analytical determination. DNA was isolated using a phenol extraction method, purified on a CsCl gradient, and carcinogen binding-assayed as previously described (Autrup *et al.*, 1977). The results are given as mean ± SD.

experimental animals with the human situation. Explant cultures of human and rat colon provide a system in which such a correlation could be made on the same biological organizational level. Both cultured human and rat colon were shown to be able to metabolize procarcinogens from various classes of chemicals, such as polynuclear aromatic hydrocarbons, *N*-nitrosamines, dialkylhydrazine, and a mycotoxin (Table VII; Autrup *et al.*, 1978a–d, 1979a). The mean level of binding of aflatoxin B_1 and 1,2-dimethylhydrazine (DMH), both of which have been shown to induce cancer of the colon in rat (Druckrey, 1972; Newberne and Rogers, 1973), was higher in rat colonic DNA than in human colonic DNA. The carcinogen–DNA adducts formed were qualitatively similar in both species (Autrup *et al.*, 1979b). The ability of colonic tissue to metabolize benzo[*a*]pyrene (BP) into ultimate carcinogenic forms could be an indicator for assessing an individual's susceptibility for colonic carcinoma caused by BP. The binding level of BP to cellular DNA showed a 100-fold interindividual variation (Autrup *et al.*, 1978a) and had a one-sided normal distribution (Fig. 5). Several factors which could, in part, explain this variability have previously been discussed (Autrup *et al.*, 1978a). When the metabolism of two different carcinogens, BP and DMH, was compared in cultured colonic tissue from the same patient, a positive correlation was observed (Fig. 6), which indicates that the same activation system is involved in the activation of carcinogens of different chemical classes.

The influence of potential inhibitors and suspected cocarcinogenic factors in colon carcinogenesis can also be studied in this experimental system.

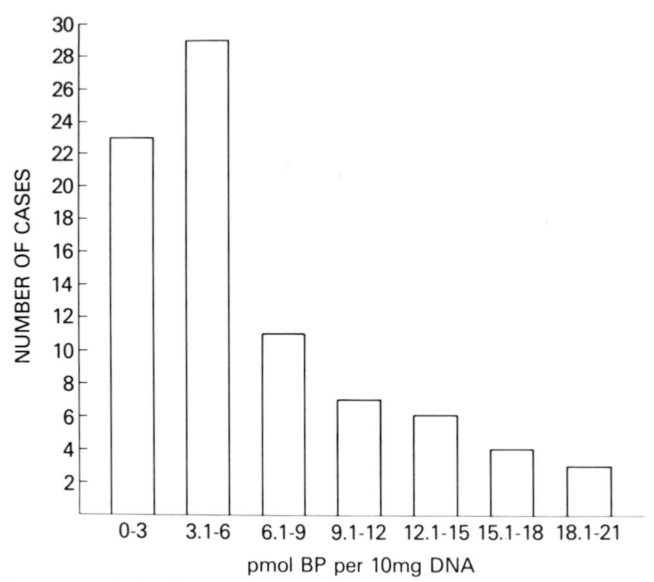

FIG. 5. Frequency distribution of binding levels of benzo[*a*]pyrene to DNA in cultured human colon (87 cases).

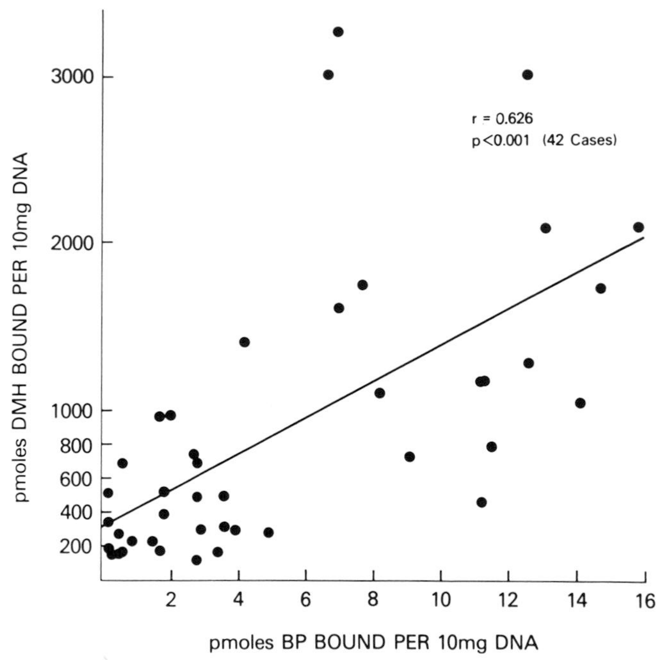

FIG. 6. Correlation between binding levels of benzo[*a*]pyrene and 1,2-dimethylhydrazine to DNA in cultured human colon (42 cases).

B. Xenotransplantation

The expression time for chemically altered cells to malignant transformation by far exceeds the time in which colonic tissue can be maintained in explant culture. A combination of the *in vitro* system with an *in vivo* system has been developed (Valario *et al.*, 1980). Using this approach we have found that adult human colonic specimens can be maintained as subcutaneous xenotransplants in athymic nude mice for periods in excess of 3 months. After an ischemic crisis and revascularization, the colonic epithelium regenerates and attains a normal morphology by approximately 7 weeks (Fig. 7). The tissue can be cultured for up to 14 days prior to implantation, which allows exposure of the tissue to carcinogens and various promotors prior to xenografting. This combined approach may provide a system in which promotion and progression in human colon carcinogenesis can be studied.

C. Perspectives

One of the difficulties in biomedical research is the lack of comparable experimental systems in which to study the morphological and biochemical responses in both animal and human tissue—normal and diseased. The recent

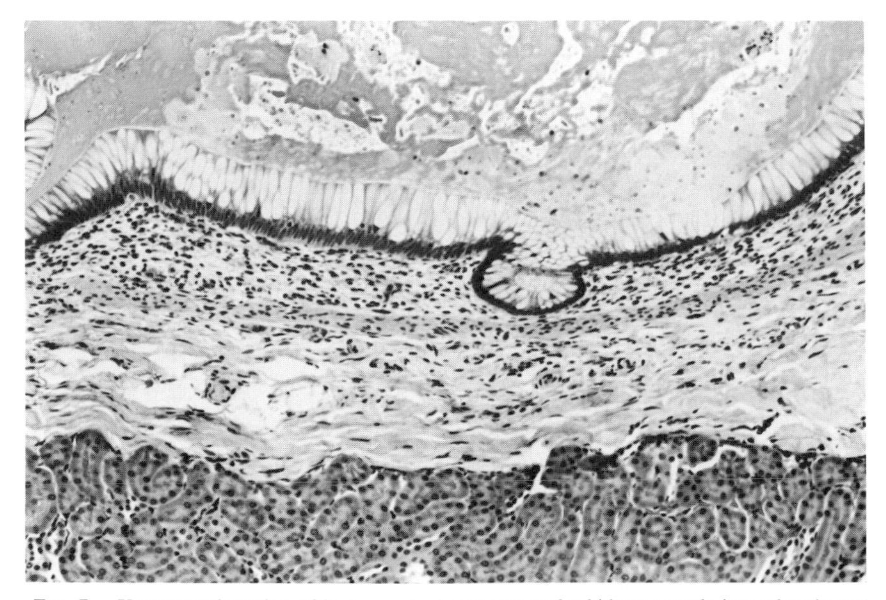

FIG. 7. Xenotransplantation of human colonic mucosa under kidney capsule into athymic nude mouse after 7 weeks. The tissue was maintained in explant culture for 4 days prior to implantation. H & E. ×330.

development of long-term cell and explant cultures of human tissues has made such comparisons possible. Using the explant culture system of human colonic tissue, we have studied the first step in the carcinogenic process, i.e., the interaction of carcinogen with cellular DNA, and work is in progress to study the persistence of such damage and the biological consequences thereof. The system could also be used to identify and to investigate the mechanism of potential cocarcinogenic and promoting compounds, and could be used to study chemoprevention of colonic cancer.

Acknowledgment

The author wishes to thank Dr. B. F. Trump, Department of Pathology, University of Maryland and Dr. L. Smith, National Naval Medical Center for supplying the colonic tissues, Dr. Y. Katoh for performing the immunostaining for the secretory component, Drs. C. Harris and G. D. Stoner for comments during the development of the culture conditions and in the preparation of this manuscript, and Dr. M. Valario, Litton Bionetics for xenotransplantation experiments. The technical assistance of Ms. R. Schwartz, Ms. M. Yamaguchi, and Mr. F. Jackson and the secretarial assistance of Ms. M. Bellman are greatly appreciated.

References

Autrup, H., Harris, C. C., Stoner, G. D., Jesudason, M. L., and Trump, B. F. (1977). *J. Natl. Cancer Inst.* **59,** 351.

Autrup, H., Barrett, L. A., Jackson, F. E., Jesudason, M. L., Stoner, G. D., Phelps, P., Trump, B. F., and Harris, C. C. (1978a). *Gastroenterology* **74,** 1248.

Autrup, H., Stoner, G. D., Jackson, F., Harris, C. C. Shamsuddin, A.K.M., Barrett, L. A., and Trump, B. F. (1978b). *In Vitro* **14**, 868.

Autrup, H., Harris, C. C., Trump, B. F., and Jeffrey, A. M. (1978c). *Cancer Res.* **38,** 3689.

Autrup, H., Harris, C. C., and Trump, B. F. (1978d). *Proc. Soc. Expl. Biol. Med.* **159,** 111.

Autrup, H., Essigmann, J. M., Croy, R. G., Trump, B. F., Wogan, G. N., and Harris, C. C. (1979a). *Cancer Res.* **39,** 694.

Autrup, H., Essigmann, J. M., and Harris, C. C. (1979b). *Fed. Proc.* **38,** 1073.

Brown, W. R. (1977). *Cancer* **40,** 2497.

Browning, T. H., and Trier, J. S. (1969). *J. Clin. Invest.* **48,** 1423.

Defries, E. A., and Frank, L. M. (1977). *J. Natl. Cancer Inst.* **58,** 1323.

Dexter, D. L., Barbosa, J. A., and Calabresi, P. (1979). *Cancer Res.* **39,** 1020.

Druckrey, E. (1972). *In* "Topics in Chemical Carcinogenesis" (W. Nakahara, S. Takeyama, T. Sugimura and S. Odashima, eds.), pp. 73–103. Univ. Park Press, Baltimore, Maryland.

Friend, C. W., Scher, W., Holland, J. G., and Sato, T. (1971). *Proc. Natl. Acad. Sci. U.S.A.* **68,** 378.

Halpin, T. C., Gregoire, R. P., and Izant, Jr., R. J. (1978). *Lancet* Sept. 16, 606.

Katoh, Y., Stoner, G. D., McIntire, K. R., Hill, T. A., Anthony, R., McDowell, E. M., Trump, B. F., and Harris, C. C. (1979). *J. Natl. Cancer Inst.* **62,** 1177.

Knowles, M. A., Hicks, R. M., Berry, R. J., and Milroy, E. (1980). *In* "Cultured Human Cells and Tissues in Biomedical Research" (C. C. Harris, B. F. Trump and G. D. Stoner, eds.), Vol. II, pp. XX. Academic Press, New York.

McDowell, E. M., and Trump, B. F. (1976). *Arch. Pathol. Lab. Med.* **100,** 405.
Mak, K. M., and Chang, W. W. (1978). *J. Natl. Cancer Inst.* **61,** 799.
Mak, K. M., Slater, G. I., and Hoff, M. B. (1979). *J. Natl. Cancer Inst.* **63,** 1305.
Martin, G. P., Marriott, C., and Kellaway, I. W. (1978). *Gut* **9,** 103.
Neutra, M. R., Grand, R. J., and Trier, J. S. (1977). *Lab. Invest.* **36,** 535.
Newberne, P. M., and Rogers, A. E. (1973). *J. Natl. Cancer Inst.* **50,** 434.
O'Gorman, T. A., and LaMont, J. T. (1978). *Cancer Res.* **38,** 2784.
Rachmilewitz, D., Ligumsky, M., Sharon, P., Karmeli, F., and Zor, U. (1978). *Gastroenterology* **75,** 929.
Schiff, L. J. (1975). *In Vitro* **11,** 46.
Shamsuddin, A.K.M., Barrett, L. A., Autrup, H., Harris, C. C., and Trump, B. F. (1978). *Pathol. Res. Pract.* **163,** 362.
Stoner, G. D., Harris, C. C., Autrup, H., Trump, B. F., Kingsbury, E. W., and Myers, G. A. (1978). *Lab. Invest.* **38,** 685.
Stoner, G. D., Katoh, Y., Foidart, J. M., Myers, G. A., and Harris, C. C. (1980). *In* "Cultured Human Cells and Tissues in Biomedical Research" (C. C. Harris, B. F. Trump and G. D. Stoner, eds.), Vol. I, pp. XX. Academic Press, New York.
Trier, J. S. (1976). *N. Engl. J. Med.* **295,** 150.
Valario, M., Fineman, E. L., Bowman, R. L., Harris, C. C., Stoner, G. D., Autrup, H., Trump, B. F., McDowell, E. M., and Jones, R. T. (1980).

Chapter 20

Establishment and Characterization of Intestinal Epithelial Cell Cultures

ANDREA QUARONI AND ROGER J. MAY

Department of Medicine,
Harvard Medical School,
and The Gastrointestinal Unit,
Massachusetts General Hospital,
Boston, Massachusetts

I. Introduction

The small intestinal epithelium constitutes a system of constant renewal. Continuous cell proliferation is limited to the lower two-thirds of the crypts: as the cells migrate out of the crypts and onto the villus, they lose their capacity to divide and differentiate into absorptive cells, and eventually are extruded into the lumen from the villous tip (Cheng and Leblond, 1974). Cell turnover is very rapid, with a mean cell duration of 2–3 days in most mammals and 3–5 days in man (Cairnie *et al.*, 1965a,b). This system constitutes an important model for the

ISBN 0-12-564140-0

study of the regulation of cell growth, induction of cellular differentiation, and characterization of cell adhesion.

Little is known about the factors which regulate the orderly arrangement of intestinal epithelial cells in different phases of maturation and control the precisely balanced ratio of proliferating and differentiated cells. They are likely to be of importance in the intestine's ability to adapt to bowel resection and to respond to diseases such as gluten-sensitive enteropathy. A large body of work has been done *in vivo* in order to characterize these factors, however, the complexity of the intestinal mucosa, with the possibility of extensive interactions among different cell types, has been a major obstacle to the interpretation of the results of many of these studies. Short- and/or long-term cultures of intestinal epithelial cells, under well-defined experimental conditions and uncontaminated by different cell types, would, therefore, be of considerable importance. Many methods have been described to obtain suspensions of isolated intestinal villus and crypt cells from different experimental animals and man (Harrer *et al.*, 1964, Sjöstrand, 1968; Harrison and Webster, 1969; Iemhoff *et al.*, 1970; Evans *et al.*, 1971; Kaur *et al.*, 1972; Weiser, 1973; Castro *et al.*, 1974; Mian and Pover, 1974; Kremski *et al.*, 1977; Perret *et al.*, 1977; Towler *et al.*, 1978). Although quite useful for different biochemical studies, these cell suspensions have proven rather disappointing in short- and long-term culture, due to the rapid autolysis of the isolated intestinal epithelial cells *in vitro*. Two reports have dealt with the establishment of monolayer cultures of epithelial cells from the small intestine (Henle and Deinhardt, 1957; Lichtenberger *et al.*, 1973). However, the lack of a precise identification of the cellular origin of the cells described in these studies does not allow an evaluation of their potential value. We have recently described a new technique which has allowed us to establish pure cultures of epithelioid cells derived from small intestinal crypt cells (Quaroni *et al.*, 1979). These cells grow rapidly, and can be maintained in culture for long periods of time without significant change in their morphological or functional properties. They appear to have retained, *in vitro,* specific regulatory properties and should provide important information about the factors controlling growth and differentiation of the intestinal epithelium *in vivo*.

II. Methods and Materials

A. Isolated Cell Suspensions

Many different techniques have been described for the preparation of isolated intestinal cell suspensions. Most of them provide a large amount of cells and usually allow a reasonable separation between villous and crypt cells. They can be grouped according to the principle employed for cell dissociation: (*a*) mechanical dissociation (Sjöstrand, 1968; Harrison and Webster, 1969; Iemhoff *et al.*, 1970; Kaur *et al.*, 1972); (*b*) use of chelating agents (Evans *et al.*, 1971; Weiser, 1973), and (*c*) enzymatic digestion, with proteolytic enzymes and/or hyaluronidase (Harrer *et al.*, 1964; Castro *et al.*, 1974; Mian and Pover, 1974; Kremski *et al.*, 1977; Towler *et al.*, 1978). These techniques will not be described here and the reader is referred to the original studies for details.

B. Monolayer Cultures

1. General Culture Conditions and Techniques

Monolayer cultures are grown in plastic petri dishes (Lux Scientific Co.), at 37°C in an atmosphere of 90% air–10% CO_2. The complete medium routinely used consists of: Dulbecco's modified Eagle medium (DMEM)[1] with 4.5 gm/liter glucose, containing 5% fetal bovine serum (irradiated) (FCS), 10 μg/ml insulin, 4 m*M* glutamine, 50 U/ml penicillin, and 50 μg/ml streptomycin. The cultures are refed twice weekly.

Cells are subcultured exclusively by using EDTA-containing buffers as follows: the medium is removed and the cell layer rinsed with 10 ml (for 100-mm dishes) of an EDTA solution (0.02% in phosphate-buffered saline (PBS,10 m*M* phosphate buffer, pH 7.2, 0.154 *M* NaCl); the dishes are incubated at 37°C for 30–60 minutes, then 10 ml of complete medium is added and the cells are suspended into the medium by gentle scraping with a rubber-tipped spatula. The cell suspension is then diluted as desired and transferred to new dishes.

Cells are stored in liquid nitrogen (in plastic vials) in 1-ml aliquots ($2–4 \times 10^6$ cells/ml) in DMEM containing 7.5% dimethyl sulfoxide and 20% FCS. Cell recovery is accomplished by quickly thawing the vial at 37°C, and transferring the cell suspension into a dish containing 10 ml complete medium prewarmed at 37°C; after 10–24 hours the medium is removed and the cells are refed with fresh medium.

Cell counts are performed either electronically with a Coulter counter or by hemocytometer counting. Growth rates are determined by plating cells in complete medium at $2–3 \times 10^4$ cells/ml in 60-mm-diameter dishes (4 ml medium/dish). After 1 day the medium is changed, and proper additions are made; cell counts are normally performed at 24-hour intervals in duplicate or triplicate. Plating efficiencies are determined by plating $10^2–10^4$ cells/100-mm dish in 10 ml of either fresh complete medium or conditioned medium obtained from confluent cultures of IEC cells 48 hours after the last medium change. Cell suspen-

[1]Abbreviations used: DMEM, Dulbecco's modified Eagle medium; FCS, fetal calf serum; EDTA, ethylenediaminotetraacetic acid; PBS, phosphate-buffered saline (10 m*M* phosphate buffer, pH 7.2, 154 m*M* NaCl); IEC, intestinal epithelial cells; CFA, complete Freund's adjuvant.

sions are microscopically checked before plating to ensure that single cells are present. After 10–14 days the dishes are rinsed twice with PBS, and the cells are fixed with absolute methanol (30 minutes) and stained with Giemsa (diluted 1:50). Colonies of at least 10 cells are counted. Cells are tested for their ability to grow in soft agar as described by MacPherson and Montagnier (1964): 10^3–10^4 cells are seeded in 60-mm dishes and incubated at 37°C for 2 weeks. The dishes are then examined with an inverted microscope and colonies of at least five cells are counted.

Karyotyping is performed as follows: cultures to be examined are exposed to colchicine (1 μg/ml) for 2 hours at 37°C, and then trypsinized, treated with a hypotonic (0.075 *M*) KC1 solution, and fixed in a 3:1 mixture of cold absolute methanol and glacial acetic acid; several drops of the cell suspensions are placed on a glass slide, air dried, stained with Giemsa (1:50), buffered to pH 6.8 with citric–phosphate buffer (18.2 m*M* citrate, 163 m*M* sodium phosphate), and examined under a microscope at a ×1000 magnification.

DNA synthesis is determined by measuring the uptake of [*methyl*-^{3}H]thymidine into cellular DNA; cells are seeded into 60-mm dishes and, when the desired cell density is reached (usually cells are tested when about 50% confluent), the medium is removed and fresh medium containing 0.5 μCi/ml [*methyl*-^{3}H]thymidine is added. Following an incubation (at 37°C), the duration of which depends on the purpose of the experiment, the medium is aspirated and each dish is sequentially washed twice with PBS, treated twice with cold 5% trichloroacetic acid (10 minutes each time), and rinsed three times with distilled water. The cells are then dissolved in 1 ml 0.1 *M* NaOH, and aliquots of the solution are used for determination of the radioactivity incorporated (by scintillation counting) and of the protein (by the method of Lowry *et al.*, 1951). Alternatively, the total amount of DNA is determined as follows: following incubation and rinsing with PBS, the cultures are trypsinized, washed twice with PBS, and extracted twice with 0.75 ml of 0.5 *N* perchloric acid at 77°C for 20 minutes. The two extracts are pooled, and aliquots are used for scintillation counting and determination of DNA by the diphenylamine reaction (Burton, 1968).

2. Establishment of Small Intestinal Epithelial Cell Lines (IEC Cells)

The following protocol has given the most consistent results. The entire small intestines of two germ-free rats (18–24 days old) are resected, thoroughly rinsed with 1 m*M* dithiothreitol in normal saline (about 50 ml/intestine), and then flushed with complete medium (100 ml/intestine). While kept in cold complete medium, they are cut into 4- to 5-mm fragments which are then rinsed three to four times with complete medium and finally divided among 18 dishes (100 mm diameter, 10 ml medium/dish). Because a large amount of debris usually accumulates on the surface of the dishes during the first period of incubation, the intestinal fragments are preincubated in standard medium overnight before transferring them to new dishes containing fresh medium and collagenase (type I, Sigma Chemical Co., 200 μg/ml). After 24 hours of incubation with collagenase the medium and the tissue fragments are carefully removed and fresh medium (containing 20–50 μg/ml collagenase) is added to the dishes which are then left undisturbed in the incubator for 1 week. The medium (always containing collagenase) is afterward changed at biweekly intervals. After 4–6 weeks colonies of both fibroblastic and epitheliod morphology are apparent in most of the dishes. Epithelioid colonies of at least 10^3 cells showing no visible contamination with fibroblasts are encircled with stainless-steel cylinders (1 cm diameter), kept in place with sterile silicon grease, and rinsed with 0.02% EDTA in PBS. After 30–60 minutes at 37°C, the cells in each cylinder are suspended in 1 ml complete medium by gentle scraping with a rubber-tipped spatula, and separately transferred to a well of a Costar plate (Cluster[24]). At least 20 colonies are normally selected for each preparation. The colonies which grow to confluency are transferred to 35-mm dishes and, subsequently, 100-mm dishes. The first confluent culture in a 100-mm dish obtained from each original colony is defined "passage No. 1" and serially subcultured in complete medium lacking collagenase.

C. Characterization of the IEC Cells

1. Cell Morphology and Ultrastructure

For light microscopic examination cells are grown on glass coverslips, washed with PBS (two to three times), fixed in absolute methanol (30 minutes), and stained with Giemsa (diluted 1:50). Cells grown in plastic dishes can be similarly treated and examined with an inverted microscope.

Samples of cells for electron microscopy are grown on plastic dishes, fixed at room temperature for 1 hour in 4% paraformaldehyde–2.5% glutaraldehyde buffered with 0.1 *M* sodium cacodylate at pH 7.4, rinsed, and postfixed in 1.3% osmium tetroxide buffered with 0.2 *M*-collidine at pH 7.4 for 1 hour at 4°C. The samples are then briefly washed with 0.2 *M* collidine buffer, pH 6.1, and stained *en bloc* in a 1% solution of uranyl acetate in the same buffer for 1 hour at 4°C. Following *en bloc* staining, samples are again washed with pH 6.1 collidine buffer, dehydrated in a graded series of alcohols, and embedded in a mixture of Epon 812 and Araldite 6005. Sections with gray interference colors are cut (we use a Porter-Blum MT-2 ultramicrotome), stained with uranyl acetate and lead citrate, and examined in the transmission electron microscope.

2. Immunochemical Characterization

a. Preparation of Antigens. Papain-solubilized sucrase-isomaltase is isolated according to Cogoli *et al.* (1972). Rat intestinal crypt cells are isolated as described by Weiser (1973); fractions 7–9 of the cell gradient are used. Crypt cells surface membrane is prepared from this cell suspension according to Weiser *et al.*, (1978). Similarly, lateral-basal membranes are isolated from villus cell fractions (fractions 2–5, Weiser, 1973), as described (Weiser *et al.*, 1978). Brush border membranes are purified from intestinal mucosal scrapings (Hopfer *et al.*, 1973)

b. Preparation of Antisera in Rabbits. Antisera are produced in white male New Zealand rabbits (2 kg). Two immunization schedules are used, depending on the sensitizing antigen: (*a*) antisera against membrane fractions and purified sucrase isomaltase are prepared as follows: for each rabbit the antigen (1–2 mg protein) in 1 ml PBS is combined with 1 ml complete Freund's adjuvant (CFA) and a stable emulsion is obtained and inoculated at multiple subcutaneous sites. The rabbit is boosted with a similarly prepared emulsion using incomplete Freund's adjuvant (containing 0.5–1 mg protein) on days 14 and 21 and bled on day 28. Subsequently, rabbits are boosted at 2-month intervals, and bled 7 days following each boost; (*b*) antisera against intact crypt cells or IEC cells are prepared as follows: cell suspensions, containing $1\text{–}2 \times 10^7$ cells in 3 ml sterile PBS, are injected intraperitoneally into each rabbit twice a week for a month; the rabbits are bled 1 week after the last injection. Twenty to fifty milliliters of blood (ear vein) are collected at each bleed, and left to clot 2 hours at 20°C and then overnight at 4°C. The serum is collected and ammonium sulfate is added to 40% saturation. The precipitate formed is dissolved in PBS, sterilized by Millipore filtration, and stored in 2-ml aliquots. The specificity of many of these antisera can be increased by absorption with isolated liver cell suspensions, kidney glomeruli, and spleen cell suspensions, as follows: (*a*) 1 ml antiserum is absorbed five times with a suspension of liver cells obtained by the collagenase perfusion technique of Witters *et al.* (1976); for each absorption, 2×10^7 liver cells are used, the cell suspension is kept at 37°C for 30 minutes with shaking and spun, and the absorbed serum is removed and then added to a fresh pellet of liver cells; (*b*) absorption with kidney glomeruli (isolated as described by Quadracci and Striker, 1973) and isolated spleen cells (2×10^7 cells/absorption) is similarly performed.

c. Preparation of Monoclonal Antibodies Direction against Brush Border Membrane Proteins. The recent adaptation of cell hybridization techniques to the construction of hybrid cell lines producing monoclonal antibodies with a desired specificity has revolutionized the approach of the production of pure, standardized antibodies (Köhler and Milstein, 1975, 1976). This technique consists of fusing mouse myeloma cells with spleen cells from mice or other animals immunized with a given antigen or antigen mixture. Hybrid lines are derived which secrete specific antibodies. Such lines can be manipulated in culture so that the multispecific response to a complex immunogen can be resolved into a set of monospecific responses by cloning. We have prepared in this way a number of monospecific antibodies directed against different proteins of the brush border membrane. Most of them have proven specific for the intestinal epithelial cells *in vivo,* and many have been shown to interact specifically with the IEC cells *in vitro*. Their preparation is as follows: BALBc mice are immunized intraperitoneally (ip) with 200 μg brush border membrane proteins (in 200 μl PBS) mixed with 200 μl CFA. A boost with 200 μg membrane proteins in 200 μl PBS (ip) is given 3 weeks later and is followed by an identical boost 6 weeks later. Four days after the second boost the mice are bled (the serum obtained is tested for the presence of specific antibodies), and the spleen is removed from each mouse. Spleen cell suspensions are obtained by teasing the spleens in serum-free DMEM, counted, and fused with myeloma cells. The myeloma cell line P3-NSI/1-Ag 4-1 is used: this line has been selected for its resistance to 8-azaguanine, and does not secrete Ig chains, but contains intracellular k chains (Cowan *et al.*, 1974). Fusion of 10^8 spleen cells with 10^7 myeloma cells is performed as described by Galfre *et al.* (1977). The cell suspension, after fusion, is divided among 48 wells (Costar, Cluster^{24} plate). Hybrids are selected with HAT medium (Littlefield, 1964). Vigorous cell growth is usually observed in 80–100% of the wells after 10–14 days. At this point cultures are first tested for production of specific antibodies (see following), and positive cultures are subcultured, in HT medium (standard medium containing 16 μM thymidine and 100 μM hypoxanthine) for 7 days, and in standard medium afterward. The cultures are assayed once a week for production of specific antibodies. Aliquots of each positive culture are frozen in liquid nitrogen. After 1–2 months, cultures are cloned either by dilution plating using 3T3 cells as feeder layers, or in soft agar as described by Cotton *et al.* (1973). Clones are tested for production of specific antibodies and, if positive, cloned a second time. Clonal lines secreting immunoglobulins specific for brush border membrane proteins are characterized in terms of the immunoglobulin type produced, and used for the preparation of large amounts of specific immunoglobulins. This is achieved by injecting the hybrid clones i.p. into mice (primed with 0.5 ml pristane 2 weeks in advance) and collecting the ascitic fluid produced, from which up to 5–10 mg specific immunoglobulin/ml can be purified. Two different assays are used to test the hybrid cultures for the production of specific immunoglobulins: (*a*) pure brush borders, prepared according to Forstner *et al.* (1968) are divided among the required number of tubes (1–2 mg protein/tube), mixed with 2.5 ml of assay buffer (PBS containing 0.2% bovine serum albumin and 0.05% sodium azide), and spun (at 2500 *g* for 10 minutes at 4°C); the supernatant is removed, and 100 μl hybridoma culture media (or fresh medium as control) is added to each pellet; following incubation at 4°C for 90 minutes, the brush borders are spun and washed twice with 2.5 ml buffer; they are then incubated with ^{125}I-labeled $F(ab)_2$ fragment of a rabbit IgG-anti-mouse IgG (about 20 μCi/μg protein): 10^5 cpm in 50 μl assay buffer is added to each tube; following incubation at 4°C for 90 minutes, the brush borders are spun, washed twice with 2.5 ml assay buffer, and counted in a gamma counter; (*b*) culture media from the hybrids are tested for immunofluorescent staining of intestinal frozen sections and IEC cell monolayers as described in the following.

d. Immunofluorescent Staining of Cells and Tissue Sections. Isolated intestinal crypt and villus cells, suspended in saline, are placed on glass slides, allowed to air dry, and fixed with ethanol (10 minutes at -40°C) followed by acetone (10 minutes at -40°C). Cells cultured *in vitro* on glass coverslips are washed two to three times with saline and fixed as previously described. Cryostat sections (2–4 μm) of rat small intestine, liver, kidney, and heart are placed on glass slides, allowed to air dry, and used shortly after preparation. The double antibody fluorescent technique is used for staining: the first antiserum (or normal rabbit serum or PBS serving as controls) at a 1:25 dilution is placed over the slides for 30 minutes at room temperature, followed by three 10-minute washes with PBS. Thereafter, fluorescein-conjugated goat anti-rabbit immunoglobulins (at a 1:25 dilution) are placed over the slides for 30 minutes at room temperature, followed again by three 10-minute PBS washes. In some cases, cells and tissues are counterstained with Evans blue (0.01% in PBS) for 15 seconds; in all cases, they are mounted in a 9:1 mixture of glycerol–PBS and examined under the fluorescence microscope (we use a Zeiss model 260 fluorescence microscope: fluorescence is excited with the output of an Osram 200 lamp filtered through a Zeiss BG_{12} interference filter; samples are observed with a barrier filter 50). Alternatively, the peroxidase–antiperoxidase staining technique can be used (Sternberger *et al.*, 1970).

D. Materials

Male germ-free rats, 18–24 days old (Charles River CD/SD GN rats) obtained from Charles River Breeding Laboratories are used for *in vitro* culture experiments. They are sacrificed within 24 hours after arrival. Isolated crypt and villous cells and intestinal mucosal scrapings, for immunization and for immunochemical staining, and for isolation of subcellular fractions, are prepared from Sprague–Dawley rats, CD strain, weighing 170–225 gm of either sex. Insulin (Iletin, 100 *U*/ml) is obtained from Lilly Laboratories. Collagenase (types I and III) is obtained from Sigma Chemical Co., and Dulbecco's modified Eagle medium, fetal bovine serum (irradiated), penicillin–streptomycin mixture, and penicillin–streptomycin–fungizone mixture are obtained from Microbiological Associates. Fluorescein-conjugated goat anti-rabbit immunoglobulins (molar F/P ratio 4.0–5.0) and goat anti-rabbit immunoglobulins are from Behring Diagnostics; goat anti-rabbit IgG, normal rabbit serum, goat serum, peroxidase–antiperoxidase-soluble complex, rabbit anti-mouse IgG (IgG fraction), mouse IgG, and fluorescein-conjugated rabbit anti-mouse IgG (molar F/P ratio 4.7) are from Miles Laboratories. [*methyl*-^{3}H]Thymidine, 20 Ci/mmole (sterile aqueous solution), is purchased from New England Nuclear.

III. Results

A. Isolated Cell Suspensions

Suspensions of isolated intestinal epithelial cells, minimally contaminated by mesenchymal and plasma elements, are relatively easy to obtain and represent an obvious source of cells for primary culture. Many reports have appeared describing their preparation, and these are summarized in Table I. Three different dissociation techniques, sometimes used in combination, have been employed: mechanical dissociation, often accompanied by dilation of the intestine with air or other means to release the crypt cells; treatment with chelating agents (EDTA and Na-citrate); and enzymatic digestion (with trypsin, Pronase, collagenase, hyaluronidase). We, as undoubtedly many other investigators, have used cell suspensions prepared according to many of these techniques in our early attempts

TABLE I

METHODS TO OBTAIN ISOLATED CELL SUSPENSIONS FROM SMALL AND LARGE INTESTINE

Animal[a]	Method	Reference	Cell viability[b]
Rat	Trypsin	Harrer *et al.* (1964)	±
Rat	Trypsin	Stern and Reilly (1965)	±
Rat	Hyaluronidase	Perris (1966)	N.T.
Rat	Mechanical dissociation	Sjöstrand (1968)	N.T.
Rat	Mechanical dissociation + EDTA	Harrison and Webster (1969)	
Rat	Mechanical dissociation	Iemhoff *et al.* (1970)	±
Chicken	Hyaluronidase	Kimmich (1970)	+
Guinea pig	EDTA	Evans *et al.* (1971)	+
Rabbit	Mechanical dissociation	Kaur *et al.* (1972)	±
Rat	Na-Citrate + EDTA	Weiser (1973)	N.T.
Rat	Hyaluronidase	Castro *et al.* (1974)	±
Pig	EDTA + UREA + hyaluronidase + trypsin	Mian and Pover (1974)	±
Mouse	Trypsin	Kremski *et al.* (1977)	N.T.
Rat	Mannitol	Perret *et al.* (1977)	+
Rat	Na-citrate + hyaluronidase	Towler *et al.* (1978)	+

[a] All studies are with small intestine, except Perret *et al.* (1977) who used large intestine.
[b] N.T., not tested; ±, limited viability; +, viable cells.

to establish short- and/or long-term cultures of intestinal epithelial cells. Our efforts have, invariably, met with a complete lack of success. Since similarly prepared cell suspensions, obtained from different tissues, have proven a much more satisfactory source of cells for culture *in vitro*, these negative results deserve a comment. In most of the studies listed in Table I the problem of the viability of the dissociated intestinal cells has been either completely ignored, or only briefly examined (by measuring the uptake of vital dyes such as trypan blue or nigrosine). When, as in a recent study by Towler *et al.* (1978), this question has been examined by using many different and more stringent criteria, the isolated intestinal epithelial cells have been found to maintain a satisfactory cellular integrity for only 30–60 minutes at 37°C. Many modifications of the incubation medium have failed to prolong cell viability. Similar results have been reported in other studies (Perris, 1966; Kimmich, 1970; Evans *et al.,* 1971). Two other observations are of interest: (*a*) in general, the techniques listed in Table I initially dissociate nonproliferative villous cells whereas crypt cells are obtained after a relatively long incubation, often under inhospitable conditions (in the presence of chelating agents, etc.); (*b*) in our studies using whole fragments of small intestine (of both rat and human origin), a large number of cells has been observed tightly adherent to the surface of the dish, many at least partly spread, within a few hours from the beginning of the incubation: we have never observed a similar phenomenon with isolated cell suspensions. On the basis of these observations, it seems fair to conclude that isolated cell suspensions of intestinal epithelial cells, irrespective of the method used for their preparation, contain cells which remain apparently intact for a short period *in vitro,* but are probably intrinsically damaged and doomed to die within a few hours from isolation. They are, however, of considerable value for short-term biochemical studies and as a source of cells for subcellular fractionations.

B. Monolayer Cultures

1. Establishment of Intestinal Epithelial Cell Lines (IEC Cells)

The general approach used for the establishment of intestinal epithelial cell cultures is to maintain whole fragments of tissue floating near the surface of the culture medium (containing 100–200 μg/ml collagenase) while slowly releasing cells which are then found attached to the surface of the dish (Quaroni *et al.,* 1979). In this way it was hoped that the trauma to the epithelial cells would be as limited as possible. Since a large amount of debris accumulates on the surface of the dishes in the first few hours of incubation, the intestinal fragments are preincubated in standard medium overnight before transferring them to new dishes containing fresh medium and collagenase. The length of the period of incubation of the tissue with collagenase does not appear to be critical importance; however, after 2–3 days of incubation, contamination of the cultures with fibroblasts increases dramatically and pure cultures of epithelial cells cannot be obtained. Higher collagenase concentrations are not beneficial and do not result in a more rapid release of epithelial cells, but in the absence of collagenase few or no cells are dissociated from the tissue and only cultures with a fibroblastic morphology are obtained. During the first 2 months in culture, and until pure epithelioid colonies are selected with cloning cylinders, the presence of a small amount of collagenase (20–50 μg/ml) is required in the culture medium. We cannot, at present, explain this requirement, particularly since the commercial preparation of collagenase used is contaminated with other enzymes and proteins (of note is the presence of a significant amount of unspecific proteolytic enzymes). It is possible, but perhaps unlikely, that its effect is due to its collagenolytic activity: once established, the intestinal epithelial cell lines neither require it nor are affected by collagenase at similar or higher concentrations. However, these cells, when seeded on dishes coated with rat tail tendon collagen, fail to adhere to it and subsequently die. Contamination by fibroblasts is usually not an overwhelming problem in primary cultures; intestinal fibroblasts have generally a longer population doubling time than epithelial cells. They are, however, difficult to eliminate completely during successive subculturing. The most convenient method to obtain pure cultures of epithelial cells is to select uncontaminated colonies from the primary cultures with cloning cylinders before confluency is achieved.

2. Characterization of the IEC Cells

The IEC cells have the typical morphologic appearance of epithelioid cells in culture. They have a large, centrally located nucleus, and grow as tightly packed colonies of polygonal cells (Fig. 1). Numerous, thin microvilli cover the perinuclear portion of the surface membrane (Quaroni *et al.,* 1979). Characteristically, even in confluent cultures, relatively large intercellular spaces divide adjacent cells; frequent pseudopods extend from the cell borders to the surface of the culture dish and to neighboring cells, establishing frequent cell-to-cell contacts. No particularly characteristic ultrastructural feature has been described for these cells: transmission electron microscopic studies have demonstrated the presence of free and bound ribosomes, an extensive Gogli complex, numerous mitochondria, and dense granules throughout the cytoplasm. Occasional junctional complexes have been observed. The IEC cells possess a normal rat diploid karotype, without evidence of genetic heterogeneity. The general growth characteristics of these cells are listed in Table II: of note are the relatively low cell density at confluency, and the rapid rate of growth of subconfluent cultures. The properties listed in Table II are typical of most ''normal'' epithelioid cells grown *in vitro.*

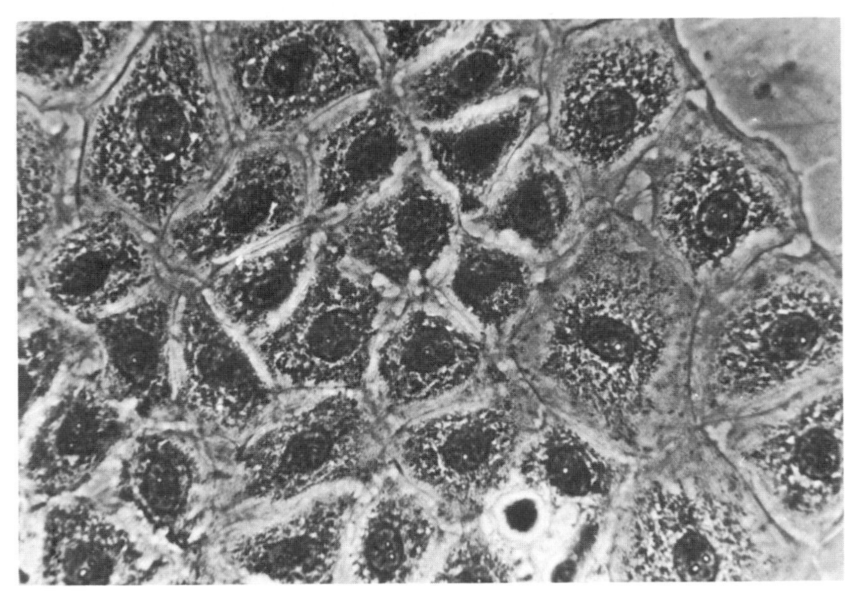

FIG. 1. Morphology of the IEC cells. These cells have a rather large centrally located nucleus and grow as tightly packed colonies of closely opposed, polygonal cells. Giemsa stained. ×250.

The problem of the cellular origin of the IEC cells is complicated by three circumstances: (*a*) the use of whole fragments of small intestine as a source of cells for primary culture; (*b*) the lack of a secure morphological, biochemical, or functional marker for undifferentiated crypt cells; and (*c*) the cellular

TABLE II

CHARACTERISTICS OF INTESTINAL EPITHELIAL CELL LINES

Saturation density	3.8–4.2 (cells × 10^4/cm^2)
Plating efficiency[a]	
In fresh medium	2–3%
In conditioned medium	7–10%
Efficiency of colony formation in soft agar[b]	0
Population doubling time[c]	20–24 hours

[a] 10^3 cells plated/100-mm dish. The complete medium contained 5% FCS. Conditioned medium was obtained from confluent IEC cell cultures 48 hours after last medium change. Cultures were fixed after 2 weeks from plating and stained with Giemsa. Colonies of at least 10 cells were counted.

[b] 10^4 cells plated in 4 ml medium/60-mm dish. After 2 weeks in culture, no colonies of at least five cells were observed.

[c] Determined during the logarithmic phase of growth. The complete medium contained 10 μg/ml insulin and 5% FCS.

heterogeneity of the small intestinal crypts. The presence or absence of cell- and/or tissue-specific antigens on the surface membrane of the IEC cells has been, therefore, the criterion used to determine their origin and identity. Antisera have been prepared against different membrane fractions of intestinal crypt and villous cells, and also against intact crypt and IEC cells. The tissue and cell

TABLE III

IMMUNOLOGICAL CHARACTERIZATION OF THE IEC CELLS

	Immunofluorescence staining of			
Antigen	Intestine	Other tissues	IEC cells	Other cell lines
Sucrase–isomaltase complex[a]	Specific (villus-crypt cell surface membrane)	–	–	–
Crypt cell surface membrane (absorbed)[b]	Specific (villus-crypt cell surface membrane)	–	Specific (perinuclear surface membrane)	–
Crypt cells, whole (intraperitoneal, absorbed)[c]	Specific (mucous, Goblet cells)	–	–	–
Villous cells, lateral-basal membrane[d]	Unspecific	Unspecific	+	+
Brush border membrane (intact)[e]	Specific (villus cells luminal membrane)	Kidney, specific	–	–
Brush border membrane (Triton X-100 solubilized)[f]	Unspecific	Unspecific	+	+
IEC cells, whole (intraperitoneal)[g]	Specific (basement membrane + lamina propria)	Specific (kidney, heart)	Specific (extracellular matrix)	Specific (extracellular matrix)

[a] Sucrase–isomatase prepared by papain solubilization (Cogoli *et al.*, 1972).

[b] Crypt cells were isolated according to Weiser (1973) and the surface membrane obtained as described (Weiser *et al.*, 1978).

[c] Crypt cell suspensions isolated as described (Weiser, 1973).

[d] Villus cells lateral-basal membranes obtained as described by Weiser *et al.* (1978).

[e] Prepared according to the method of Hopfer *et al.* (1973).

[f] Brush border membranes, prepared as described by Hopfer *et al.* (1973), suspended in a solution containing 1% Triton X-100 (2–3 mg protein/ml), incubated 4 hours at 4°C. The suspension was then spun, and the supernatant (containing 80–90% of the protein) used for immunization.

[g] IEC cells (from confluent cultures) injected intraperitoneally at 5–10 × 10^6 cells/injection.

specificity of these antisera has been determined by immunofluorescent staining of different tissue sections (including intestine, heart, kidney, and liver), and various established cell lines. A full account of the results of these studies, which are summarized in Table III, has been reported (Quaroni *et al.*, 1979). Two findings are of major interest: (*a*) immunization with crypt cells surface membrane has produced an antiserum which, after extensive absorptions, has been

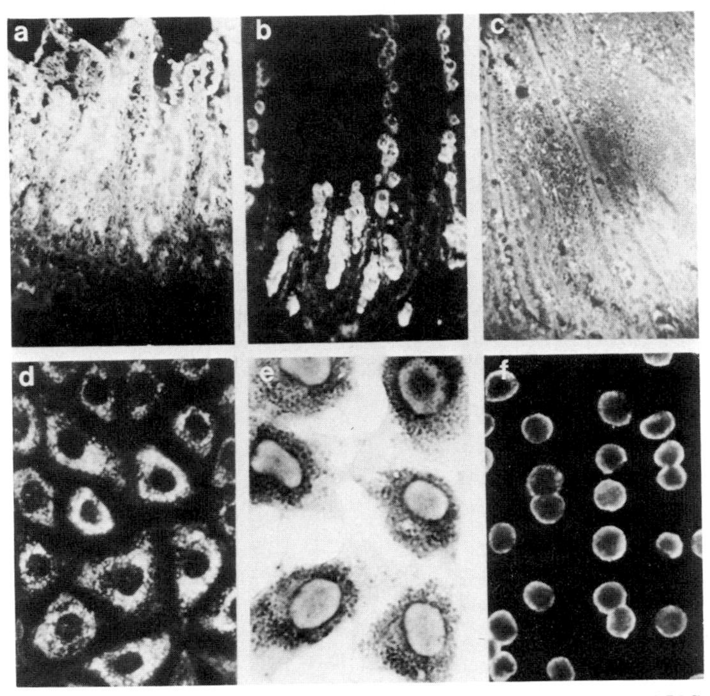

FIG. 2. Immunochemical staining of small intestinal frozen sections (a–c) and IEC cells (d–f) with anticrypt cells surface membrane antiserum. (a) Unabsorbed antiserum: the fluorescence is strongest in correspondence with the luminal membrane of the epithelial cells, but is also present throughout the entire section, with the exception of the muscularis mucosa (not counterstained). ×200. (b) After absorption of the antiserum with liver cell suspensions, kidney glomeruli, and spleen cells, immunofluorescence staining is limited to the luminal membrane of the epithelial cells both in the crypts and on the villi (not counterstained). ×500. (c) Absorption of the antiserum with isolated crypt cell suspensions totally abolished the immunofluorescent staining (counterstained with Evans blue). ×500. (d) Immunofluorescent staining of a confluent monolayer of IEC cells with the absorbed antiserum. Fluorescence is confined to a perinuclear region, with the supranuclear region and peripheral portions of the cells negative (not counterstained). ×800. (e) The uneven distribution of the specific antigens and the patchy appearance of the stain is also apparent after immunochemical staining by the PAP technique (lightly counterstained with hematoxylin). ×1200. (f) Immunofluorescent staining of IEC cells pretreated with EDTA before fixation. The fluorescence is present as a ring around the cells, demonstrating the presence of specific antigens on the surface membrane (counterstained with Evans blue). ×800. From Quaroni *et al.* (1979).

proven to recognize antigens totally specific for intestinal (crypt and villus) epithelial cells; the presence of these antigens on the surface membrane of the IEC cells (see Fig. 2) demonstrates their origin from the intestinal epithelium; (*b*) antisera recognizing antigens specific for differentiated villous cells (anti-sucrase-isomaltase and anti-brush border membrane antisera) have failed to stain the IEC cells, identifying them as derived from crypt and not villus cells. Recent observations, obtained by using monospecific antibodies produced by hybridomas *in vitro* (see Section IV), seem however to contradict the latter findings, and suggest that antigens specifically present in the luminal membrane of the differentiated villous cells may be expressed by the IEC cells.

3. Regulation of IEC Cell Growth

Most aspects of the regulation of IEC cell proliferation are still undefined. Preliminary results, however, seem to suggest that these cells possess a number of intriguing regulatory properties (see Table IV). The growth curve of the IEC cells (see Figs. 3 and 4) can be divided into three distinct periods: (*a*) during the first 2 days after plating, cell growth is very limited or absent; (*b*) a phase of logarithmic growth follows, characterized by a population doubling time of 20–24 hours, which lasts until the cell density reaches $2\text{–}2.5 \times 10^4$ cells/cm^2; (*c*) finally, at higher cell densities, proliferation rapidly declines and is not stimulated by frequent changes of medium. These different periods may reflect significant differences in the properties of the IEC cells with respect to their interaction with various hormones, growth factors, etc. Cell density appears to play a key role in the regulation of IEC cell proliferation. Fetal calf serum, although required for the adhesion of the IEC cells to the growth surface and their prolifera-

TABLE IV

Factors Regulating Growth and Function of the IEC Cells

Stimulators
Insulin
Epidermal growth factor
Inhibitors
Fetal calf serum at concentrations >5–7.5%
Adult rat serum (inhibitory effect absent in serum from vitamin D-deficient rats)
Specific inhibitor extracted from villous cells
Glucocorticoids
Partially reversible inhibition of growth
Induction of synthesis and/or release of a mitotic inhibitor
Striking morphological changes
Synthesis of new membrane glycoproteins
Increased transport of sugars

tion, has significant inhibitory effects on cell growth at concentrations higher than 5–7.5% (see Fig. 3). Among the many substances tried, only insulin and epidermal growth factor appear to stimulate cell proliferation in subconfluent cultures significantly; nevertheless, these two peptides have no detectable effect on confluent cultures. The presence, in animal sera, of inhibitory or stimulatory factors specific for intestinal epithelial cells has been proposed in many studies, and is one of the favored explanations of the intestinal hyperplasia which follows partial small intestinal resection (Tutton, 1973; Sassier and Bergeron, 1977). In this respect it is of interest that adult rat serum has been found to inhibit almost completely IEC cell proliferation, without having a similar effect on fibroblastic cell lines established from rat small intestine. Little or no inhibition was found using serum from rats kept on a vitamin D-deficient diet: this difference is, however, apparently not directly related to the presence of vitamin D or its metabolites in the serum (Weiser and Quaroni, 1979).

Evidence from studies *in vivo* strongly supports the concept of a feedback regulation of crypt cell proliferation by the differentiated villous cells (Quastler,

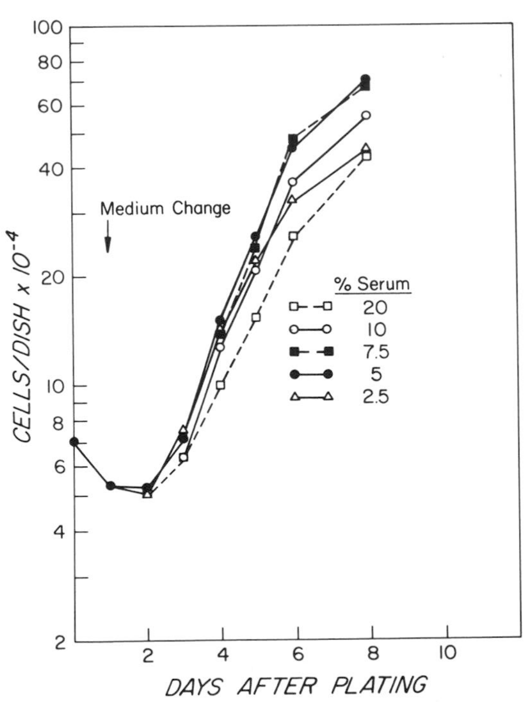

FIG. 3. Growth curves of the IEC cells in media containing different fetal calf serum concentrations. Population doubling time (in 5% FCS) during the logarithmic phase of growth was 22 hours. Note the significant inhibitory effect of FCS at concentrations higher than 7.5%.

TABLE V

CHARACTERIZATION OF THE GROWTH INHIBITOR EXTRACTED FROM INTESTINAL VILLOUS CELLS

High molecular weight (nondialyzable)
Not cytotoxic
No increase in ^{51}Cr release
Inhibition reversible
No change in plating efficiency
Inhibits DNA synthesis and cell growth
Specific for IEC cells—no effect on a large panel of other cell lines in culture

1956; Rijke *et al.*, 1974). Consistent with this hypothesis is the finding that soluble fractions obtained from villous cell homogenates inhibit growth of the IEC cells. Preincubation of the IEC cells with this extract produces a subsequent inhibition of DNA synthesis that is (*a*) reversible; (*b*) not associated with cytotoxicity; and (*c*) apparently cell-type specific (R. J. May and A. Quaroni, in preparation). The results of these studies are summarized in Table V.

4. EFFECT OF GLUCOCORTICOIDS

Glucocorticoids play a fundamental role in the postnatal development of the structure and functions of the small intestinal mucosa (Moog, 1971). They are, therefore, a priori, the most likely factor capable of inducing differentiated functions in the cultured intestinal epithelial cells. Preliminary observations have shown that indeed hydrocortisone has striking effects on the IEC cells, namely: (*a*) a complete inhibition of growth has been observed (Fig. 4) after addition of hydrocortisone to the culture medium; the effect is concentration dependent (maximum inhibition is obtained with 0.5 μg/ml) and is reversible (only in the first week after addition of the hormone); (*b*) prolonged culture of the IEC cells in the presence of hydrocortisone results in striking morphological changes (Fig. 5); the cells become much larger and tightly adherent to each other and their surface, as seen by SEM, is covered with a large number of microvilli characteristically grouped in distinct patches; (*c*) the addition of hydrocortisone to the culture medium results in a different pattern of fucosylated membrane glycoproteins and in the synthesis and release of different soluble proteins.

5. SYNTHESIS OF EXTRACELLULAR MATRIX-INTERACTION OF THE IEC CELLS WITH COLLAGEN

The continuous migration of the intestinal epithelial cells *in vivo,* from the crypt region to the tip of the villus, is an intrinsic property of these cells, and is

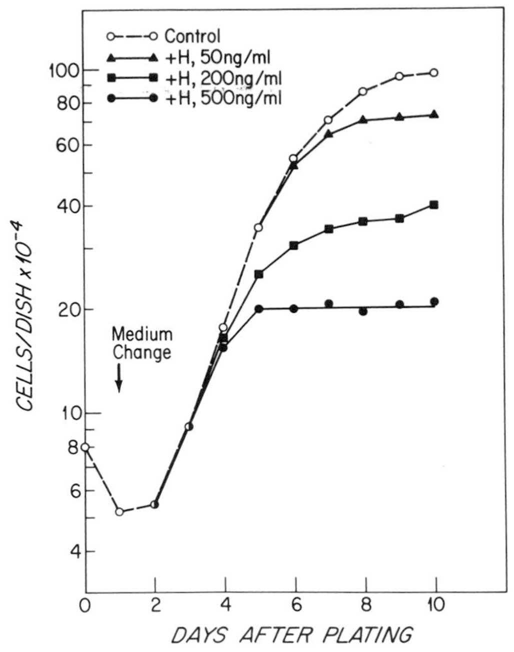

FIG. 4. Suppression of IEC cells proliferation by hydrocortisone. The cells were seeded on day 0 in standard medium. One day later, the medium was changed: one group of cultures received standard fresh medium, while three other groups were treated with medium containing three different concentrations of hydrocortisone (50, 200, and 500 ng/ml, respectively). At daily intervals, cell numbers were determined in triplicate cultures from each group. The symbols represent the mean cell numbers.

likely to be mediated by specific interactions between the basal plasma membrane of the epithelial cells and the underlying basement membrane. A gradient of adhesivity of the intestinal epithelial cells to the basement membrane, with the villus cells the least adhesive, has been demonstrated by Weiser (1973). We have demonstrated (Quaroni *et al.*, 1978) using immunofluorescence, radioimmunoassay, and collagen binding, that the IEC cells synthesize and release large amounts of fibronectin into the culture medium. In IEC cell monolayers, fibronectin has been found present exclusively in regions of cell-to-cell contact, and to be absent underneath the cells where contact with the growth surface is made. This distribution is quite different from that described for fibroblasts and other cell types. In microscopic sections of small intestine, the highest concentration of fibronectin was detected in the region of the crypts, in correspondence with the basement membrane underlying the epithelial cells. The amount of fibronectin present in the intestinal basement membrane, as judged by the intensity of fluorescence, was much greater in the crypt than in the villus region, apparently paralleling the gradient in adhesivity of the epithelial cells. We have also shown (A. Quaroni and R. L. Trelstad, submitted) that the IEC cells synthesize and deposit different types of collagen in the extracellular matrix. Synthesis of collagen has been found strikingly dependent on the presence of ascorbate in the culture medium. In its presence, at least four different collagen types have identified: type I, α_1 (I) trimer, and two type IV-like procollagens containing approximately half of the radioactive proline incorporated into collageneous proteins present in the medium. This last two collagen types did not correspond to any known collagen, and may be specific components of the intestinal basement membrane. Preliminary results have also shown that the IEC cells have unusual properties in their interaction with collagen; when seeded on dishes coated with type I collagen obtained from rat tail tendons, they cannot adhere and subsequently die (Fig. 6). This effect is apparently specific for these cells and suggests that the IEC cells have specific adhesive properties.

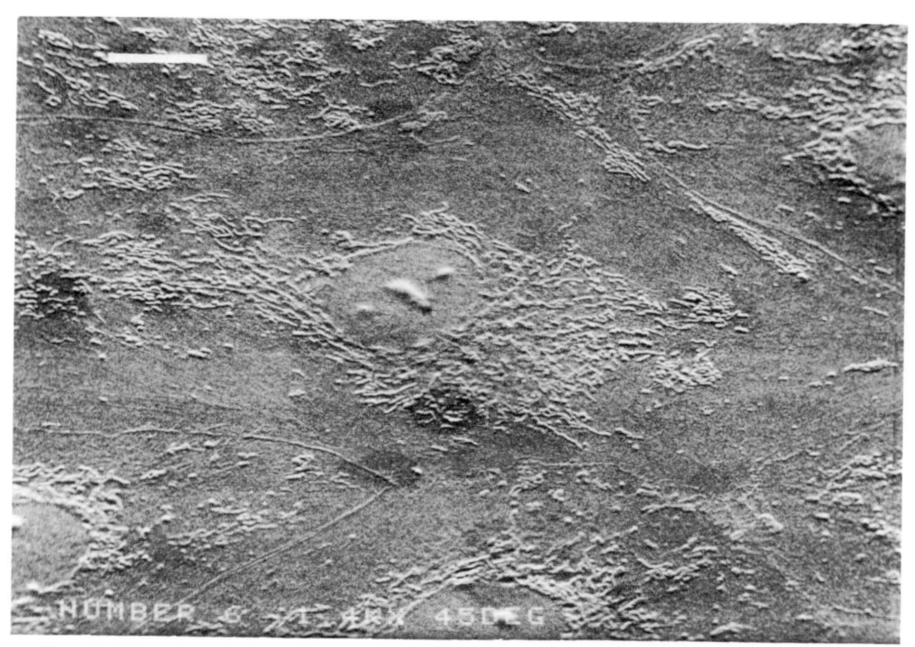

FIG. 5. Scanning electron micrograph of hydrocortisone-treated IEC cells. The cells were treated with medium containing 0.5 μg/ml hydrocortisone for 7 days prior to examination. Note the close apposition of adjacent cells to each other, leaving little or no intercellular space, and the clustering of the microvilli into distinct patches. Large areas of the cell surface are totally free of microvilli. Bar=5 μm. ×4000.

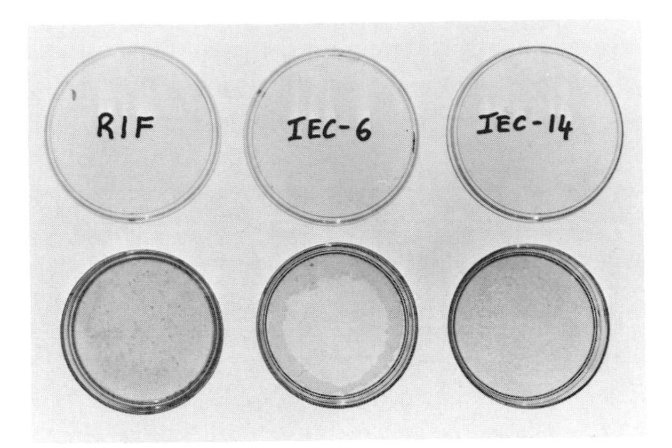

FIG. 6. Lack of binding and growth of the IEC cells on the central region of the dish, where rat tail tendon collagen has been deposited. Cells on the peripheral regions of the dish, where no collagen is present, have reached a normal cell density. This difference is not observed with a fibroblastic (RIF) and a transformed epithelioid cell line (IEC-14) established from rat small intestine (Quaroni *et al.*, 1979), seeded on similarly coated dishes.

C. Attempts to Derive Cultures of Epithelial Cells from Human Intestine

Our attempts to culture epithelial cells from human small intestine have followed a protocol very similar to that employed for the establishment of rat intestinal epithelial cell lines. Complete DMEM was modified so as to contain 10% FCS and supplemented with additional antibiotics and fungizone. Specimens of human small intestine were received from the operating room and quickly cut into tissue fragments; free cells were then obtained by a series of collagenase digestions.

To date, primary cultures of human intestinal epithelium have been attempted from tissue from five ileal and two jejunal specimens. Routinely, the first tissue-digestion is initiated within 30 minutes of surgical excision of the bowel. With each collagenase digestion, large numbers of cells are released from the tissue fragments, and many attach to the dish and begin to spread. However, within 48–72 hours of release, the majority of the cells die and are found floating in the medium or adherent to the dish as membrane "ghosts." Eventually, cultures from all specimens have proved unsuccessful. From one ileal specimen, rare viable cells of unknown identity were observed at 7–14 days, and a small number of fibroblasts, at 28 days; however neither cell type produced viable colonies. Cell viability has not been improved by the use of alternate media (RPMI-1640, Ham's F-10), heat-inactivated serum, or epidermal growth factor. Surprisingly, bacterial and fungal overgrowth of cultures has been quite rare.

A comparison of the techniques used in the rat and human intestinal cell cultures indicates three factors that may be responsible for the failure of the latter: (*a*) differences in the age of the tissue donor (human, 17–70 years vs rat, 18–22 days); (*b*) possible differences in nutritional requirements (human cells vs rat cells); and (*c*) differences in bacterial contamination of tissue (human gut vs germ-free rat gut). Age-related variations in intestinal epithelial cell growth have not been shown to be of clinical significance in man and probably do not play a role in the failure of these cultures. Because cell cultures of other human tissues have been obtained with conventional culture media and FCS, it seems possible but unlikely that human intestinal epithelial cells would possess unique nutritional requirements. In our studies, the relatively rapid death of human intestinal cells, that had initially attached and spread, suggests a specific cytotoxic factor, such as bacterial contamination. Bacterial counts in the normal human jejunum may occasionally approach 10^4 organisms/ml and 10^5–10^8/ml in the ileum. It is well established that *Escherichia coli* bind to human epithelial cells by a mannose-specific lectin (Eshdat *et al.*, 1978). Despite the presence of antibiotics in the culture medium, enteric bacteria could very well be adherent to the human intestinal epithelial cells and could predispose to cytotoxicity by several mechanisms: (*a*) cell-mediated or antibody-dependent cytotoxicity directed against adherent bacterial antigens; (*b*) bacterial lectin-mediated cytotoxicity manifested uniquely in isolated cells; (*c*) chemical cytotoxicity due to local release of bacterial toxins (e.g., lecithinase). Efforts in the future will be directed toward documenting the presence of such bacteria through several techniques (dye staining, electron microscopy, etc.). If enteric bacteria are adherent to the cells, they might be eluted off the cells by the preincubation of the tissue fragments with mannose and other simple sugars prior to collagenase digestion. If immunologic factors are producing cytotoxicity, it may be necessary to deplete the FCS or human serum of immunoglobulin and/or complement prior to its use in culture medium. As necessary, the culture medium will be modified to improve cell viability.

IV. Discussion and Perspectives

Intensive studies *in vivo* have been directed toward elucidating the regulatory mechanisms operative in the small intestinal mucosa and the complex interrelationships among its cellular components. In contrast the problem of establishing monolayer cultures of intestinal epithelial cells has received little attention, judging from the extremely limited literature on this subject. Intestinal crypt cells are characterized, *in vivo*, by a very rapid rate of growth which should facilitate their growth *in vitro*. However, the very ease with which suspensions of isolated crypt and villous cells are obtained by using different dissociation techniques may have represented an unsuspected obstacle to the establishment of monolayer cultures of these cells. Evidence presented in some carefully designed studies (Kimmich, 1970; Evans *et al.*, 1971; Towler *et al.*, 1978) seems to demonstrate convincingly that suspensions of isolated intestinal cells undergo relatively rapid degeneration *in vitro* and are not a good source of cells for long-term culture. This spontaneous tendency to autolysis is probably not an intrinsic property of the intestinal epithelial cells but rather a function of the dissociation techniques employed. Organ cultures of small intestinal fragments or explants have been shown to remain well preserved for up to 48 hours (Trier, 1976). Our approach to establishing monolayer cultures of intestinal crypt cells has been designed to limit, as much as possible, the trauma to the epithelial cells. It is based on two simple observations: (*a*) when whole fragments of small intestine are suspended in culture medium containing relatively low amounts of collagenase, large numbers of cells are easily released which attach and fairly rapidly spread onto the surface of the dish; (*b*) although many morphologically distinct cells can be observed during the first days of culture, only three cell types appear to grow consistently and establish large colonies: the IEC cells, fibroblasts, and large, pleomorphic epithelioid cells which are usually a rather minor component and have never been successfully subcultured. An extensive characterization of the IEC cells, based on morphological and immunological observations, has established their origin from intestinal crypt cells (Quaroni *et al.*, 1979). The key finding, in this respect, has been the demonstration that the surface membrane of IEC cells possesses antigens which are specific for intestinal epithelial cells *in vivo*. This observation, by itself, gives no indication of the potential research value of these cells. It is increasingly evident that cells need a period of adaptation before their establishment as long-term monolayer cultures, and as a result often lose regulatory or differentiated properties specific for the cells from which they are derived. The crucial question is whether IEC cells are truly representative of the intestinal crypt cells *in vitro,* or are partially or totally "dedifferentiated" cells of little or no value. This question can be answered by determining how well their *in vitro* properties correlate with *in vivo* observations about the following phenomena: (*a*) *The regulation of proliferation of the intestinal crypt cells*. These cells are characterized, *in vivo,* by one of the shortest cell cycle times known for normal cells; yet their rate of growth is perfectly coordinated with the loss of differentiated cells at the villous tips, and malignancy is a rare event in the small intestine. Many models and regulatory factors have been proposed as important in the control of this well-balanced cell growth: hormones, specific serum factors, a feedback regulation of crypt cell growth by the differentiated villous cells. (b) *The induction and modulation of intestinal cell differentiation*. The factors which trigger the differentiation of the epithelial cells in the crypts and coordinate such a complex set of morphological and functional changes are at present unknown and solely a matter of speculation. No obvious structural or histological change is detectable in the region of the crypt where differentiation appears to start. This transition zone is, in addition, displaced toward the mouth of the crypt and the base of the villus in conditions requiring increased cell proliferation. Hormones, like glucocorticoids (Moog, 1971) and thyroxine are likely to be of considerable importance in the modulation of cell differentiation. (*c*) *The process of migration of the intestinal epithelial cells along the villous axis*. The process does not appear to be due to pressure from the proliferating cells in the crypts; in conditions of dramatically reduced crypt cellularity differentiated cells continue to migrate and are lost into the lumen. Rather, this phenomena is likely to be regulated by specific changes in the adhesive properties of the intestinal cells and, possibly, in the composition of the intestinal basement membrane.

Our knowledge of the growth regulation and other properties of the IEC cells is still too preliminary to determine their usefulness in the study of these phenomena clearly. The indications from past results, however, suggest that these cells have retained, *in vitro,* specific regulatory functions which may be related to the previously mentioned properties of the intestinal cells *in vivo*.

The considerable inhibitory effect of adult rat serum on their proliferation appears to be due to the presence of specific factor(s) which are, at least in part, under vitamin D control. The inhibitory activity of intestinal villous cell extracts is apparently also specific for the IEC cells and is consistent with the proposed role of the differentiated villus cells in the feedback regulation of crypt cell proliferation. The strong inhibition of cell proliferation induced by glucocorticoids is of considerable interest in view of their proposed role in the modulation of intestinal cell differentiation, but is difficult to relate to a similar effect *in vivo* since the information available is, in this respect, quite contradictory. Other aspects of the regulation of IEC cell proliferation, like the key role played by cell density and the stimulatory effects of insulin and epidermal growth factor, are not specific for these cells and may be general properties of epithelial cells in monolayer culture.

The problem of the induction in the IEC cells of differentiated properties typical of absorptive villous cells is completely open to investigation and presents many aspects which still require definition. The major question, in this respect, is how differentiation can be defined *in vitro*. The striking morphological, functional, and enzymatic changes which accompany intestinal cell differentiation *in vivo* are probably induced and modulated by a complex set of different factors: hormones, epithelial–mesenchymal interactions, luminal content, pancreatic enzymes, enzyme substrates, mucous, bile salts, and other unknown influences. Clearly, a cell growing in monolayer culture cannot be reasonably expected to transform itself into a columnar cell with a well-defined brush border, terminal web, and other characteristic ultrastructural features which are observed *in vivo*. What is, therefore, the minimal requirement defining cell differentiation in the IEC cells? Cessation of growth? Morphological and ultrastructural changes? Expression of specific enzymatic activities? Obviously, one would like to observe as many of these changes as possible. The considerable morphological, structural, and biosynthetic changes which accompany the arrest of growth induced by glucocorticoids may represent the expression of differentiated properties. The recent finding, that monospecific antibodies directed against brush border membrane proteins appear to recognize antigens expressed by the IEC cells, may open new exciting possibilities in the study of cell differentiation in the IEC cells. The use of these monoclonal antibodies, whose antigenic specificity can be exactly defined, will allow a more complete and unequivocable investigation at the cellular level.

The synthesis of basement membrane components (fibronectin, type IV-like procollagen) by the IEC cells and their apparently specific requirements for cell adhesion are among the more clearly defined properties of these cells. They open a number of possibilities for the investigation of the changes in adhesive properties of the epithelial cells *in vivo,* and the study of the synthesis of the intestinal epithelial basement membrane. It will be important to determine if the different collagens synthesized by the IEC cells and deposited in their extracellular matrix have a specific distribution in the intestinal mucosa and, in particular, if a compositional gradient of collagens can be observed along the villus axis, as has been found for fibronectin.

In conclusion, the results so far obtained seem to suggest that the intestinal epithelial cells, once established as monolayer cultures, retain functions specific for the cells from which they are derived. They may provide important insights into the factors controlling growth, differentiation, and migration of the intestinal epithelium *in vivo*.

Acknowledgments

We especially thank Dr. Kurt J. Isselbacher for his support and helpful criticism of this work. This work was supported by grants AM-01392, AM-25956, AM-20309, AA-0225, and CA-26675 of the National Institutes of Health.

References

Autrup, H., Barrett, L. A., Jackson, F. E., Jesudason, M. L., Stoner, G., Phelps, P., Trump, B. F., and Harris, C. C. (1978). *Gastroenterology* **74,** 1248.

Burton, K. (1968). *In* "Methods in Enzymology," Vol. XII B, pp. 163–166. Academic Press, New York.

Cairnie, A. B., Lamerton, L. F., and Steel, G. G. (1965a). *Expl. Cell Res.* **39,** 528.

Cairnie, A. B., Lamerton, L. F., and Steel, G. G. (1965b). *Expl. Cell Res.* **39,** 539.

Castro, G. A., Roy, S. A., and Stockstill, A. D. (1974). *Expl. Parasitol.* **36,** 307.

Cheng, H., and Leblond, C. P. (1974). *Am. J. Anat.* **141,** 461.
Cogoli, A., Mosimann, H., Vock, C., von Balthazar, A.-K., and Semenza, G. (1972). *Eur. J. Biochem.* **30,** 7.
Cotton, R.G.H., Secher, D. S., and Milstein, C. (1973). *Eur. J. Immunol.* **3,** 135.
Cowan, N. J., Secher, D. S., and Milstein, C. (1974). *J. Mol. Biol.* **90,** 691.
Eshdat, Y., Ofek, I., Yashouv-Gan, Y., Sharon, N., Mirelman, D. (1978). *Biochem. Biophys. Res. Commun.* **85,** 1551.
Evans, E. M., Wrigglesworth, J. M., Burdett, K., and Pover, W.F.R. (1971). *J. Cell Biol.* **51,** 452.
Forstner, G. G., Sabesin, S. M., and Isselbacher, K. J. (1968). *Biochem. J.* **106,** 381.
Galfre, G., Howe, S. C., Milstein, C., Butcher, G. W., and Howard, J. C. (1977). *Nature (London)* **266,** 550.
Harrer, G. S., Stern, B. K., and Reilly, R. W. (1964). *Nature (London)* **203,** 319.
Harrison, D. D., and Webster, H. L. (1969). *Exp. Cell Res.* **55,** 257.
Henle, G., and Deinhardt, F. (1957). *J. Immunol.* **79,** 54.
Hopfer, U., Nelson, K., Perrotto, J., and Isselbacher, K. J. (1973). *J. Biol. Chem.* **248,** 25.
Iemhoff, W.G.J., Van Den Berg, J.W.D., DePijper, A. M., and Hülsmann, W. C. (1970). *Biochim. Biophys. Acta* **215,** 229.
Kaur, J., McGhee, J. R., and Burrows, W. (1972). *J. Immunol.* **108,** 387.
Kimmich, G. A. (1970). *Biochemistry* **9,** 3659.
Köhler, G., and Milstein, C. (1975). *Nature (London)* **256,** 495.
Köhler, G., and Milstein, C. (1976). *Eur. J. Immunol.* **6,** 511.
Kremski, V. C., Varani, L., DeSaive, C., Miller, P., and Nicolini, C. (1977). *J. Histochem. Cytochem.* **25,** 554.
Lichtenberger, L., Miller, L. R., Erwin, D. N., and Johnson, L. R. (1973). *Gastroenterology,* **65,** 242.
Littlefield, J. W. (1964). *Science* **145,** 709.
Lowry, O. H., Rosebrough, N. J., Farr, A. L., and Randall, R. J. (1951). *J. Biol. Chem.* **193,** 265.
MacPherson, I., and Montagnier, L. (1964). *Virology* **23,** 291.
Mian, N., and Pover, W.F.R. (1974). *Biomedicine* **20,** 186.
Moog, F. (1971) *In* "Hormones in Development" (M. Hamburg and E.J.W. Barrington, eds.), pp. 143–160. Appleton, New York.
Perret, V., Lev, R., and Pigman, W. (1977). *Gut* **18,** 382.
Perris, A. D. (1966). *Can. J. Biochem.* **44,** 687.
Quadracci, L. J., and Striker, G. E. (1973). *In* "Tissue Culture, Methods and Applications" (P. F. Kruse, Jr. and M. K. Patterson, Jr., eds.), pp. 82–86. Academic Press, New York.
Quaroni, A., Isselbacher, K. J., and Ruoslahti, E. (1978). *Proc. Natl. Acad. Sci. U.S.A.* **75,** 5548.
Quaroni, A., Wands, J., Trelstad, R. L., and Isselbacher, K. J. (1979). *J. Cell Biol.* **80,** 248.
Quastler, H. (1956). *Radiat. Res.* **4,** 303.
Rijke, R.P.C., Van der Meer-Fieggen, W., and Galjaard, H. (1974). *Cell Tissue Kinet.* **7,** 577.
Sassier, P., and Bergeron, M. (1977). *Cell Tissue Kinet.* **10,** 223.
Sjöstrand, F. S. (1968). *J. Ultrastruct. Res.* **22,** 424.
Stern, B. K., and Reilly, R. W. (1965). *Nature (London)* **205,** 563.
Sternberger, L. A., Hardy, P. H., Jr., Cuculis, J. J., and Meyer, H. G. (1970). *J. Histochem. Cytochem.* **18,** 315.
Towler, C. M., Pugh-Humphreys, G. P., and Porteous, J. W. (1978). *J. Cell Sci.* **29,** 53.
Trier, J. S. (1976). *N. Engl. J. Med.* **295,** 150.
Tutton, P.J.M. (1973). *Cell Tissue Kinet.* **6,** 211.
Weiser, M. M. (1973). *J. Biol. Chem.* **248,** 2536.
Weiser, M. M., and Quaroni, A. (1979). *Biochem. Biophys. Res. Commun.* **90,** 788.
Weiser, M. M., Neumeier, M., Quaroni, A., and Kirsch, K. (1978). *J. Cell. Biol.* **77,** 722.
Witters, L. A., Alberico, L., and Avruch, J. (1976). *Biochem. Biophys. Res. Commun.* **79,** 997.

Chapter 21

Culture of Human Pancreatic Ducts

RAYMOND T. JONES AND BENJAMIN F. TRUMP

Department of Pathology and Maryland Institute for Emergency Medical Services, University of Maryland School of Medicine, Baltimore, Maryland

GARY D. STONER

Human Tissue Studies Section, Laboratory of Experimental Pathology, National Cancer Institute, Bethesda, Maryland

---

I. Introduction

There is a paucity of studies involving cultured human pancreatic ductal and exocrine tissues in the literature. A few attempts have been made to establish pancreatic cell lines from fetal tissue (Hay, 1975; Wellman *et al.*, 1975) and

ISBN 0-12-564140-0

Carrel and Lindberg (1938) described *in vitro* culture of cat pancreas for 4 days. The development of several clonal lines of epithelial cells from adult bovine pancreatic duct was described by Stoner *et al.* (1978). The small number of studies involving cultured human pancreatic tissues as well as pancreatic tissues from experimental animals is perhaps due to the misconception that pancreatic tissue is extremely labile, undergoing autolysis at a very rapid rate which therefore makes it difficult to obtain and culture pancreatic tissues. However, a study by Jones and Trump (1975) demonstrated that rat pancreatic slices maintained ultrastructural evidence of viability for up to 3 hours under complete ischemic conditions at 37°C. This study led to the development of a bovine pancreatic ductal system which was the model for the successful long-term culture of human pancreatic ductal explants (Jones *et al.*, 1977). The present chapter will discuss the culturing of ductal elements of the human pancreas (culture of human islets is discussed in Chapter 6).

II. Materials and Methods

A. Ductal Explants

1. Source

Human pancreas was obtained either at immediate autopsy (Trump *et al.*, 1975), at routine autopsy (within 3 hours of death), or at surgery and returned to the laboratory. Under sterile conditions and using sterile techniques, the main duct was opened the entire length of the pancreas, dissected out and placed in cold L-15 [Grand Island Biological Co. (GIBCO), Grand Island, NY] medium. Some interlobular ducts remained attached to the main duct and were removed with it.

2. Culture Conditions

Aftter the glandular part of the pancreas was dissected away from the duct, the duct was cut into pieces approximately 1 cm long. These pieces were placed in 60-mm plastic petri dishes (Falcon Plastics, Oxnard, CA). Three pieces of duct were placed with the epithelial surface toward the gas-medium interface in each dish and 5 ml of supplemented CMRL 1066 (GIBCO, Grand Island, NY) was added. The CMRL 1066 was supplemented with 1 μg insulin/ml, 0.1 μg hydrocortisone hemisuccinate/ml, 2 m*M* L-glutamine, 100 U penicillin G/ml, 100 μg streptomycin/ml, and 5% heat-inactivated fetal bovine serum. The petri dishes were maintained in a controlled atmosphere chamber (Bellco Glass Co.,

Vinland, NJ) which was gassed with a mixture of 45% 0_2, 50% N_2, 5% CO_2. The chambers were placed on rocker platforms (Bellco Glass Co., Vinland, NJ) in a 37°C incubator and rocked 10 times/minute. All explants and isolated cells (see Section II,B) were treated for morphologic study by techniques previously described (Jones *et al.*, 1977).

B. Ductal Cells

1. Source and Isolation

Three organs removed at immediate autopsy were used for the isolation of human pancreatic ductal cells. Two isolation techniques were used, one a modification of a procedure for isolation of bovine pancreatic ductal cells described by Jones *et al.* (1979). This technique is as follows: The main pancreatic duct was cannulated at both the head and tail with an 18-gauge needle at the head and a 22-guage needle at the tail. Both needles had the pointed ends removed. Pieces of silicon tubing were connected to the needles through which 100 ml of Eagle's minimal essential medium (MEM-F-14 GIBCO, Grand Island, NY) was perfused through the duct (inlet at the tail, outlet at the head) to wash cell debris and mucus from the duct as well as to locate branches of the duct which needed to be clamped with hemostats. Following the wash, 50 ml of enzyme solution [0.1% Worthington Type III collagenase + 0.1% dispase in Hank's balanced salt solution (GIBCO)] was continually perfused through the duct for 40 minutes. The enzyme solution was then washed out of the duct with 150 ml of MEM-F-12 cell culture medium (GIBCO, Grand Island, NY) and collected. The cells were then centrifuged at 600 *g* for 5 minutes and washed twice with MEM-F-12. The second technique to isolate human pancreatic ductal epithelial cells was by permitting epithelial cells to grow out from explants onto culture dishes and removing the explants from the dishes before outgrowth of fibroblasts. In this manner, epithelial cultures could be obtained which were relatively free of fibroblasts.

2. Culture Conditions

The cells were grown in 250-ml Falcon plastic flasks containing 12 ml of MEM-F-12 supplemented as follows: 10% heat-inactivated fetal bovine serum, 100 U/ml penicillin G, 100 μg/ml streptomycin, 0.25 μ/ml fungizone, 1 μg/ml insulin, 0.292 mg/ml L-glutamine, 2.2 gm/liter $NaHCO_3$, and 10 m*M* HEPES buffer. The cells were maintained at 37°C, 95% relative humidity with 5% CO_2 in air.

C. Xenotransplantation Studies

Human pancreatic ductal explants were cultured for 1 week under conditions described in Section II,A,2 prior to xenotransplantation into athymic "nude" mice. Since this is the subject of a manuscript in preparation by Valerio (1980), no further methods will be described in the present chapter.

III. Results

A. Explants

Human pancreatic duct obtained for explant culture consists of the main duct which is composed of columnar epithelial cells (Fig. 1) as well as a few secondary ducts which are attached to the main duct. By transmission microscopy, ducts obtained at autopsy or surgery show evidence of cell injury (Fig. 2). The dilated endoplasmic reticulum, swollen mitochondria, clumped chromatin, and blebbing of the plasma membrane (all examples of reversible cell injury) were all repaired by the culture conditions (Fig. 3). However, numerous autophagic vacuoles were seen in epithelial cells of the explants following 3 weeks in culture (Fig. 4).

Alcian-blue/periodic acid–Schiff (PAS)-positive material could still be seen in epithelial cells of the explants and in lumens of the secondary ducts at 5 weeks in culture (Fig. 5). Any acinar or islet cells which were attached to the main duct prior to the culturing of the explants were lost during culture. Ductal epithelium grew over the necrotic tissue as well as onto the petri dish. This could be seen quite dramatically by scanning electron microscopy (Fig. 6).

Viable human pancreatic ductal cells were maintained *in vitro* by means of this explant procedure for up to 2 months (Fig. 7).

Human pancreatic ductal explants have been successfully xenotransplanted into athymic "nude" mice. The success of this technique can be seen in Fig. 8, a plastic embedded section of a human pancreatic duct after 123 days in a nude mouse. There is excellent preservation of the architecture of the duct. Normal appearing epithelium with goblet cells have been maintained in this system.

B. Cells

To date, only preliminary data have been obtained from ongoing studies of human pancreatic duct cells in culture. Cells isolated by the digestion technique described and grown in tissue culture for 18 days are shown in Fig. 9. These epithelial-like cells have material suggestive of mucus present in the culture.

We have also cultured human pancreatic ductal epithelial-like cells that grew

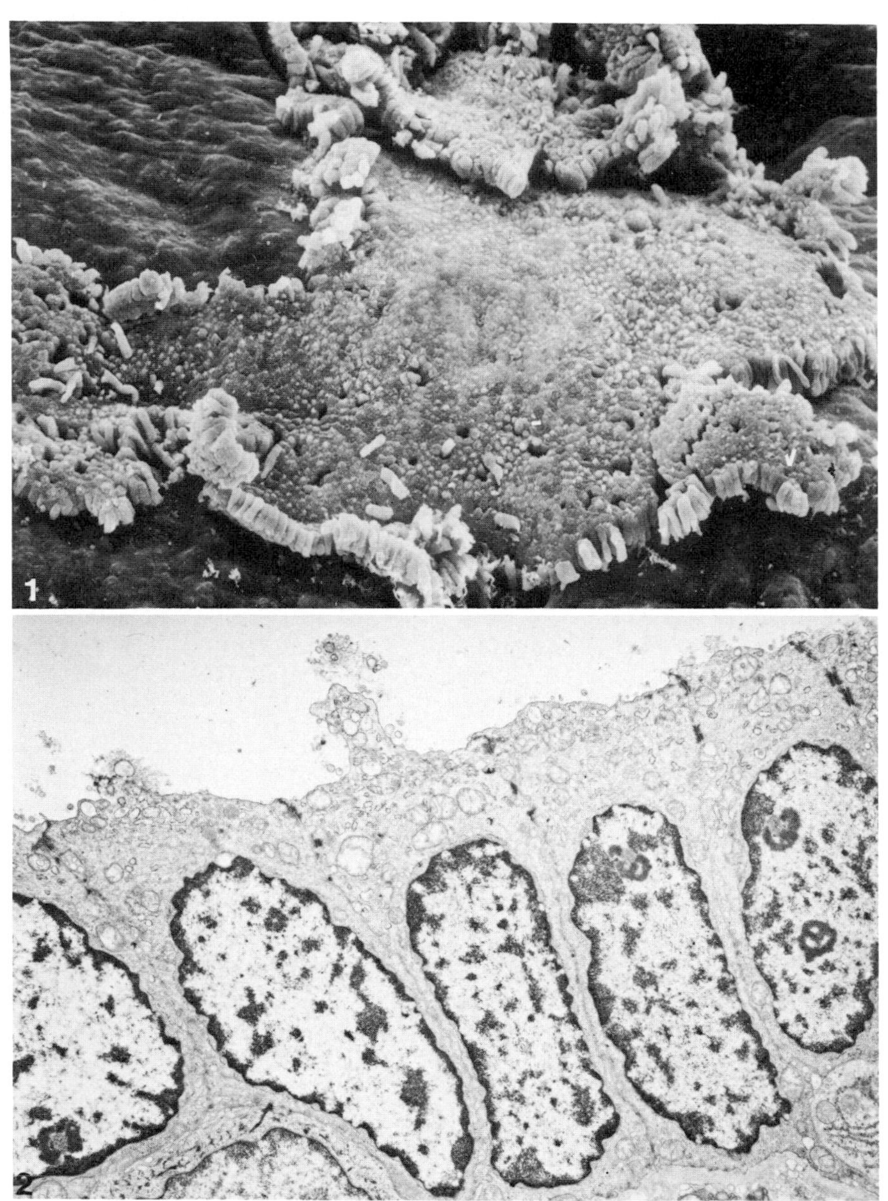

FIG 1. Scanning electron micrograph (SEM) of a piece of human pancreatic duct (HPD) fixed at immediate autopsy. Note the blebbing of some of the columinar epithelial cells and areas where cells have sloughed. Courtesy of Dr. H. Sanefugi.

FIG. 2. Transmission electron micrograph (TEM) of a HPD fixed prior to culture. Cells show evidence of reversible injury: dilated endoplasmic reticulum, swollen mitochondria, blebbing of the plasma membrane, and chromatin clumping.

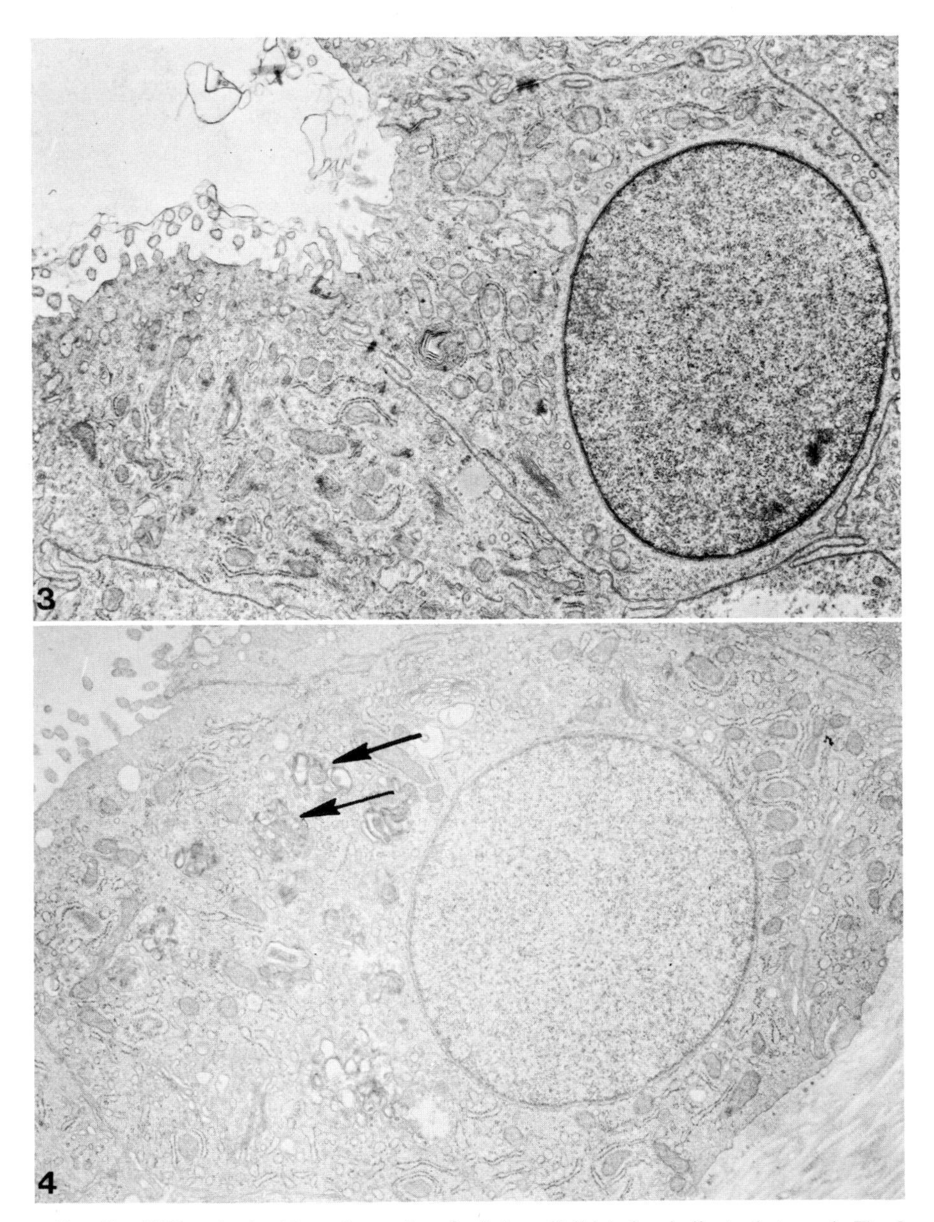

FIG. 3. HPD maintained in explant culture for 8 days. Cell injuries similar to that seen in Fig. 2 have been reversed by the culture conditions.

FIG. 4. HPD cultured for 3 weeks. Numerous autophagic vacuoles are present (arrows).

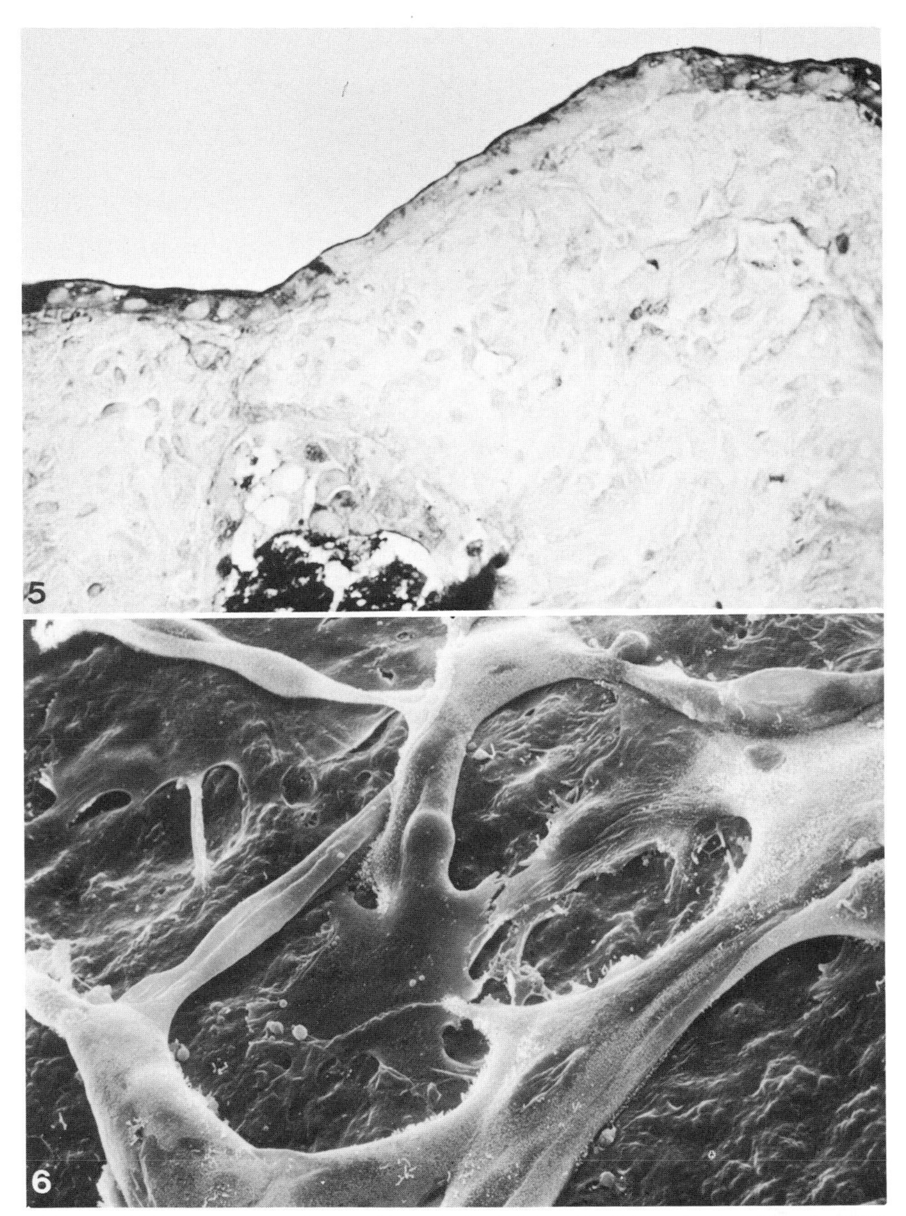

FIG. 5. Light micrograph of a paraffin-embedded alcian blue/PAS-stained section of a HPD which was cultured for 5 weeks. Alcian blue-positive material is on the apex of the ductal epithelial cells as well as within a lumen of a secondary duct.

FIG. 6. SEM of a HPD explant also cultured for 5 weeks. The epithelial cells have migrated over the substrate. Numerous microvilli and cilia can be seen in the epithelial cells. Courtesy of Dr. H. Sanefugi.

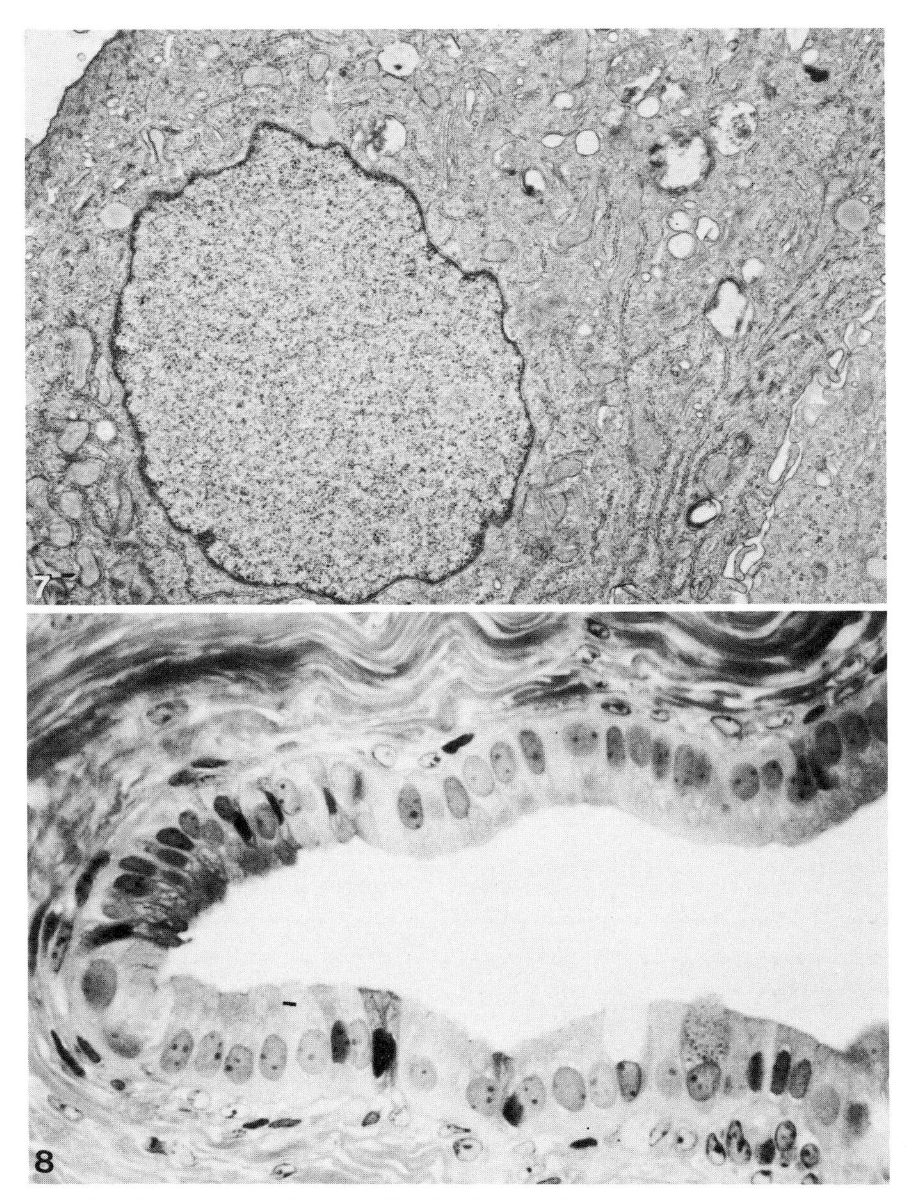

FIG. 7. HPD cultured for 2 months. Note that the organelles are within normal limits.

FIG. 8. Plastic-embedded toluidine blue-stained section of a HPD which was xenotransplanted into a nude mouse for 123 days. The ductal epithelial cells show excellent preservation of their normal characteristics. Note the presence of goblet cells. Courtesy of Dr. M. G. Valerio.

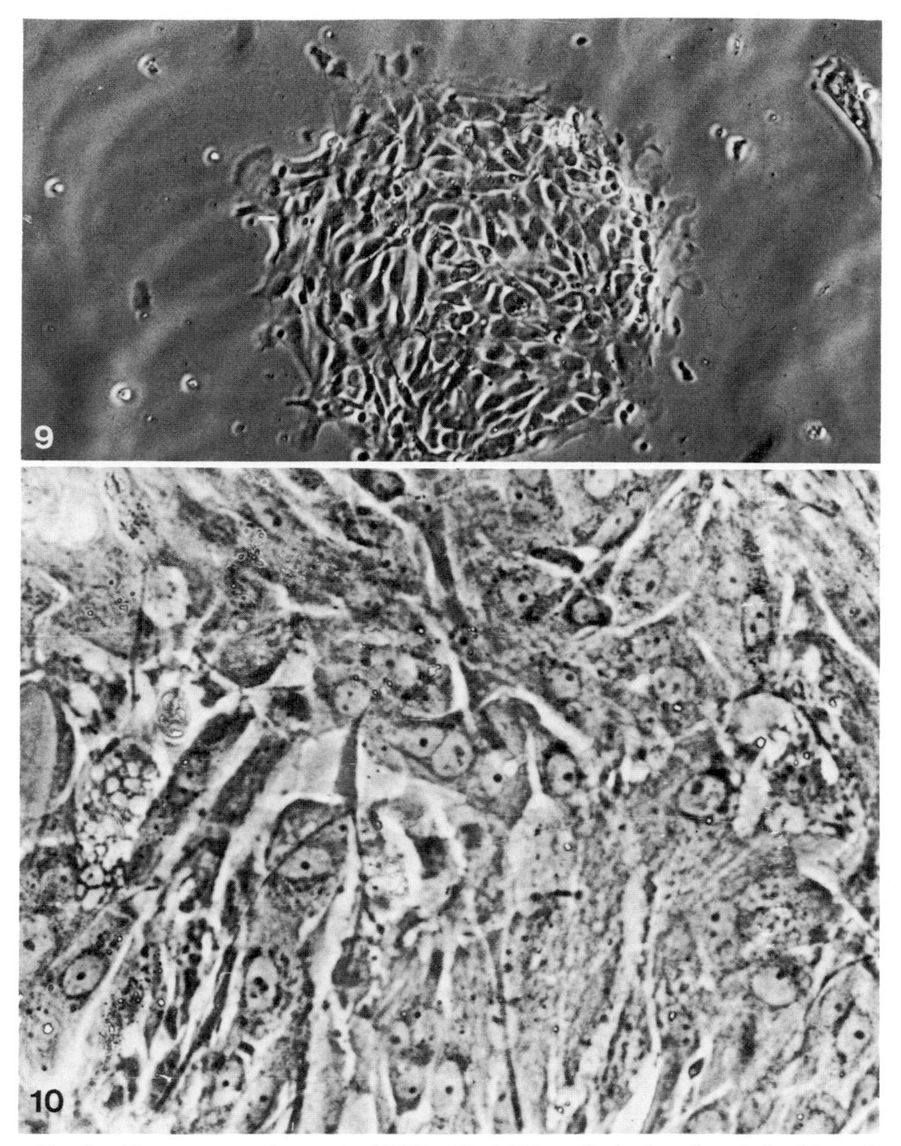

FIG. 9. Phase contrast micrograph of HPD epithelial-like cells that have been isolated by techniques described in the text and maintained in primary culture for 40 hours.

FIG. 10. Phase contrast micrograph of HPD cells that migrated onto a petri dish from an explant.

out from explants in primary culture. These cells are similar in appearance (Fig. 10) to those which were obtained by the digestion technique.

IV. Discussion

The observations presented here demonstrate that it is possible to culture human pancreatic ductal tissues obtained at autopsy within 4 hours of somatic death for extended periods since many injuries to the tissues are reversed by the culture conditions. Pancreatic ductal explants maintain excellent ultrastructural preservation for up to 60 days and we feel that it is possible to culture human pancreatic duct for at least twice this long since we have been able to culture bovine pancreatic ductal explants for longer than 4 months.

The present studies have also demonstrated the feasibility of obtaining epithelial-like cell cultures from human pancreatic duct by either an enzyme digestion technique or from outgrowths from explant cultures. Although these cell isolation and culture studies are preliminary, we feel these techniques will enable one to obtain cell cultures of pancreatic ductal epithelial cells for a variety of investigations.

V. Perspectives

A. Current Applications and Findings

The utilization of human pancreatic ductal tissue in long-term organ and cell cultures has made it possible to study the effects of chemical carcinogens on this important target tissue by our laboratories (Jones *et al,* 1977; Harris *et al.,* 1977), and hopefully these techniques will make it possible to study the pathogenesis of human pancreatic cancer under *in vitro* conditions.

B. Future Studies

It is our opinion that both explant and cell culture methods for human pancreatic ducts described in this chapter make it possible for functional studies to be carried out in order to determine the role of these tissues in disease processes and in the normal physiology of digestion. For example, the preparations that are discussed in this chapter could be used to study diseases such as cystic fibrosis.

The ease with which human pancreatic ductal explants can be xenotransplanted into nude mice and the excellent maintenance of normal cellular structure also makes it possible to utilize this technique to study the pathophysiology of human pancreatic ducts.

It is also our opinion that the methodology now exists for culturing human pancreatic exocrine tissue (as single cells or as acini) for several weeks. Rat acinar cells have been isolated and cultured by Oliver (1978) for up to 4 weeks. Hopefully, the long-term culture of human exocrine pancreatic tissue with preservation of secretory function will soon be a reality.

Acknowledgments

These studies were supported in part by Public Health Service Contracts NO 1-CP-43237 and NO 1-CP-75947 from the Division of Cancer Cause and Prevention, National Cancer Institute. This is publication #774 from the Department of Pathology of the University of Maryland School of Medicine.

References

Carrel, A., and Lindberg, C. A. (1938). "The Culture of Organs," p. 151. Hoeber, New York.

Harris, C. C., Autrup, H., Stoner, G., Wang, S. K., Leutz, J. L., Gelboin, H. V., Selkirk, J. K., Connor, R. J., Barrett, L. A., Jones, R. T., McDowell, E. M., and Trump, B. F. (1977). *Cancer Res.* **37**, 3349.

Hay, R. J. (1975). *Adv. Exp. Med. Biol.* **53,** 23.

Jones, R. T., Barrett, L. A., van Haaften, C., Harris, C. C., and Trump, B. F. (1977). *J. Natl. Cancer Inst.* **58,** 557.

Jones, R. T., and Trump, B. F. (1975). *Virchows Arch. B. Cell Pathol.* **19,** 325.

Jones, R. T., Lakshmanan, M., and Stoner, G. D. (1979). *In Vitro* **15,** 215.

Oliver, C. (1978). *J. Cell Biol.* **79,** 68a.

Stoner, G. D., Harris, C. C., Bostwick, D. G., Jones, R. T., Trump, B. F., Kingsbury, E. W., Fineman, E., and Newkirk, C. (1978). *In Vitro* **14,** 581.

Trump, B. F., Valigorsky, J. M., Jones, R. T., Mergner, W. J., Garcia, J. H., and Cowley, R. A. (1975). *Hum. Pathol.* **6,** 499.

Valerio, M. G. (1980). Unpublished studies.

Wellman, K., Amsterdam, D., and Valk, B. (1975). *Arch. Pathol.* **99,** 424.

Chapter 22

Methodology and Utility of Primary Cultures of Hepatocytes from Experimental Animals

HENRY C. PITOT AND ALPHONSE E. SIRICA

Departments of Oncology and Pathology,
McArdle Laboratory for Cancer Research,
The Medical School,
University of Wisconsin,
Madison, Wisconsin

I. Introduction

Whereas the tissue culture of nonepithelial cells such as fibroblasts, as well as of primitive epithelium, has been developed to a significant degree, the sustained culture of highly differentiated glandular epithelium from the mammal has not yet achieved a similar degree of refinement. As with most techniques in cell culture, early attempts to culture highly differentiated organs led to a rapid overgrowth from a relative minority of primitive cells present in all tissues. Usually these cells grow out as fibroblasts or as simple epithelium (Alexander and Grisham,

ISBN 0-12-564140-0

1970). Also, although some differentiated characteristics of the major epithelial components *in vivo* could be demonstrated for short periods in culture, little or no replication of such tissues could be shown such as that which can be produced *in vivo* with appropriate stimuli, e.g., subtotal hepatic resection.

The culture of adult hepatocytes followed the pattern described until the development of a reproducible, relatively nondestructive technique for the separation of hepatocytes into individual cells or small clusters of less than 10 cells. This critical point was established initially by the investigations of Howard *et al.* (1967) and further refined by Berry and Friend (1969) as well as by the original authors (Howard *et al.*, 1973). The development of such a technique, which allowed for the preparation of isolated hepatocytes with better than a 75% yield, opened the door for the establishment of primary cultures of adult hepatocytes from several different species. This chapter will attempt to describe the methods employed for the culture of fetal, neonatal, and adult hepatocytes, as well as cell lines derived from liver, and to point out some of the experimental systems to which such cultures have been applied.

II. Methods of Liver Cell Culture

A. Methods of Liver Cell Culture Not Involving Hepatic Perfusion

1. Fetal and Neonatal Hepatocyte Culture

Methods for the culture of fetal and neonatal hepatocytes have involved the mincing of the freshly removed tissue and/or enzymatic dissociation of the tissue. In the former case, one of the earlier techniques used was that of Alexander and Grisham (1970), who minced livers of 1- to 5-day-old rats and then explanted several of the fragments onto collagenized coverslips maintained in a gas-tight chamber. Morphologic, histochemical, and autoradiographic studies could be carried out in this system. Rose *et al.* (1968) had also described an elegant circumfusion system with multiple chambers useful for culturing minced fragments of embryonic mice. In addition, MacDonnell *et al.* (1975) used cut fragments of fetal liver for the study of enzyme release from the explants, as well as the hormonal regulation of tyrosine aminotransferase. This latter enzyme was also studied in similar cultures by Wicks (1968).

The enzymatic isolation of fetal hepatocytes was used extensively by Leffert and his associates (e.g., Leffert and Paul, 1972; Leffert and Sell, 1974). These authors investigated a number of characteristics of short-term cultures of fetal rat hepatocytes that were isolated by the use of collagenase and then maintained in a medium deficient in arginine. In addition, Buckley and Walton (1974) have employed trypsin to isolate hepatocytes from chick embryo liver in order to study the morphology of such cells in culture. More recently, Acosta *et al.* (1978) have used collagenase and hyaluronidase in solution to dissociate cells from the livers of 7- to 10-day-old rats. When these liver cells were maintained in an arginine-deficient medium, they were found to express the L-isozyme of pyruvate kinase, which is characteristic of the adult hepatocyte.

2. Culture of Hepatocytes from Adult Animals

Methods similar to those described for fetal and neonatal tissues have also been used in the preparation of hepatic cells from the adult rodent. In all these cases, enzymatic dissociation with either trypsin or collagenase ± hyaluronidase was the principal technique employed to isolate liver cells. Both normal liver (Takaoka *et al.*, 1975; Gebhardt *et al.*, 1978) and regenerating liver (Hays *et al.*, 1975) have been used as the starting tissues, and yields of up to 30 million cells from adult rat liver have been reported with this technique (Gebhardt *et al.*, 1978). Other methods to separate the cells from liver have included chelating agents and agitation of the tissue in the presence of digestive enzymes.

Brief perfusion of the liver followed by enzymatic digestion with mincing, dissection, and/or agitation has also been employed to isolate liver cells of adult animals. In an early investigation by Dickson (1971), a special "filter-well" culture was used in a study of the labeling of intermediates with radioactive substrates. Recently, Leffert *et al.* (1977) used a short (2–3 minutes) perfusion of the liver *in situ* followed by dissection and agitation in the presence of collagenase. This method was similar to that described for fetal tissue (Leffert and Paul, 1972), although the use of short-term perfusion was omitted in the latter case.

The principal difficulties with all the methods described are the very low yield of cells obtained (less than 5% of the total liver cell number) and the question of selection of those cells most able to withstand the rigors of isolation. It was not until the establishment of hepatic perfusion as a reproducible method for obtaining high cell yield that primary liver cell cultures became practical on a large scale.

B. Methods of Primary Adult Hepatocyte Culture with Hepatic Perfusion

1. Liver Perfusion Methods

As indicated in Section I, the basis for the preparation of viable hepatocytes satisfactory for cell culture was the report by Howard *et al.* (1967) that superseded earlier studies in which chelating agents were used in the perfusion medium in the absence of enzymes (Jacob and Bhargava, 1962; Takeda *et al.*, 1964). The refinement of this new technique by Berry and Friend (1969) and again by Howard *et al.* (1973) has been adopted by many laboratories with some variation for the preparation of suspensions of hepatocytes to be explanted directly to cell culture. Some variations in this procedure have been described by Fry *et al.* (1976) and reviewed by Wagle (1975). Such cell preparations have been separated into size classes using ficoll gradients (Drochmans *et al.*, 1975), and the separation of centrolobular and perilobular hepatocytes has also been accomplished following suitable preparation of the animal (Wanson *et al.*, 1975). In addition, some studies have been undertaken to determine the optimal concentrations of ions and chelators, as well as optimal characteristics of the enzymatic requirements for obtaining hepatocyte suspensions (Seglen, 1973a,b). Although the collagenase perfusion techniques have been applied mainly to the rat, they have also been used to isolate hepatic parenchymal cells from guinea pig liver (Elliott and Pogson, 1977), mouse liver (Renton *et al.*, 1978), frog liver (Stanchfield and Yager, 1978), and eel liver (Hayashi and Ooshiro, 1978).

Just as brief perfusions were used in conjunction with mincing and enzymatic digestion, prolonged hepatic perfusion has been followed by mincing of the perfused liver in the solution of collagenase and hyaluronidase for both rodent and avian liver (Capuzzi *et al.*, 1971; Horsfall and Ketterer, 1976). Studies have also demonstrated (Kreusch *et al.*, 1977) that hepatocytes can be selectively destroyed by treatment of cell suspensions with proteolytic enzymes such as Pronase. This treatment results in a cell suspension that contains a significant yield of sinusoidal cells but no hepatocytes. Thus, this finding may explain the extremely low hepatocyte yields obtained by investigators who employed comparable proteolytic enzymes such as trypsin to digest fragments of liver.

2. Short-Term Suspension Cultures of Hepatocytes

The preparation of metabolically active hepatocyte suspensions in a good yield has led to the use of such preparations for short-term biochemical investigations. Many of the characteristics of short-term cultures have been reviewed by Jeejeebhoy and Phillips (1976). As they have pointed out, a variety of different media have been used, varying from a simple salt solution such as Hanks to the use of tissue culture medium such as Eagle's, to medium containing a variety of macromolecules such as gelatin or albumin. Unlike the prolonged cell culture systems described later, incubation is carried out in various containers usually immersed in a shaking water bath. Gasing with 95% 0_2/5% CO_2 has been used by many investigators employing the short-term culture systems. In addition, incubation in rat serum has been shown to improve the response of isolated liver cells to hormones, especially glucagon (Siess and Wieland, 1975).

One of the difficulties of using hepatocyte suspensions is the leakage of enzymes from the dispersed liver cells (Takeda *et al.*, 1964). Although these authors demonstrated that cortical steroids had a preventive effect on the leakage of certain enzymes, this problem remains and drastically limits the use of suspensions for any studies other than extremely short-term ones (probably less than 4 hours).

3. Prolonged Maintenance of Hepatocytes in Primary Culture

One of the first techniques with hepatic perfusion for the preparation of rat hepatocytes in a primary culture was that of Bissell *et al.* (1973). These investigators plated cells directly onto plastic petri dishes and found that albumin synthesis continued for at least 6 days. A similar system, in which the induction of tyrosine aminotransferase activity by corticosteroids could be demonstrated, was reported by Bonney *et al.* (1974). Comparable systems were then reported by Laishes and Williams (1976) and, more recently in the case of mouse liver, by Renton *et al.* (1978).

In most of the studies mentioned, the hepatocytes survived for approximately 1 week, and during this period an extensive and continuous loss of hepatocytes from the cultured plate surface was noted. In 1975, Michalopoulos and Pitot (1975) used the collagenase perfusion technique described by Bonney *et al.* (1974) to plate isolated adult rat hepatocytes onto collagen gels prepared from rat tail collagen. The advantage of this system was that the viability and function of the cells could be maintained for at least 3 weeks as evidenced by the corticosteroid induction of tyrosine aminotransferase. Michalopoulos and Pitot contrasted their findings with the plating of hepatic cells on plastic dishes or on collagen-coated plates. Recently, other supports for hepatocytes have also been described (Savage and Bonney, 1978; Seglen and Fossa, 1978), although without further extension of viability. In comparison, Sirica *et al.* (1979) modified the technique of Michalopoulos and Pitot (1975) by plating the hepatocytes on a substratum of nylon mesh coated with a thin collagen gel (Fig. 1). In addition to the extended cell longevity this technique allowed for the removal of the cells from the substratum by gentle treatment with a dilute collagenase solution and provided for their replating. Unfortunately, none of these techniques for culturing adult rat liver cells obtained by perfusion results in any significant degree of hepatocyte replication. The study of Sirica *et al.* (1979) did demonstrate that up to 9% of the cells were in DNA synthesis after 1 week in culture, but extensive examination did not reveal any significant degree of mitotic activity in such cultures, nor was there any sustained increase in DNA synthesis. Although there had been previous reports of cell replicative activity in adult hepatocyte cultures (Leffert *et al.*, 1977), such studies have not been carefully controlled to determine exactly which cells are replicating, hepatocytes or littoral cells. Thus the

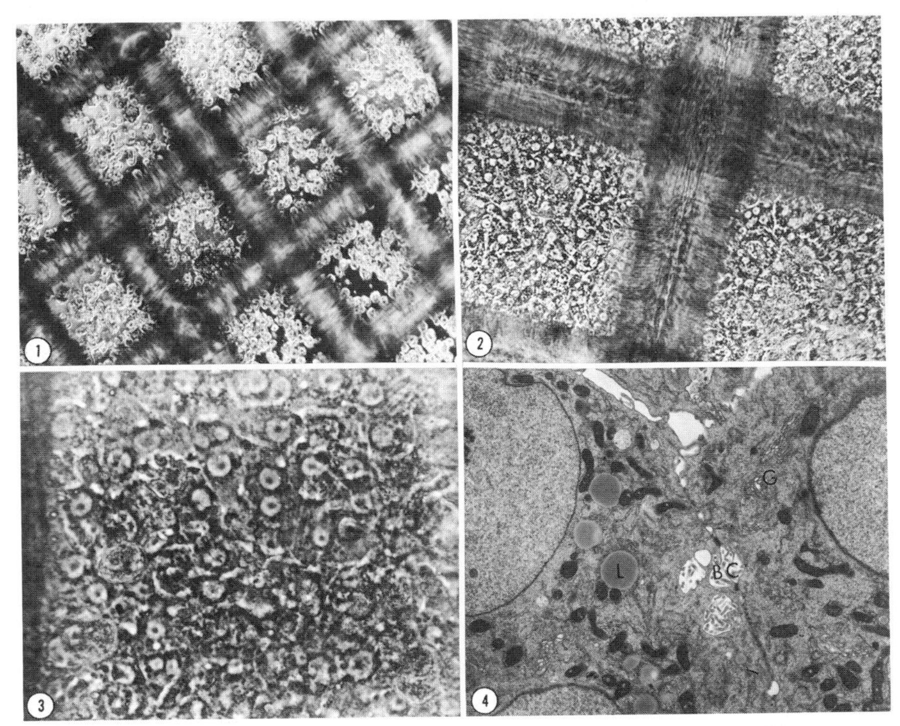

FIG. 1. Primary culture of adult rat hepatocytes on collagen gel–nylon meshes. Phase contrast micrographs of (1) 4-hour culture (original magnification ×40); (2) 6-day-old culture (original magnification ×100); (3) 10-day-old culture (original magnification ×200); (4) electron micrograph of hepatocytes in a 10-day-old culture: L, lipid droplet; BC, bile canaliculus; G, Golgi (original magnification ×6000). From Sirica *et al.* (1979).

mass culture of replicating hepatocytes remains as a goal for future investigations.

C. Culture of Hepatic Cell Strains and Lines

It has not yet been possible to obtain reproducible and substantial levels of mitotic activity of adult hepatocytes in culture, but a number of reports have demonstrated the production of continuous lines or strains of cells from adult liver, as well as from hepatomas. Perhaps one of the most noteworthy hepatoma cell lines is the H-4-II-E, first established at the McArdle Laboratory from the Reuber hepatoma (Pitot *et al.*, 1963). Since then, other strains of hepatoma cells have been derived from several transplanted Morris hepatomas (Kido *et al.*, 1977; Richardson *et al.*, 1969).

The establishment of cell lines from adult rat liver cells has been reported by a number of authors (Gerschenson *et al.*, 1974a; Williams and Gunn, 1974; Chessebeuf *et al.*, 1974). In at least one of these cell lines (Gerschenson *et al.*, 1974b), the induction of tyrosine aminotransferase by insulin and corticosteroids was retained, although the karyotype of the cell line was distinctly abnormal. In another study, Bausher and Schaeffer (1974) described a diploid rat liver cell line that was relatively resistant to the toxic action of the hepatocarcinogen, aflatoxin B_1, and that exhibited tyrosine aminotransferase activity by a histochemical method, although no evidence for its hormonal regulation was presented. Epithelial cell cultures established from normal rat liver have also been "transformed" in cell culture (Borenfreund *et al.*, 1975).

Virtually without exception, the light microscopic and ultrastructural morphology of cell strains derived from liver is significantly different from that of adult liver *in vivo* or of that reported for primary hepatocytes cultured *en masse*. Many of the cell lines obtained from normal adult liver resemble endothelium in culture, especially those cell lines that exhibit virtually no hepatic functions. Therefore, with the exception of the hepatoma cell lines, many of which can produce neoplasms when reinoculated *in vivo,* cell lines derived from normal adult liver may not be parenchymal cells at all. The line described by Gerschenson *et al.* (1974a,b) appears to be an exception, although its neoplastic potential has not been reported.

III. Utilization of Primary Cultures of Adult Hepatocytes

In view of the relatively complex metabolic machinery of the liver, the culture of hepatic parenchymal cells has been viewed as a major tool in the study of a variety of metabolic mechanisms in a highly differentiated cell outside of the "internal milieu" of the organism. Because of the ability to control the environment of cells in cell culture, systems of hepatocytes in culture have been looked at longingly by many investigators. Here we will briefly summarize some of the biochemical approaches that have been used to study primary cultures of mammalian hepatocytes.

A. Studies on Intermediary Metabolism and Nucleic Acid Synthesis

Although freshly isolated hepatocytes prepared by the liver perfusion techniques have been employed for a number of studies in intermediary metabolism (Howard and Pesch, 1968; Story *et al.*, 1976; Shanmugam and Bhargava, 1968; Schreiber and Schreiber, 1972) and in liver cell lines and strains (Beck and Bollack, 1974; Munns *et al.*, 1976; Levine *et al.*, 1978; Smith *et al.*, 1978), studies on intermediary metabolism in primary cultures of hepatocytes have not been so extensive. In general, the most interesting effects have been alterations in the metabolic characteristics of hepatocytes in culture. Walker *et al.* (1972) demonstrated that isozymes of several glycolytic enzymes of adult hepatocytes that had been cultured for several days resembled those of the fetal liver more than those of the adult liver. Studies by Leffert *et al.* (1978) showed a similar appearance of the fetal form of pyruvate kinase and the production of α_1-fetoprotein in adult hepatocytes prepared by methods previously described (Leffert *et al.*, 1977). More recently, Sirica *et al.* (1979) have demonstrated, with mass cultures of primary hepatocytes on nylon meshes coated with collagen gel, several fetal characteristics of the cells. These fetal characteristics, which became apparent after 3 days in culture, included the production of α_1-fetoprotein and the expression of the fetal isozymic form of fructose diphosphate aldolase, as well as the appearance of the rat fetal liver cell enzyme, γ-glutamyl transpeptidase (Figs. 2 and 3). Parallel to this appearance of fetal characteristics was the gradual loss of certain adult characteristics including the production of albumin (Sirica *et al.*, 1979), glycogen synthesis (Walker, 1977), some enzymes of the

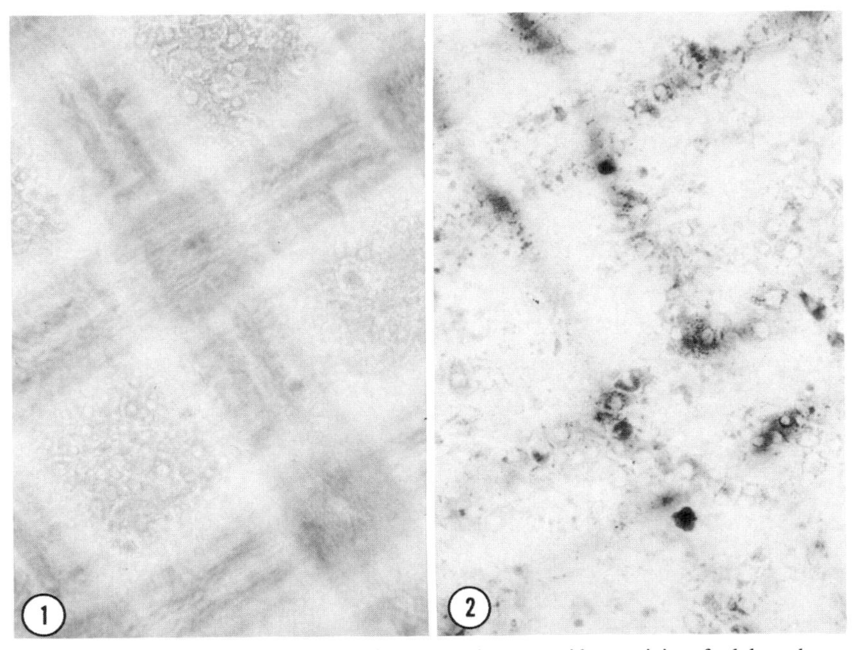

FIG. 2. Histochemical demonstration of γ-glutamyl transpeptidase activity of adult rat hepatocytes on collagen gel-nylon meshes. (1) Two-day-old cultures: no enzyme activity; (2) 9-day-old culture: cytoplasmic enzyme activity (original magnification ×100). From Sirica *et al.* (1979).

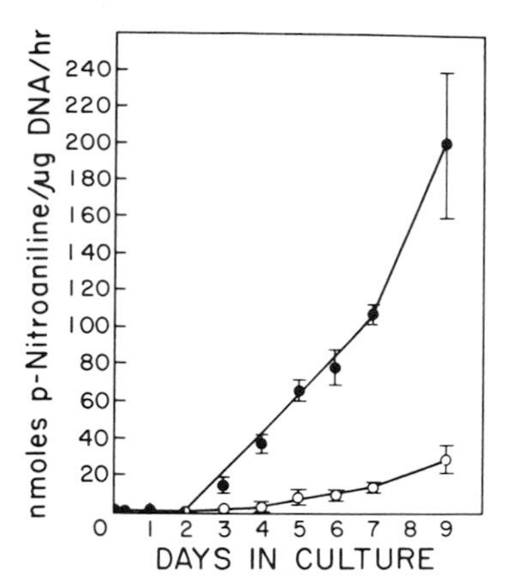

FIG. 3. γ-Glutamyl transpeptidase activity of adult rat hepatocytes in primary culture. Enzyme activity was measured as the release of *p*-nitroaniline from the substrate L-γ-glutamyl-*p*-nitroanilide in the presence (●) and absence (○) of the γ-glutamyl group acceptor glycylglycine. From Sirica *et al.* (1979).

urea cycle (Lin and Snodgrass, 1975), and the synthesis and secretion of α-2u-globulin (Haars and Pitot, 1979).

In contrast, when hepatocytes are cultured from fetal liver, adult characteristics appear in these cells after they are cultured for various time periods, including the development of gluconeogenesis (Coufalik and Monder, 1976) and the appearance of the synthesis of adult serum proteins (Grieninger and Granick, 1975) in cultured embryonic chicken liver cells.

Studies on DNA synthesis in primary cultures of hepatic cells have been somewhat varied in their results. In fetal rat hepatocytes, DNA synthesis was induced by serum (Paul and Walter, 1974) as well as by several hormones (Leffert, 1974). In adults, culture of liver cells 15 hours after partial hepatectomy resulted in an increase in [^{3}H]Tdr incorporation in a manner similar to that seen after the operation *in vivo*. However, according to autoradiographic studies, only 1–2% of the parenchymal cells in the culture were labeled (Yager *et al.*, 1975). Insulin and epidermal growth factor were found to stimulate a large number of hepatocytes to enter DNA synthesis in primary cultures of adult rat liver (Richman *et al.*, 1976). Sirica *et al.*, (1979) also found that, with insulin alone, up to 9% of the adult rat hepatocytes maintained on nylon mesh–collagen gel substrata exhibited DNA synthesis by 8 days in culture. Thus, it would appear that in hepatic cells in culture DNA synthesis is stimulated both by hormones and also by explantation to certain substrata. None of the reports mentioned gives evidence for a substantial increase in the number of adult hepatocytes in the cell cultures.

B. Protein Secretion by Cultures of Hepatocytes

Several studies have demonstrated that isolated rat hepatocytes in suspension secrete or release plasma proteins including albumin (Weigand and Otto, 1974), fibrinogen (Crane and Miller, 1977), and very low-density lipoproteins (Jeejeebhoy *et al.*, 1975). Ammonia appears to inhibit such secretion in isolated rat hepatocytes (Seglen and Reith, 1977). Chen and Feigelson (1978a) have further demonstrated the synthesis and secretion as well as the hormonal regulation of α-2u-globulin synthesis in isolated rat hepatocytes.

In embryonic chick hepatocytes in culture, albumin and several other plasma proteins are synthesized and secreted at rather substantial levels (Grieninger and Granick, 1978). Serum albumin is also synthesized and secreted by one or more mouse hepatoma strains (Bernhardt *et al.*, 1973). In adult rat hepatocytes on nylon mesh–collagen gel substrata, several serum proteins including albumin, transferrin, α_1-acid glycoprotein, and α-2u-globulin have been shown to be synthesized and secreted (Sirica *et al.*, 1979; Haars and Pitot, 1979). In addition, these studies have also demonstrated the secretion of α_1-fetoprotein by these cells (Fig. 4). Thus, cultured hepatocytes either in short-term suspensions or in longer term cultures of fetal and adult cells may be quite useful for the investigation of the synthesis and secretion of serum proteins.

C. Studies on the Regulation of Genetic Expression in Cultured Hepatocytes

Since the hepatic cell represents one of the most responsive tissues in the body to environmental stimuli, it is not surprising that many investigations of cultured hepatocytes have been directed toward the regulation of metabolism. Even in

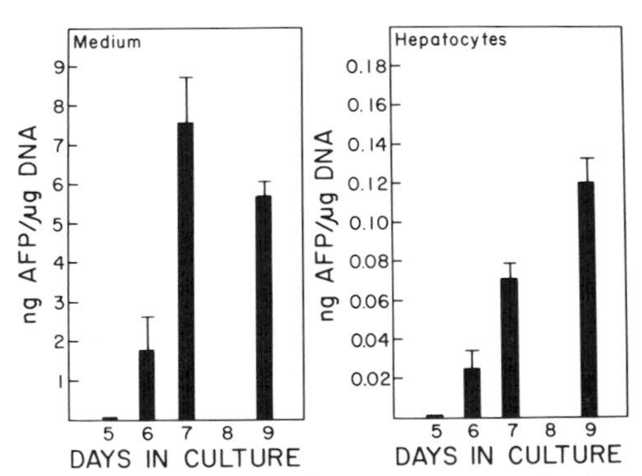

FIG. 4. α_1-Fetoprotein production by hepatocytes on collagen gel–nylon meshes. α_1-Fetoprotein was determined by radioimmunoassay in cell-free culture medium and in hepatocytes collected from culture at 24-hour intervals. From Sirica *et al.* (1979).

isolated rat liver cells in very short-term study, both the adaptive synthesis of enzymes (Lakshmanan *et al.*, 1975; Henderson, 1971), messenger RNA synthesis (Chen and Feigelson, 1978b), and the hormonal control of enzyme activity (Feliu *et al.*, 1976) have been investigated. However, because of the relatively short-lived nature of hepatic cell suspensions, as well as the questions of viability and cell damage in these preparations, many of the more recent investigations have been directed toward hepatocytes maintained in cell culture.

Perhaps the most well-known example of the study of the regulation of genetic expression in liver cell cultures is that of the enzyme tyrosine aminotransferase, whose induction by corticosteroids was first described more than 15 years ago by Pitot *et al.* (1963). Since that time numerous investigations of this enzyme, both in hepatoma cells (Thompson *et al.*, 1966; Reel *et al.*, 1970) and in primary cultures of normal adult cells (Bonney *et al.*, 1974; Michalopoulos *et al.*, 1978a), have been reported. In addition, other enzymes have been induced by corticosteroids in hepatoma cell lines (Lee and Kenney, 1970; Haggerty *et al.*, 1973), as well as in primary adult liver cell cultures (Oliver *et al.*, 1978; Spence and Pitot, 1979). Enzyme repression has also been described in a strain of cells obtained from liver (Kitajima *et al.*, 1975).

In addition to the corticosteroid regulation of enzyme synthesis, insulin, glucagon, and T_3 have been shown to regulate enzyme synthesis (Goodridge and Adelman, 1976), amino acid transport (Kletzien *et al.*, 1976), and protein secretion (Tarlow *et al.*, 1977). Since several laboratories have demonstrated a requirement for insulin in the culture of hepatocytes, the regulation by this hormone of metabolic function is not surprising.

D. Drug and Carcinogen Metabolism by Primary Cultures of Hepatocytes

Because the liver is the principal site of xenobiotic metabolism in the mammalian organism, a number of investigations have been directed toward a study of this type of metabolism in hepatic cell cultures. Among the earliest such investigations were those Gielen and Nebert (1971), who investigated the regulation of the cytochrome P-450 system in fetal cultures from rat liver. Although they showed the regulation of the cytochrome P-450-dependent enzyme, aryl hydrocarbon benzo[*a*]pyrene hydroxylase, using polycyclic hydrocarbons and phenobarbital, the molecular species of cytochrome responding was like that of P-448, which is usually associated only with polycyclic hydrocarbon induction *in vivo*. However, the studies by Michalopoulos *et al.* (1976) on primary cultures of adult hepatocytes on collagen gel demonstrated the induction of cytochrome P-450 by phenobarbital. But even this latter situation does not appear to replicate exactly the pattern of drug metabolism control that phenobarbital exhibits *in vivo* (Fahl *et al.*, 1979).

The studies by Michalopoulos and his associates also demonstrated that the cytrochrome P-450 levels in cultured hepatocytes rapidly decreased after explantation. More recently, Decad *et al.* (1977) have maintained higher levels of cytochrome P-450 for somewhat longer periods in short-term cultures of hepatocytes with a complex mixture of various hormones, vitamins, and metabolites in the medium. In addition, Guzelian *et al.* (1977) have also demonstrated the presence and regulation of a number of drug-metabolizing enzymes in primary cultures of adult hepatocytes on plastic plates.

Isolated hepatocytes are capable of metabolizing carcinogens and converting such molecules to active forms that react with DNA and other macromolecules (Burke *et al.*, 1977; King *et al.*, 1976). This characteristic, both in hepatocyte suspensions and in primary cultures, has been used to study DNA repair (induced by such molecules in the cultured cells) as a potential system for monitoring environmental carcinogens (Michalopoulos *et al.*, 1978b; Williams, 1977).

IV. Perspectives and Conclusions

From this brief review of the methodologies and uses of cultured hepatocytes, the reader must be struck by the variety of culture techniques employed in attempts to maintain and grow these highly differentiated cells in artificial environments suitable for experimental manipulation. Although a major advantage of such primary adult liver cell cultures is the ability to maintain these cells in serum-free medium of known molecular composition, there still remains the problem of the inability of the cells to continue through the cell cycle from G_2 into the mitotic phase. While some claims have been made for hepatocyte replication in cell culture, there is no proof that such cells undergoing replication are in fact parenchymal hepatocytes (Leffert *et al.*, 1977).

This latter point is important since there are presently many "liver cell strains" and comparable cell lines with essentially little or no hepatocyte function but which are still utilized as models for liver cells in culture. Only cells capable of critical hepatocyte functions, such as the corticosteroid induction of tyrosine aminotransferase, the phenobarbital induction of cytochrome P-450, and/or the complete urea cycle, should be considered as bona fide hepatocyte-derived cells. Clearly, epithelial cell appearance in culture is an insufficient criterion for hepatocyte derivation of a cell line.

One of the most critical areas in which hepatocytes have been useful is the study of xenobiotic metabolism. The use of hepatic cell cultures in monitoring environmental carcinogens is potentially great, since most known chemical carcinogens are metabolized to active forms in the liver. Thus the requirement for an S-9 fraction as seen in the Ames system (Ames *et al.*, 1973) is built into the primary hepatocyte culture, although as yet not at the same level as that found *in vivo*. This latter problem is still to be overcome, but the studies of Decad *et al.* (1977) and others on the hormonal requirements of liver cells in culture (Hayashi *et al.*, 1978) offer promise that this difficulty will be surmounted before long.

Related to the hormonal regulation of liver cell metabolism is the question of the regulation of DNA synthesis in these cells. Clearly, primary adult rat liver cells are incapable of completing the cell cycle *in vitro,* but several systems have demonstrated an increase in DNA synthesis in parenchymal liver cells in culture (Richman *et al.*, 1976; Sirica *et al.*, 1979). The hormonal regulation of DNA synthesis in liver is well known. Thus, under some suitable culture conditions the replication of hepatocytes *in vitro* may become a reality.

Finally, the interesting phenomenon of the "fetalization" of adult hepatocytes in culture as described by Sirica *et al.* (1979) and Leffert *et al.* (1978) has yet to be explained. Pitot and Sirica (1979) have suggested that this effect is due to the lack of hormones in the culture medium; the hepatocytes revert to their fetal characteristics since the hormonal environment of the adult is required to maintain the highly differentiated biochemistry of the hepatic parenchymal cell. It is of interest, as noted earlier, that fetal cells in culture tend to acquire adult cell characteristics (Grieninger and Granick, 1975). Thus, it is possible that hepatic cells tend toward some "equilibrium differentiated state" in environments that do not possess all of the hormonal and substrate characteristics of the *in vivo* situation. Whether this phenomenon can be explained in such simplistic terms or represents some other major mechanism of differentiation remains to be determined.

Acknowledgments

The authors wish to express their sincere appreciation to Dr. Carol Sattler for the preparation of the electron micrographic Fig. 1(4) and to Mr. Gerald Sattler for assistance in the micrography.

References

Acosta, D., Anuforo, D. C., and Smith, R. E. (1978). *In Vitro* **14,** 428.
Alexander, R. W., and Grisham, J. W. (1970). *Lab. Invest.* **22,** 50.
Ames, B. N., Lee, F. D., and Durston, W. E. (1973). *Proc. Natl. Acad. Sci. U.S.A.* **70,** 782.
Bausher, J., and Schaeffer, W. I. (1974). *In Vitro* **9,** 286.
Beck, J., and Bollack, C. (1974). *FEBS Lett.* **47,** 314.
Bernhardt, H. P., Darlington, G. J., and Ruddle, F. H. (1973). *Dev. Biol.* **35,** 83.
Berry, M. N., and Friend, D. S. (1969). *J. Cell Biol.* **43,** 506.
Bissell, D. M., Hammaker, L. E., and Meyer, U. A. (1973). *J. Cell Biol.* **59,** 722.
Bonney, R. J., Becker, J. E., Walker, P. R., and Potter, V. R. (1974). *In Vitro* **9,** 399.

Borenfreund, E., Higgins, P. J., Steinglass, M., and Bendich, A. (1975). *J. Nat. Cancer Inst.* **55,** 375.
Buckley, I. K., and Walton, J. R. (1974). *Tissue Cell* **6,** 641.
Burke, M. D., Vadi, H., Jernstrom, B., and Orrenius, S. (1977). *J. Biol. Chem.* **252,** 6424.
Capuzzi, D. M., Rothman, V., and Margolis, S. (1971). *Biochem. Biophys. Res. Commun.* **45,** 421.
Chen, C-L., and Feigelson, P. (1978a). *Biochemistry* **17,** 5308.
Chen, C-L., and Feigelson, P. (1978b). *J. Biol. Chem.* **253,** 7880.
Chessebeuf, M., Olsson, A., Bournot, P., Desgres, J., Guiguet, M., Maume, G., Maume, B. F., Perissel, B., and Padieu, P. (1974). *Biochimie* **56,** 1365.
Coufalik, A., and Monder, C. (1976). *Proc. Soc. Exp. Biol. Med.* **152,** 603.
Crane, L. J., and Miller, D. L. (1977). *J. Cell Biol.* **72,** 11.
Decad, G. M., Hsieh, D.P.H., and Byard, J. L. (1977). *Biochem. Biophys. Res. Commun.* **78,** 279.
Dickson, J. A. (1971). *Exp. Cell Res.* **64,** 17.
Drochmans, P., Wanson, J-C., and Mosselmans, R. (1975). *J. Cell Biol.* **66,** 1.
Elliott, K.R.F., and Pogson, C. I. (1977). *Mol. Cell. Biochem.* **16,** 23.
Fahl, W. E., Michalopoulos, G., Sattler, G. L., Jefcoate, C. R., and Pitot, H. C. (1979). *Arch. Biochem. Biophys.* **192,** 61.
Feliu, J. E., Hue, L., and Hers, H.-G. (1976). *Proc. Natl. Acad. Sci. U.S.A.* **73,** 2762.
Fry, J. R., Jones, C. A., Wiebkin, P., Bellemann, P., and Bridges, J. W. (1976). *An. Biochem.* **71,** 341.
Gebhardt, R., Bellemann, P., and Mecke, D. (1978). *Exp. Cell Res.* **112,** 431.
Gerschenson, L. E., Berliner, J., and Davidson, M. D. (1974a) *In* "Methods in Enzymology" (S. Fleischer and L. Packer, eds.), Vol. 32, pp. 733–740. Academic Press, New York.
Gerschenson, L. E., Davidson, M. B., and Anderson, M. (1974b). *Eur. J. Biochem.* **41,** 139.
Gielen, J. E., and Nebert, D. W. (1971). *J. Biol. Chem.* **246.** 5189.
Goodridge, A. G., and Adelman, T. G. (1976). *J. Biol. Chem.* **251,** 3027.
Grieninger, G., and Granick, S. (1975). *Proc. Natl. Acad. Sci. U.S.A.* **72,** 5007.
Grieninger, G., and Granick, S. (1978). *J. Exp. Med.* **147,** 1806.
Guzelian, P. S., Bissell, D. M., and Meyer, U. A. (1977). *Gastroenterology* **72,** 1232.
Haars, L., and Pitot, H. C. (1979). *J. Biol. Chem.* **254,** 9401.
Haggerty, D. F., Young, P. L., Popjack, G., and Carnes, W. H. (1973). *J. Biol. Chem.* **248,** 223.
Hayashi, S., and Ooshiro, Z. (1978). *Bull. Jpn. Soc. Sci. Fish.* **44,** 499.
Hayashi, I., Larner, J., and Sato, G. (1978). *In Vitro* **14,** 23.
Hays, D. M., Sera, Y., Koga, Y., and Hays, E. F. (1975). *Proc. Soc. Exp. Biol. Med.* **148,** 596.
Henderson, P. T. (1971). *Life Sci.* **10,** 655.
Horsfall, A. C., and Ketterer, B. (1976). *Br. J. Cancer* **33,** 96.
Howard, R. B., and Pesch, L. A. (1968). *J. Biol. Chem.* **243,** 3105.
Howard, R. B., Christensen, A. K., Gibbs, F. A., and Pesch, L. A. (1967). *J. Cell Biol.* **35,** 675.
Howard, R. B., Lee, J. C., and Pesch, L. A. (1973). *J. Cell Biol.* **57,** 642.
Jacob, S. T., and Bhargava, P. M. (1962). *Exp. Cell Res.* **27,** 453.
Jeejeebhoy, K. N., and Phillips, M. J. (1976). *Gastroenterology* **71,** 1086.
Jeejeebhoy, K. N., Ho, J., Breckenridge, C., Bruce-Robertson, A., Steiner, G., and Jeejeebhoy, J. (1975). *Biochem. Biophys. Res. Commun.* **66,** 1147.
Kido, H., Sanada, Y., Katunuma, N., and Morris, H. P. (1977). *Gann* **68,** 691.
King, C. M., Traub, N. R., Cardon, R. A., and Howard, R. B. (1976). *Cancer Res.* **36,** 2374.
Kitajima, K., Tashiro, S.-I., and Numa, S. (1975). *Eur. J. Biochem.* **54,** 375.
Kletzien, R. F., Pariza, M. W., Becker, J. E., Potter, V. R., and Richard, F. R. (1976). *J. Biol. Chem.* **251,** 3014.
Kreusch, J., Hoffmann, F., Maier, K. P., and Decker, K. (1977). *Hoppe-Seyler's Z. Physiol. Chem.* **358,** 1513.
Laishes, B. A., and Williams, G. M. (1976). *In Vitro* **12,** 521.

Lakshmanan, M. R., Nepokroeff, C. M., Kim, M., and Porter, J. W. (1975). *Arch. Biochem. Biophys.* **169,** 737.
Lee, J.-L., and Kenney, F. T. (1970). *Biochem. Biophys. Res. Commun.* **40,** 469.
Leffert, H. L. (1974). *J. Cell Biol.* **62,** 792.
Leffert, H. L., and Paul, D. (1972). *J. Cell Biol.* **52,** 559.
Leffert, H. L., and Sell, S. (1974). *J. Cell Biol.* **61,** 823.
Leffert, H. L., Moran, T., Boorstein, R., and Koch, K. S. (1977). *Nature (London)* **267,** 58.
Leffert, H., Moran, T., Sell, S., Skelly, H., Ibsen, K., Mueller, M., and Arias, I. (1978). *Proc. Natl. Acad. Sci. U.S.A.* **75,** 1834.
Levine, G. A., Bissell, M. J., and Bissell, D. M. (1978). *J. Biol. Chem.* **253,** 5985.
Lin, R. C., and Snodgrass, P. J. (1975). *Biochem. Biophys. Res. Commun.* **64,** 725.
MacDonnell, P. C., Ryder, E., Delvalle, J. A., and Greengard, O. (1975). *Biochem. J.* **150,** 269.
Michalopoulos, G., and Pitot, H. C. (1975). *Exp. Cell Res.* **94,** 70.
Michalopoulos, G., Sattler, C. A., Sattler, G. L., and Pitot, H. C. (1976). *Science* **193,** 907.
Michalopoulos, G., Sattler, G. L., and Pitot, H. C. (1978a). *Cancer Res.* **38,** 1550.
Michalopoulos, G., Sattler, G. L., O'Connor, L., and Pitot, H. C. (1978b). *Cancer Res.* **38,** 1866.
Munns, T. W., Johnson, M.F.M., Liszewski, M. K., and Olson, R. E. (1976). *Proc. Natl. Acad. Sci. U.S.A.* **73,** 2803.
Oliver, I. T., Edwards, A. M., and Pitot, H. C. (1978). *Eur. J. Biochem.* **87,** 221.
Paul, D., and Walter, S. (1974). *Proc. Soc. Exp. Biol. Med.* **145,** 456.
Pitot, H. C., and Sirica, A. E. (1979). *Differentiation* (in press).
Pitot, H. C., Peraino, C., Morris, P. A., and Potter, V. R. (1963). *Natl. Cancer Inst. Monogr.* **13,** 229.
Reel, J. R., Lee, K.-L., and Kenney, F. T. (1970). *J. Biol. Chem.* **245,** 5800.
Renton, K. W., Deloria, L. B., and Mannering, G. J. (1978). *Mol. Pharmacol.* **14,** 672.
Richardson, U. I., Tashjian, A. H., Jr., and Levine, L. (1969). *J. Cell Biol.* **40,** 236.
Richman, R. A., Claus, T. H., Pilkis, S. J., and Friedman, D. L. (1976). *Proc. Natl. Acad. Sci. U.S.A.* **73,** 3589.
Rose, G. G., Kumegawa, M., and Cattoni, M. (1968). *J. Cell Biol.* **39,** 430.
Savage, C. R., and Bonney, R. J. (1978). *Exp. Cell Res.* **114,** 307.
Schreiber, G., and Schreiber, M. (1972). *J. Biol. Chem.* **247,** 6340.
Seglen, P. O. (1973a). *Exp. Cell Res.* **76,** 25.
Seglen, P. O. (1973b). *Exp. Cell Res.* **82,** 391.
Seglen, P. O., and Fossa, J. (1978). *Exp. Cell Res.* **116,** 199.
Seglen, P. O., and Reith, A. (1977). *Biochim. Biophys. Acta.* **496,** 29.
Shanmugam, G., and Bhargava, P. M. (1968). *Biochem. J.* **108,** 741.
Siess, E. A., and Wieland, O. H. (1975). *Biochem. Biophys. Res. Commun.* **64,** 323.
Sirica, A. E., Richards, W., Tsukada, Y., Sattler, C. A., and Pitot, H. C. (1979). *Proc. Natl. Acad. Sci. U.S.A.* **76,** 283.
Smith, J. E., Borek, C., and Goodman, D. S. (1978). *Cell* **15,** 865.
Spence, J. T., and Pitot, H. C. (1979). *Fed. Proc.* **38,** 504.
Stanchfield, J. E., and Yager, J. D. (1978). *Exp. Cell Res.* **116,** 239.
Story, D. L., O'Donnell, J. A., Dong, F. M., and Freedland, R. A. (1976). *Biochem. Biophys. Res. Commun.* **73,** 799.
Takaoka, T., Yasumoto, S., and Katsuta, H. (1975). *Jpn. J. Exp. Med.* **45,** 317.
Takeda, Y., Ichihara, A., Tanioka, H., and Inoue, H. (1964). *J. Biol. Chem.* **239,** 3590.
Tarlow, D. M., Watkins, P. A., Reed, R. E., Miller, R. S., Zwergel, E. E., and Lane, M. D. (1977). *J. Cell Biol.* **73,** 332.
Thompson, E. B., Tompkins, G. M., and Curan, J. F. (1966). *Proc. Natl. Acad. Sci. U.S.A.* **56,** 296.
Wagle, S. R. (1975). *Life Sci.* **17,** 827.

Walker, P. R. (1977). *J. Cell. Physiol.* **91,** 169.
Walker, P. R., Bonney, R. J., Becker, J. E., and Potter, V. R. (1972). *In Vitro* **8,** 107.
Wanson, J.-C., Drochmans, P., May, C., Penasse, W., and Popowski, A. (1975). *J. Cell Biol.* **66,** 23.
Weigand, K., and Otto, I. (1974). *FEBS Lett.* **46,** 127.
Wicks, W. D. (1968). *J. Biol. Chem.* **243,** 900.
Williams, G. M. (1977). *Cancer Res.* **37,** 1845.
Williams, G. M., and Gunn, J. M. (1974). *Exp. Cell Res.* **89,** 139.
Yager, J. D., Pariza, M. W., Becker, J. E., and Potter, V. R. (1975). *In* "Liver Regeneration after Experimental Injury" (R. Lesch and W. Reutter, eds.), pp. 148–151. Stratton, New York.

Chapter 23

Human Liver Cells in Culture[1]

WARREN I. SCHAEFFER AND DANA J. KESSLER

Department of Medical Microbiology,
College of Medicine,
University of Vermont,
Burlington, Vermont

I. Introduction

The culture of human hepatic cells suffers from two major types of difficulties: one is semantic and the other is technical. It will be necessary to iterate these in order to understand the way in which we have evaluated our work and that of others.

First, from the point of view of semantics, what do we mean when we refer to a culture of hepatocytes? We feel that it is essential, as will be discussed later, in depth, that the answer to this question be approached from a functional point of view. That is, the cultures produced should possess one or more liver-specific biochemical characteristics. Further, if one is going to call the culture a "hepatocyte culture" then the culture should not merely *contain* hepatocytes, it should be totally *composed* of hepatocytes, otherwise the culture would more properly

[1]This project was supported in part by NIAID Contract NO1-AI-72500 and NCI grant CA12056-08.

ISBN 0-12-564140-0

be termed a "liver cell culture," one *derived* from liver tissue. Hepatocyte refers to the *specific* liver cell, whose functional capability endows the liver with its characteristics. We have purposely omitted any consideration of structure from this discussion at this point as it will be discussed in depth later. Suffice it to say that structure, even when typical, is not sufficiently specific to unequivocally identify a hepatocyte and when atypical, is absolutely useless.

From a technical viewpoint, there have been no systematic studies related to the development of media specifically for the cultivation of human hepatocytes. This may possibly explain the current lack of an unequivocally identified diploid human hepatocyte culture. As will be discussed, morphological appearance of cultivated human liver cells is a poor criterion for hepatocyte identification and the morphological variability may reflect suboptimal cultural conditions for such cells. Furthermore, the current inability to maintain morphological homogeneity may also reflect this problem.

We will, therefore, within this framework, present the results which we have obtained in our attempts to cultivate human hepatocytes.

II. Methods and Materials

A. Source of Tissue

In all cases, our source of tissue has been from postmortem samples obtained from newborns less than 1 week old. Tissue samples were obtained as soon as possible, but always less than 12 hours following death. No samples were taken from infants in which the suspected cause of death was an infectious agent. In one case, tissue samples were obtained from twin fetuses at approximately 20 weeks of gestation. Tissue samples were obtained, at autopsy, and placed in culture medium (see following) containing twice the normal (see following) concentration of gentamicin (transport medium). Usually no more than 1 hour of time elapsed from the time the samples were obtained until processing of the samples for culture was begun.

B. Processing the Tissue for Culture

The tissue samples, weighing approximately 7 gm, were washed with transport medium and were then transferred to a 100-mm petri plate containing 2 ml Dulbecco's phosphate-buffered saline (PBS). The liver samples were then minced into 1 mm^3 pieces using two scalpels worked in opposition to one another taking care to make clean cuts and to avoid tearing the tissue. Obvious nonparenchymal elements such as fatty tissue were dissected away during the mincing process. When mincing was completed, the PBS was removed and the minced tissue was washed twice with 3 to 5 ml of additional PBS to remove red blood cells. The tissue mince was then transferred to 50-ml Ehrlenmeyer flasks (2–3 gm of tissue to each flask). A sufficient amount of 0.1% collagenase (usually 10 ml) was added and the flask was placed into a rotary shaking waterbath adjusted to 130 revolutions per minute. (It is important to maintain the mince in suspension but not to produce foaming in the process.) At 25, 50, and 75 minutes, the flasks were removed from the bath and the enzyme solution containing the disaggregated cells was removed to a 30-ml centrifuge tube containing an equal quantity of ice-cold fetal bovine serum. The serum was used to increase the viscosity of the suspending fluid and to coat the now fragile hepatocyte cells in order to protect them during the centrifugation process and succeeding steps. Unless centrifuged immediately, the cell suspension was maintained at 4°C. Centrifugation was performed by spinning at 126 *g* for 10 minutes in a Sorvall GLC-1 centrifuge using a swinging bucket rotor. The supernate was discarded and the cell pellet was gently resuspended in complete medium maintained at 37°C and recentrifuged to effect a second wash. One-half of the cell suspension was then planted in each of two T-25 flasks.

C. Culture Medium

The culture medium used was the Kaighn modification of Ham's F-12, F-12K, supplemented with 10% fetal bovine serum (always pretested for plating efficiency and growth promotion), 50 μg/ml gentamicin, and 0.22 μg/ml hydrocortisone.

When enrichment for hepatocytes was desired, cells were planted in F-12K, containing dialyzed serum and lacking arginine (hepatocyte cells possess enzymes of the urea cycle and can synthesize arginine *de novo*). Other arginine-deficient media tested were: Dulbecco's High Glucose MEM and RPMI-1640. Only the arginine-deficient F-12K and RPMI-1640 proved successful, therefore, F-12K was used routinely thereafter. Cells were grown in the arginine-deficient medium for 2 weeks and the medium was replaced once during this period. After selection the cells were refed with F-12K and were cloned by low-density plating. Clones were selected for epithelioid morphology.

D. Identification of Hepatocyte Products

1. Detection of Albumin by the Enzyme-Linked Immunosorbent Assay (ELISA) Engvall and Perlman (1972)

The indirect method used (described in the following) was modified by Boraker (personal communication).

a. Materials. The first five reagents listed are common to all assays:

1. Carbonate coating buffer—pH 8
   1.59 gm sodium carbonate ($Na2CO_3$)
   2.93 gm sodium bicarbonate ($NaHCO_3$)
   0.2 gm sodium azide (NaN_3)
   Make up to 1 liter in distilled water. Store at 4°C
   Stable for 2 weeks
2. PBS–Tween 20 (PBS-T)
   100 ml calcium and magnesium-free, phosphate buffered saline (PBS-CMF) (10×)
   0.5 ml Tween 20
   0.2 gm NaN_3
   Bring up to 1 liter with distilled water—pH 7.4
   Store at 4°C
3. Diethanolamine buffer (10%)
   97 ml diethanolamine
   800 ml distilled water
   0.2 gm NaN_3
   100 mg magnesium chloride ($MgCl_2 \cdot 6H_2O$)
   Adjust to pH 9.8 with 1 *M* HCl
   Make up to 1 liter with distilled water. Store at 4°C in dark (wrap bottle in tin foil—should be at room temperature before use
4. 3 *M* sodium hydroxide (NaOH)
5. Substrate—*p*-nitrophenyl phosphate
   Use Sigma 104 tablets (5 mg tablets)
   Dissolve one tablet in 5 ml of diethanolamine buffer
   Must be used the same day as that on which it was prepared
   Use at room temperature

For assay of human albumin in cell culture medium:

6. Human albumin
   Stock solution of 5 mg/ml
   Dilute in PBS-T (10 μg, 100 ng, 10 ng, 1 ng, 100 pg, 10 pg, 1 pg)
   This serves as basis for the standard curve of human albumin and these various dilutions can also be employed in running controls
7. Media from cell cultures
   The media are concentrated 10× (Millipore-immersible molecular separator kit)
   The media are then diluted in PBS-T (1:10, 1:100, 1:1000, 1:10000)
8. Goat anti-human albumin
   Miles-Yeda, Ltd. #61-015 4.3 mg AB/ml
   Dilute 1:500 in carbonate coating buffer. (This serves as solid phase for the indirect assay or "sandwich.")

9. Rabbit anti-human albumin
   Miles-Yeda, Ltd. 5 mg AB/ml
   Absorbed against bovine serum albumin (5 mg/ml). The precipitated antigen–antibody complex is discarded, and the supernatant is saved.
   Dilute supernatant 1:250 in PBS-T
10. Goat anti-rabbit immunoglobulin conjugated with alkaline phosphatase (G × R Ig–alkaline phosphatase)
    (Supplied by Dr. Boraker)
    Dilute 1:2000 in PBS-T
11. Normal goat serum
    (Supplied by Dr. Boraker)
    Dilute 1:100 in carbonate coating buffer
    This serves as the solid phase control
12. Normal rabbit serum
    (Supplied by Dr. Boraker)
    Dilute 1:100 in PBS-T
    This serves as the control for rabbit anti-human albumin
13. Microtiter plates—96 well micro-ELISA microtiter plates (Dynatech Labs) with "peel off" covers

b. Methods. The following methods are used:

1. For testing cell culture media and establishing a standard curve for human albumin, 200 ul of goat anti-human albumin (G × human albumin) is pipetted into the wells of the microtiter plate designated for this purpose. The plastic serves as the "solid phase" and binds the other layers of the "sandwich" to the well. Binding to the plastic occurs within 15 minutes.

2. The G × human albumin is carefully aspirated off and the wells are washed three times in PBS-T for 3 minutes/wash. Thorough washing is a very important step and must not be skipped or abbreviated. Note: great care must be exercised in aspirating any reagent from the plate as a mere scratch to the surface may remove anything that has been bound. A wash bottle filled with PBS-T may be used to quickly fill the wells for any wash.

3. Next, the second layer of the "sandwich" is added to the wells. This is the material to be tested, either culture fluid or human albumin for the standard curve. All reagents subsequent to the coating step are diluted in PBS-T; the Tween 20 serves to prevent nonspecific binding of the subsequent reagents. Two hundred microliters of either solution is added to each well and incubation is carried out for a minimum of 6 hours to overnight at 4°C.

4. The wells are then washed again in PBS-T (three times for 3 minutes each) and then 200 ul rabbit anti-human albumin (R × human albumin) is added and allowed to incubate at room temperature 2–3 hours.

5. The R × human albumin is aspirated, the plate is washed in PBS-T (three times for 3 minutes each), and 200 μl G × R-Ig-alkaline phosphatase (1:2000) is added. This requires an incubation period of 8 hours to overnight at 4°C.

6. Again, the plate is washed three times for 3 minutes per wash in PBS-T.

7. Two hundred microliters *p*-nitrophenyl phosphate in 10% diethanolamine is added to each well at 15-second intervals. The timing is critical, as the reaction may go to completion very quickly (in our assays of human albumin, the reaction was halted at approximately 3.5 minutes). To stop the reaction, 50 μl of 3 *N* NaOH is added to the well at 15-second intervals corresponding to the pattern in which the substrate was added.

8. The "peel off" microtiter cover is replaced on the plate. The plate is inverted in order to evenly distribute the NaOH and ensure that the reaction has been completely halted.

9. This laboratory has a Beckman DB-G spectrophotometer modified to aspirate the sample into the cuvette and automatically discard the sample at the end of the reading. The samples are read at 405 nm.

Note: The reaction is critical. *p*-Nitrophenyl phosphate is dephosphorylated by the active alkaline phosphatase and gives rise to a phenol derivative which displays a bright yellow color. The reaction should be halted before the solution becomes too "optically dense" to read. The substrate incubation time should be empirically adjusted to give optical density readings of approximately 0.6 in 15–30 minutes. A control for all components of the system (nonspecific binding, cross-reactivity of the various antibodies used with one another, efficiency of washing medium components, etc.) must be run to assure that the final reaction received is specific.

2. Albumin Detection by Immunoelectrophoresis

This procedure was accomplished by standard methodology (Kabat and Mayer, 1964) applied to culture medium which was concentrated 10 ×.

3. Detection of Albumin and Transferrin by Agar Overlay of Cell Clones (Immunooverlay)

This procedure was developed and preliminarily described by Sammons *et al.* (1978). The methodology essentially consists of incorporating antibody to human albumin or transferrin into agar which is then spread over clones of mixed liver cells or hepatocytes. The cells are then incubated for 24 hours. The agar overlay is then removed and any antigen–antibody complexes are detected by staining with Coomassie blue. Since this procedure is nondestructive, the cells from which this overlay is removed can be refed and reincubated. By premarking the agar overlay prior to its removal, cell clones producing the desired product can be identified, removed, and grown in separate cultures.

4. Growth Curves and the Effects of Growth Factors

Growth curves were conducted by planting 3–4 x 10^4 cells in each well of 24-well plastic multidishes (Costar). Each day, for 7 days, two or three wells are washed three times with PBS. At the end of the growth period the amount of protein in each well is quantitated by the method of Oyama and Eagle (1956). To examine the effect of epidermal growth factor (EGF), insulin, glucagon, and chorionic gonadotropin on the growth of the liver cell cultures, the particular factor was incorporated into the growth medium and growth curves were conducted. The levels tested were as follows: insulin, 30 U/ml; glucagon, 10 μg/ml; chorionic gonadotropin, 15 U/ml; EGF, 100, 10, and 1.0 ng/ml.

III. Results

We have been involved, for a number of years, in the cultivation of rat liver cells and have definitively demonstrated (Bausher and Schaeffer, 1974; Schaeffer and Heintz, 1978) that the culture which we have developed, RL-PR-C, was derived from a hepatocyte. During the course of this study, however, it rapidly became apparent that, with human liver, we were dealing with a tissue which, when disaggregated and placed in culture, would behave quite differently from rat liver tissue. That is, we were unable to find areas of typical epithelioid cells such as we detected in our early passage, mixed cultures of rat liver (Fig. 1). Instead, what is seen are cells which can be generally classified as intermediate between epithelioid and fibroblastic cells. They are best described as being stellate or polygonal. Figure 2 shows one such cell juxtaposed to a classical spindle-shaped or fibroblastic cell and Fig. 3 shows a grouping of such cells. As one became more familiar with these cultures and was willing to examine them without being prejudiced as to how liver cells "should" look, other morphological variations in cell type became apparent. For example, a large, polygonal cell type with granular cytoplasm was observed (Fig. 4) as well as a large polygonal cell with a relatively clearer cytoplasm (Fig. 5). Binucleate cells of *both* types are also found (Fig. 4). Figure 3 shows an example of the final cell type seen which is a smaller polygonal cell with a clear cytoplasm, approximately one-half to one-third the size of the larger cell types. These results correspond exceedingly well to those of Zuckerman *et al.* (1968), Guillouzo *et al.* (1972), and LeGuilly *et al.* (1973), who have examined cells which proliferated from explanted biopsy material obtained from *adult* humans.

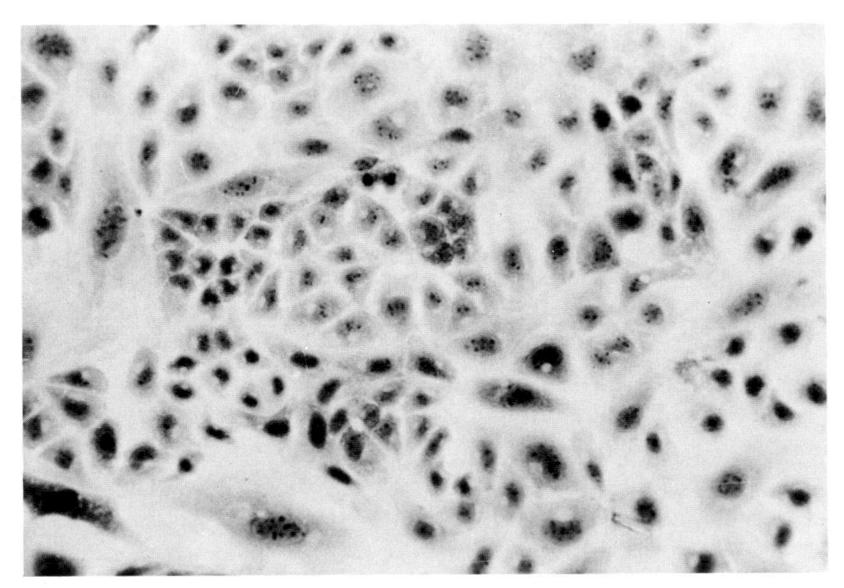

FIG. 1. Early passage (Passage 9) uncloned culture of rat liver cells derived from a 3-day-old rat. Note the clearly visible clusters of epithelioid cells and the morphologically heterogeneous nature of the monolayer. The cells were stained with Giemsa and were photographed with a 10× objective lens.

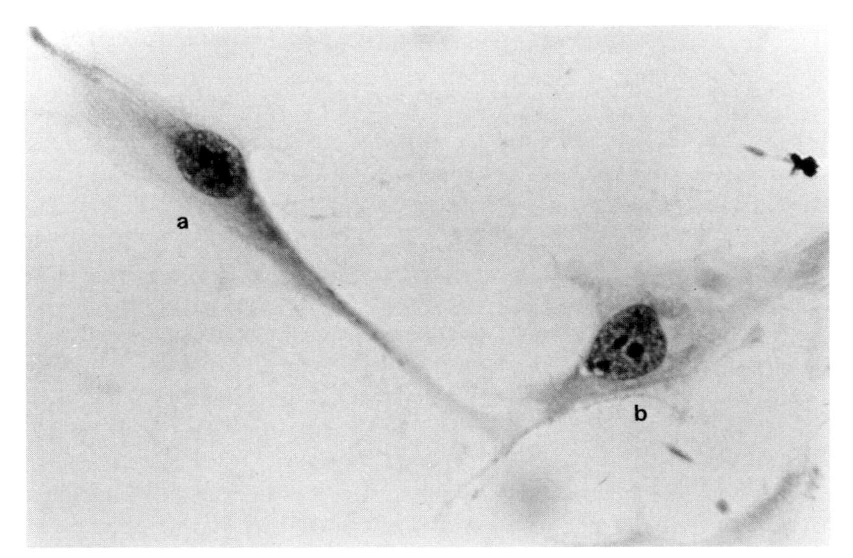

FIG. 2. Two morphologically distinct cells from an early passage (Passage 4) culture of human liver. This photograph shows clearly the difference between a fibroblastic cell (a) and the stellate or polygonal cell (b). These cells were stained with Giemsa and were photographed with a 40× objective lens.

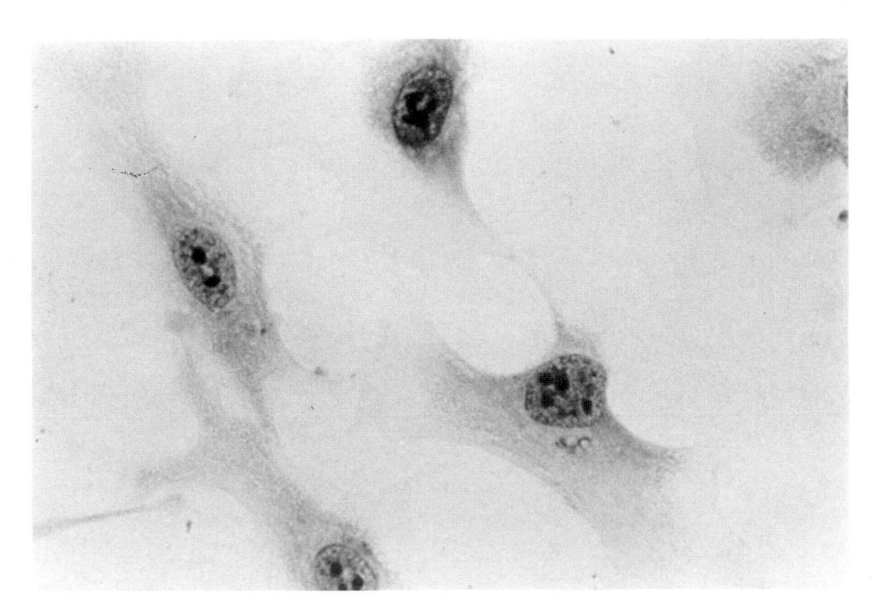

FIG. 3. A grouping of stellate or polygonal cells with a clear cytoplasm in culture of human liver. Note that these cells are clearly morphologically distinct from fibroblastic cells but are not of a classically epithelioid morphology. These cells were stained with Giemsa and were photographed with a 40× objective lens.

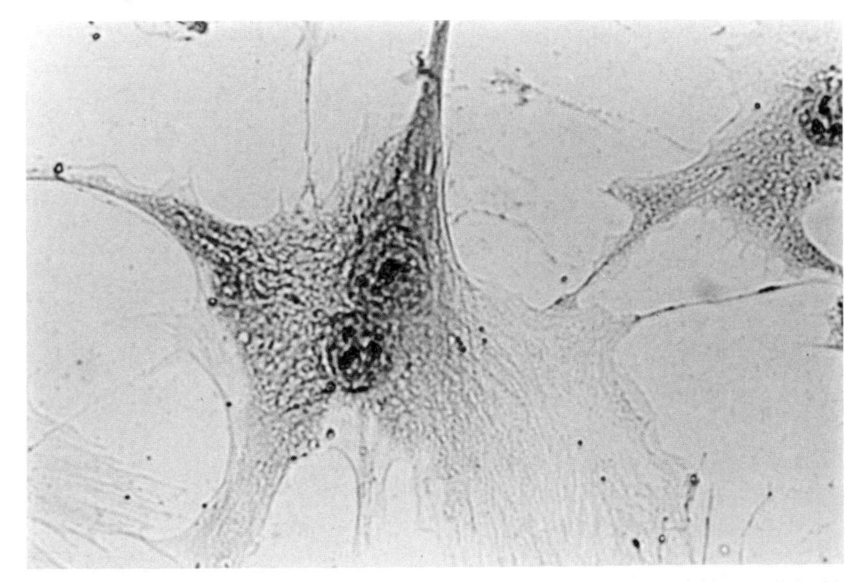

FIG. 4. An example of the large, granular, polygonal cells found in cultures of human liver. Note that one particular cell is binucleate, a characteristic often found in hepatocyte cells. The cells were stained with Giemsa and were photographed with a 40× objective lens.

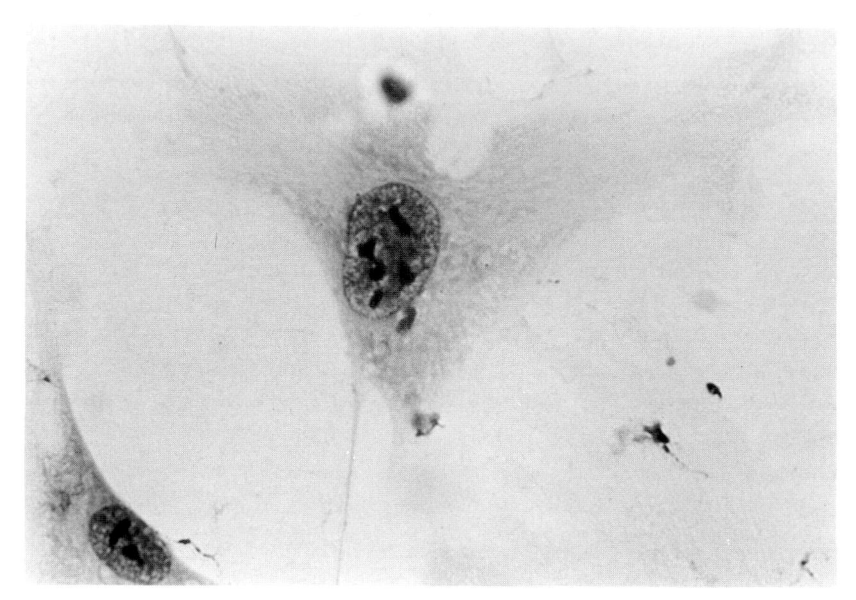

FIG. 5. An example of the large, polygonal cells with a clear cytoplasm found in cultures of human liver. This preparation was stained with Giemsa and photographed with a 40× objective lens.

Initially, when scanning these cultures for "epithelioid" patches of cells, we tended to dismiss all of these cell types as merely variants of fibroblast cells which will frequently appear somewhat polygonal when plated at low density. However, careful examination of more densely packed cultures clearly showed a morphologically heterogeneous monolayer. Figure 6 a and b illustrates this point. In those areas which are relatively "pure" with respect to morphological type (Fig. 6a), the fibroblastic cells align themselves along their long axes into a characteristic "streaming" pattern. However, in more mixed areas such alignment is impossible and the monolayer appears heterogeneous with spacing apparent even at relatively higher densities (Fig. 6b). Eventually, when confluency is reached, these heterogeneous areas appear "criss-crossed" and this provides a selective mechanism wherein one can presume that putative hepatocyte cells are present. When the cells in these areas are selectively removed and replated the density of polygonal types is increased and cloning becomes more feasible. Thus far, we do not have a cloned human hepatocyte culture, however, we have one which is just at the stage of "scaling up" from micro-dishes.

In order to increase the potential for obtaining hepatocyte cells we examined the possibility of enhancing growth by incorporating epidermal growth factor, glucagon, insulin, or chorionic gonadotropin in the medium. We were unable to

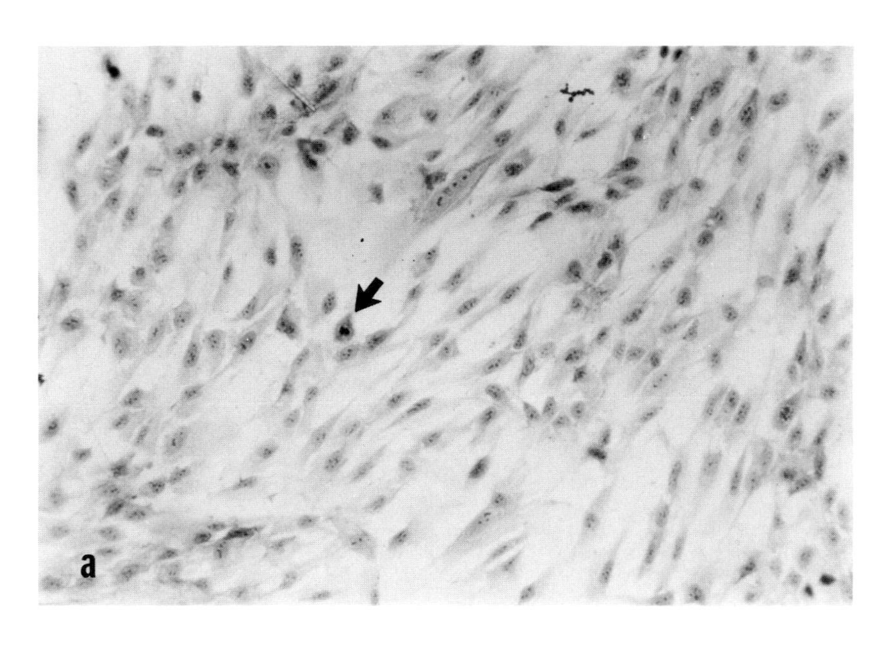

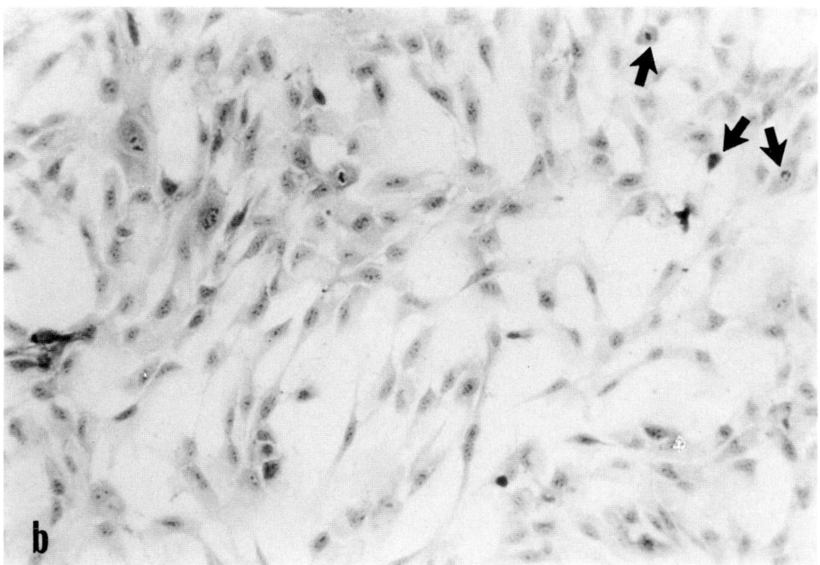

FIG. 6. (a) An example of a relatively pure fibroblastic area of a human liver culture at Passage 4. Note the overall parallel alignment of the cells along their long axes. (b) An example of a morphologically heterogenous area of the same slide as in (a). Note the inability of the fibroblastic cells to align themselves as in (a) and the more randomized appearance of the culture. In both areas it is possible to discern all of the cell types mentioned in the text and several cells in various stages of mitosis (arrows). These cultures were stained with Giemsa and were photographed with a 10× objective lens.

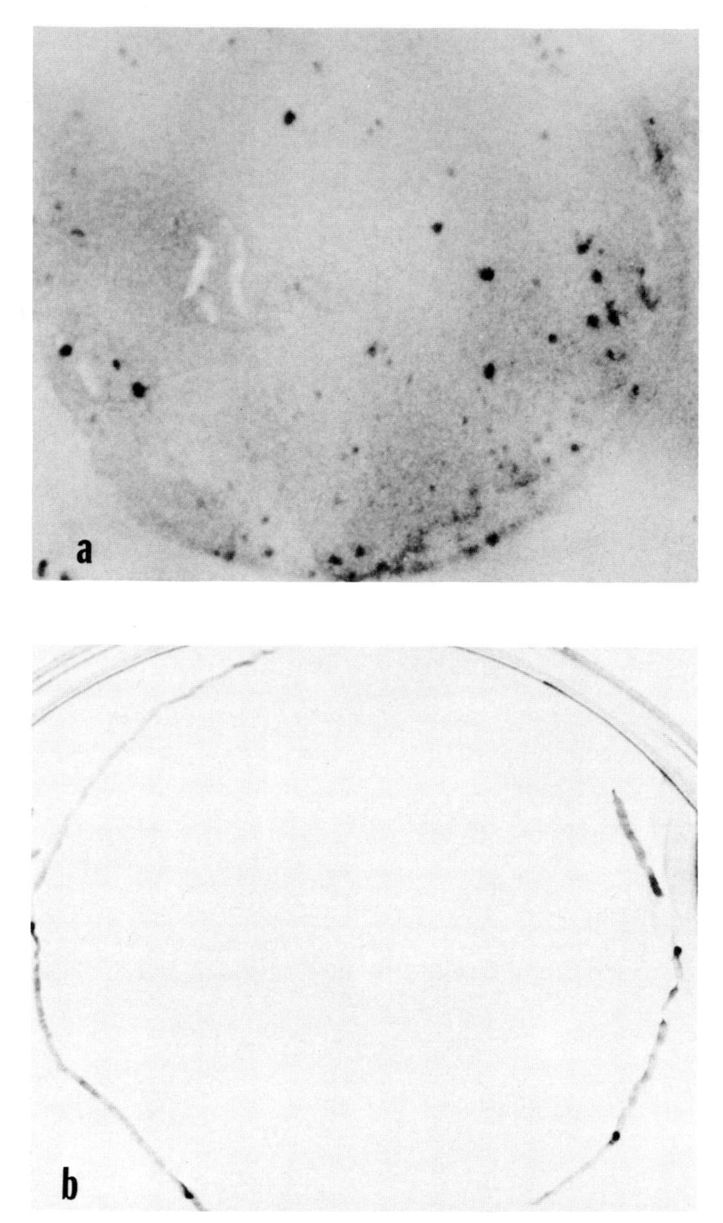

FIG. 7. (a) An immunooverlay containing antibody to human albumin which was removed from a culture of human liver cells plated at clonal density. Note the dark, Coomassie blue-stained areas of antigen–antibody complexes. Compare this photograph with the control immunooverlay(b). (b) This was prepared by overlaying agar, containing antibody to human albumin, onto a culture of human liver fibroblasts plated at clonal density. It was necessary to outline the agar overlay in order to provide sufficient contrast for photography, there being no retention of the dye.

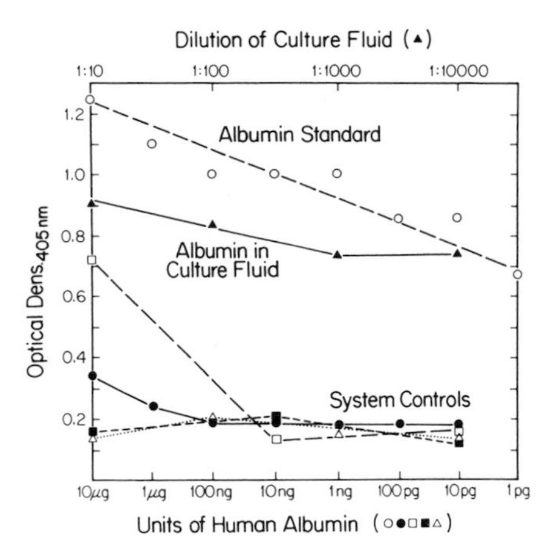

FIG. 8. This graph depicts the results of an ELISA for albumin production run on media recovered from a culture of human liver cells. Note the specificity of the reaction and the fact that the sensitivity extends to 10 pg. ○, Human albumin standard curve; ▲, albumin in culture medium; ●, binding control; □ and ■, bovine serum albumin controls; △, plastic control.

discern any enhanced growth or plating efficiency of either selected or nonselected cells using these hormones.

Another procedure which we used to increase the numbers of putative hepatocytes has been to plant cell suspensions in media lacking arginine, an amino acid required by most cells. Hepatocytes possess an active urea cycle and are able to synthesize arginine *de novo*. Nonhepatocytic cells do not contain enzymes of the urea cycle and. therefore, cannot proliferate under these conditions. Care must be taken to plant such cultures at a fairly low density in order to avoid metabolic cooperation, whereby hepatocyte cells could provide arginine to nonhepatocytic cells *via* membrane contact or secretion of arginine into surrounding medium.

Putative hepatocyte cells can also be located functionally, as opposed to morphologically, using the method of Sammons *et al.* (1978), namely, the "immunooverlay." This is an excellent, nondestructive means of identifying cloned, functional, hepatocyte cells. Figure 7a illustrates a stained "immunooverlay" and Fig. 7b illustrates a control culture of liver fibroblasts. It should be emphasized that since every hepatocyte cell will not, necessarily, be totipotent with respect to serum protein biosynthesis, one should screen for at least three liver-specific products (see Section IV) or some hepatocyte clones might be overlooked. We have found albumin and transferrin to be produced in the same culture but by different clones. We have also confirmed the presence of albumin in our cultures by the use of the ELISA method. Figure 8 demonstrates the results of this assay.

IV. Discussion and Perspectives

It must be stated, at the outset, that the cultivation of human liver cells has been a somewhat perplexing problem. This is especially true when one compares the cultivation of human liver cells with the cultivation of rat liver cells. Rat liver cells "obey" all of the rules defined by cell culturists which permit one to use morphology as a first approximation of cell type. That is, in primary cultures of newborn or fetal rat liver, one can distinguish between epithelial and fibroblastic clusters by light microscopy. These morphological types are retained upon subcultivation and, therefore, it is possible to clone for a particular morphological cell type that bears the "textbook" shape. These clones can then be tested for ultrastructural and biochemical characteristics typical of the hepatocyte. From the results of this study, it is apparent, however, that human liver tissue does not follow the dogma of "form following function." Our data are supported by the earlier study of Kaighn and Prince (1971) and by the studies of Guillouzo *et al.* (1972) and LeGuilly *et al.* (1973). Although there were significant differences in the methodology of Guillouzo *et al.* (1972) and LeGuilly *et al.* (1973) and ours (they used biopsies from adult patients and cultured these only as primary explants), they were able to describe the following five different morphological cell types by phase microscopy: fibroblasts, macrophages, a granular epithelioid cell characteristic of *in vivo* hepatocytes, a clear epithelioid cell of undetermined origin, and a polygonal cell, also of undetermined origin. When they examined culture fluids for the presence of 10 different plasma proteins, and correlated this with the predominant morphological cell type(s) in each of the cultures, they found that all 10 plasma proteins were never found simultaneously in a single culture. Orosomucoid, α_1-antitrypsin, ceruloplasmin, and plasminogen were produced by the granular and clear epithelioid cells; transferrin and hemopexin were produced by the polygonal cell type, and albumin, β-lipoprotein, α_2-macroglobulin, and haptoglobulin were produced by all three cell types.

One question which remains following the studies of human liver cultivation is: Excluding the fibroblast and macrophage elements, which are relatively easy to identify, are the remaining epithelioid cell types bona fide components of liver *in vivo* which can only be seen when the tissue is cultured or are these cell types derived by a process of dedifferentiation from the granular epithelioid cell characteristic of the *in vivo* liver? This question remains to be answered. From our data and those of Guillouzo *et al.* (1972), we would support the latter.

This then brings us to another question whose answer is of urgency: In the cultivation of cells in which classical morphology may not be exhibited and in which the tissue, *in vivo,* is capable of a repertoire of product synthesis considered to be organ specific, what sorts of criteria should be used to identify the cells *in vitro*? There appears to be little doubt, considering human liver cell cultivation, that morphological criteria, as a first approximation, are of little use. We have shown, supported by the work to others, that cells of intermediate morphology can be mistaken for contaminating fibroblasts or vice versa. Morphology can be of some help in indicating morphological diversity. This can be seen in areas of confluency. In areas comprised exclusively of fibroblasts, cells assume the parallel orientation and swirling so characteristic of such cultures, whereas mixtures of fibroblasts and epithelioid cells exhibit a more randomized appearance characterized more by a ''criss-crossed'' look. Selection of such ''patches'' for subcultivation can lead to separation of hepatocytes. Still, not all nonfibroblastic cells may be hepatocytes either. They could be aberrant fibroblasts or aberrant, nonhepatocytic epithelial cells. At this point, only biochemical parameters can suffice to make the distinctions complete.

As to what biochemical parameters to choose, one can, from the data presented earlier, make some judicious choices. Probably due to a ''division of labor'' in the intact organ, the hepatocytes in culture exhibit a diversity of functional capability with respect to synthesis of serum proteins. Of those tested, albumin, β-lipoprotein, α_2-macroglobulin, and haptoglobulin are apparently made by all three of the putative hepatocytes in culture, and, therefore, represent reasonable choices to begin testing. One should be aware of the fact that of the four proteins listed, α_2-macroglobulin, haptoglobulin, and β-lipoproteins can be synthesized by other than liver tissue, *in vivo,* as indicated by Schultze and Heremans (1966). One should also recall our data and those of Kaighn and Prince (1971) which suggest that not all *clones* will necessarily produce all of the four serum proteins listed. It must be remembered that LeGuilly *et al.* (1973) examined culture fluid from *mixed* cultures of the three morphologically distinct putative hepatocytes. As a result, any one of the predominant cell types, in any given culture, that is, granular epithelial, clear epithelial, or polygonal, may be composed of ''individuals'' from disparate areas of the liver whose functional repertoire may be somewhat different from others of the same type. Alternatively, the dedifferentiation which may take place to yield the various cell types could lead to the loss, or gain, of biosynthetic capability. Thus, it is always safest to screen for more than one product from any given clone. It is incumbent upon the investigator to prove that the cell type cultured is hepatocytic before calling it such. The mere cultivation of cells derived from liver does not make the culture a hepatocyte culture.

Perhaps some of the difficulties listed, such as variation in morphology and biosynthetic capability, will be overcome once truly suitable or ideal culture media are defined for any given cell type. Until such time as this and other cultural conditions become more standardized, variation will have to be expected. It is akin to listening to Rudolph Serkin and Artur Rubinstein playing the same piece of music; the notes are the same but the sound is different.

One further note with respect to recognition of cell types involves some recent findings which suggest the need for studies of the biochemical potential of cells both *in vivo* and *in vitro*. Humphries *et al.* (1976) have recently found mouse globin messenger RNA in the nuclear and cytoplasmic RNA not only from a fetal liver and reticulocytes but in low amounts in nonerythroid tissue of adult brain, liver, cultured lymphoma, and untransformed and transformed fibroblast lines. They used cDNA probes in locating the RNA sequences. Norling *et al.* (1978) have described the production of large chondroitin sulfate proteoglycan molecules in cultured human glial cells. The properties of the proteoglycans produced in the glial cultures suggest that its production may not be a feature only of cartilage cells. Newsome *et al.* (1976) and Smith *et al.* (1976) demonstrated the production, by the neural retina of the developing chick eye, of type II collagen. It had been initially thought that type II collagen production was restricted to hyaline cartilage.

These studies, both *in vivo* and *in vitro,* indicate that quite possibly we have overlooked the fact that even those tissues and cells considered to be differentiated may, in fact, have the ability to continue to express areas of the genome normally considered to be terminally repressed. When cells are removed from the *in vivo* situation and cultured *in vitro* in the absence of homeostatic control, there is an even higher probability of the expression of a greater percentage of the total genome. ''Dedifferentiation'' has long been recognized by those using cell cultures. However, it now appears as though it may be necessary to reexamine this process to see if certain cell types, in fact, were ever totally differentiated to the extent that one can absolutely relate a given function *only* with a particular tissue type. Dedifferentiation might even have to include the acquisition of additional synthetic capabilities.

The studies cited suggest the need for a systematic appraisal of the biochemical potential of cells. All cells of a given individual, after all, have derived from the single fertilized ovum and, therefore, presumably, possess the same genome. The extremely powerful tools of molecular biology will now permit a deeper probing of the functional capability of cells and this should add measurably to our understanding of the behavior of cells both *in vivo* and *in vitro*.

References

Bausher, J., and Schaeffer, W. I. (1974). *In Vitro* **9,** 286.

Boraker, D. (1979). Personal Communication.

Engvall, E., and Perlman, P. (1972). *J. Immunol.* **109,** 129.

Guillouzo, A., Oudea, P., LeGuilly, Y., Oudea, M. C., Lenoir, P., and Bourel, M. (1972). *Exp. Molc. Pathol.* **16,** 1.

Humphries, S., Windass, J., and Williamson, R. (1976). *Cell* **7,** 267.

Kabat, E. A., and Mayer, M. M. (1964). ''Experimental Immunochemistry,'' pp. 622–657. Thomas, Springfield, Illinois.

Kaighn, M. E., and Prince, A. M. (1971). *Proc. Natl. Acad. Sci. U.S.A.* **68,** 2396.

LeGuilly, Y., Launois, B., Lenoir, P., and Bourel, M. (1973). *Biomedicine* **19,** 248.

Newsome, D. A., Linsenmayer, T. F., and Trelstad, R. L. (1976). *J. Cell Biol.* **71,** 59.
Norling, B., Glimelius, B., Westermark, B., and Wasteson, A. (1978). *Biochem. Biophys. Res. Commun.* **84,** 914.
Oyama, V. I., and Eagle, H. (1956). *Proc. Soc. Exp. Biol. Med.* **91,** 305.
Sammons, D. W., Sanchez, E., and Darlington, G. J. (1978). *In Vitro* **14,** 347, and personal communication.
Schaeffer, W. I., and Heintz, N. H. (1978). *In Vitro* **14,** 418.
Schultze, H. E., and Heremans, J. F. (1966). "Molecular Biology of Human Proteins," Vol. I, p. 321. Elsevier, Amsterdam.
Smith, G. N., Linsenmayer, T. F., and Newsome, D. A. (1976). *Proc. Natl. Acad. Sci. U.S.A.* **73,** 4420.
Zuckerman, A. J., Rees, K. R., Inman, D. R., and Robb, I. A. (1968). *Br. J. Exp. Pathol.* **49,** 33.

Chapter 24

Overview

JERRY S. TRIER

Division of Gastroenterology,
Department of Medicine,
Peter Bent Brigham Hospital
and
Department of Medicine,
Harvard Medical School,
Boston, Massachusetts

That the technical aspects of *in vitro* culture of tissues and cells from the alimentary tract has posed major problems has been clearly stated in the foregoing chapters. As a result, culture of alimentary tract mucosal explants and epithelial cells has, in general, been less successful than culture of tissues and cells from most other organ systems. With the exception of the esophageal epithelium which, as pointed out by Hillman *et al.* (Chapter 16), survives nicely in explant culture, the characteristics of the epithelia of the alimentary tract are such that it is not surprising that it has not been easy to develop culture methods that ensure consistent, long-term survival.

In alimentary tract organs, mesenchymal cells are abundant and are intimately associated with epithelial cells. The problem of their tendency to overgrow epithelial cell cultures is well recognized, but potentially solvable as indicated by Quaroni and May (Chapter 20) and Schaeffer and Kessler (Chapter 23). In addition, the epithelial cell populations lining the stomach, small intestine, and colon are extremely heterogeneous, ranging from many rapidly renewing undifferentiated cells in the intestinal crypts and the base of the gastric foveoli to highly differentiated cells including the absorptive cells on intestinal villi and on the colonic luminal surface and the secretory cells in the glands of the gastric mucosa and in the intestinal crypts. The optimal requirements for *in vitro* maintenance and growth of these various cell types that populate the epithelial lining of the stomach and intestine may thus differ substantively. For example, explant or cell culture conditions ideal for maintaining proliferating crypt cells might

ISBN 0-12-564140-0

preclude proper differentiation of crypt cells to mature villus absorptive cells, whereas conditions optimal for maintaining differentiated absorptive cells might not permit survival of proliferating crypt cells.

Moreover, the milieu from which alimentary mucosa and its epithelial elements are obtained may be hostile to tissue and cell survival *in vitro,* a point which is underscored by the rapidity with which gastric and intestinal epithelial cells undergo degeneration after death. In adults, gastric mucosa is bathed with acid and proteases, small intestinal mucosa is bathed with proteases, phospholipases, detergent bile salts, and some bacteria, whereas the colonic mucosa is in constant contact with enormous numbers of microbial organisms. It is understandable that when viability of differentiated epithelial cell elements is achieved *in vivo,* continued production of acid and pepsinogens by gastric glandular cells and peptidases by villus absorptive cells may create an unfavorable microenvironment for cell survival, especially if removal, neutralization, or inactivation of these cell products cannot be readily achieved. Perhaps these inherent difficulties have discouraged some investigators from venturing into this research area for the amount of published work relating to alimentary tract tissue and cell culture, both methodological and experimental, is limited.

Although much remains to be accomplished, as the preceding eight chapters indicate, progress is being made toward the development of reliable *in vitro* explant and cell culture methods for non-neoplastic alimentary tract tissues. Explants of gastric, small intestinal, and colonic mucosa can now be maintained consistently *in vitro* for 24 to 48 hours with minimal alteration of morphology and continued activity of most functional parameters assessed to date. Thus, a wide range of short-term studies of gastric and intestinal mucosal function and metabolism would now seem possible. As described by Autrup (Chapter 19), it now appears feasible to culture explants of normal human colonic mucosa for as long as several weeks. Further refinement of Autrup's method to permit consistent survival of explants and to minimize epithelial cell dedifferentiation may well be possible. Moreover, his methods should be tested to see if they are applicable to longer term explant culture of human gastric mucosa and human small intestinal mucosa. Availability of such relatively long-term culture methods for mucosa of the tubular alimentary tract should permit not only studies of the effects of carcinogens on the mucosa and their metabolism by the mucosa, but also experiments that might clarify fundamental mechanisms that regulate epithelial cell renewal, migration and differentiation, and alimentary tract metabolism in health and disease.

The development of stable epithelial cell culture lines which survive for long periods from non-neoplastic gastric mucosa and intestinal mucosa has been uniformly unsuccessful until the recent important, encouraging, and provocative studies reported by Quaroni and his associates. These investigators present impressive morphological and immunological evidence that the IEC cell line which they have established is, indeed, derived from the undifferentiated crypt epithelial cells of mouse intestine. As they point out, a major challenge now is to determine whether this cell line retains sufficient properties of undifferentiated crypt cells to permit the study of factors which may regulate proliferation and, possibly, migration of these cells, as well as their eventual maturation into cells which have characteristics of differentiated villus absorptive cells or differentiated intestinal secretory cells, including goblet, Paneth, and, perhaps, even endocrine cells. Another important question which might now be approached is clarification of the potential influence of intestinal mesenchymal cells on proliferation and differentiation of the intestinal epithelium. It may be feasible to devise methods by which Quaroni's IEC cells and a mesenchymal cell line derived from intestinal mucosa are utilized in proximate or coculture experiments to explore this question.

As pointed out by Pitot and Sirica (Chapter 22), *in vitro* maintenance for relatively long periods of differentiated non-neoplastic mammalian hepatocytes in defined media on collagen gels is now possible. Using such methods, the feasibility of examining at least some of the complex metabolic, synthetic, and excretory functions of hepatocytes has now been demonstrated. Although limited DNA synthesis occurs in some of these cells, especially following stimulation by tropic substances such as insulin and epidermal growth factor, the evidence that monolayers of non-neoplastic mammalian hepatocytes actually replicate is limited and not definitive. In that regard, the evidence presented by Schaeffer and Kessler, that cultures of human cells derived from human hepatocytes may change their morphological appearance to resemble mesenchymal cells and yet contain and produce substances characteristic of hepatocytes, is of interest, although the cultures in which these observations have been made are not yet pure but also contain mesenchymal cells. Clearly, the development of conditions which permit pure culture of non-neoplastic hepatocytes which replicate yet retain the functional characteristics of differentiated hepatocytes remains a major challenge to workers in this field.

In conclusion, only relatively recently have investigators invested substantial effort in developing methods for the *in vitro* culture of mucosal explants and epithelial cells from normal alimentary tract tissues. The significant progress made during the past decade, as reviewed in the preceding chapters, is encouraging. From the technical point of view, alimentary tract cell and explant culture is still in its infancy. However, the prognosis for developing reliable *in vitro* culture methods that will permit study of fundamental biological processes in alimentary tract tissues under precisely controlled conditions appears excellent.

Index

C

D

CONTENTS OF PREVIOUS VOLUMES

(Volumes I–XX edited by David M. Prescott)

Volume I

Volume II

Volume III

Volume IV

Volume V

Volume VI

Volume VII

Volume VIII

Volume IX

Volume X

Volume XI

Volume XII

Volume XIII

Volume XIV

Volume XV

Volume XVI

Volume XVII

Volume XVIII

Volume XIX

Volume XX

Volume 21A